David M. _____

CAST IRON PIPE RESEARCH ASSOCIATION

HANDBOOK

Ductile Iron Pipe
Cast Iron Pipe

Fifth Edition

CAST IRON PIPE RESEARCH ASSOCIATION
1301 W. 22nd St.
Oak Brook, Illinois 60521

2

ISBN — 0-916 442-01-2

Library of Congress No. 75-405-82

Printed in the United States of America

PREFACE

The Cast Iron Pipe Research Association (CIPRA) is a non-profit corporation whose members are manufacturers of gray and ductile cast iron pressure pipe. Throughout its 60 years of operation, the Association has served its members, the engineering profession, private and public utilities, industry and the public by providing sound engineering and research information at the highest level of integrity, thus setting an example of good business conduct. Results have included steady improvements in cast and ductile iron pipe, significant contributions to public health and safety through research and education.

CIPRA's staff of professional engineers and researchers provide services to consulting engineers, utilities and industry through the following activities:

1. Participation in national standards development by American National Standards Institute (ANSI), American Water Works Association (AWWA), and American Society for Testing and Materials (ASTM).

2. Presentation of engineering seminars on design, specifications and pipeline construction, corrosion control, and special applications and problems.

3. Provision of field services, such as soil and environmental investigations, flow tests, consultation with engineers on piping problems, including pipe and system design.

4. Publication of research and engineering information in technical papers, brochures, manuals and the Ductile Iron Pipe News.

Engineering services are available by request through member companies or by direct request to the President of CIPRA.

This Handbook is furnished to design engineers and purchasers of cast and ductile iron pipe as a complete reference for use in development of piping design and specifications. Unabridged ANSI and AWWA Standards are included, together with other helpful reference material.

Membership Directory

CAST IRON PIPE RESEARCH ASSOCIATION

AMERICAN CAST IRON PIPE COMPANY
General Office, Birmingham, Alabama

ATLANTIC STATES CAST IRON PIPE COMPANY
General Office, Phillipsburg, New Jersey

CLOW CORPORATION
CAST IRON PIPE & FOUNDRY DIVISION, General Office, Bensenville, Illinois

McWANE CAST IRON PIPE COMPANY
General Offices, Birmingham, Alabama

PACIFIC STATES CAST IRON PIPE COMPANY
General Office, Provo, Utah

UNITED STATES PIPE AND FOUNDRY COMPANY
General Office, Birmingham, Alabama

4

TABLE OF CONTENTS

Preface

7

Cast Iron Pipe Installed 1664 Versailles

8

SECTION I
EVOLUTION AND HISTORY OF
GRAY AND DUCTILE CAST IRON PIPE

Early Uses of Cast Iron as Pipe

Man's ability to cast pipe probably developed from or coincidentally with the manufacture of cannons, which is reported as early as the year 1313. There is an official record of cast iron pipe manufactured at Siegerland, Germany, in 1455 for installation at the Dillenburg Castle.

A Revolutionary Experiment

In the year 1664, King Louis XIV of France ordered the construction of a cast iron pipe main extending fifteen miles from a pumping station at Marly-on-Seine to Versailles to supply water for the fountains and town. This cast iron pipe is still functioning after more than 300 years of continuous service. When the line was begun (1664), the production of iron required the use of expensive charcoal for the reduction of the iron ore; however, by 1738 success had been achieved in the production of lower cost iron through the use of coke instead of charcoal. Immediately thereafter, the more progressive cities installed cast iron mains.

Invention of Bell & Spigot Joint

The joints of these early cast iron lines were of the flanged type, with lead gaskets. These joints were used until Sir Thomas Simpson, engineer of the Chelsea Water Company, London, invented the bell and spigot joint in 1785. The bell and spigot joint is assembled by caulking yarn or braided hemp into the base of the annular bell cavity, then pouring molten lead into the remaining space inside the bell. Upon solidification, the lead is compacted by caulking, thus effecting a water-tight seal. This joint was used extensively until recent years. Many of the original bell and spigot lines are still in use and apparently will serve for many more years.

Extensive Use of Cast Iron Pipe

There are more than 170 water and gas utilities in the United States that have cast iron distribution mains with continuous service records of more than 100 years.

Since the introduction of cast iron in this country, shortly after 1816, various other materials have been offered as suitable for water distribution mains. That none of these has proved able to supplant cast iron in the confidence and preference of water works engineers is demonstrated by a survey of the kinds of pipe in service in water distribution systems of the 100 largest U.S. cities. This survey revealed that cast iron pipe represented more than 90 percent of the pipe in service in these 100 cities.

Because of such strong evidence and the cumulative experience of generations of water works engineers, cast iron pipe remains more firmly entrenched than ever in its acknowledged position as the standard material for water distribution mains.

Development of Joints

Flanged. Originally, cast iron pipe was made with flanged joints, using lead gaskets. Much improved joints of this type are still used for many above ground plant installations using rubber, fiber, metal or other types of gaskets.

Bell and Spigot. The bell and spigot joint was developed in 1785 and was extensively used for new installations until the 1930's.

Mechanical Joint. The mechanical joint was developed for gas industry use in the late 1920's, but has since been used extensively in the water industry. This joint utilizes the basic principle of the stuffing box.

Roll-on Joint. The roll-on joint was developed in 1937 and was in use approximately 20 years. Assembly of the joint involved a compressed rubber gasket rolled under a restriction ring, followed by caulked square braided jute, with the remainder of the joint packed with a bituminous compound.

Push-on Joint. The push-on joint was developed in 1956 and represented an extremely important advance in the water distribution field. This joint consists of a single rubber gasket placed in a groove inside the bell end of the pipe. By pushing the plain end of the pipe through the lubricated rubber gasket, the gasket is compressed and the joint becomes pressure tight. Assembly of the push-on joint is simple and fast. Large bell holes are not required for this joint, and it may be assembled under wet-trench conditions or even under water.

Special Joints. Several special joints are available. These joints include ball and socket for submarine or stream crossings, plain end coupled, threaded and coupled, and various restrained type joints (see Section VI).

Ductile Iron Pipe

The advent of ductile iron pipe in 1948 was the most significant development in the pressure pipe industry. Quickly recognized as a pipe material with all of the good qualities of gray cast iron pipe plus additional strength coupled with ductility, it was first used for special and severe conditions of high pressure, water hammer and excessive external loads. More than 25 years of experience have proved it to be completely trouble-free as an underground pressure pipe material, and today it is used in the transportation of raw and potable water, sewage, gas, slurries and process chemicals.

Ductile iron pipe is designed as a flexible conduit and separate and distinct American National Standards were developed and have been available since 1965. Casting processes are similar to those for gray cast iron pipe; however, ductile iron pipe requires significant refinement in casting, higher quality raw materials, treatment with special additives, and a higher level of quality control.

It has an ultimate tensile strength of 60,000 psi, a yield strength of 42,000 psi with 10 percent elongation, and is available with standard joints and many types of special joints.

The high level of operational dependability of ductile iron pipe stems from its major advantages of machineability, impact and corrosion resistance and high strength, making it a pipe material that is rugged and durable with high resistance to impact requiring virtually no maintenance.

Manufacture of Ductile and Cast Iron Pipe

The centrifugal casting methods used in manufacturing ductile and cast iron pipe have been in the process of commercial development and refinement since 1925. The steady improvements that have led to the present state of the art have been covered by literally hundreds of patents and technical papers, and represent the ingenuity of many dedicated engineers, metallurgists, and foundrymen.

In the centrifugal casting process, a controlled amount of molten metal having the proper characteristics is introduced into a rotating metal or sand-lined mold, fitted with a socket core, in such way as to distribute the metal over the interior of the mold surface by centrifugal force. This force holds the metal in place until solidification occurs. Pipe removed from the mold is furnace-annealed to produce the prescribed physical and mechanical properties and eliminate any casting stresses that may have been present.

After cleaning, hydrostatic testing, dimensional gaging, weighing, coating, lining and marking, the pipe is ready for shipment.

Microstructures

Gray Iron

Ductile Iron

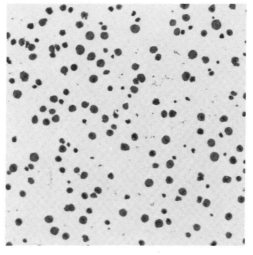

SECTION II

Gray Cast Iron Pipe

Introduction

Cast iron pipe is widely used for water transmission and distribution, gravity sewers and force mains, water and sewage treatment plants and industrial piping. Its characteristics make it an ideal piping material for underground, underwater and aboveground installations. It is available with a wide variety of joints, each suited for specific service conditions. Linings are available for protection against aggressive liquid; and external corrosion control measures can be provided where severe conditions prevail (see Section VIII).

Advantages of cast iron pipe include machineability, which provides ease of field cutting and tapping; convenient pipe lengths for ease of installation; and inherent corrosion resistance, assuring long service life.

Cast iron pressure pipe is designed in accordance with ANSI Standard A21.1 (AWWA C101) and manufactured in accordance with ANSI Standards A21.6 (AWWA C106).

Metallurgy of Gray Cast Iron Pipe

Structure

Gray cast iron is essentially an alloy of iron and carbon containing appropriate amounts of silicon, manganese, phosphorus and sulfur. In gray cast iron, a major part of the carbon content occurs as free carbon or graphite in the form of flakes interspersed throughout the metal. The engineering properties specific to gray cast iron are principally due to the presence of these graphite flakes.

The excellent corrosion resistance of cast iron pipe in underground service is well known. The corrosion products of cast iron pipe are tightly adherent and help protect the metal beneath. The appreciable amount of graphite, about 10 percent by volume, together with relatively inert iron oxides and phosphides, causes gray cast iron to be characteristically corrosion resistant.

In corrosive environments where the metallic content of cast iron pipe is reduced by corrosion, the corrosion products of cast iron form an interlocking mat of graphite, phosphides and iron oxides, which is dense and strong enough to enable the pipe to continue to serve indefinitely as an effective conduit in many instances.

Machineability of any metal structure is important, particularly where it must be drilled, tapped, or cut with ordinary tools. At a given hardness level cast iron is more easily machined than most other metals because the graphite flakes break up the chips and lubricate the cutting tool.

Chemistry

Carbon.   Carbon in cast iron may vary from about 3.00 to 3.75 percent. In general, the carbon content is adjusted to suit the particular method of manufacture and the cooling rate of a given size of casting.

Silicon.   Silicon promotes graphitization, and, therefore, is a very useful element to control the properties of gray cast iron. The proper use of silicon permits a wide range of different castings to be made with uniform strength and hardness properties.

Phosphorus.   Phosphorus in cast iron pipe ranges up to a maximum of 0.90 percent. Phosphorus increases the fluidity of the molten iron and is useful for wear and corrosion resistance.

Manganese and Sulfur.   Manganese content is related to the content of sulfur and other elements in the control of physical properties. Sulfur is limited to 0.12 percent maximum in the current pipe standards.

Several other elements may be present and affect the physical properties. For this reason, the acceptance test requirements of cast iron pipe standards are based on the physical properties.

Quality Control

Modern metallurgical control enables the foundry to produce quality castings with the desired combination of properties. Tests carried out as a guide to metallurgical control include: frequent chemical analyses for each mix used in the cupola, chill tests for graphitizing tendency, Talbot strips from the pipe wall, ring tests on rings cut from the pipe, full length bursting tests of pipe, impact tests, direct tensile tests, and others. One of the routine tests of the finished product is the hydrostatic test to which every length is subjected. Correlation of the values obtained from all of these tests with service performance of the castings have enabled the cast iron pressure pipe industry to produce progressively better and more reliable cast iron pipe and fittings.

ANSI A21.1–1967
(AWWA C101–67)
(Formerly AWWA H1–67)
Reaffirmed w/o revision 1972

AMERICAN NATIONAL STANDARD

for

THICKNESS DESIGN OF CAST-IRON PIPE

With Tables of Pipe Thicknesses

In Four Parts

Sec. 1–1—Thickness Tables for Standard Conditions

Sec. 1–2—General Procedure for Thickness Determination

Sec. 1–3—Design Theory—Determination of Net Thickness, Earth Load, and Truck Superload

Sec. 1–4—Thickness Determination for Pipe on Piers or Piling Aboveground or Underground

SPONSORS

AMERICAN GAS ASSOCIATION

AMERICAN SOCIETY FOR TESTING AND MATERIALS

AMERICAN WATER WORKS ASSOCIATION

NEW ENGLAND WATER WORKS ASSOCIATION

Approved by The American National Standard Institute, Nov. 10, 1967
Reaffirmed without revision Jul. 6, 1972

NOTICE

This Standard has been especially printed by the American Water Works Association for incorporation into this volume. It is current as of December 1, 1975. It should be noted, however, that all AWWA Standards are updated at least once in every five years. Therefore, before applying this Standard it is suggested that you confirm its currency with the American Water Works Association.

AMERICAN WATER WORKS ASSOCIATION
6666 West Quincy Avenue, Denver, Colorado 80235

American National Standard

An American National Standard implies a consensus of those substantially concerned with its scope and provisions. An American National Standard is intended as a guide to aid the manufacturer, the consumer, and the general public. The existence of an American National Standard does not in any respect preclude anyone, whether he has approved the standard or not, from manufacturing, marketing, purchasing, or using products, processes, or procedures not conforming to the standard. American National Standards are subject to periodic review, and users are cautioned to obtain the latest editions. Producers of goods made in conformity with an American National Standard are encouraged to state on their own responsibility in advertising and promotion material or on tags or labels that the goods are produced in conformity with particular American National Standards.

CAUTION NOTICE. This American National Standard may be revised or withdrawn at any time. The procedures of the American National Standards Institute require that action be taken to reaffirm, revise, or withdraw this standard no later than five (5) years from the date of publication. Purchasers of American National Standards may receive current information on all standards by calling or writing the American National Standards Institute, 1430 Broadway, New York, N.Y. 10018, (212) 868–1220.

Committee Personnel

Subcommittee 1—Pipe, which reviewed and recommended reaffirmation of this standard without revision, had the following personnel at that time:

EDWARD C. SEARS, *Chairman*
WALTER AMORY, *Vice-Chairman*

User Members	*Producer Members*
FRANK E. DOLSON	CARL A. HENRIKSON
JACK D. HASLER	THOMAS D. HOLMES
GEORGE F. KEENAN	THOMAS P. NORWOOD
LEONARD ORLANDO, JR.	SIDNEY P. TEAGUE
JAMES D. POWERS	
MILES R. SUCHOMEL	

Standards Committee A21—Cast Iron Pipe and Fittings, which reviewed and approved this standard, had the following personnel at the time of approval:

WALTER AMORY, *Chairman*
CARL A. HENRIKSON, *Vice-Chairman*
JAMES B. RAMSEY, *Secretary*

Organization Represented	*Name of Representative*
American Gas Association	LEONARD ORLANDO, JR.
American Society of Civil Engineers	KENNETH W. HENDERSON
American Society of Mechanical Engineers	CHARLES R. VELZY
American Society for Testing and Materials	(Vacant)
American Water Works Association	VANCE C. LISCHER
Cast Iron Pipe Research Association	CARL A. HENRIKSON
	EDWARD C. SEARS
	W. HARRY SMITH
Individual Producers	FRANK J. CAMEROTA
	WILLIAM T. MAHER
Manufacturers' Standardization Society of the Valve and Fittings Industry	ABRAHAM FENSTER
New England Water Works Association	WALTER AMORY
Naval Facilities Engineering Command	STANLEY C. BAKER
Underwriters' Laboratories, Inc.	MILES R. SUCHOMEL
Canadian Standards Association	W. F. SEMENCHUK *

* Liaison representative without vote.

iii

Table of Contents

1
9

Foreword

This foreword is provided for information only and is not a part of ANSI A21.1–1967 (AWWA C101–67).

History of Standard

On Sep. 10, 1902, NEWWA adopted a "Standard Specification for Cast-Iron Pipe and Special Castings," covering bell-and-spigot pit-cast pipe and fittings of ten thickness classes. The thickness classes were based on allowable internal pressures varying by increments of 50 ft of head.

On May 12, 1908, AWWA adopted a "Standard Specification for Cast-Iron Pipe and Special Castings," covering bell-and-spigot pit-cast pipe and fittings of eight classes, A through H, with allowable working pressures varying by increments of 100 ft of head from 100 to 800 ft. Dimensions and weights were given for pipe and fittings.

In 1926, ASA Sectional Committee (now ANSI Standards Committee) A21 on Cast-Iron Pipe and Fittings was organized under the sponsorship of AGA, ASTM, AWWA, and NEWWA. The present scope of Committee A21 is:

Standardization of specifications for cast-iron and ductile-iron pressure pipe for gas, water, and other liquids, and fittings for use with such pipe. These specifications to include design, dimensions, materials, coatings, linings, joints, accessories, and methods of inspection and test.

Sectional Committee A21 sponsored many tests of pipe and fittings; these included subjection of pipe to combined earth load and internal pressures (which form the basis of pipe-thickness design), corrosion tests, measurement of hydraulic friction loss in fittings, and tests of bursting strengths of pipe and fittings. After exhaustive study of the test results and other research, the committee in 1939 issued A21.1, "American Standard Practice Manual for the Computation of Strength and Thickness of Cast-Iron Pipe." The manual included nomograms and thickness tables for pit-cast pipe with 11/31 * iron strength. As stated in the preface to that manual, however, the design method was applicable to pipe of any iron strength.

Discussions and interpretations [1,2] of the method of design of cast-iron pipe were published in 1939 and presented to AWWA and AGA. As a result of these publications and because of the general acceptance of A21.1, a substantial volume of cast-iron pipe was designed by the new method and furnished to manufacturers' standards between 1939 and 1953. A standard (A21.2) for pit-cast pipe with 11/31 iron strength also was issued in 1939. Work on standards for centrifugally cast pipe with 18/40 iron strength was started after the design was completed in 1939, but, owing to the intervention of World War II and other causes, they were not formally issued until 1953.

---

* The first figure designates the bursting tensile strength in units of 1,000 psi and the second figure designates the ring modulus of rupture in units of 1,000 psi.

vii

In 1957, a revision of A21.1 was issued. In that revision, designated ASA A21.1–1957 (AWWA H1–57) the major change was the addition of a method for computing earth loads on pipe laid under embankments and of nomograms and thickness tables for centrifugally cast pipe with 18/40 iron strength.

In 1958, Sectional Committee A21 was reorganized. Subcommittees were established to study each group of standards in accordance with the review and revision policy of ASA (now ANSI). The scope of Subcommittee 1—Pipe is:

To include the periodic review of all current A21 standards for pipe, the preparation of revisions or new standards when needed, as well as other matters pertaining to pipe standards.

As a result of the work of Subcommittee No. 1 on this assignment, revisions of the cast-iron pipe standards A21.6 (AWWA C106), A21.7, A21.8 (AWWA C108), and A21.9 were issued in 1962 and again in 1970. Revision of A21.1–1957 (AWWA H1–57) was delayed in order for the subcommittee to carry out a new assignment from the sectional committee to develop a design standard and pipe standards for ductile-iron pipe. Subsequently, the subcommittee completed its study and a revision of A21.1 was issued in 1967.

In 1971, Subcommittee No. 1 reviewed the 1967 edition and submitted a recommendation to Committee A21 that the standard be reaffirmed without change. Therefore, this reaffirmed edition is unchanged from the 1967 edition, except for the updating of this foreword. For convenience, the major features listed in the foreword of the 1967 edition are repeated here.

Major Features of 1967 Revision

Although ANSI A21.1–1967 (AWWA H1–67) contained no changes in the basic design method, a number of revisions were incorporated to simplify the design procedure and to reflect changes in technology. Major features of the 1967 revision are discussed.

1. *Format.* This revision is divided into four major sections: Sec. 1–1 gives thickness tables for standard conditions; Sec. 1–2 gives the general procedure for thickness determination; Sec. 1–3 gives design theory and provides methods for determining pipe thicknesses, earth loads, and truck superloads for both standard and special conditions; and Sec. 1–4 gives the design procedure for a special installation condition, pipe on piers aboveground or underground.

2. *Iron strength.* Thickness tables, nomograms, and other data for pit-cast pipe (11/31 iron strength) were deleted, as this type of pipe is seldom furnished today.

Centrifugally cast pipe covered under ANSI Standards A21.6 (AWWA C106), A21.7, A21.8 (AWWA C108), and A21.9 are specified to have an iron strength of not less than 18/40. Advances in production technology have enabled the manufacturers to furnish pipe with greater strength, and pipe with 21/45 iron strength has been available for many years. Thus, for the convenience of users, tables and figures in the standard cover pipe with iron strengths of both 18/40 and 21/45.

3. *Laying conditions.* Laying conditions C, D, and E (also called "field conditions" in the 1957 revision) were deleted. Conditions C and D were for pipe laid on blocks, a method sometimes used in the 1920s and 1930s. This method has been recognized as undesirable and is seldom used today. Condition E was for pipe laid with special bedding but without tamping the backfill. It was felt that this combination would rarely be used today; when special bedding is used, the backfill is almost always tamped, giving laying condition F, which was retained in this revision.

4. *Earth loads.* Formulas and procedures were added (Sec. 1–3) for determining earth loads for standard and special conditions. Earth loads shown in Table 1–8 are the same as those given in the 1957 revision, except that loads for 20 and 24 ft of cover were added.

5. *Allowance for truck superloads.* Formulas and procedures were added (Sec. 1–3) for determining truck superloads for standard and special conditions. These procedures may be used to compute truck superloads for unpaved roads, flexible pavement, or rigid pavement; one truck or two passing trucks; and any wheel load and impact factor, including AASHO truck loadings. Truck superloads for standard conditions are shown in Table 1–8 and are the same as given in the 1957 revision, except that loads for 20 and 24 ft of cover were added.

6. *Allowance for surge pressure.* The allowances for surge pressure (water hammer), shown in Table 1–10, are unchanged from those given in the 1957 revision.

7. *Allowance for corrosion.* A standard allowance for soil corrosion of 0.08 in., based on judgment and experience of early engineers, was used in the 1939 manual and continued in the 1957 revision. The allowance of 0.08 in. was also retained in this edition. It is very conservative for many soils, and has proved to be adequate in most. In areas suspected or known to be highly corrosive, however, the designer should take special precautions. Where unusually corrosive soil conditions are anticipated, a soil survey is recommended.

8. *Pipe on piers.* A new section (Sec. 1–4) was added to provide procedures for computing thicknesses of pipe installed on piers or piling, a condition sometimes encountered in laying pipe in unstable soil, across streams or swamps, either aboveground or underground, and in installing pipe on bridges and other aboveground structures.

Standard Thicknesses of Pit-Cast Pipe

As stated in the foregoing, thickness tables for pit-cast pipe were deleted in the 1967 edition. During the review of this standard in 1971, it was determined that some standards and codes refer to the table of standard thickness classes for pit-cast pipe shown as Table 10 in the 1957 edition of A21.1. For reference, the contents of that table are reproduced on the next page:

Standard Thickness Classes for Pit-Cast Pipe *

Pipe Size in.	Pipe-Wall Thickness (in.) for Standard Thickness Class No.:													
	1	2	3	4	5	6	7	8	9	10	11	12	13	14
3	0.37	0.40	0.43	0.46	0.50	0.54	0.58	0.63	0.68	0.73	0.79	0.85	0.92	0.99
4	0.40	0.43	0.46	0.50	0.54	0.58	0.63	0.68	0.73	0.79	0.85	0.92	0.99	1.07
6	0.43	0.46	0.50	0.54	0.58	0.63	0.68	0.73	0.79	0.85	0.92	0.99	1.07	1.16
8	0.46	0.50	0.54	0.58	0.63	0.68	0.73	0.79	0.85	0.92	0.99	1.07	1.16	1.25
10	0.50	0.54	0.58	0.63	0.68	0.73	0.79	0.85	0.92	0.99	1.07	1.16	1.25	1.35
12	0.54	0.58	0.63	0.68	0.73	0.79	0.85	0.92	0.99	1.07	1.16	1.25	1.35	1.46
14	0.54	0.58	0.63	0.68	0.73	0.79	0.85	0.92	0.99	1.07	1.16	1.25	1.35	1.46
16	0.58	0.63	0.68	0.73	0.79	0.85	0.92	0.99	1.07	1.16	1.25	1.35	1.46	1.58
18	0.63	0.68	0.73	0.79	0.85	0.92	0.99	1.07	1.16	1.25	1.35	1.46	1.58	1.71
20	0.66	0.71	0.77	0.83	0.90	0.97	1.05	1.13	1.22	1.32	1.43	1.54	1.66	1.79
24	0.74	0.80	0.86	0.93	1.00	1.08	1.17	1.26	1.36	1.47	1.59	1.72	1.86	2.01
30	0.87	0.94	1.02	1.10	1.19	1.29	1.39	1.50	1.62	1.75	1.89	2.04	2.20	2.38
36	0.97	1.05	1.13	1.22	1.32	1.43	1.54	1.66	1.79	1.93	2.08	2.25	2.43	2.62
42	1.07	1.16	1.25	1.35	1.46	1.58	1.71	1.85	2.00	2.16	2.33	2.52	2.72	2.94
48	1.18	1.27	1.37	1.48	1.60	1.73	1.87	2.02	2.18	2.35	2.54	2.74	2.96	3.20
54	1.30	1.40	1.51	1.63	1.76	1.90	2.05	2.21	2.39	2.58	2.79	3.01	3.25	3.51
60	1.39	1.50	1.62	1.75	1.89	2.04	2.20	2.38	2.57	2.78	3.00	3.24	3.50	3.78

2
3

* Each class is made 8 per cent heavier than the preceding class, starting with the thinnest, i.e., minimum thickness, as the base class.

References

1. WIGGIN, T. H.; ENGER, M. L.; & SCHLICK, W. J. A Proposed New Method for Determining Barrel Thickness of Cast-Iron Pipe. *Jour. AWWA*, 31:841 (May 1939).
2. MOORE, W. D. Discussion of the New Law of Design of Cast-Iron Pipe. *Jour. AWWA*, 31:1655 (Oct. 1939).
3. SCHLICK, W. J. Supporting Strength of Cast-Iron Pipe for Water and Gas. *Iowa State College Eng. Exp. Sta. Bul.* 146 (1940).

Bibliography

MARSTON, ANSON. The Theory of External Loads on Closed Conduits in the Light of the Latest Experiments. *Iowa State College Eng. Exp. Sta. Bul.* 96 (1930).

SCHLICK, W. J. Loads on Pipe in Wide Ditches. *Iowa State College Eng. Exp. Sta. Bul.* 108 (1932).

SCHLICK, W. J. & MOORE, B. A. Strength and Elastic Properties of Cast-Iron. *Iowa State College Eng. Exp. Sta. Bul.* 127 (1936).

SPANGLER, M. G. *Soil Engineering.* International Textbook Co., Scranton, Pa. (2nd ed., 1960).

SPANGLER, M. G. The Supporting Strength of Rigid Pipe Culverts. *Iowa State College Eng. Exp. Sta. Bul.* 112 (1933).

SPANGLER, M. G., ET AL. Experimental Determination of Static and Impact Loads Transmitted to Culverts. *Iowa State College Eng. Exp. Sta. Bul.* 79 (1926).

SPANGLER, M. G. & SCHLICK, W. J. Negative Projecting Conduits. *Iowa State College Eng. Report* No. 14 (1952-53).

Standard Specifications for Highway Bridges. American Association of State Highway Officials, Washington, D.C. (8th ed., 1961).

Vertical Pressure on Culverts Under Wheel Loads on Concrete Pavement Slabs. Bul. ST 65. Portland Cement Assn., Chicago, Ill. (1951).

American National Standard for

Thickness Design of Cast-Iron Pipe

2
4

Sec. 1–1—Thickness Tables for Standard Conditions

Sec. 1–1.1—General

Tables 1–1, 1–2, and 1–3, as applicable, permit the direct determination of the required thickness of cast-iron pipe limited to the following conditions:

a. Vertical-sided trench of width at top of pipe not greater than the nominal pipe diameter plus 2 ft

b. Unit weight of soil 120 lb/cu ft

c. $K\mu = 0.1924$, $K\mu' = 0.130$ (see Sec. 1–3.2 for definition)

d. Truck superload based on two passing trucks with adjacent wheels 3 ft apart, 9,000 lb wheel load, unpaved road or flexible pavement, 1.50 impact factor

e. Surge allowances as shown in Table 1–10

f. Iron strengths of 18/40 and 21/45 *

g. The three most common laying conditions:

A—Pipe laid on flat-bottom trench, backfill not tamped

B—Pipe laid on flat-bottom trench, backfill tamped

F—Pipe bedded in gravel or sand, backfill tamped

h. Allowances for casting tolerance as shown in Table 1–6

i. A corrosion allowance of 0.08 in.

Sec. 1–1.2—Trench Load and Internal Pressure

The required thickness of cast-iron pressure pipe is determined from a consideration of trench load and internal pressure in combination. Trench load is considered to consist of the earth load on the pipe plus any superload resulting from traffic over the trench. Internal pressure is considered to consist of the design working pressure plus an additional allowance for surge pressure. Two different combinations of trench load and internal pressure are considered in the design:

Case 1. Trench load (earth load but no truck superload) in combination with internal pressure (working pressure plus surge pressure) and with 2.5 factor of safety applied to both trench load and internal pressure

Case 2. Trench load (earth load plus truck superload) in combination with internal pressure (working pressure but no surge pressure) and with

---

* The first figure designates the bursting tensile strength (S) in units of 1,000 psi and the second figure designates the ring modulus of rupture (R) in units of 1,000 psi.

1

a 2.5 factor of safety applied to both trench load and internal pressure.

Sec. 1–1.3—Traffic Superload and Surge Pressure

In designing water pipe it is customary to assume that neither traffic superload nor surge pressure will occur in important magnitude simultaneously. Thus, calculations for the required thickness of water pipe are made for both conditions independently, and the greater of the two thicknesses thus determined is chosen as the net thickness.

In designing gas pipe the procedure is the same, except that surge pressure is not a factor and only Case 2 is considered.

Sec. 1–1.4—Corrosion Allowance and Casting Tolerance

To the net thickness determined as explained above, a corrosion allowance and a casting tolerance are added to obtain the calculated thickness shown in Tables 1–1, 1–2, and 1–3. The standard thickness class and/or the nominal thickness for this class shown in Tables 1–1, 1–2, and 1–3 are used for specifying and ordering pipe.

For other than standard conditions the formulas, tables, and diagrams in Sec. 1–2 may be used. The design theory on which Tables 1–1, 1–2, and 1–3 are based is presented in Sec. 1–3. Procedures for determining the net thickness of pipe on piers or piling above- or belowground are presented in Sec. 1–4.

2
5

2

TABLE 1-1

Schedule of Barrel Thickness for Water Pipe of 18/40 Iron Strength

Laying Condition	Depth of Cover ft	Thickness Specifications	Internal Pressure—psi						
			50	100	150	200	250	300	350
			Barrel Thicknesses—in.						

Three-Inch Water Pipe

Laying Condition	Depth of Cover ft	Thickness Specifications	50	100	150	200	250	300	350
A	2½	Calculated Thickness	.21*	.21*	.22*	.22*	.22*	.22*	.23*
		Use { Thickness Class	22	22	22	22	22	22	22
		Thickness	.32	.32	.32	.32	.32	.32	.32
	3½	Calculated Thickness	.21*	.21*	.22*	.22*	.22*	.22*	.23*
		Use { Thickness Class	22	22	22	22	22	22	22
		Thickness	.32	.32	.32	.32	.32	.32	.32
	5	Calculated Thickness	.21*	.21*	.22*	.22*	.23	.23	.24
		Use { Thickness Class	22	22	22	22	22	22	22
		Thickness	.32	.32	.32	.32	.32	.32	.32
	8	Calculated Thickness	.22*	.23*	.23*	.24	.24	.24	.25
		Use { Thickness Class	22	22	22	22	22	22	22
		Thickness	.32	.32	.32	.32	.32	.32	.32
	12	Calculated Thickness	.24*	.25*	.25*	.26	.26	.27	.28
		Use { Thickness Class	22	22	22	22	22	22	22
		Thickness	.32	.32	.32	.32	.32	.32	.32
	16	Calculated Thickness	.27	.27	.28	.28	.28	.29	.29
		Use { Thickness Class	22	22	22	22	22	22	22
		Thickness	.32	.32	.32	.32	.32	.32	.32
B	2½	Calculated Thickness	.21*	.21*	.22*	.22*	.22*	.22*	.23*
		Use { Thickness Class	22	22	22	22	22	22	22
		Thickness	.32	.32	.32	.32	.32	.32	.32
	3½	Calculated Thickness	.21*	.21*	.22*	.22*	.22*	.22*	.23*
		Use { Thickness Class	22	22	22	22	22	22	22
		Thickness	.32	.32	.32	.32	.32	.32	.32
	5	Calculated Thickness	.21*	.21*	22*	.22*	.23	.23	.24
		Use { Thickness Class	22	22	22	22	22	22	22
		Thickness	.32	.32	.32	.32	.32	.32	.32
	8	Calculated Thickness	.22*	.22*	.23*	.23*	.24	.24	.24
		Use { Thickness Class	22	22	22	22	22	22	22
		Thickness	.32	.32	.32	.32	.32	.32	.32
	12	Calculated Thickness	.24*	.24*	.25	.25	.26	.26	.27
		Use { Thickness Class	22	22	22	22	22	22	22
		Thickness	.32	.32	.32	.32	.32	.32	.32
	16	Calculated Thickness	.26	.26	.27	.27	.28	.28	.29
		Use { Thickness Class	22	22	22	22	22	22	22
		Thickness	.32	.32	.32	.32	.32	.32	.32
F	2½	Calculated Thickness	.21*	.21*	.21*	.21*	.22*	.22*	.23
		Use { Thickness Class	22	22	22	22	22	22	22
		Thickness	.32	.32	.32	.32	.32	.32	.32
	3½	Calculated Thickness	.21*	.21*	.21*	.21*	.22*	.22*	.23
		Use { Thickness Class	22	22	22	22	22	22	22
		Thickness	.32	.32	.32	.32	.32	.32	.32
	5	Calculated Thickness	.21*	.21*	.21*	.22*	.22*	.22*	.23
		Use { Thickness Class	22	22	22	22	22	22	22
		Thickness	.32	.32	.32	.32	.32	.32	.32
	8	Calculated Thickness	.21	.21	.22	.22	.23	.23	.24
		Use { Thickness Class	22	22	22	22	22	22	22
		Thickness	.32	.32	.32	.32	.32	.32	.32
	12	Calculated Thickness	.23	.23	.23	.24	.24	.24	.25
		Use { Thickness Class	22	22	22	22	22	22	22
		Thickness	.32	.32	.32	.32	.32	.32	.32
	16	Calculated Thickness	.24	.24	.25	.25	.26	.26	.27
		Use { Thickness Class	22	22	22	22	22	22	22
		Thickness	.32	.32	.32	.32	.32	32	.32

* Asterisk following total calculated thickness indicates that truck superload (Case 2) is the controlling factor. When total calculated thickness is not followed by asterisk, surge pressure (Case 1) is the controlling factor. See Sec. 1-2.1.

TABLE 1–1 (*Continued*)

Schedule of Barrel Thickness for Water Pipe of 18/40 Iron Strength

Laying Condition	Depth of Cover ft	Thickness Specifications	Internal Pressure—*psi*						
			50	100	150	200	250	300	350
			Barrel Thicknesses—*in.*						
Four-Inch Water Pipe									
A	2½	Calculated Thickness	.23*	.23*	.24*	.24*	.24*	.25*	.27
		Use { Thickness Class	22	22	22	22	22	22	22
		{ Thickness	.35	.35	.35	.35	.35	.35	.35
	3½	Calculated Thickness	.23*	.23*	.23*	.24	.24	.25	.27
		Use { Thickness Class	22	22	22	22	22	22	22
		{ Thickness	.35	.35	.35	.35	.35	.35	.35
	5	Calculated Thickness	.23*	.24*	.24*	.25	.26	.27	.28
		Use { Thickness Class	22	22	22	22	22	22	22
		{ Thickness	.35	.35	.35	.35	.35	.35	.35
	8	Calculated Thickness	.25*	.26	.27	.28	.28	.29	.30
		Use { Thickness Class	22	22	22	22	22	22	22
		{ Thickness	.35	.35	.35	.35	.35	.35	.35
	12	Calculated Thickness	.29	.29	.30	.30	.31	.31	.32
		Use { Thickness Class	22	22	22	22	22	22	22
		{ Thickness	.35	.35	.35	.35	.35	.35	.35
	16	Calculated Thickness	.31	.31	.32	.32	.33	.33	.34
		Use { Thickness Class	22	22	22	22	22	22	22
		{ Thickness	.35	.35	.35	.35	.35	.35	.35
B	2½	Calculated Thickness	.23*	.23*	.23*	.23*	.24*	.25	.27
		Use { Thickness Class	22	22	22	22	22	22	22
		{ Thickness	.35	.35	.35	.35	.35	.35	.35
	3½	Calculated Thickness	.23*	.23*	.23*	.24	.25	.26	.27
		Use { Thickness Class	22	22	22	22	22	22	22
		{ Thickness	.35	.35	.35	.35	.35	.35	.35
	5	Calculated Thickness	.23	.24	.24	.25	.26	.27	.28
		Use { Thickness Class	22	22	22	22	22	22	22
		{ Thickness	.35	.35	.35	.35	.35	.35	.35
	8	Calculated Thickness	.25	.26	.27	.27	.28	.29	.29
		Use { Thickness Class	22	22	22	22	22	22	22
		{ Thickness	.35	.35	.35	.35	.35	.35	.35
	12	Calculated Thickness	.28	.28	.29	.30	.30	.31	.32
		Use { Thickness Class	22	22	22	22	22	22	22
		{ Thickness	.35	.35	.35	.35	.35	.35	.35
	16	Calculated Thickness	.30	.31	.31	.32	.32	.33	.33
		Use { Thickness Class	22	22	22	22	22	22	22
		{ Thickness	.35	.35	.35	.35	.35	35	.35
F	2½	Calculated Thickness	.23*	.23*	.23*	.24*	.24	.25	.26
		Use { Thickness Class	22	22	22	22	22	22	22
		{ Thickness	.35	.35	.35	.35	.35	.35	.35
	3½	Calculated Thickness	.23*	.23*	.23	.24	.24	.25	.26
		Use { Thickness Class	22	22	22	22	22	22	22
		{ Thickness	.35	.35	.35	.35	.35	.35	.35
	5	Calculated Thickness	.23	.23	.23	.24	.24	.25	.26
		Use { Thickness Class	22	22	22	22	22	22	22
		{ Thickness	.35	.35	.35	.35	.35	.35	.35
	8	Calculated Thickness	.24	.24	.25	.25	.25	.26	.27
		Use { Thickness Class	22	22	22	22	22	22	22
		{ Thickness	.35	.35	.35	.35	.35	.35	.35
	12	Calculated Thickness	.25	.26	.27	.28	.28	.29	.30
		Use { Thickness Class	22	22	22	22	22	22	22
		{ Thickness	.35	.35	.35	.35	.35	.35	.35
	16	Calculated Thickness	.28	.28	.29	.29	.30	.30	.31
		Use { Thickness Class	22	22	22	22	22	22	22
		{ Thickness	.35	.35	.35	. 35	.35	.35	.35

* Asterisk following total calculated thickness indicates that truck superload (Case 2) is the controlling factor. When total calculated thickness is not followed by asterisk, surge pressure (Case 1) is the controlling factor. See Sec. 1–2.1.

TABLE 1-1 (*Continued*)

Schedule of Barrel Thickness for Water Pipe of 18/40 Iron Strength

Laying Condition	Depth of Cover ft	Thickness Specifications	Internal Pressure—psi						
			Barrel Thicknesses—in.						
			50	100	150	200	250	300	350
		Six-Inch Water Pipe							
A	2½	Calculated Thickness	.29*	.29*	.30*	.31*	.32*	.33*	.34
		Use Thickness Class	22	22	22	22	22	22	22
		Thickness	.38	.38	.38	.38	.38	.38	.38
	3½	Calculated Thickness	.28*	.28*	.29*	.30*	.31*	.32*	.34
		Use Thickness Class	22	22	22	22	22	22	22
		Thickness	.38	.38	.38	.38	.38	.38	.38
	5	Calculated Thickness	.29*	.29*	.30*	.31*	.33	.34	.35
		Use Thickness Class	22	22	22	22	22	22	22
		Thickness	.38	.38	.38	.38	.38	.38	.38
	8	Calculated Thickness	.32	.33	.34	.35	.36	.37	.38
		Use Thickness Class	22	22	22	22	22	22	22
		Thickness	.38	.38	.38	.38	.38	.38	.38
	12	Calculated Thickness	.36	.37	.38	.38	.39	.40	.41
		Use Thickness Class	22	22	22	22	22	23	23
		Thickness	.38	.38	.38	.38	.38	.41	.41
	16	Calculated Thickness	.39	.40	.41	.42	.42	.43	.44
		Use Thickness Class	22	23	23	23	23	24	24
		Thickness	.38	.41	.41	.41	.41	.44	.44
B	2½	Calculated Thickness	.28*	.28*	.29*	.30*	.31*	.32*	.33*
		Use Thickness Class	22	22	22	22	22	22	22
		Thickness	.38	.38	.38	.38	.38	.38	.38
	3½	Calculated Thickness	.27*	.28*	.28*	.29*	.30*	.32	.34
		Use Thickness Class	22	22	22	22	22	22	22
		Thickness	.38	.38	.38	.38	.38	.38	.38
	5	Calculated Thickness	.28*	.29	.30	.31	.32	.33	.34
		Use Thickness Class	22	22	22	22	22	22	22
		Thickness	38	.38	.38	.38	.38	.38	.38
	8	Calculated Thickness	.31	.32	.33	.34	.35	.36	.37
		Use Thickness Class	22	22	22	22	22	22	22
		Thickness	.38	.38	.38	.38	.38	.38	.38
	12	Calculated Thickness	.35	.35	.36	.37	.38	.39	.40
		Use Thickness Class	22	22	22	22	22	22	23
		Thickness	.38	.38	.38	.38	.38	.38	.41
	16	Calculated Thickness	.38	.38	.39	.40	.41	.42	.43
		Use Thickness Class	22	22	22	23	23	23	24
		Thickness	.38	.38	.38	.41	.41	.41	.44
F	2½	Calculated Thickness	.26*	.27*	.28	.29	.30	.32	.34
		Use Thickness Class	22	22	22	22	22	22	22
		Thickness	.38	.38	.38	.38	.38	.38	.38
	3½	Calculated Thickness	.27	.27	.28	.29	.30	.32	.34
		Use Thickness Class	22	22	22	22	22	22	22
		Thickness	.38	.38	.38	.38	.38	.38	.38
	5	Calculated Thickness	.28	.28	.29	.29	.30	.32	.35
		Use Thickness Class	22	22	22	22	22	22	22
		Thickness	.38	.38	.38	.38	.38	.38	.38
	8	Calculated Thickness	.29	.30	.31	.32	.33	.34	.35
		Use Thickness Class	22	22	22	22	22	22	22
		Thickness	.38	.38	.38	.38	.38	.38	.38
	12	Calculated Thickness	.32	.33	.34	.35	.36	.37	.38
		Use Thickness Class	22	22	22	22	22	22	22
		Thickness	.38	.38	.38	38	.38	.38	.38
	16	Calculated Thickness	.35	.35	.36	.37	.38	.39	.40
		Use Thickness Class	22	22	22	22	22	22	23
		Thickness	.38	.38	.38	.38	.38	.38	.41

* Asterisk following total calculated thickness indicates that truck superload (Case 2) is the controlling factor. When total calculated thickness is not followed by asterisk, surge pressure (Case 1) is the controlling factor. See Sec. 1-2.1.

TABLE 1-1 (*Continued*)

Schedule of Barrel Thickness for Water Pipe of 18/40 Iron Strength

Laying Condition	Depth of Cover ft	Thickness Specifications	Internal Pressure—psi						
			50	100	150	200	250	300	350
			Barrel Thicknesses—in.						
Eight-Inch Water Pipe									
A	2½	Calculated Thickness	.34*	.35*	.36*	.37*	.38*	.39*	.40*
		Use { Thickness Class	22	22	22	22	22	22	22
		{ Thickness	.41	.41	.41	.41	.41	.41	.41
	3½	Calculated Thickness	.33*	.34*	.34*	.35*	.36*	.39	.41
		Use { Thickness Class	22	22	22	22	22	22	22
		{ Thickness	.41	.41	.41	.41	.41	.41	.41
	5	Calculated Thickness	.34*	.35*	.36*	.37*	.39	.41	.43
		Use { Thickness Class	22	22	22	22	22	22	23
		{ Thickness	.41	.41	.41	.41	.41	.41	.44
	8	Calculated Thickness	.38	.39	.40	.41	.43	.44	.46
		Use { Thickness Class	22	22	22	22	23	23	24
		{ Thickness	.41	.41	.41	.41	.44	.44	.48
	12	Calculated Thickness	.43	.44	.45	.46	.47	.49	.50
		Use { Thickness Class	23	23	23	24	24	24	25
		{ Thickness	.44	.44	.44	.48	.48	.48	.52
	16	Calculated Thickness	.46	.47	.48	.49	.51	.52	.53
		Use { Thickness Class	24	24	24	24	25	25	25
		{ Thickness	.48	.48	.48	.48	.52	.52	.52
B	2½	Calculated Thickness	.32*	.33*	.34*	.35*	.36*	.38*	.39*
		Use { Thickness Class	22	22	22	22	22	22	22
		{ Thickness	.41	.41	.41	.41	.41	.41	.41
	3½	Calculated Thickness	.32*	.33*	.33*	.34*	.36*	.38	.40
		Use { Thickness Class	22	22	22	22	22	22	22
		{ Thickness	.41	.41	.41	.41	.41	.41	.41
	5	Calculated Thickness	.33*	.34*	.35	.36	.38	.40	.42
		Use { Thickness Class	22	22	22	22	22	22	22
		{ Thickness	.41	.41	.41	.41	.41	.41	.41
	8	Calculated Thickness	.36	.37	.38	.40	.41	.43	.45
		Use { Thickness Class	22	22	22	22	22	23	23
		{ Thickness	.41	.41	.41	.41	.41	.44	.44
	12	Calculated Thickness	.41	.42	.43	.44	.46	.47	.48
		Use { Thickness Class	22	22	23	23	24	24	24
		{ Thickness	.41	.41	.44	.44	.48	.48	.48
	16	Calculated Thickness	.44	.45	.46	.47	.48	.50	.51
		Use { Thickness Class	23	23	24	24	24	25	25
		{ Thickness	.44	.44	.48	.48	.48	.52	.52
F	2½	Calculated Thickness	.30*	.31*	.32*	.33*	.35	.37	.39
		Use { Thickness Class	22	22	22	22	22	22	22
		{ Thickness	.41	.41	.41	.41	.41	.41	.41
	3½	Calculated Thickness	.29*	.30*	.31	.33	.35	.37	.39
		Use { Thickness Class	22	22	22	22	22	22	22
		{ Thickness	.41	.41	.41	.41	.41	.41	.41
	5	Calculated Thickness	.30*	.31*	.32	.34	.36	.38	.40
		Use { Thickness Class	22	22	22	22	22	22	22
		{ Thickness	.41	.41	.41	.41	.41	.41	.41
	8	Calculated Thickness	.34	.35	.36	.37	.39	.41	.43
		Use { Thickness Class	22	22	22	22	22	22	23
		{ Thickness	.41	.41	.41	.41	.41	.41	.44
	12	Calculated Thickness	.38	.39	.40	.41	.43	.44	.46
		Use { Thickness Class	22	22	22	22	23	23	24
		{ Thickness	.41	.41	.41	.41	.44	.44	.48
	16	Calculated Thickness	.40	.41	.42	.43	.45	.46	.48
		Use { Thickness Class	22	22	22	23	23	24	24
		{ Thickness	.41	.41	.41	.44	.44	.48	.48

* Asterisk following total calculated thickness indicates that truck superload (Case 2) is the controlling factor. When total calculated thickness is not followed by asterisk, surge pressure (Case 1) is the controlling factor. See Sec. 1-2.1.

TABLE 1–1 (*Continued*)

Schedule of Barrel Thickness for Water Pipe of 18/40 Iron Strength

Laying Condition	Depth of Cover ft	Thickness Specifications	Internal Pressure—psi						
			50	100	150	200	250	300	350
			Barrel Thicknesses—in.						

Ten-Inch Water Pipe

Laying Condition	Depth of Cover ft	Thickness Specifications	50	100	150	200	250	300	350
A	2½	Calculated Thickness	.40*	.41*	.42*	.43*	.45*	.46*	.48*
		Use { Thickness Class	22	22	22	22	22	23	23
		{ Thickness	.44	.44	.44	.44	.44	.48	.48
	3½	Calculated Thickness	.39*	.40*	.41*	.42*	.44*	.46	.49
		Use { Thickness Class	22	22	22	22	22	23	23
		{ Thickness	.44	.44	.44	.44	.44	.48	.48
	5	Calculated Thickness	.40*	.41*	.42*	.44	.46	.48	.51
		Use { Thickness Class	22	22	22	22	23	23	24
		{ Thickness	.44	.44	.44	.44	.48	.48	.52
	8	Calculated Thickness	.44	.46	.47	.49	.51	.53	.55
		Use { Thickness Class	22	23	23	23	24	24	25
		{ Thickness	.44	.48	.48	.48	.52	.52	.56
	12	Calculated Thickness	.51	.52	.53	.55	.56	.58	.60
		Use { Thickness Class	24	24	24	25	25	26	26
		{ Thickness	.52	.52	.52	.56	.56	.60	.60
	16	Calculated Thickness	.53	.54	.56	.57	.59	.60	.62
		Use { Thickness Class	24	25	25	25	26	26	26
		{ Thickness	.52	.56	.56	.56	.60	.60	.60
B	2½	Calculated Thickness	.38*	.39*	.40*	.41*	.43*	.45*	.47*
		Use { Thickness Class	22	22	22	22	22	22	23
		{ Thickness	.44	.44	.44	.44	.44	.44	48
	3½	Calculated Thickness	.37*	.38*	.39*	.40*	.42*	.44	.48
		Use { Thickness Class	22	22	22	22	22	22	23
		{ Thickness	.44	.44	.44	.44	.44	.44	.48
	5	Calculated Thickness	.38*	.39*	.40*	.42	.45	.47	.50
		Use { Thickness Class	22	22	22	22	22	23	24
		{ Thickness	.44	.44	.44	.44	.44	.48	.52
	8	Calculated Thickness	.41*	.44	.45	.47	.49	.51	.53
		Use { Thickness Class	22	22	22	23	23	24	24
		{ Thickness	.44	.44	.44	.48	.48	.52	.52
	12	Calculated Thickness	.48	.49	.51	.52	.54	.56	.58
		Use { Thickness Class	23	23	24	24	25	25	26
		{ Thickness	.48	.48	.52	.52	.56	.56	.60
	16	Calculated Thickness	.50	.52	.53	.54	.56	.58	.60
		Use { Thickness Class	24	24	24	25	25	26	26
		{ Thickness	.52	.52	.52	.56	.56	.60	.60
F	2½	Calculated Thickness	.35*	.36*	.37*	.38*	.41	.44	.47
		Use { Thickness Class	22	22	22	22	22	22	23
		{ Thickness	.44	.44	.44	.44	.44	.44	.48
	3½	Calculated Thickness	.34*	.35*	.36*	.38	.41	.44	.47
		Use { Thickness Class	22	22	22	22	22	22	23
		{ Thickness	.44	.44	.44	.44	.44	.44	.48
	5	Calculated Thickness	.35*	.36*	.38	.40	.42	.46	.48
		Use { Thickness Class	22	22	22	22	22	23	23
		{ Thickness	.44	.44	.44	.44	.44	.48	.48
	8	Calculated Thickness	.39	.41	.42	.44	.46	.48	.51
		Use { Thickness Class	22	22	22	22	23	23	24
		{ Thickness	.44	.44	.44	.44	.48	.48	.52
	12	Calculated Thickness	.44	.45	.46	.48	.50	.52	.54
		Use { Thickness Class	22	22	23	23	24	24	25
		{ Thickness	.44	.44	.48	.48	.52	.52	.56
	16	Calculated Thickness	.46	.47	.49	.51	.52	.54	.56
		Use { Thickness Class	23	23	23	24	24	25	25
		{ Thickness	.48	.48	.48	.52	.52	.56	.56

* Asterisk following total calculated thickness indicates that truck superload (Case 2) is the controlling factor. When total calculated thickness is not followed by asterisk, surge pressure (Case 1) is the controlling factor. See Sec. 1–2.1.

TABLE 1–1 (*Continued*)

Schedule of Barrel Thickness for Water Pipe of 18/40 Iron Strength

Laying Condi-tion	Depth of Cover *ft*	Thickness Specifications	Internal Pressure—*psi*						
			50	100	150	200	250	300	350
			Barrel Thicknesses—*in.*						

Twelve-Inch Water Pipe

Laying Condi-tion	Depth of Cover *ft*	Thickness Specifications	50	100	150	200	250	300	350
A	2½	Calculated Thickness	.45*	.46*	.47*	.49*	.51*	.53*	.55*
		Use {Thickness Class	22	22	22	22	23	23	24
		Thickness	.48	.48	.48	.48	.52	.52	.56
	3½	Calculated Thickness	.43*	.45*	.46*	.48*	.50*	.51	.55
		Use {Thickness Class	22	22	22	22	23	23	24
		Thickness	.48	.48	.48	.48	.52	.52	.56
	5	Calculated Thickness	.45*	.46*	.47*	.49	.52	.54	.57
		Use {Thickness Class	22	22	22	22	23	24	24
		Thickness	.48	.48	.48	.48	.52	.56	.56
	8	Calculated Thickness	.50	.52	.53	.55	.57	.60	.62
		Use {Thickness Class	23	23	23	24	24	25	25
		Thickness	.52	.52	.52	.56	.56	.60	.60
	12	Calculated Thickness	.56	.57	.59	.60	.62	.64	.67
		Use {Thickness Class	24	24	25	25	25	26	26
		Thickness	.56	.56	.60	.60	.60	.65	.65
	16	Calculated Thickness	.59	.60	.61	.63	.65	.67	.69
		Use {Thickness Class	25	25	25	26	26	26	27
		Thickness	.60	.60	.60	.65	.65	.65	.70
B	2½	Calculated Thickness	.42*	.44*	.45*	.46*	.48*	.50*	.53*
		Use {Thickness Class	22	22	22	22	22	23	23
		Thickness	.48	.48	.48	.48	.48	.52	.52
	3½	Calculated Thickness	.41*	.42*	.44*	.45*	.47*	.51	.55
		Use {Thickness Class	22	22	22	22	22	23	24
		Thickness	.48	.48	.48	.48	.48	.52	.56
	5	Calculated Thickness	.42*	.43*	.45*	.47	.50	.53	.56
		Use {Thickness Class	22	22	22	22	23	23	24
		Thickness	.48	.48	.48	.48	.52	.52	.56
	8	Calculated Thickness	.48	.49	.51	.53	.55	.57	.60
		Use {Thickness Class	22	22	23	23	24	24	25
		Thickness	.48	.48	.52	.52	.56	.56	.60
	12	Calculated Thickness	.53	.54	.56	.57	.59	.61	.64
		Use {Thickness Class	23	24	24	24	25	25	26
		Thickness	.52	.56	.56	.56	.60	.60	.65
	16	Calculated Thickness	.55	.57	.58	.60	.62	.64	.66
		Use {Thickness Class	24	24	25	25	25	26	26
		Thickness	.56	.56	.60	.60	.60	.65	.65
F	2½	Calculated Thickness	.39*	.40*	.41*	.43*	.45*	.49	.53
		Use {Thickness Class	22	22	22	22	22	22	23
		Thickness	.48	.48	.48	.48	.48	.48	.52
	3½	Calculated Thickness	.38*	.39*	.40*	.42*	.45	.50	.53
		Use {Thickness Class	22	22	22	22	22	23	23
		Thickness	.48	.48	.48	.48	.48	.52	.52
	5	Calculated Thickness	.39*	.40*	.41*	.43*	.46	.50	.54
		Use {Thickness Class	22	22	22	22	22	23	24
		Thickness	.48	.48	.48	.48	.48	.52	.56
	8	Calculated Thickness	.44	.45	.47	.49	.52	.55	.58
		Use {Thickness Class	22	22	22	22	23	24	25
		Thickness	.48	.48	.48	.48	.52	.56	.60
	12	Calculated Thickness	.48	.49	.51	.53	.55	.58	.61
		Use {Thickness Class	22	22	23	23	24	25	25
		Thickness	.48	.48	.52	.52	.56	.60	.60
	16	Calculated Thickness	.50	.52	.53	.55	.57	.59	.62
		Use {Thickness Class	23	23	23	24	24	25	25
		Thickness	.52	.52	.52	.56	.56	.60	.60

3
1

* Asterisk following total calculated thickness indicates that truck superload (Case 2) is the controlling factor. When total calculated thickness is not followed by asterisk, surge pressure (Case 1) is the controlling factor. See Sec. 1–2.1.

TABLE 1–1 (Continued)
Schedule of Barrel Thickness for Water Pipe of 18/40 Iron Strength

Laying Condition	Depth of Cover ft	Thickness Specifications	Internal Pressure—psi						
			50	100	150	200	250	300	350
			Barrel Thicknesses—in.						

Fourteen-Inch Water Pipe

Laying Condition	Depth of Cover ft	Thickness Specifications	50	100	150	200	250	300	350
A	2½	Calculated Thickness	.51*	.52*	.54*	.55*	.58*	.60*	.62*
		Use Thickness Class	22	22	23	23	24	24	25
		Thickness	.51	.51	.55	.55	.59	.59	.64
	3½	Calculated Thickness	.50*	.51*	.53*	.54*	.57*	.60	.64
		Use Thickness Class	22	22	23	23	24	24	25
		Thickness	.51	.51	.55	.55	.59	.59	.64
	5	Calculated Thickness	.51*	.53*	.54*	.56*	.59	.62	.66
		Use Thickness Class	22	23	23	23	24	25	25
		Thickness	.51	.55	.55	.55	.59	.64	.64
	8	Calculated Thickness	.58	.60	.62	.64	.66	.69	.72
		Use Thickness Class	24	24	25	25	25	26	27
		Thickness	.59	.59	.64	.64	.64	.69	.75
	12	Calculated Thickness	.63	.65	.67	.69	.71	.74	.76
		Use Thickness Class	25	25	26	26	26	27	27
		Thickness	.64	.64	.69	.69	.69	.75	.75
	16	Calculated Thickness	.67	.68	.70	.72	.74	.76	.79
		Use Thickness Class	26	26	26	27	27	27	28
		Thickness	.69	.69	.69	.75	.75	.75	.81
B	2½	Calculated Thickness	.47*	.49*	.51*	.52*	.54*	.57*	.62
		Use Thickness Class	21	21	22	22	23	24	25
		Thickness	.48	.48	.51	.51	.55	.59	.64
	3½	Calculated Thickness	.46*	.47*	.49*	.51*	.54	.57	.63
		Use Thickness Class	21	21	22	22	23	24	25
		Thickness	.48	.48	.51	.51	.55	.59	.64
	5	Calculated Thickness	.48*	.49*	.52	.54	.57	.61	.65
		Use Thickness Class	21	22	22	23	24	24	25
		Thickness	.48	.51	.51	.55	.59	.59	.64
	8	Calculated Thickness	.54	.56	.58	.61	.63	66	.69
		Use Thickness Class	23	23	24	24	25	25	26
		Thickness	.55	.55	.59	.59	.64	.64	.69
	12	Calculated Thickness	.59	.61	.63	.65	.68	.70	.73
		Use Thickness Class	24	24	25	25	26	26	27
		Thickness	.59	.59	.64	.64	.69	.69	.75
	16	Calculated Thickness	.62	.64	.66	.68	.70	.73	.75
		Use Thickness Class	25	25	25	26	26	27	27
		Thickness	.64	.64	.64	.69	.69	.75	.75
F	2½	Calculated Thickness	.44*	.45*	.47*	.49*	.52	.57	.61
		Use Thickness Class	21	21	21	21	22	24	24
		Thickness	.48	.48	.48	.48	.51	.59	.59
	3½	Calculated Thickness	.44*	.45*	.47	.50	.53	.57	.62
		Use Thickness Class	21	21	21	22	23	24	25
		Thickness	.48	.48	.48	.51	.55	.59	.64
	5	Calculated Thickness	.45	.47	.49	.52	.55	.59	.64
		Use Thickness Class	21	21	21	22	23	24	25
		Thickness	.48	.48	.48	.51	.55	.59	.64
	8	Calculated Thickness	.50	.52	.54	.57	.60	.64	.66
		Use Thickness Class	22	22	23	24	24	25	25
		Thickness	.51	.51	.55	.59	.59	.64	.64
	12	Calculated Thickness	.54	.56	.58	.61	.63	.66	.69
		Use Thickness Class	23	23	24	24	25	25	26
		Thickness	.55	.55	.59	.59	.64	.64	.69
	16	Calculated Thickness	.57	.58	.60	.63	.65	.68	.71
		Use Thickness Class	24	24	24	25	25	26	26
		Thickness	.59	.59	.59	.64	.64	.69	.69

* Asterisk following total calculated thickness indicates that truck superload (Case 2) is the controlling factor. When total calculated thickness is not followed by asterisk, surge pressure (Case 1) is the controlling factor. See Sec. 1–2.1.

TABLE 1–1 (*Continued*)

Schedule of Barrel Thickness for Water Pipe of 18/40 Iron Strength

Laying Condition	Depth of Cover ft	Thickness Specifications	Internal Pressure—*psi*						
			50	100	150	200	250	300	350
			Barrel Thicknesses—*in.*						
Sixteen-Inch Water Pipe									
A	2½	Calculated Thickness	.54*	.56*	.58*	.60*	.62*	.65*	.68*
		Use Thickness Class	22	23	23	23	24	24	25
		Thickness	.54	.58	.58	.58	.63	.63	.68
	3½	Calculated Thickness	.54*	.55*	.57*	.59*	.61	.65	.70
		Use Thickness Class	22	22	23	23	24	24	25
		Thickness	.54	.54	.58	.58	.63	.63	.68
	5	Calculated Thickness	.56*	.57*	.59*	.61*	.64	.68	.72
		Use Thickness Class	23	23	23	24	24	25	26
		Thickness	.58	.58	.58	.63	.63	.68	.73
	8	Calculated Thickness	.62	.64	.66	.69	.72	.75	.78
		Use Thickness Class	24	24	25	25	26	26	27
		Thickness	.63	.63	.68	.68	.73	.73	.79
	12	Calculated Thickness	.68	.70	.72	.74	.77	.80	.83
		Use Thickness Class	25	25	26	26	27	27	28
		Thickness	.68	.68	.73	.73	.79	.79	.85
	16	Calculated Thickness	.72	.74	.76	.78	.81	.83	.86
		Use Thickness Class	26	26	27	27	27	28	28
		Thickness	.73	.73	.79	.79	.79	.85	.85
B	2½	Calculated Thickness	.50*	.52*	.54*	.56*	.59*	.62*	.65*
		Use Thickness Class	21	22	22	23	23	24	25
		Thickness	.50	.54	.54	.58	.58	.63	.68
	3½	Calculated Thickness	.50*	.52*	.53*	.56*	.59	.63	.68
		Use Thickness Class	21	22	22	23	23	24	25
		Thickness	.50	.54	.54	.58	.58	.63	.68
	5	Calculated Thickness	.52*	.53*	.55*	.58	.62	.66	.70
		Use Thickness Class	22	22	22	23	24	25	25
		Thickness	.54	.54	.54	.58	.63	.68	.68
	8	Calculated Thickness	.58	.60	.62	.65	.68	.71	.75
		Use Thickness Class	23	23	24	24	25	26	26
		Thickness	.58	.58	.63	.63	.68	.73	.73
	12	Calculated Thickness	.63	.65	.68	.70	.73	.76	.79
		Use Thickness Class	24	24	25	25	26	27	27
		Thickness	.63	.63	.68	.68	.73	.79	.79
	16	Calculated Thickness	.67	.69	.71	.73	.76	.79	.82
		Use Thickness Class	25	25	26	26	27	27	28
		Thickness	.68	.68	.73	.73	.79	.79	.85
F	2½	Calculated Thickness	.47*	.48*	.50*	.52*	.56	.61	.67
		Use Thickness Class	21	21	21	22	23	24	25
		Thickness	.50	.50	.50	.54	.58	.63	.68
	3½	Calculated Thickness	.46*	.47*	.50	.53	.57	.62	.67
		Use Thickness Class	21	21	21	22	23	24	25
		Thickness	.50	.50	.50	.54	.58	.63	.68
	5	Calculated Thickness	.48*	.50	.52	.56	.59	.64	.68
		Use Thickness Class	21	21	22	23	23	24	25
		Thickness	.50	.50	.54	.58	.58	.63	.68
	8	Calculated Thickness	.53	.55	.58	.61	.64	.68	.72
		Use Thickness Class	22	22	23	24	24	25	26
		Thickness	.54	.54	.58	.63	.63	.68	.73
	12	Calculated Thickness	.57	.59	.62	.65	.68	.71	.75
		Use Thickness Class	23	23	24	24	25	26	26
		Thickness	.58	.58	.63	.63	.68	.73	.73
	16	Calculated Thickness	.61	.63	.65	.68	.71	.74	.77
		Use Thickness Class	24	24	24	25	26	26	27
		Thickness	.63	.63	.63	.68	.73	.73	.79

3
3

* Asterisk following total calculated thickness indicates that truck superload (Case 2) is the controlling factor. When total calculated thickness is not followed by asterisk, surge pressure (Case 1) is the controlling factor. See Sec. 1–2.1.

TABLE 1-1 (*Continued*)

Schedule of Barrel Thickness for Water Pipe of 18/40 Iron Strength

Laying Condition	Depth of Cover ft	Thickness Specifications	Internal Pressure—psi						
			50	100	150	200	250	300	350
			Barrel Thicknesses—in.						

Eighteen-Inch Water Pipe

Laying Condition	Depth of Cover ft	Thickness Specifications	50	100	150	200	250	300	350
A	2½	Calculated Thickness	.58*	.60*	.62*	.65*	.67*	.70*	.74*
		Use { Thickness Class	22	22	23	23	24	24	25
		Thickness	.58	.58	.63	.63	.68	.68	.73
	3½	Calculated Thickness	.58*	.59*	.62*	.64*	.67*	.70	.76
		Use { Thickness Class	22	22	23	23	24	24	26
		Thickness	.58	.58	.63	.63	.68	.68	.79
	5	Calculated Thickness	.60*	.62*	.64*	.67	.70	.75	.79
		Use { Thickness Class	22	23	23	24	24	25	26
		Thickness	.58	.63	.63	.68	.68	.73	.79
	8	Calculated Thickness	.67	.69	.72	.74	.77	.81	.85
		Use { Thickness Class	24	24	25	25	26	26	27
		Thickness	.68	.68	.73	.73	.79	.79	.85
	12	Calculated Thickness	.73	.75	.78	.80	.83	.86	.90
		Use { Thickness Class	25	25	26	26	27	27	28
		Thickness	.73	.73	.79	.79	.85	.85	.92
	16	Calculated Thickness	.77	.80	.82	.85	.88	.91	.95
		Use { Thickness Class	26	26	27	27	27	28	28
		Thickness	.79	.79	.85	.85	.85	92	.92
B	2½	Calculated Thickness	.54*	.56*	.58*	.60*	.63*	.67	.73
		Use { Thickness Class	21	22	22	22	23	24	25
		Thickness	.54	.58	.58	.58	.63	.68	.73
	3½	Calculated Thickness	.53*	.55*	.57*	.60	.64	.69	.75
		Use { Thickness Class	21	21	22	22	23	24	25
		Thickness	.54	.54	.58	.58	.63	.68	.73
	5	Calculated Thickness	.55*	.58*	.60	.63	.67	.72	.77
		Use { Thickness Class	21	22	22	23	24	25	26
		Thickness	.54	.58	.58	.63	.68	.73	.79
	8	Calculated Thickness	.62	.64	.67	.70	.74	.77	.82
		Use { Thickness Class	23	23	24	24	25	26	27
		Thickness	.63	.63	.68	.68	.73	.79	.85
	12	Calculated Thickness	.68	.70	.73	.75	.78	.82	.86
		Use { Thickness Class	24	24	25	25	26	27	27
		Thickness	.68	.68	.73	.73	.79	.85	.85
	16	Calculated Thickness	.72	.74	.76	.79	.82	.85	.89
		Use { Thickness Class	25	25	26	26	27	27	28
		Thickness	.73	.73	.79	.79	.85	.85	.92
F	2½	Calculated Thickness	.50*	.51*	.53*	.56*	.61	.67	.73
		Use { Thickness Class	21	21	21	22	23	24	25
		Thickness	.54	.54	.54	.58	.63	.68	.73
	3½	Calculated Thickness	.49*	.51	.53	.57	.62	.67	.73
		Use { Thickness Class	21	21	21	22	23	24	25
		Thickness	.54	.54	.54	.58	.63	.68	.73
	5	Calculated Thickness	.51*	.53	.57	.60	.65	.70	.75
		Use { Thickness Class	21	21	22	22	23	24	25
		Thickness	.54	.54	.58	.58	.63	.68	.73
	8	Calculated Thickness	.57	.59	.62	.65	.69	.74	.78
		Use { Thickness Class	22	22	23	23	24	25	26
		Thickness	.58	.58	.63	.63	.68	.73	.79
	12	Calculated Thickness	.61	.64	.67	.70	.73	.77	.81
		Use { Thickness Class	23	23	24	24	25	26	26
		Thickness	.63	.63	.68	.68	.73	.79	.79
	16	Calculated Thickness	.65	.67	.69	.72	.76	.80	.84
		Use { Thickness Class	23	24	24	25	26	26	27
		Thickness	.63	.68	.68	.73	.79	.79	.85

* Asterisk following total calculated thickness indicates that truck superload (Case 2) is the controlling factor. When total calculated thickness is not followed by asterisk, surge pressure (Case 1) is the controlling factor.   See Sec. 1-2.1.

TABLE 1-1 (*Continued*)

Schedule of Barrel Thickness for Water Pipe of 18/40 Iron Strength

Laying Condition	Depth of Cover ft	Thickness Specifications	Internal Pressure—*psi*						
			50	100	150	200	250	300	350
			Barrel Thicknesses—*in.*						

Twenty-Inch Water Pipe

Laying Condition	Depth of Cover ft	Thickness Specifications	50	100	150	200	250	300	350
A	2½	Calculated Thickness	.62*	.65*	.67*	.70*	.73*	.76*	.80*
		Use {Thickness Class	22	23	23	24	24	25	25
		{Thickness	.62	.67	.67	.72	.72	.78	.78
	3½	Calculated Thickness	.62*	.64*	.66*	.69*	.72*	.76*	.80
		Use {Thickness Class	22	22	23	23	24	25	26
		{Thickness	.62	.62	.67	.67	.72	.78	.84
	5	Calculated Thickness	.65*	.67*	.69*	.72*	.75*	.80	.85
		Use {Thickness Class	23	23	23	24	25	25	26
		{Thickness	.67	.67	.67	.72	.78	.78	.84
	8	Calculated Thickness	.71	.73	.76	.79	.83	.87	.91
		Use {Thickness Class	24	24	25	25	26	26	27
		{Thickness	.72	.72	.78	.78	.84	.84	.91
	12	Calculated Thickness	.78	.80	.83	.86	.89	.93	.97
		Use {Thickness Class	25	25	26	26	27	27	28
		{Thickness	.78	.78	.84	.84	.91	.91	.98
	16	Calculated Thickness	.83	.86	.88	.91	.94	.97	1.01
		Use {Thickness Class	26	26	27	27	27	28	28
		{Thickness	.84	.84	.91	.91	.91	.98	.98
B	2½	Calculated Thickness	.57*	.59*	.62*	.65*	.68*	.72	.78
		Use {Thickness Class	21	21	22	23	23	24	25
		{Thickness	.57	.57	.62	.67	.67	.72	.78
	3½	Calculated Thickness	.57*	.58*	.62*	.64*	.68	.74	.80
		Use {Thickness Class	21	21	22	22	23	24	25
		{Thickness	.57	.57	.62	.62	.67	.72	.78
	5	Calculated Thickness	.59*	.61*	.64*	.67	.72	.77	.82
		Use {Thickness Class	21	22	22	23	24	25	26
		{Thickness	.57	.62	.62	.67	.72	.78	.84
	8	Calculated Thickness	.65	.68	.71	.74	.78	.82	.87
		Use {Thickness Class	23	23	24	24	25	26	26
		{Thickness	.67	.67	.72	.72	.78	.84	.84
	12	Calculated Thickness	.71	.74	.77	.80	.84	.87	.92
		Use {Thickness Class	24	24	25	25	26	26	27
		{Thickness	.72	.72	.78	.78	.84	.84	.91
	16	Calculated Thickness	.76	.78	.81	.84	.87	.91	.95
		Use {Thickness Class	25	25	26	26	26	27	28
		{Thickness	.78	.78	.84	.84	.84	.91	.98
F	2½	Calculated Thickness	.51*	.53*	.57*	.60*	.64	.70	.77
		Use {Thickness Class	21	21	21	22	22	24	25
		{Thickness	.57	.57	.57	.62	.62	.72	.78
	3½	Calculated Thickness	.52	.54	.57	.61	.66	.72	.78
		Use {Thickness Class	21	21	21	22	23	24	25
		{Thickness	.57	.57	.57	.62	.67	.72	.78
	5	Calculated Thickness	.53	.56	.60	.64	.69	.74	.80
		Use {Thickness Class	21	21	22	22	23	24	25
		{Thickness	.57	.57	.62	.62	.67	.72	.78
	8	Calculated Thickness	.60	.62	.65	.69	.73	.78	.83
		Use {Thickness Class	22	22	23	23	24	25	26
		{Thickness	.62	.62	.67	.67	.72	.78	.84
	12	Calculated Thickness	.65	.68	.71	.74	.78	.82	.87
		Use {Thickness Class	23	23	24	24	25	26	26
		{Thickness	.67	.67	.72	.72	.78	.84	.84
	16	Calculated Thickness	.69	.71	.74	.77	.81	.85	.90
		Use {Thickness Class	23	24	24	25	26	26	27
		{Thickness	.67	.72	.72	.78	.84	.84	.91

* Asterisk following total calculated thickness indicates that truck superload (Case 2) is the controlling factor When total calculated thickness is not followed by asterisk, surge pressure (Case 1) is the controlling factor. See Sec. 1–2.1.

TABLE 1–1 (*Continued*)

Schedule of Barrel Thickness for Water Pipe of 18/40 Iron Strength

Laying Condition	Depth of Cover ft	Thickness Specifications	Internal Pressure—*psi*						
			50	100	150	200	250	300	350
			Barrel Thicknesses—*in.*						

Twenty-four-Inch Water Pipe

Laying Condition	Depth of Cover ft	Thickness Specifications	50	100	150	200	250	300	350
A	2½	Calculated Thickness	.69*	.72*	.75*	.78*	.82*	.86*	.91*
		Use Thickness Class	22	23	23	24	25	25	26
		Thickness	.68	.73	.73	.79	.85	.85	.92
	3½	Calculated Thickness	.69*	.71*	.74*	.78*	.81*	.86	.93
		Use Thickness Class	22	23	23	24	24	25	26
		Thickness	.68	.73	.73	.79	.79	.85	.92
	5	Calculated Thickness	.72*	.75*	.78*	.81*	.85*	.91	.96
		Use Thickness Class	23	23	24	24	25	26	27
		Thickness	.73	.73	.79	.79	.85	.92	.99
	8	Calculated Thickness	.80	.82	.86	.90	.94	.99	1.04
		Use Thickness Class	24	25	25	26	26	27	28
		Thickness	.79	.85	.85	.92	.92	.99	1.07
	12	Calculated Thickness	.88	.91	.94	.98	1.02	1.06	1.10
		Use Thickness Class	25	26	26	27	27	28	28
		Thickness	.85	.92	.92	.99	.99	1.07	1.07
	16	Calculated Thickness	.94	.97	1.00	1.03	1.07	1.11	1.15
		Use Thickness Class	26	27	27	28	28	28	29
		Thickness	.92	.99	.99	1.07	1.07	1.07	1.16
B	2½	Calculated Thickness	.63*	.65*	.68*	.72*	.76*	.82	.90
		Use Thickness Class	21	21	22	23	24	25	26
		Thickness	.63	.63	.68	.73	.79	.85	.92
	3½	Calculated Thickness	.62*	.65*	.68*	.71*	.77	.84	.91
		Use Thickness Class	21	21	22	23	24	25	26
		Thickness	.63	.63	.68	.73	.79	.85	.92
	5	Calculated Thickness	.65*	.67*	.71*	.76	.81	.88	.94
		Use Thickness Class	21	22	23	24	24	25	26
		Thickness	.63	.68	.73	.79	.79	.85	.92
	8	Calculated Thickness	.72	.75	.79	.83	.88	.94	.99
		Use Thickness Class	23	23	24	25	25	26	27
		Thickness	.73	.73	.79	.85	.85	.92	.99
	12	Calculated Thickness	.79	.82	.86	.90	.94	.99	1.04
		Use Thickness Class	24	25	25	26	26	27	28
		Thickness	.79	.85	.85	.92	.92	.99	1.07
	16	Calculated Thickness	.85	.88	.91	.95	.99	1.03	1.08
		Use Thickness Class	25	25	26	26	27	28	28
		Thickness	.85	.85	.92	.92	.99	1.07	1.07
F	2½	Calculated Thickness	.56*	.58*	.63*	.67*	.73	.81	.89
		Use Thickness Class	21	21	21	22	23	24	26
		Thickness	.63	.63	.63	.68	.73	.79	.92
	3½	Calculated Thickness	.56*	.59	.63	.69	.75	.82	.90
		Use Thickness Class	21	21	21	22	23	25	26
		Thickness	.63	.63	.63	.68	.73	.85	.92
	5	Calculated Thickness	.59	.63	.67	.72	.78	.84	.92
		Use Thickness Class	21	21	22	23	24	25	26
		Thickness	.63	.63	.68	.73	.79	.85	.92
	8	Calculated Thickness	.66	.69	.73	.77	.83	.89	.95
		Use Thickness Class	22	22	23	24	25	26	26
		Thickness	.68	.68	.73	.79	.85	.92	.92
	12	Calculated Thickness	.72	.75	.79	.83	.88	.93	.99
		Use Thickness Class	23	23	24	25	25	26	27
		Thickness	.73	.73	.79	.85	.85	.92	.99
	16	Calculated Thickness	.77	.80	.83	.87	.91	.96	1.02
		Use Thickness Class	24	24	25	25	26	27	27
		Thickness	.79	.79	.85	.85	.92	.99	.99

* Asterisk following total calculated thickness indicates that truck superload (Case 2) is the controlling factor. When total calculated thickness is not followed by asterisk, surge pressure (Case 1) is the controlling factor. See Sec. 1–2.1.

TABLE 1-1 (*Continued*)

Schedule of Barrel Thickness for Water Pipe of 18/40 Iron Strength

Laying Condition	Depth of Cover ft	Thickness Specifications	Internal Pressure—psi						
			50	100	150	200	250	300	350
			Barrel Thicknesses—in.						

Thirty-Inch Water Pipe

Laying Condition	Depth of Cover ft	Thickness Specifications	50	100	150	200	250	300	350
A	2½	Calculated Thickness	.83*	.86*	.90*	.94*	.98*	1.04*	1.10
		Use { Thickness Class	23	23	24	24	25	26	26
		{ Thickness	.85	.85	.92	.92	.99	1.07	1.07
	3½	Calculated Thickness	.82*	.85*	.89*	.93*	.98*	1.04	1.12
		Use { Thickness Class	23	23	24	24	25	26	27
		{ Thickness	.85	.85	.92	.92	.99	1.07	1.16
	5	Calculated Thickness	.86*	.90*	.93*	.97*	1.03	1.10	1.18
		Use { Thickness Class	23	24	24	25	26	26	27
		{ Thickness	.85	.92	.92	.99	1.07	1.07	1.16
	8	Calculated Thickness	.94*	.98	1.02	1.07	1.12	1.19	1.25
		Use { Thickness Class	24	25	25	26	27	27	28
		{ Thickness	.92	.99	.99	1.07	1.16	1.16	1.25
	12	Calculated Thickness	1.05	1.08	1.12	1.17	1.22	1.28	1.33
		Use { Thickness Class	26	26	27	27	28	28	29
		{ Thickness	1.07	1.07	1.16	1.16	1.25	1.25	1.35
	16	Calculated Thickness	1.13	1.17	1.20	1.25	1.29	1.35	1.40
		Use { Thickness Class	27	27	27	28	28	29	29
		{ Thickness	1.16	1.16	1.16	1.25	1.25	1.35	1.35
B	2½	Calculated Thickness	.74*	.77*	.81*	.85*	.90*	.99	1.09
		Use { Thickness Class	21	22	22	23	24	25	26
		{ Thickness	.73	.79	.79	.85	.92	.99	1.07
	3½	Calculated Thickness	.73*	.76*	.80*	.85*	.92	1.01	1.10
		Use { Thickness Class	21	22	22	23	24	25	26
		{ Thickness	.73	.79	.79	.85	.92	.99	1.07
	5	Calculated Thickness	.77*	.80*	.84*	.90	.97	1.05	1.13
		Use { Thickness Class	22	22	23	24	25	26	27
		{ Thickness	.79	.79	.85	.92	.99	1.07	1.16
	8	Calculated Thickness	.84	.88	.93	.98	1.05	1.11	1.18
		Use { Thickness Class	23	23	24	25	26	26	27
		{ Thickness	.85	.85	.92	.99	1.07	1.07	1.16
	12	Calculated Thickness	.93	.97	1.01	1.06	1.12	1.18	1.25
		Use { Thickness Class	24	25	25	26	27	27	28
		{ Thickness	.92	.99	.99	1.07	1.16	1.16	1.25
	16	Calculated Thickness	1.00	1.04	1.08	1.13	1.18	1.24	1.30
		Use { Thickness Class	25	26	26	27	27	28	29
		{ Thickness	.99	1.07	1.07	1.16	1.16	1.25	1.35
F	2½	Calculated Thickness	.66*	.70*	.74*	.79*	.88	.98	1.08
		Use { Thickness Class	21	21	21	22	23	25	26
		{ Thickness	.73	.73	.73	.79	.85	.99	1.07
	3½	Calculated Thickness	.66*	.70*	.74*	.80	.90	1.00	1.10
		Use { Thickness Class	21	21	21	22	24	25	26
		{ Thickness	.73	.73	.73	.79	.92	.99	1.07
	5	Calculated Thickness	.70*	.73*	.79	.86	.93	1.01	1.11
		Use { Thickness Class	21	21	22	23	24	25	26
		{ Thickness	.73	.73	.79	.85	.92	.99	1.07
	8	Calculated Thickness	.77	.81	.86	.93	1.00	1.07	1.15
		Use { Thickness Class	22	22	23	24	25	26	27
		{ Thickness	.79	.79	.85	.92	.99	1.07	1.16
	12	Calculated Thickness	.85	.89	.94	.99	1.05	1.12	1.20
		Use { Thickness Class	23	24	24	25	26	27	27
		{ Thickness	.85	.92	.92	.99	1.07	1.16	1.16
	16	Calculated Thickness	.91	.95	.99	1.04	1.10	1.16	1.23
		Use { Thickness Class	24	24	25	26	26	27	28
		{ Thickness	.92	.92	.99	1.07	1.07	1.16	1.25

* Asterisk following total calculated thickness indicates that truck superload (Case 2) is the controlling factor. When total calculated thickness is not followed by an asterisk, surge pressure (Case 1) is the controlling factor. See Sec. 1–2.1.

TABLE 1–1 (*Continued*)

Schedule of Barrel Thickness for Water Pipe of 18/40 Iron Strength

Laying Condition	Depth of Cover ft	Thickness Specifications	Internal Pressure—psi						
			50	100	150	200	250	300	350
			Barrel Thicknesses—in.						
		Thirty-six-Inch Water Pipe							
A	2½	Calculated Thickness	.94*	.98*	1.02*	1.07*	1.13*	1.19*	1.27
		Use {Thickness Class	23	24	24	25	25	26	27
		{Thickness	.94	1.02	1.02	1.10	1.10	1.19	1.29
	3½	Calculated Thickness	.93*	.97*	1.02*	1.07*	1.12*	1.19	1.31
		Use {Thickness Class	23	23	24	25	25	26	27
		{Thickness	.94	.94	1.02	1.10	1.10	1.19	1.29
	5	Calculated Thickness	.98*	1.02*	1.07*	1.12*	1.17*	1.26	1.35
		Use {Thickness Class	24	24	25	25	26	27	28
		{Thickness	1.02	1.02	1.10	1.10	1.19	1.29	1.39
	8	Calculated Thickness	1.07*	1.11*	1.16	1.22	1.28	1.36	1.44
		Use {Thickness Class	25	25	26	26	27	28	28
		{Thickness	1.10	1.10	1.19	1.19	1.29	1.39	1.39
	12	Calculated Thickness	1.19	1.24	1.28	1.34	1.40	1.47	1.54
		Use {Thickness Class	26	27	27	28	28	29	29
		{Thickness	1.19	1.29	1.29	1.39	1.39	1.50	1.50
	1	Calculated Thickness	1.29	1.33	1.38	1.43	1.49	1.55	1.62
		Use {Thickness Class	27	27	28	28	29	29	30
		{Thickness	1.29	1.29	1.39	1.39	1.50	1.50	1.62
B	2½	Calculated Thickness	.82*	.86*	.91*	.97*	1.03*	1.14	1.26
		Use {Thickness Class	21	22	23	23	24	25	27
		{Thickness	.81	.87	.94	.94	1.02	1.10	1.29
	3½	Calculated Thickness	.81*	.85*	.90*	.96*	1.04	1.15	1.28
		Use {Thickness Class	21	22	22	23	24	26	27
		{Thickness	.81	.87	.87	.94	1.02	1.19	1.29
	5	Calculated Thickness	.86*	.90*	.94*	1.01	1.10	1.20	1.30
		Use {Thickness Class	22	22	23	24	25	26	27
		{Thickness	.87	.87	.94	1.02	1.10	1.19	1.29
	8	Calculated Thickness	.94	.99	1.05	1.11	1.18	1.27	1.36
		Use {Thickness Class	23	24	24	25	26	27	28
		{Thickness	.94	1.02	1.02	1.10	1.19	1.29	1.39
	12	Calculated Thickness	1.05	1.09	1.14	1.20	1.27	1.34	1.43
		Use {Thickness Class	24	25	25	26	27	28	28
		{Thickness	1.02	1.10	1.10	1.19	1.29	1.39	1.39
	16	Calculated Thickness	1.12	1.17	1.22	1.27	1.34	1.40	1.48
		Use {Thickness Class	25	26	26	27	28	28	29
		{Thickness	1.10	1.19	1.19	1.29	1.39	1.39	1.50
F	2½	Calculated Thickness	.73*	.79*	.84*	.89	1.01	1.13	1.25
		Use {Thickness Class	21	21	22	22	24	25	27
		{Thickness	.81	.81	.87	.87	1.02	1.10	1.29
	3½	Calculated Thickness	.73*	.78*	.83*	.92	1.03	1.14	1.26
		Use {Thickness Class	21	21	21	23	24	25	27
		{Thickness	.81	.81	.81	.94	1.02	1.10	1.29
	5	Calculated Thickness	.78*	.82*	.89	.97	1.07	1.17	1.28
		Use {Thickness Class	21	21	22	23	25	26	27
		{Thickness	.81	.81	.87	.94	1.10	1.19	1.29
	8	Calculated Thickness	.86	.91	.97	1.04	1.13	1.22	1.33
		Use {Thickness Class	22	23	23	24	25	26	27
		{Thickness	.87	.94	.94	1.02	1.10	1.19	1.29
	12	Calculated Thickness	.95	1.00	1.06	1.12	1.20	1.28	1.37
		Use {Thickness Class	23	24	25	25	26	27	28
		{Thickness	.94	1.02	1.10	1.10	1.19	1.29	1.39
	16	Calculated Thickness	1.02	1.07	1.12	1.18	1.25	1.33	1.42
		Use {Thickness Class	24	25	25	26	27	27	28
		{Thickness	1.02	1.10	1.10	1.19	1.29	1.29	1.39

*Asterisk following total calculated thickness indicates that truck superload (Case 2) is the controlling factor. When total calculated thickness is not followed by asterisk, surge pressure (Case 1) is the controlling factor. See Sec. 1–2.1.

TABLE 1–1 (*Continued*)
Schedule of Barrel Thickness for Water Pipe of 18/40 Iron Strength

Laying Condition	Depth of Cover ft	Thickness Specifications	Internal Pressure—psi						
			50	100	150	200	250	300	350
			Barrel Thicknesses—in.						

Forty-two-Inch Water Pipe

Laying Condition	Depth of Cover ft	Thickness Specifications	50	100	150	200	250	300	350
A	2½	Calculated Thickness	1.04*	1.08*	1.14*	1.20*	1.27*	1.34*	1.43
		Use Thickness Class	23	23	24	25	26	26	27
		Thickness	1.05	1.05	1.13	1.22	1.32	1.32	1.43
	3½	Calculated Thickness	1.04*	1.08*	1.14*	1.20*	1.27*	1.33	1.46
		Use Thickness Class	23	23	24	25	26	26	27
		Thickness	1.05	1.05	1.13	1.22	1.32	1.32	1.43
	5	Calculated Thickness	1.09*	1.14*	1.19*	1.25*	1.32*	1.43	1.52
		Use Thickness Class	24	24	25	25	26	27	28
		Thickness	1.13	1.13	1.22	1.22	1.32	1.43	1.54
	8	Calculated Thickness	1.19*	1.24*	1.30	1.37	1.45	1.53	1.63
		Use Thickness Class	25	25	26	26	27	28	29
		Thickness	1.22	1.22	1.32	1.32	1.43	1.54	1.66
	12	Calculated Thickness	1.33	1.38	1.44	1.50	1.58	1.65	1.74
		Use Thickness Class	26	27	27	28	28	29	30
		Thickness	1.32	1.43	1.43	1.54	1.54	1.66	1.79
	16	Calculated Thickness	1.46	1.51	1.56	1.62	1.68	1.75	1.83
		Use Thickness Class	27	28	28	29	29	30	30
		Thickness	1.43	1.54	1.54	1.66	1.66	1.79	1.79
B	2½	Calculated Thickness	.89*	.94*	1.00*	1.07*	1.14	1.28	1.42
		Use Thickness Class	21	22	22	23	24	26	27
		Thickness	.90	.97	.97	1.05	1.13	1.32	1.43
	3½	Calculated Thickness	.89*	.94*	1.00*	1.07*	1.17	1.30	1.43
		Use Thickness Class	21	22	22	23	24	26	27
		Thickness	.90	.97	.97	1.05	1.13	1.32	1.43
	5	Calculated Thickness	.94*	.98*	1.04*	1.13	1.23	1.34	1.47
		Use Thickness Class	22	22	23	24	25	26	27
		Thickness	.97	.97	1.05	1.13	1.22	1.32	1.43
	8	Calculated Thickness	1.04	1.09	1.16	1.23	1.32	1.42	1.53
		Use Thickness Class	23	24	24	25	26	27	28
		Thickness	1.05	1.13	1.13	1.22	1.32	1.43	1.54
	12	Calculated Thickness	1.15	1.21	1.27	1.33	1.42	1.51	1.60
		Use Thickness Class	24	25	26	26	27	28	29
		Thickness	1.13	1.22	1.32	1.32	1.43	1.54	1.66
	16	Calculated Thickness	1.25	1.30	1.36	1.43	1.50	1.58	1.67
		Use Thickness Class	25	26	26	27	28	28	29
		Thickness	1.22	1.32	1.32	1.43	1.54	1.54	1.66
F	2½	Calculated Thickness	.80*	.86*	.92*	1.00	1.15	1.28	1.42
		Use Thickness Class	21	21	21	22	24	26	27
		Thickness	.90	.90	.90	.97	1.13	1.32	1.43
	3½	Calculated Thickness	.80*	.86*	.92*	1.04	1.17	1.29	1.43
		Use Thickness Class	21	21	21	23	24	26	27
		Thickness	.90	.90	.90	1.05	1.13	1.32	1.43
	5	Calculated Thickness	.85*	.90*	.98	1.08	1.19	1.32	1.45
		Use Thickness Class	21	21	22	23	25	26	27
		Thickness	.90	.90	.97	1.05	1.22	1.32	1.43
	8	Calculated Thickness	.94	1.00	1.08	1.16	1.26	1.36	1.49
		Use Thickness Class	22	22	23	24	25	26	28
		Thickness	.97	.97	1.05	1.13	1.22	1.32	1.54
	12	Calculated Thickness	1.05	1.10	1.17	1.25	1.34	1.43	1.54
		Use Thickness Class	23	24	24	25	26	27	28
		Thickness	1.05	1.13	1.13	1.22	1.32	1.43	1.54
	16	Calculated Thickness	1.13	1.18	1.24	1.32	1.40	1.49	1.59
		Use Thickness Class	24	25	25	26	27	28	28
		Thickness	1.13	1.22	1.22	1.32	1.43	1.54	1.54

* Asterisk following total calculated thickness indicates that truck superload (Case 2) is the controlling factor. When total calculated thickness is not followed by asterisk, surge pressure (Case 1) is the controlling factor. See Sec. 1–2.1.

TABLE 1–1 (*Continued*)

Schedule of Barrel Thickness for Water Pipe of 18/40 Iron Strength

Laying Condition	Depth of Cover ft	Thickness Specifications	Internal Pressure—*psi*						
			50	100	150	200	250	300	350
			Barrel Thicknesses—*in.*						

Forty-eight-Inch Water Pipe

Laying Condition	Depth of Cover ft	Thickness Specifications	50	100	150	200	250	300	350
A	2½	Calculated Thickness	1.14*	1.19*	1.25*	1.32*	1.40*	1.49*	1.61
		Use { Thickness Class	23	24	24	25	26	26	27
		Thickness	1.14	1.23	1.23	1.33	1.44	1.44	1.56
	3½	Calculated Thickness	1.14*	1.20*	1.26*	1.33*	1.40*	1.50	1.66
		Use { Thickness Class	23	24	24	25	26	27	28
		Thickness	1.14	1.23	1.23	1.33	1.44	1.56	1.68
	5	Calculated Thickness	1.22*	1.27*	1.32*	1.39*	1.47	1.58	1.71
		Use { Thickness Class	24	24	25	26	26	27	28
		Thickness	1.23	1.23	1.33	1.44	1.44	1.56	1.68
	8	Calculated Thickness	1.33*	1.38*	1.44	1.53	1.61	1.71	1.83
		Use { Thickness Class	25	25	26	27	27	28	29
		Thickness	1.33	1.33	1.44	1.56	1.56	1.68	1.81
	12	Calculated Thickness	1.49	1.55	1.61	1.69	1.77	1.85	1.96
		Use { Thickness Class	26	27	27	28	29	29	30
		Thickness	1.44	1.56	1.56	1.68	1.81	1.81	1.95
	16	Calculated Thickness	1.63	1.68	1.74	1.81	1.89	1.98	2.08
		Use { Thickness Class	28	28	28	29	30	30	
		Thickness	1.68	1.68	1.68	1.81	1.95	1.95	
B	2½	Calculated Thickness	.97*	1.03*	1.09*	1.17*	1.28	1.43	1.59
		Use { Thickness Class	21	22	22	23	25	26	27
		Thickness	.98	1.06	1.06	1.14	1.33	1.44	1.56
	3½	Calculated Thickness	.98*	1.03*	1.10*	1.17	1.31	1.45	1.61
		Use { Thickness Class	21	22	23	23	25	26	27
		Thickness	.98	1.06	1.14	1.14	1.33	1.44	1.56
	5	Calculated Thickness	1.03*	1.09*	1.15*	1.25	1.37	1.50	1.65
		Use { Thickness Class	22	22	23	24	25	27	28
		Thickness	1.06	1.06	1.14	1.23	1.33	1.56	1.68
	8	Calculated Thickness	1.13	1.20	1.28	1.37	1.47	1.58	1.72
		Use { Thickness Class	23	24	25	25	26	27	28
		Thickness	1.14	1.23	1.33	1.33	1.44	1.56	1.68
	12	Calculated Thickness	1.28	1.34	1.41	1.49	1.58	1.69	1.80
		Use { Thickness Class	25	25	26	26	27	28	29
		Thickness	1.33	1.33	1.44	1.44	1.56	1.68	1.81
	16	Calculated Thickness	1.38	1.43	1.50	1.58	1.67	1.77	1.87
		Use { Thickness Class	25	26	27	27	28	29	29
		Thickness	1.33	1.44	1.56	1.56	1.68	1.81	1.81
F	2½	Calculated Thickness	.88*	.93*	1.00*	1.12	1.27	1.43	1.59
		Use { Thickness Class	21	21	21	23	24	26	27
		Thickness	.98	.98	.98	1.14	1.23	1.44	1.56
	3½	Calculated Thickness	.88*	.94*	1.01*	1.14	1.29	1.44	1.60
		Use { Thickness Class	21	21	21	23	25	26	27
		Thickness	.98	.98	.98	1.14	1.33	1.44	1.56
	5	Calculated Thickness	.93*	.99*	1.08	1.20	1.33	1.48	1.63
		Use { Thickness Class	21	21	22	24	25	26	28
		Thickness	.98	.98	1.06	1.23	1.33	1.44	1.68
	8	Calculated Thickness	1.03	1.10	1.19	1.28	1.40	1.52	1.67
		Use { Thickness Class	22	23	24	25	26	27	28
		Thickness	1.06	1.14	1.23	1.33	1.44	1.56	1.68
	12	Calculated Thickness	1.15	1.22	1.29	1.39	1.48	1.60	1.73
		Use { Thickness Class	23	24	25	26	26	27	28
		Thickness	1.14	1.23	1.33	1.44	1.44	1.56	1.68
	16	Calculated Thickness	1.25	1.31	1.38	1.46	1.56	1.67	1.78
		Use { Thickness Class	24	25	25	26	27	28	29
		Thickness	1.23	1.33	1.33	1.44	1.56	1.68	1.81

* Asterisk following total calculated thickness indicates that truck superload (Case 2) is the controlling factor. When total calculated thickness is not followed by asterisk, surge pressure (Case 1) is the controlling factor. See Sec. 1–2.1.

TABLE 1–2

Schedule of Barrel Thickness for Water Pipe of 21/45 Iron Strength

Laying Condi-tion	Depth of Cover ft	Thickness Specifications	Internal Pressure—psi						
			50	100	150	200	250	300	350
			Barrel Thickness—in.						

Three-Inch Water Pipe

Laying Condi-tion	Depth of Cover ft	Thickness Specifications	50	100	150	200	250	300	350
A	2½	Calculated Thickness	0.20*	0.20*	0.20*	0.20*	0.21*	0.21*	0.21
		Use {Thickness Class	22	22	22	22	22	22	22
		{Thickness	0.32	0.32	0.32	0.32	0.32	0.32	0.32
	3½	Calculated Thickness	0.20*	0.20*	0.20*	0.21*	0.21*	0.22*	0.22
		Use {Thickness Class	22	22	22	22	22	22	22
		{Thickness	0.32	0.32	0.32	0.32	0.32	0.32	0.32
	5	Calculated Thickness	0.21	0.21	0.21	0.22	0.22	0.23	0.23
		Use {Thickness Class	22	22	22	22	22	22	22
		{Thickness	0.32	0.32	0.32	0.32	0.32	0.32	0.32
	8	Calculated Thickness	0.22	0.23	0.23	0.23	0.24	0.24	0.25
		Use {Thickness Class	22	22	22	22	22	22	22
		{Thickness	0.32	0.32	0.32	0.32	0.32	0.32	0.32
	12	Calculated Thickness	0.24	0.25	0.25	0.25	0.26	0.26	0.26
		Use {Thickness Class	22	22	22	22	22	22	22
		{Thickness	0.32	0.32	0.32	0.32	0.32	0.32	0.32
	16	Calculated Thickness	0.26	0.26	0.27	0.27	0.27	0.28	0.28
		Use {Thickness Class	22	22	22	22	22	22	22
		{Thickness	0.32	0.32	0.32	0.32	0.32	0.32	0.32
B	2½	Calculated Thickness	0.19*	0.20*	0.20*	0.20*	0.21	0.22	0.22
		Use {Thickness Class	22	22	22	22	22	22	22
		{Thickness	0.32	0.32	0.32	0.32	0.32	0.32	0.32
	3½	Calculated Thickness	0.19*	0.20*	0.20	0.21	0.21	0.22	0.22
		Use {Thickness Class	22	22	22	22	22	22	22
		{Thickness	0.32	0.32	0.32	0.32	0.32	0.32	0.32
	5	Calculated Thickness	0.21	0.21	0.21	0.21	0.22	0.22	0.23
		Use {Thickness Class	22	22	22	22	22	22	22
		{Thickness	0.32	0.32	0.32	0.32	0.32	0.32	0.32
	8	Calculated Thickness	0.22	0.22	0.22	0.23	0.23	0.24	0.24
		Use {Thickness Class	22	22	22	22	22	22	22
		{Thickness	0.32	0.32	0.32	0.32	0.32	0.32	0.32
	12	Calculated Thickness	0.24	0.24	0.24	0.25	0.25	0.25	0.26
		Use {Thickness Class	22	22	22	22	22	22	22
		{Thickness	0.32	0.32	0.32	0.32	0.32	0.32	0.32
	16	Calculated Thickness	0.25	0.26	0.26	0.26	0.27	0.27	0.27
		Use {Thickness Class	22	22	22	22	22	22	22
		{Thickness	0.32	0.32	0.32	0.32	0.32	0.32	0.32
F	2½	Calculated Thickness	0.19*	0.19*	0.19	0.20	0.20	0.21	0.22
		Use {Thickness Class	22	22	22	22	22	22	22
		{Thickness	0.32	0.32	0.32	0.32	0.32	0.32	0.32
	3½	Calculated Thickness	0.19*	0.19	0.19	0.20	0.20	0.21	0.22
		Use {Thickness Class	22	22	22	22	22	22	22
		{Thickness	0.32	0.32	0.32	0.32	0.32	0.32	0.32
	5	Calculated Thickness	0.19	0.20	0.20	0.21	0.21	0.22	0.22
		Use {Thickness Class	22	22	22	22	22	22	22
		{Thickness	0.32	0.32	0.32	0.32	0.32	0.32	0.32
	8	Calculated Thickness	0.21	0.21	0.21	0.22	0.22	0.23	0.23
		Use {Thickness Class	22	22	22	22	22	22	22
		{Thickness	0.32	0.32	0.32	0.32	0.32	0.32	0.32
	12	Calculated Thickness	0.22	0.22	0.23	0.23	0.24	0.24	0.25
		Use {Thickness Class	22	22	22	22	22	22	22
		{Thickness	0.32	0.32	0.32	0.32	0.32	0.32	0.32
	16	Calculated Thickness	0.24	0.24	0.24	0.25	0.25	0.25	0.26
		Use {Thickness Class	22	22	22	22	22	22	22
		{Thickness	0.32	0.32	0.32	0.32	0.32	0.32	0.32

4
1

* Asterisk following total calculated thickness indicates that truck superload (Case 2) is the controlling factor. When total calculated thickness is not followed by asterisk, surge pressure (Case 1) is the controlling factor. See Sec. 1–2.1.

TABLE 1-2 (*Continued*)

Schedule of Barrel Thickness for Water Pipe of 21/45 Iron Strength

Laying Condition	Depth of Cover ft	Thickness Specifications	Internal Pressure—psi						
			50	100	150	200	250	300	350
			Barrel Thickness—in.						

Four-Inch Water Pipe

Laying Condition	Depth of Cover ft	Thickness Specifications	50	100	150	200	250	300	350
A	2½	Calculated Thickness	0.22*	0.23*	0.23*	0.24*	0.24*	0.25*	0.25*
		Use { Thickness Class	22	22	22	22	22	22	22
		Thickness	0.35	0.35	0.35	0.35	0.35	0.35	0.35
	3½	Calculated Thickness	0.22*	0.22*	0.23*	0.23*	0.24	0.25	0.26
		Use { Thickness Class	22	22	22	22	22	22	22
		Thickness	0.35	0.35	0.35	0.35	0.35	0.35	0.35
	5	Calculated Thickness	0.23*	0.23*	0.24*	0.24*	0.25	0.26	0.26
		Use { Thickness Class	22	22	22	22	22	22	22
		Thickness	0.35	0.35	0.35	0.35	0.35	0.35	0.35
	8	Calculated Thickness	0.25	0.26	0.26	0.27	0.27	0.28	0.28
		Use { Thickness Class	22	22	22	22	22	22	22
		Thickness	0.35	0.35	0.35	0.35	0.35	0.35	0.35
	12	Calculated Thickness	0.28	0.28	0.29	0.29	0.29	0.30	0.31
		Use { Thickness Class	22	22	22	22	22	22	22
		Thickness	0.35	0.35	0.35	0.35	0.35	0.35	0.35
	16	Calculated Thickness	0.30	0.30	0.31	0.31	0.32	0.32	0.33
		Use { Thickness Class	22	22	22	22	22	22	22
		Thickness	0.35	0.35	0.35	0.35	0.35	0.35	0.35
B	2½	Calculated Thickness	0.22*	0.22*	0.23*	0.23*	0.24*	0.24*	0.24
		Use { Thickness Class	22	22	22	22	22	22	22
		Thickness	0.35	0.35	0.35	0.35	0.35	0.35	0.35
	3½	Calculated Thickness	0.22*	0.22*	0.22*	0.23*	0.24	0.24	0.25
		Use { Thickness Class	22	22	22	22	22	22	22
		Thickness	0.35	0.35	0.35	0.35	0.35	0.35	0.35
	5	Calculated Thickness	0.22	0.23	0.23	0.24	0.25	0.25	0.26
		Use { Thickness Class	22	22	22	22	22	22	22
		Thickness	0.35	0.35	0.35	0.35	0.35	0.35	0.35
	8	Calculated Thickness	0.25	0.25	0.25	0.26	0.27	0.28	0.28
		Use { Thickness Class	22	22	22	22	22	22	22
		Thickness	0.35	0.35	0.35	0.35	0.35	0.35	0.35
	12	Calculated Thickness	0.27	0.27	0.28	0.28	0.29	0.29	0.30
		Use { Thickness Class	22	22	22	22	22	22	22
		Thickness	0.35	0.35	0.35	0.35	0.35	0.35	0.35
	16	Calculated Thickness	0.29	0.29	0.30	0.30	0.31	0.31	0.32
		Use { Thickness Class	22	22	22	22	22	22	22
		Thickness	0.35	0.35	0.35	0.35	0.35	0.35	0.35
F	2½	Calculated Thickness	0.21*	0.21*	0.21*	0.22*	0.23	0.24	0.25
		Use { Thickness Class	22	22	22	22	22	22	22
		Thickness	0.35	0.35	0.35	0.35	0.35	0.35	0.35
	3½	Calculated Thickness	0.20*	0.21	0.21	0.22	0.23	0.24	0.25
		Use { Thickness Class	22	22	22	22	22	22	22
		Thickness	0.35	0.35	0.35	0.35	0.35	0.35	0.35
	5	Calculated Thickness	0.21	0.22	0.22	0.23	0.24	0.25	0.25
		Use { Thickness Class	22	22	22	22	22	22	22
		Thickness	0.35	0.35	0.35	0.35	0.35	0.35	0.35
	8	Calculated Thickness	0.23	0.24	0.24	0.25	0.25	0.26	0.27
		Use { Thickness Class	22	22	22	22	22	22	22
		Thickness	0.35	0.35	0.35	0.35	0.35	0.35	0.35
	12	Calculated Thickness	0.25	0.26	0.26	0.27	0.27	0.28	0.28
		Use { Thickness Class	22	22	22	22	22	22	22
		Thickness	0.35	0.35	0.35	0.35	0.35	0.35	0.35
	16	Calculates Thickness	0.27	0.27	0.28	0.28	0.29	0.29	0.30
		Use { Thickness Class	22	22	22	22	22	22	22
		Thickness	0.35	0.35	0.35	0.35	0.35	0.35	0.35

* Asterisk following total calculated thickness indicates that truck superload (Case 2) is the controlling factor. When total calculated thickness is not followed by asterisk, surge pressure (Case 1) is the controlling factor.   See Sec .1-2.1.

TABLE 1-2 (*Continued*)

Schedule of Barrel Thickness for Water Pipe of 21/45 Iron Strength

Laying Condition	Depth of Cover ft	Thickness Specifications	Internal Pressure—psi						
			50	100	150	200	250	300	350
			Barrel Thickness—in.						

Six-Inch Water Pipe

Laying Condition	Depth of Cover ft	Thickness Specifications	50	100	150	200	250	300	350
A	2½	Calculated Thickness	0.28*	0.28*	0.29*	0.30*	0.30*	0.31*	0.32*
		Use {Thickness Class	21	21	21	21	21	21	21
		Thickness	0.35	0.35	0.35	0.35	0.35	0.35	0.35
	3½	Calculated Thickness	0.27*	0.27*	0.28*	0.29*	0.29*	0.30*	0.32
		Use {Thickness Class	21	21	21	21	21	21	21
		Thickness	0.35	0.35	0.35	0.35	0.35	0.35	0.35
	5	Calculated Thickness	0.28*	0.28*	0.29	0.30	0.31	0.32	0.33
		Use {Thickness Class	21	21	21	21	21	21	21
		Thickness	0.35	0.35	0.35	0.35	0.35	0.35	0.35
	8	Calculated Thickness	0.31	0.31	0.32	0.33	0.34	0.35	0.36
		Use {Thickness Class	21	21	21	21	21	21	21
		Thickness	0.35	0.35	0.35	0.35	0.35	0.35	0.35
	12	Calculated Thickness	0.34	0.35	0.36	0.36	0.37	0.38	0.39
		Use {Thickness Class	21	21	21	21	22	22	22
		Thickness	0.35	0.35	0.35	0.35	0.38	0.38	0.38
	16	Calculated Thickness	0.38	0.38	0.39	0.39	0.40	0.41	0.42
		Use {Thickness Class	22	22	22	22	23	23	23
		Thickness	0.38	0.38	0.38	0.38	0.41	0.41	0.41
B	2½	Calculated Thickness	0.27*	0.27*	0.28*	0.29*	0.29*	0.30*	0.31*
		Use {Thickness Class	21	21	21	21	21	21	21
		Thickness	0.35	0.35	0.35	0.35	0.35	0.35	0.35
	3½	Calculated Thickness	0.26*	0.27*	0.27*	0.28*	0.29*	0.30	0.31
		Use {Thickness Class	21	21	21	21	21	21	21
		Thickness	0.35	0.35	0.35	0.35	0.35	0.35	0.35
	5	Calculated Thickness	0.27*	0.27*	0.28*	0.29	0.30	0.31	0.32
		Use {Thickness Class	21	21	21	21	21	21	21
		Thickness	0.35	0.35	0.35	0.35	0.35	0.35	0.35
	8	Calculated Thickness	0.30	0.30	0.31	0.32	0.33	0.34	0.35
		Use {Thickness Class	21	21	21	21	21	21	21
		Thickness	0.35	0.35	0.35	0.35	0.35	0.35	0.35
	12	Calculated Thickness	0.33	0.34	0.34	0.35	0.36	0.37	0.38
		Use {Thickness Class	21	21	21	21	21	22	22
		Thickness	0.35	0.35	0.35	0.35	0.35	0.38	0.38
	16	Calculated Thickness	0.36	0.37	0.37	0.38	0.39	0.39	0.40
		Use {Thickness Class	21	22	22	22	22	22	23
		Thickness	0.35	0.38	0.38	0.38	0.38	0.38	0.41
F	2½	Calculated Thickness	0.25*	0.26*	0.26*	0.27*	0.28*	0.29*	0.30
		Use {Thickness Class	21	21	21	21	21	21	21
		Thickness	0.35	0.35	0.35	0.35	0.35	0.35	0.35
	3½	Calculated Thickness	0.24*	0.25*	0.25	0.26	0.28	0.29	0.31
		Use {Thickness Class	21	21	21	21	21	21	21
		Thickness	0.35	0.35	0.35	0.35	0.35	0.35	0.35
	5	Calculated Thickness	0.25	0.26	0.27	0.28	0.29	0.30	0.31
		Use {Thickness Class	21	21	21	21	21	21	21
		Thickness	0.35	0.35	0.35	0.35	0.35	0.35	0.35
	8	Calculated Thickness	0.28	0.28	0.29	0.30	0.31	0.32	0.33
		Use {Thickness Class	21	21	21	21	21	21	21
		Thickness	0.35	0.35	0.35	0.35	0.35	0.35	0.35
	12	Calculated Thickness	0.31	0.31	0.32	0.33	0.34	0.34	0.35
		Use {Thickness Class	21	21	21	21	21	21	21
		Thickness	0.35	0.35	0.35	0.35	0.35	0.35	0.35
	16	Calculated Thickness	0.33	0.34	0.34	0.35	0.36	0.37	0.38
		Use {Thickness Class	21	21	21	21	21	22	22
		Thickness	0.35	0.35	0.35	0.35	0.35	0.38	0.38

* Asterisk following total calculated thickness indicates that truck superload (Case 2) is the controlling factor. When total calculated thickness is not followed by asterisk, surge pressure (Case 1) is the controlling factor. See Sec. 1-2.1.

TABLE 1-2 (*Continued*)

Schedule of Barrel Thickness for Water Pipe of 21/45 Iron Strength

Laying Condition	Depth of Cover ft	Thickness Specifications	Internal Pressure—psi						
			50	100	150	200	250	300	350
			Barrel Thickness—in.						

Eight-Inch Water Pipe

Laying Condition	Depth of Cover ft	Thickness Specifications	50	100	150	200	250	300	350
A	2½	Calculated Thickness	0.33*	0.33*	0.34*	0.35*	0.36*	0.37*	0.38*
		Use { Thickness Class	20	20	20	20	20	21	21
		Thickness	0.35	0.35	0.35	0.35	0.35	0.38	0.38
	3½	Calculated Thickness	0.32*	0.32*	0.33*	0.34*	0.35*	0.36*	0.38
		Use { Thickness Class	20	20	20	20	20	20	21
		Thickness	0.35	0.35	0.35	0.35	0.35	0.35	0.38
	5	Calculated Thickness	0.32*	0.33*	0.34*	0.35	0.36	0.38	0.39
		Use { Thickness Class	20	20	20	20	20	21	21
		Thickness	0.35	0.35	0.35	0.35	0.35	0.38	0.38
	8	Calculated Thickness	0.36	0.37	0.38	0.39	0.40	0.41	0.43
		Use { Thickness Class	20	21	21	21	22	22	23
		Thickness	0.35	0.38	0.38	0.38	0.41	0.41	0.44
	12	Calculated Thickness	0.41	0.42	0.43	0.44	0.45	0.46	0.47
		Use { Thickness Class	22	22	23	23	23	24	24
		Thickness	0.41	0.41	0.44	0.44	0.44	0.48	0.48
	16	Calculated Thickness	0.44	0.45	0.46	0.47	0.47	0.48	0.50
		Use { Thickness Class	23	23	24	24	24	24	25
		Thickness	0.44	0.44	0.48	0.48	0.48	0.48	0.52
B	2½	Calculated Thickness	0.31*	0.32*	0.33*	0.34*	0.35*	0.36*	0.37*
		Use { Thickness Class	20	20	20	20	20	20	21
		Thickness	0.35	0.35	0.35	0.35	0.35	0.35	0.38
	3½	Calculated Thickness	0.30*	0.31*	0.32*	0.33*	0.34*	0.35	0.37
		Use { Thickness Class	20	20	20	20	20	20	21
		Thickness	0.35	0.35	0.35	0.35	0.35	0.35	0.38
	5	Calculated Thickness	0.31*	0.32*	0.33	0.34	0.35	0.37	0.38
		Use { Thickness Class	20	20	20	20	20	21	21
		Thickness	0.35	0.35	0.35	0.35	0.35	0.38	0.38
	8	Calculated Thickness	0.35	0.35	0.36	0.37	0.39	0.40	0.41
		Use { Thickness Class	20	20	20	21	21	22	22
		Thickness	0.35	0.35	0.35	0.38	0.38	0.41	0.41
	12	Calculated Thickness	0.39	0.40	0.41	0.42	0.43	0.44	0.45
		Use { Thickness Class	21	22	22	22	23	23	23
		Thickness	0.38	0.41	0.41	0.41	0.44	0.44	0.44
	16	Calculated Thickness	0.42	0.43	0.44	0.44	0.45	0.46	0.48
		Use { Thickness Class	22	23	23	23	23	24	24
		Thickness	0.41	0.44	0.44	0.44	0.44	0.48	0.48
F	2½	Calculated Thickness	0.29*	0.30*	0.30*	0.31*	0.32*	0.34	0.36
		Use { Thickness Class	20	20	20	20	20	20	20
		Thickness	0.35	0.35	0.35	0.35	0.35	0.35	0.35
	3½	Calculated Thickness	0.28*	0.29*	0.30*	0.31*	0.33	0.35	0.36
		Use { Thickness Class	20	20	20	20	20	20	20
		Thickness	0.35	0.35	0.35	0.35	0.35	0.35	0.35
	5	Calculated Thickness	0.29*	0.30	0.31	0.32	0.34	0.36	0.37
		Use { Thickness Class	20	20	20	20	20	20	21
		Thickness	0.35	0.35	0.35	0.35	0.35	0.35	0.38
	8	Calculated Thickness	0.32	0.33	0.34	0.35	0.36	0.38	0.39
		Use { Thickness Class	20	20	20	20	20	21	21
		Thickness	0.35	0.35	0.35	0.35	0.35	0.38	0.38
	12	Calculated Thickness	0.36	0.37	0.38	0.39	0.40	0.41	0.42
		Use { Thickness Class	20	21	21	21	22	22	22
		Thickness	0.35	0.38	0.38	0.38	0.41	0.41	0.41
	16	Calculated Thickness	0.38	0.39	0.40	0.41	0.42	0.43	0.44
		Use { Thickness Class	21	21	22	22	22	23	23
		Thickness	0.38	0.38	0.41	0.41	0.41	0.44	0.44

* Asterisk following total calculated thickness indicates that truck superload (Case 2) is the controlling factor When total calculated thickness is not followed by asterisk, surge pressure (Case 1) is the controlling factor. See Sec. 1-2.1.

TABLE 1–2 (*Continued*)

Schedule of Barrel Thickness for Water Pipe of 21/45 Iron Strength

Laying Condi- tion	Depth of Cover ft	Thickness Specifications	Internal Pressure—*psi*						
			50	100	150	200	250	300	350
			Barrel Thickness—*in.*						

Ten-Inch Water Pipe

Laying Condi- tion	Depth of Cover ft	Thickness Specifications	50	100	150	200	250	300	350
A	2½	Calculated Thickness	0.38*	0.39*	0.40*	0.41*	0.42*	0.44*	0.45*
		Use {Thickness Class	20	20	21	21	21	22	22
		{Thickness	0.38	0.38	0.41	0.41	0.41	0.44	0.44
	3½	Calculated Thickness	0.37*	0.38*	0.39*	0.40*	0.41*	0.43*	0.44
		Use {Thickness Class	20	20	20	21	21	22	22
		{Thickness	0.38	0.38	0.38	0.41	0.41	0.44	0.44
	5	Calculated Thickness	0.38*	0.39*	0.40*	0.41	0.43	0.44	0.46
		Use {Thickness Class	20	20	21	21	22	22	23
		{Thickness	0.38	0.38	0.41	0.41	0.44	0.44	0.48
	8	Calculated Thickness	0.42	0.44	0.45	0.46	0.47	0.49	0.51
		Use {Thickness Class	21	22	22	23	23	23	24
		{Thickness	0.41	0.44	0.44	0.48	0.48	0.48	0.52
	12	Calculated Thickness	0.48	0.49	0.50	0.51	0.53	0.54	0.56
		Use {Thickness Class	23	23	24	24	24	25	25
		{Thickness	0.48	0.48	0.52	0.52	0.52	0.56	0.56
	16	Calculated Thickness	0.51	0.52	0.53	0.54	0.55	0.56	0.58
		Use {Thickness Class	24	24	24	25	25	25	26
		{Thickness	0.52	0.52	0.52	0.56	0.56	0.56	0.60
B	2½	Calculated Thickness	0.36*	0.37*	0.38*	0.39*	0.41*	0.42*	0.44*
		Use {Thickness Class	20	20	20	20	21	21	22
		{Thickness	0.38	0.38	0.38	0.38	0.41	0.41	0.44
	3½	Calculated Thickness	0.35*	0.36*	0.37*	0.39*	0.40*	0.41	0.44
		Use {Thickness Class	20	20	20	20	21	21	22
		{Thickness	0.38	0.38	0.38	0.38	0.41	0.41	0.44
	5	Calculated Thickness	0.36*	0.37*	0.38*	0.40	0.41	0.43	0.46
		Use {Thickness Class	20	20	20	21	21	22	23
		{Thickness	0.38	0.38	0.38	0.41	0.41	0.44	0.48
	8	Calculated Thickness	0.40	0.42	0.43	0.44	0.46	0.47	0.49
		Use {Thickness Class	21	21	22	22	23	23	23
		{Thickness	0.41	0.41	0.44	0.44	0.48	0.48	0.48
	12	Calculated Thickness	0.46	0.47	0.48	0.49	0.50	0.52	0.53
		Use {Thickness Class	23	23	23	23	24	24	24
		{Thickness	0.48	0.48	0.48	0.48	0.52	0.52	0.52
	16	Calculated Thickness	0.48	0.49	0.50	0.51	0.52	0.54	0.55
		Use {Thickness Class	23	23	24	24	24	25	25
		{Thickness	0.48	0.48	0.52	0.52	0.52	0.56	0.56
F	2½	Calculated Thickness	0.33*	0.34*	0.35*	0.37*	0.38*	0.40	0.43
		Use {Thickness Class	20	20	20	20	20	21	22
		{Thickness	0.38	0.38	0.38	0.38	0.38	0.41	0.44
	3½	Calculated Thickness	0.33*	0.33*	0.35*	0.36*	0.38	0.40	0.43
		Use {Thickness Class	20	20	20	20	20	21	22
		{Thickness	0.38	0.38	0.38	0.38	0.38	0.41	0.44
	5	Calculated Thickness	0.33*	0.34	0.36	0.38	0.40	0.42	0.44
		Use {Thickness Class	20	20	20	20	21	21	22
		{Thickness	0.38	0.38	0.38	0.38	0.41	0.41	0.44
	8	Calculated Thickness	0.37	0.38	0.40	0.41	0.43	0.45	0.47
		Use {Thickness Class	20	20	21	21	22	22	23
		{Thickness	0.38	0.38	0.41	0.41	0.44	0.44	0.48
	12	Calculated Thickness	0.42	0.43	0.44	0.45	0.47	0.48	0.50
		Use {Thickness Class	21	22	22	22	23	23	24
		{Thickness	0.41	0.44	0.44	0.44	0.48	0.48	0.52
	16	Calculated Thickness	0.44	0.45	0.46	0.47	0.48	0.50	0.52
		Use {Thickness Class	22	22	23	23	23	24	24
		{Thickness	0.44	0.44	0.48	0.48	0.48	0.52	0.52

4
5

* Asterisk following total calculated thickness indicates that truck superload (Case 2) is the controlling factor. When total calculated thickness is not followed by asterisk, surge pressure (Case 1) is the controlling factor. See Sec. 1–2.1.

TABLE 1-2 (*Continued*)

Schedule of Barrel Thickness for Water Pipe of 21/45 Iron Strength

Laying Condition	Depth of Cover ft	Thickness Specifications	Internal Pressure—psi						
			50	100	150	200	250	300	350
			Barrel Thickness—in.						

Twelve-Inch Water Pipe

Laying Condition	Depth of Cover ft	Thickness Specifications	50	100	150	200	250	300	350
A	2½	Calculated Thickness	0.43*	0.44*	0.45*	0.47*	0.48*	0.49*	0.51*
		Use Thickness Class	21	21	21	22	22	22	23
		Thickness	0.44	0.44	0.44	0.48	0.48	0.48	0.52
	3½	Calculated Thickness	0.42*	0.43*	0.44*	0.45*	0.47*	0.48*	0.50*
		Use Thickness Class	20	21	21	21	22	22	23
		Thickness	0.41	0.44	0.44	0.44	0.48	0.48	0.52
	5	Calculated Thickness	0.43*	0.44*	0.45*	0.46*	0.48	0.50	0.52
		Use Thickness Class	21	21	21	22	22	23	23
		Thickness	0.44	0.44	0.44	0.48	0.48	0.52	0.52
	8	Calculated Thickness	0.48	0.49	0.50	0.52	0.54	0.55	0.57
		Use Thickness Class	22	22	23	23	24	24	24
		Thickness	0.48	0.48	0.52	0.52	0.56	0.56	0.56
	12	Calculated Thickness	0.53	0.54	0.56	0.57	0.58	0.60	0.62
		Use Thickness Class	23	24	24	24	25	25	25
		Thickness	0.52	0.56	0.56	0.56	0.60	0.60	0.60
	16	Calculated Thickness	0.56	0.57	0.58	0.60	0.61	0.63	0.64
		Use Thickness Class	24	24	25	25	25	26	26
		Thickness	0.56	0.56	0.60	0.60	0.60	0.65	0.65
B	2½	Calculated Thickness	0.41*	0.42*	0.43*	0.44*	0.46*	0.47*	0.49*
		Use Thickness Class	20	20	21	21	22	22	22
		Thickness	0.41	0.41	0.44	0.44	0.48	0.48	0.48
	3½	Calculated Thickness	0.39*	0.40*	0.42*	0.43*	0.45*	0.46*	0.49
		Use Thickness Class	20	20	20	21	21	22	22
		Thickness	0.41	0.41	0.41	0.44	0.44	0.48	0.48
	5	Calculated Thickness	0.40*	0.41*	0.42*	0.44*	0.46	0.48	0.51
		Use Thickness Class	20	20	20	21	22	22	23
		Thickness	0.41	0.41	0.41	0.44	0.48	0.48	0.52
	8	Calculated Thickness	0.45	0.46	0.48	0.49	0.51	0.53	0.55
		Use Thickness Class	21	22	22	22	23	23	24
		Thickness	0.44	0.48	0.48	0.48	0.52	0.52	0.56
	12	Calculated Thickness	0.50	0.51	0.52	0.54	0.56	0.57	0.59
		Use Thickness Class	23	23	23	24	24	24	25
		Thickness	0.52	0.52	0.52	0.56	0.56	0.56	0.60
	16	Calculated Thickness	0.52	0.54	0.55	0.56	0.58	0.60	0.61
		Use Thickness Class	23	24	24	24	25	25	25
		Thickness	0.52	0.56	0.56	0.56	0.60	0.60	0.60
F	2½	Calculated Thickness	0.37*	0.38*	0.39*	0.41*	0.42*	0.44*	0.48
		Use Thickness Class	20	20	20	20	20	21	22
		Thickness	0.41	0.41	0.41	0.41	0.41	0.44	0.48
	3½	Calculated Thickness	0.36*	0.37*	0.38*	0.40*	0.42	0.45	0.48
		Use Thickness Class	20	20	20	20	20	21	22
		Thickness	0.41	0.41	0.41	0.41	0.41	0.44	0.48
	5	Calculated Thickness	0.37*	0.38*	0.39*	0.42	0.44	0.47	0.49
		Use Thickness Class	20	20	20	20	21	22	22
		Thickness	0.41	0.41	0.41	0.41	0.44	0.48	0.48
	8	Calculated Tjickness	0.41	0.43	0.44	0.46	0.48	0.50	0.53
		Use Thickness Class	20	21	21	22	22	23	23
		Thickness	0.41	0.44	0.44	0.48	0.48	0.52	0.52
	12	Calculated Thickness	0.45	0.47	0.48	0.50	0.52	0.54	0.56
		Use Thickness Class	21	22	22	23	23	24	24
		Thickness	0.44	0.48	0.48	0.52	0.52	0.56	0.56
	16	Calculated Thickness	0.48	0.49	0.50	0.52	0.54	0.55	0.57
		Use Thickness Class	22	22	23	23	24	24	24
		Thickness	0.48	0.48	0.52	0.52	0.56	0.56	0.56

* Asterisk following total calculated thickness indicates that truck superload (Case 2) is the controlling factor. When total calculated thickness is not followed by asterisk, surge pressure (Case 1) is the controlling factor. See Sec. 1-2.1.

TABLE 1–2 (*Continued*)
Schedule of Barrel Thickness for Water Pipe of 21/45 Iron Strength

Laying Condition	Depth of Cover ft	Thickness Specifications	Internal Pressure—*psi*						
			50	100	150	200	250	300	350
			Barrel Thickness—*in.*						

Fourteen-Inch Water Pipe

Laying Condition	Depth of Cover ft	Thickness Specifications	50	100	150	200	250	300	350
A	2½	Calculated Thickness	0.48*	0.50*	0.51*	0.53*	0.54*	0.56*	0.58*
		Use {Thickness Class	21	22	22	23	23	23	24
		Thickness	0.48	0.51	0.51	0.55	0.55	0.55	0.59
	3½	Calculated Thickness	0.47*	0.49*	0.50*	0.52*	0.53*	0.55*	0.58
		Use {Thickness Class	21	21	22	22	23	23	24
		Thickness	0.48	0.48	0.51	0.51	0.55	0.55	0.59
	5	Calculated Thickness	0.49*	0.51*	0.52*	0.53*	0.55*	0.58	0.60
		Use {Thickness Class	21	22	22	23	23	24	24
		Thickness	0.48	0.51	0.51	0.55	0.55	0.59	0.59
	8	Calculated Thickness	0.55	0.56	0.58	0.60	0.62	0.64	0.66
		Use {Thickness Class	23	23	24	24	25	25	25
		Thickness	0.55	0.55	0.59	0.59	0.64	0.64	0.64
	12	Calculated Thickness	0.60	0.61	0.63	0.65	0.67	0.68	0.71
		Use {Thickness Class	24	24	25	25	26	26	26
		Thickness	0.59	0.59	0.64	0.64	0.69	0.69	0.69
	16	Calculated Thickness	0.63	0.65	0.66	0.68	0.69	0.71	0.73
		Use {Thickness Class	25	25	25	26	26	26	27
		Thickness	0.64	0.64	0.64	0.69	0.69	0.69	0.75
B	2½	Calculated Thickness	0.45*	0.47*	0.48*	0.50*	0.51*	0.53*	0.56*
		Use {Thickness Class	21	21	21	22	22	23	23
		Thickness	0.48	0.48	0.48	0.51	0.51	0.55	0.55
	3½	Calculated Thickness	0.44*	0.46*	0.47*	0.49*	0.51*	0.53	0.57
		Use {Thickness Class	21	21	21	21	22	23	24
		Thickness	0.48	0.48	0.48	0.48	0.51	0.55	0.59
	5	Calculated Thickness	0.46*	0.47*	0.49*	0.50	0.53	0.56	0.59
		Use {Thickness Class	21	21	21	22	23	23	24
		Thickness	0.48	0.48	0.48	0.51	0.55	0.55	0.59
	8	Calculated Thickness	0.52	0.53	0.55	0.57	0.59	0.61	0.64
		Use {Thickness Class	22	23	23	24	24	24	25
		Thickness	0.51	0.55	0.55	0.59	0.59	0.59	0.64
	12	Calculated Thickness	0.56	0.57	0.59	0.61	0.63	0.65	0.67
		Use {Thickness Class	23	24	24	24	25	25	26
		Thickness	0.55	0.59	0.59	0.59	0.64	0.64	0.69
	16	Calculated Thickness	0.59	0.60	0.62	0.64	0.65	0.67	0.70
		Use {Thickness Class	24	24	25	25	25	26	26
		Thickness	0.59	0.59	0.64	0.64	0.64	0.69	0.69
F	2½	Calculated Thickness	0.42*	0.43*	0.44*	0.46*	0.48*	0.51	0.55
		Use {Thickness Class	21	21	21	21	21	22	23
		Thickness	0.48	0.48	0.48	0.48	0.48	0.51	0.55
	3½	Calculated Thickness	0.41*	0.42*	0.44*	0.45*	0.48	0.52	0.56
		Use {Thickness Class	21	21	21	21	21	22	23
		Thickness	0.48	0.48	0.48	0.48	0.48	0.51	0.55
	5	Calculated Thickness	0.42*	0.43*	0.45	0.48	0.51	0.54	0.57
		Use {Thickness Class	21	21	21	21	22	23	24
		Thickness	0.48	0.48	0.48	0.48	0.51	0.55	0.59
	8	Calculated Thickness	0.47	0.49	0.51	0.53	0.55	0.58	0.61
		Use {Thickness Class	21	21	22	23	23	24	24
		Thickness	0.48	0.48	0.51	0.55	0.55	0.59	0.59
	12	Calculated Thickness	0.51	0.53	0.54	0.56	0.59	0.61	0.63
		Use {Thickness Class	22	23	23	23	24	24	25
		Thickness	0.51	0.55	0.55	0.55	0.59	0.59	0.64
	16	Calculated Thickness	0.54	0.55	0.57	0.59	0.61	0.63	0.65
		Use {Thickness Class	23	23	24	24	24	25	25
		Thickness	0.55	0.55	0.59	0.59	0.59	0.64	0.64

4
7

* Asterisk following total calculated thickness indicates that truck superload (Case 2) is the controlling factor. When total calculated thickness is not followed by asterisk, surge pressure (Case 1) is the controlling factor. See Sec. 1–2.1.

TABLE 1-2 (*Continued*)
Schedule of Barrel Thickness for Water Pipe of 21/45 Iron Strength

Laying Condition	Depth of Cover ft	Thickness Specifications	Internal Pressure—psi						
			50	100	150	200	250	300	350
			Barrel Thickness—in.						

Sixteen-Inch Water Pipe

Laying Condition	Depth of Cover ft	Thickness Specifications	50	100	150	200	250	300	350
A	2½	Calculated Thickness	0.52*	0.53*	0.55*	0.57*	0.59*	0.61*	0.63*
		Use {Thickness Class	22	22	22	23	23	24	24
		Thickness	0.54	0.54	0.54	0.58	0.58	0.63	0.63
	3½	Calculated Thickness	0.51*	0.53*	0.54*	0.56*	0.58*	0.60*	0.63*
		Use {Thickness Class	21	22	22	23	23	23	24
		Thickness	0.50	0.54	0.54	0.58	0.58	0.58	0.63
	5	Calculated Thickness	0.53*	0.55*	0.56*	0.58*	0.60*	0.63	0.66
		Use {Thickness Class	22	22	23	23	23	24	25
		Thickness	0.54	0.54	0.58	0.58	0.58	0.63	0.68
	8	Calculated Thickness	0.59	0.61	0.62	0.64	0.67	0.69	0.72
		Use {Thickness Class	23	24	24	24	25	25	26
		Thickness	0.58	0.63	0.63	0.63	0.68	0.68	0.73
	12	Calculated Thickness	0.64	0.66	0.68	0.70	0.72	0.74	0.77
		Use {Thickness Class	24	25	25	25	26	26	27
		Thickness	0.63	0.68	0.68	0.68	0.73	0.73	0.79
	16	Calculated Thickness	0.68	0.70	0.71	0.73	0.75	0.77	0.80
		Use {Thickness Class	25	25	26	26	26	27	27
		Thickness	0.68	0.68	0.73	0.73	0.73	0.79	0.79
B	2½	Calculated Thickness	0.48*	0.50*	0.51*	0.53*	0.55*	0.58*	0.60*
		Use {Thickness Class	21	21	21	22	22	23	23
		Thickness	0.50	0.50	0.50	0.54	0.54	0.58	0.58
	3½	Calculated Thickness	0.48*	0.49*	0.51*	0.53*	0.55*	0.57	0.61
		Use {Thickness Class	21	21	21	22	22	23	24
		Thickness	0.50	0.50	0.50	0.54	0.54	0.58	0.63
	5	Calculated Thickness	0.49*	0.51*	0.53*	0.54	0.57	0.60	0.64
		Use {Thickness Class	21	21	22	22	23	23	24
		Thickness	0.50	0.50	0.54	0.54	0.58	0.58	0.63
	8	Calculated Thickness	0.55	0.57	0.59	0.61	0.63	0.66	0.69
		Use {Thickness Class	22	23	23	24	24	25	25
		Thickness	0.54	0.58	0.58	0.63	0.63	0.68	0.68
	12	Calculated Thickness	0.60	0.62	0.63	0.65	0.68	0.70	0.73
		Use {Thickness Class	23	24	24	24	25	25	26
		Thickness	0.58	0.63	0.63	0.63	0.68	0.68	0.73
	16	Calculated Thickness	0.63	0.65	0.67	0.69	0.71	0.73	0.76
		Use {Thickness Class	24	24	25	25	26	26	27
		Thickness	0.63	0.63	0.68	0.68	0.73	0.73	0.79
F	2½	Calculated Thickness	0.44*	0.46*	0.47*	0.49*	0.52*	0.55	0.60
		Use {Thickness Class	21	21	21	21	22	22	23
		Thickness	0.50	0.50	0.50	0.50	0.54	0.54	0.58
	3½	Calculated Thickness	0.44*	0.45*	0.47*	0.49*	0.52	0.56	0.60
		Use {Thickness Class	21	21	21	21	22	23	23
		Thickness	0.50	0.50	0.50	0.50	0.54	0.58	0.58
	5	Calculated Thickness	0.45*	0.47*	0.49	0.52	0.55	0.58	0.62
		Use {Thickness Class	21	21	21	22	22	23	24
		Thickness	0.50	0.50	0.50	0.54	0.54	0.58	0.63
	8	Calculated Thickness	0.50	0.52	0.54	0.57	0.59	0.62	0.66
		Use {Thickness Class	21	22	22	23	23	24	25
		Thickness	0.51	0.54	0.54	0.58	0.5S	0.63	0.68
	12	Calculated Thickness	0.55	0.56	0.58	0.61	0.63	0.66	0.69
		Use {Thickness Class	22	23	23	24	24	25	25
		Thickness	0.54	0.58	0.58	0.63	0.63	0.68	0.68
	16	Calculated Thickness	0.58	0.59	0.61	0.63	0.66	0.68	0.71
		Use {Thickness Class	23	23	24	24	25	25	26
		Thickness	0.58	0.58	0.63	0.63	0.68	0.68	0.73

* Asterisk following total calculated thickness indicates that truck superload (Case 2) is the controlling factor. When total calculated thickness is not followed by asterisk, surge pressure (Case 1) is the controlling factor. See Sec. 1–2.1.

TABLE 1–2 (*Continued*)

Schedule of Barrel Thickness for Water Pipe of 21/45 Iron Strength

Laying Condition	Depth of Cover ft	Thickness Specifications	Internal Pressure—psi						
			50	100	150	200	250	300	350
			Barrel Thickness—in.						

Eighteen-Inch Water Pipe

Laying Condition	Depth of Cover ft	Thickness Specifications	50	100	150	200	250	300	350
A	2½	Calculated Thickness	0.56*	0.57*	0.59*	0.61*	0.63*	0.66*	0.68*
		Use Thickness Class	22	22	22	23	23	24	24
		Thickness	0.58	0.58	0.58	0.63	0.63	0.68	0.68
	3½	Calculated Thickness	0.55*	0.57*	0.58*	0.60*	0.63*	0.65*	0.68
		Use Thickness Class	21	22	22	22	23	23	24
		Thickness	0.54	0.58	0.58	0.58	0.63	0.63	0.68
	5	Calculated Thickness	0.57*	0.59*	0.61*	0.63*	0.65	0.68	0.72
		Use Thickness Class	22	22	23	23	23	24	25
		Thickness	0.58	0.58	0.63	0.63	0.63	0.68	0.73
	8	Calculated Thickness	0.63*	0.65	0.67	0.69	0.72	0.75	0.78
		Use Thickness Class	23	23	24	24	25	25	26
		Thickness	0.63	0.63	0.68	0.68	0.73	0.73	0.79
	12	Calculated Thickness	0.69	0.71	0.73	0.75	0.78	0.80	0.83
		Use Thickness Class	24	25	25	25	26	26	27
		Thickness	0.68	0.73	0.73	0.73	0.79	0.79	0.85
	16	Calculated Thickness	0.73	0.75	0.77	0.79	0.82	0.84	0.87
		Use Thickness Class	25	25	26	26	27	27	27
		Thickness	0.73	0.73	0.79	0.79	0.85	0.85	0.85
B	2½	Calculated Thickness	0.51*	0.53*	0.55*	0.57*	0.59*	0.62*	0.65
		Use Thickness Class	21	21	21	22	22	23	23
		Thickness	0.54	0.54	0.54	0.58	0.58	0.63	0.63
	3½	Calculated Thickness	0.51*	0.53*	0.54*	0.56*	0.59*	0.62	0.67
		Use Thickness Class	21	21	21	22	22	23	24
		Thickness	0.54	0.54	0.54	0.58	0.58	0.63	0.68
	5	Calculated Thickness	0.53*	0.54*	0.56*	0.59	0.62	0.66	0.70
		Use Thickness Class	21	21	22	22	23	24	24
		Thickness	0.54	0.54	0.58	0.58	0.63	0.68	0.68
	8	Calculated Thickness	0.58	0.60	0.62	0.65	0.68	0.71	0.74
		Use Thickness Class	22	22	23	23	24	25	25
		Thickness	0.58	0.58	0.63	0.63	0.68	0.73	0.73
	12	Calculated Thickness	0.64	0.66	0.68	0.70	0.73	0.76	0.79
		Use Thickness Class	23	24	24	24	25	26	26
		Thickness	0.63	0.68	0.68	0.68	0.73	0.79	0.79
	16	Calculated Thickness	0.68	0.70	0.72	0.74	0.76	0.79	0.82
		Use Thickness Class	24	24	25	25	26	26	27
		Thickness	0.68	0.68	0.73	0.73	0.79	0.79	0.85
F	2½	Calculated Thickness	0.47*	0.49*	0.50*	0.53*	0.55*	0.60	0.65
		Use Thickness Class	21	21	21	21	21	22	23
		Thickness	0.54	0.54	0.54	0.54	0.54	0.58	0.63
	3½	Calculated Thickness	0.46*	0.48*	0.50*	0.52*	0.56	0.61	0.66
		Use Thickness Class	21	21	21	21	22	23	24
		Thickness	0.54	0.54	0.54	0.54	0.58	0.63	0.68
	5	Calculated Thickness	0.48*	0.50*	0.52	0.56	0.59	0.63	0.68
		Use Thickness Class	21	21	21	22	22	23	24
		Thickness	0.54	0.54	0.54	0.58	0.58	0.63	0.68
	8	Calculated Thickness	0.54	0.56	0.58	0.61	0.64	0.67	0.71
		Use Thickness Class	21	22	22	23	23	24	25
		Thickness	0.54	0.58	0.58	0.63	0.63	0.68	0.73
	12	Calculated Thickness	0.58	0.60	0.63	0.65	0.68	0.71	0.74
		Use Thickness Class	22	22	23	23	24	25	25
		Thickness	0.58	0.58	0.63	0.63	0.68	0.73	0.73
	16	Calculated Thickness	0.61	0.63	0.66	0.68	0.71	0.74	0.77
		Use Thickness Class	23	23	24	24	25	25	26
		Thickness	0.63	0.63	0.68	0.68	0.73	0.73	0.79

* Asterisk following total calculated thickness indicates that truck superload (Case 2) is the controlling factor. When total calculated thickness is not followed by asterisk, surge pressure (Case 1) is the controlling factor. See Sec. 1–2.1.

TABLE 1–2 (*Continued*)

Schedule of Barrel Thickness for Water Pipe of 21/45 Iron Strength

Laying Condition	Depth of Cover ft	Thickness Specifications	Internal Pressure—*psi*						
			50	100	150	200	250	300	350
			Barrel Thickness—*in.*						

Twenty-Inch Water Pipe

Laying Condition	Depth of Cover ft	Thickness Specifications	50	100	150	200	250	300	350
A	2½	Calculated Thickness	0.59*	0.61*	0.63*	0.65*	0.68*	0.71*	0.74*
		Use {Thickness Class	21	22	22	23	23	24	24
		{Thickness	0.57	0.62	0.62	0.67	0.67	0.72	0.72
	3½	Calculated Thickness	0.59*	0.61*	0.63*	0.65*	0.68*	0.70*	0.73*
		Use {Thickness Class	21	22	22	23	23	24	24
		{Thickness	0.57	0.62	0.62	0.67	0.67	0.72	0.72
	5	Calculated Thickness	0.62*	0.64*	0.65*	0.67*	0.70*	0.73	0.77
		Use {Thickness Class	22	22	23	23	24	24	25
		{Thickness	0.62	0.62	0.67	0.67	0.72	0.72	0.78
	8	Calculated Thickness	0.67*	0.69	0.71	0.74	0.77	0.80	0.83
		Use {Thickness Class	23	23	24	24	25	25	26
		{Thickness	0.67	0.67	0.72	0.72	0.78	0.78	0.84
	12	Calculated Thickness	0.74	0.76	0.78	0.80	0.83	0.86	0.89
		Use {Thickness Class	24	25	25	25	26	26	27
		{Thickness	0.72	0.78	0.78	0.78	0.84	0.84	0.91
	16	Calculated Thickness	0.78	0.80	0.83	0.85	0.87	0.90	0.93
		Use {Thickness Class	25	25	26	26	26	27	27
		{Thickness	0.78	0.78	0.84	0.84	0.84	0.91	0.91
B	2½	Calculated Thickness	0.55*	0.56*	0.58*	0.61*	0.63*	0.66*	0.70*
		Use {Thickness Class	21	21	21	22	22	23	24
		{Thickness	0.57	0.57	0.57	0.62	0.67	0.67	0.72
	3½	Calculated Thickness	0.54*	0.56*	0.58*	0.60*	0.63*	0.66*	0.71*
		Use {Thickness Class	21	21	21	22	22	23	24
		{Thickness	0.57	0.57	0.57	0.62	0.62	0.67	0.72
	5	Calculated Thickness	0.57*	0.58*	0.60*	0.63*	0.66	0.70	0.74
		Use {Thickness Class	21	21	22	22	23	24	24
		{Thickness	0.57	0.57	0.62	0.62	0.67	0.72	0.72
	8	Calculated Thickness	0.62	0.64	0.66	0.69	0.72	0.76	0.79
		Use {Thickness Class	22	22	23	23	24	24	25
		{Thickness	0.62	0.62	0.67	0.67	0.72	0.78	0.78
	12	Calculated Thickness	0.68	0.70	0.72	0.75	0.77	0.81	0.84
		Use {Thickness Class	23	24	24	25	25	26	26
		{Thickness	0.67	0.72	0.72	0.78	0.78	0.84	0.84
	16	Calculated Thickness	0.72	0.74	0.76	0.78	0.81	0.84	0.87
		Use {Thickness Class	24	24	25	25	25	26	26
		{Thickness	0.72	0.72	0.78	0.78	0.84	0.84	0.84
F	2½	Calculated Thickness	0.50*	0.52*	0.54*	0.56*	0.59*	0.64	0.69
		Use {Thickness Class	21	21	21	21	21	22	23
		{Thickness	0.57	0.57	0.57	0.57	0.57	0.62	0.67
	3½	Calculated Thickness	0.49*	0.51*	0.53*	0.56*	0.60	0.65	0.70
		Use {Thickness Class	21	21	21	21	22	23	24
		{Thickness	0.57	0.57	0.57	0.57	0.62	0.67	0.72
	5	Calculated Thickness	0.51*	0.53	0.55	0.59	0.63	0.67	0.72
		Use {Thickness Class	21	21	21	21	22	23	24
		{Thickness	0.57	0.57	0.57	0.57	0.62	0.67	0.72
	8	Calculated Thickness	0.56	0.59	0.61	0.64	0.68	0.72	0.76
		Use {Thickness Class	21	21	22	22	23	24	25
		{Thickness	0.57	0.57	0.62	0.62	0.67	0.72	0.78
	12	Calculated Thickness	0.61	0.64	0.66	0.69	0.72	0.75	0.79
		Use {Thickness Class	22	22	23	23	24	25	25
		{Thickness	0.62	0.62	0.67	0.67	0.72	0.78	0.78
	16	Calculated Thickness	0.65	0.67	0.70	0.72	0.75	0.78	0.82
		Use {Thickness Class	23	23	24	24	25	25	26
		{Thickness	0.67	0.67	0.72	0.72	0.78	0.78	0.84

* Asterisk following total calculated thickness indicates that truck superload (Case 2) is the controlling factor. When total calculated thickness is not followed by asterisk, surge pressure (Case 1) is the controlling factor. See Sec. 1–2.1.

TABLE 1–2 (*Continued*)
Schedule of Barrel Thickness for Water Pipe of 21/45 Iron Strength

Laying Condition	Depth of Cover ft	Thickness Specifications	50	100	150	200	250	300	350
			Internal Pressure—*psi* — Barrel Thickness—*in.*						

Twenty-four-Inch Water Pipe

Laying Condition	Depth of Cover ft	Thickness Specifications	50	100	150	200	250	300	350
A	2½	Calculated Thickness	0.66*	0.68*	0.71*	0.73*	0.76*	0.80*	0.84*
		Use Thickness Class	22	22	23	23	24	24	25
		Thickness	0.68	0.68	0.73	0.73	0.79	0.79	0.85
	3½	Calculated Thickness	0.66*	0.68*	0.70*	0.73*	0.76*	0.79*	0.83*
		Use Thickness Class	22	22	22	23	24	24	25
		Thickness	0.68	0.68	0.68	0.73	0.79	0.79	0.85
	5	Calculated Thickness	0.69*	0.71*	0.74*	0.76*	0.79*	0.83	0.88
		Use Thickness Class	22	23	23	24	24	25	25
		Thickness	0.68	0.73	0.73	0.79	0.79	0.85	0.85
	8	Calculated Thickness	0.75*	0.77	0.80	0.83	0.87	0.91	0.95
		Use Thickness Class	23	24	24	25	25	26	26
		Thickness	0.73	0.79	0.79	0.85	0.85	0.92	0.92
	12	Calculated Thickness	0.83	0.85	0.88	0.91	0.94	0.98	1.01
		Use Thickness Class	25	25	25	26	26	27	27
		Thickness	0.85	0.85	0.85	0.92	0.92	0.99	0.99
	16	Calculated Thickness	0.89	0.91	0.94	0.96	0.99	1.03	1.06
		Use Thickness Class	26	26	26	27	27	28	28
		Thickness	0.92	0.92	0.92	0.99	0.99	1.07	1.07
B	2½	Calculated Thickness	0.60*	0.62*	0.65*	0.67*	0.71*	0.74*	0.80
		Use Thickness Class	21	21	21	22	23	23	24
		Thickness	0.63	0.63	0.68	0.68	0.73	0.73	0.79
	3½	Calculated Thickness	0.59*	0.62*	0.64*	0.67*	0.70*	0.75	0.81
		Use Thickness Class	21	21	21	22	22	23	24
		Thickness	0.63	0.63	0.63	0.68	0.68	0.73	0.79
	5	Calculated Thickness	0.62*	0.65*	0.67*	0.70	0.74	0.79	0.85
		Use Thickness Class	21	21	22	22	23	24	25
		Thickness	0.63	0.63	0.68	0.68	0.73	0.79	0.85
	8	Calculated Thickness	0.68	0.71	0.74	0.77	0.81	0.85	0.90
		Use Thickness Class	22	23	23	24	24	25	26
		Thickness	0.68	0.73	0.73	0.79	0.79	0.85	0.92
	12	Calculated Thickness	0.75	0.78	0.80	0.83	0.87	0.91	0.95
		Use Thickness Class	23	24	24	25	25	26	26
		Thickness	0.73	0.79	0.79	0.85	0.85	0.92	0.92
	16	Calculated Thickness	0.80	0.82	0.85	0.88	0.91	0.95	0.99
		Use Thickness Class	24	25	25	25	26	26	27
		Thickness	0.79	0.85	0.85	0.85	0.92	0.92	0.99
F	2½	Calculated Thickness	0.54*	0.57*	0.59*	0.62*	0.66*	0.72	0.79
		Use Thickness Class	21	21	21	21	22	23	24
		Thickness	0.63	0.62	0.63	0.63	0.68	0.73	0.79
	3½	Calculated Thickness	0.54*	0.56*	0.59*	0.62*	0.67	0.73	0.80
		Use Thickness Class	21	21	21	21	22	23	24
		Thickness	0.63	0.63	0.63	0.63	0.68	0.72	0.79
	5	Calculated Thickness	0.57*	0.59*	0.62	0.66	0.71	0.76	0.82
		Use Thickness Class	21	21	21	22	23	24	25
		Thickness	0.63	0.63	0.63	0.68	0.73	0.79	0.85
	8	Calculated Thickness	0.62	0.65	0.68	0.72	0.76	0.81	0.86
		Use Thickness Class	21	21	22	23	24	24	25
		Thickness	0.63	0.63	0.68	0.73	0.79	0.79	0.85
	12	Calculated Thickness	0.68	0.71	0.74	0.77	0.81	0.85	0.90
		Use Thickness Class	22	23	23	24	24	25	26
		Thickness	0.68	0.73	0.73	0.79	0.79	0.85	0.92
	16	Calculated Thickness	0.72	0.75	0.78	0.81	0.85	0.89	0.93
		Use Thickness Class	23	23	24	24	25	25	26
		Thickness	0.73	0.73	0.79	0.79	0.85	0.92	0.92

* Asterisk following total calculated thickness indicates that truck superload (Case 2) is the controlling factor. When total calculated thickness is not followed by asterisk, surge pressure (Case 1) is the controlling factor. See Sec. 1–2.1.

5
1

TABLE 1-2 (*Continued*)
Schedule of Barrel Thickness for Water Pipe of 21/45 Iron Strength

Laying Condition	Depth of Cover ft	Thickness Specifications	Internal Pressure—*psi*						
			50	100	150	200	250	300	350
			Barrel Thickness—*in.*						

Thirty-Inch Water Pipe

Laying Condition	Depth of Cover ft	Thickness Specifications	50	100	150	200	250	300	350
A	2½	Calculated Thickness	0.79*	0.81*	0.84*	0.88*	0.92*	0.96*	1.01*
		Use {Thickness Class	22	22	23	23	24	25	25
		{Thickness	0.79	0.79	0.85	0.85	0.92	0.99	0.99
	3½	Calculated Thickness	0.78*	0.81*	0.84*	0.87*	0.91*	0.95*	1.00
		Use {Thickness Class	22	22	23	23	24	24	25
		{Thickness	0.79	0.79	0.85	0.85	0.92	0.92	0.99
	5	Calculated Thickness	0.82*	0.85*	0.88*	0.91*	0.94	1.00	1.06
		Use {Thickness Class	23	23	23	24	24	25	26
		{Thickness	0.85	0.85	0.85	0.92	0.92	0.99	1.0,
	8	Calculated Thickness	0.89*	0.92	0.95	0.99	1.04	1.09	1.14
		Use {Thickness Class	24	24	24	25	26	26	27
		{Thickness	0.92	0.92	0.92	0.99	1.07	1.07	1.16
	12	Calculated Thickness	0.99	1.02	1.05	1.09	1.13	1.17	1.22
		Use {Thickness Class	25	25	26	26	27	27	28
		{Thickness	0.99	0.99	1.07	1.07	1.16	1.16	1.25
	16	Calculated Thickness	1.06	1.09	1.12	1.16	1.20	1.24	1.29
		Use {Thickness Class	26	26	27	27	27	28	25
		{Thickness	1.07	1.07	1.16	1.16	1.16	1.25	1.29
B	2½	Calculated Thickness	0.70*	0.73*	0.76*	0.80*	0.84*	0.89*	0.88
		Use {Thickness Class	20	21	22	22	23	24	25
		{Thickness	0.68	0.73	0.79	0.79	0.85	0.92	0.99
	3½	Calculated Thickness	0.69*	0.72*	0.75*	0.79*	0.83*	0.90	0.98
		Use {Thickness Class	20	21	21	22	23	24	25
		{Thickness	0.68	0.73	0.73	0.79	0.85	0.92	0.99
	5	Calculated Thickness	0.73*	0.76*	0.79*	0.83*	0.88	0.95	1.02
		Use {Thickness Class	21	22	22	23	23	24	25
		{Thickness	0.73	0.79	0.79	0.85	0.85	0.92	0.99
	8	Calculated Thickness	0.79	0.83	0.86	0.91	0.96	1.01	1.07
		Use {Thickness Class	22	23	23	24	25	25	26
		{Thickness	0.79	0.85	0.85	0.92	0.99	0.99	1.07
	12	Calculated Thickness	0.88	0.91	0.95	0.99	1.03	1.08	1.14
		Use {Thickness Class	23	24	24	25	26	26	27
		{Thickness	0.85	0.92	0.92	0.99	1.07	1.07	1.16
	16	Calculated Thickness	0.94	0.97	1.01	1.05	1.09	1.13	1.18
		Use {Thickness Class	24	25	25	26	26	27	27
		{Thickness	0.92	0.99	0.99	1.07	1.07	1.16	1.16
F	2½	Calculated Thickness	0.64*	0.67*	0.70*	0.74*	0.79*	0.87	0.96
		Use {Thickness class	20	20	20	21	22	23	25
		{Thickness	0.68	0.68	0.68	0.73	0.79	0.85	0.99
	3½	Calculated Thickness	0.63*	0.66*	0.69*	0.73*	0.81	0.88	0.97
		Use {Thickness Class	20	20	20	21	22	23	25
		{Thickness	0.68	0.68	0.68	0.73	0.79	0.85	0.99
	5	Calculated Thickness	0.66*	0.69*	0.73	0.78	0.85	0.92	0.99
		Use {Thickness Class	20	20	21	22	23	24	25
		{Thickness	0.68	0.68	0.73	0.79	0.85	0.92	0.99
	8	Calculated Thickness	0.72	0.76	0.80	0.85	0.90	0.96	1.03
		Use {Thickness Class	21	22	22	23	24	25	26
		{Thickness	0.73	0.79	0.79	0.85	0.92	0.99	1.07
	12	Calculated Thickness	0.80	0.84	0.87	0.91	0.96	1.02	1.08
		Use {Thickness Class	22	23	23	24	25	25	26
		{Thickness	0.79	0.85	0.85	0.92	0.99	0.99	1.07
	16	Calculated Thickness	0.85	0.89	0.92	0.96	1.01	1.06	1.12
		Use {Thickness Class	23	24	24	25	25	26	27
		{Thickness	0.85	0.92	0.92	0.99	0.99	1.07	1.91

* Asterisk following total calculated thickness indicates that truck superload (Case 2) is the controlling factor. When total calculated thickness is not followed by asterisk, surge pressure (Case 1) is the controlling factor. See Sec. 1-2.1.

TABLE 1–2 (*Continued*)

Schedule of Barrel Thickness for Water Pipe of 21/45 Iron Strength

Laying Condition	Depth of Cover ft	Thickness Specifications	Internal Pressure—psi						
			50	100	150	200	250	300	350
			Barrel Thickness—in.						

Thirty-six-Inch Water Pipe

Laying Condition	Depth of Cover ft	Thickness Specifications	50	100	150	200	250	300	350
A	2½	Calculated Thickness	0.89*	0.92*	0.96*	1.00*	1.05*	1.10*	1.16*
		Use { Thickness Class	22	23	23	24	24	25	26
		Thickness	0.87	0.94	0.94	1.02	1.02	1.10	1.19
	3½	Calculated Thickness	0.88*	0.92*	0.95*	1.00*	1.04*	1.10*	1.15
		Use { Thickness Class	22	23	23	24	24	25	26
		Thickness	0.87	0.94	0.94	1.02	1.02	1.10	1.19
	5	Calculated Thickness	0.93*	0.96*	1.00*	1.04*	1.09*	1.14	1.22
		Use { Thickness Class	23	23	24	24	25	25	26
		Thickness	0.94	0.94	1.02	1.02	1.10	1.10	1.19
	8	Calculated Thickness	1.01*	1.04*	1.08	1.13	1.18	1.24	1.31
		Use { Thickness Class	24	24	25	25	26	27	27
		Thickness	1.02	1.02	1.10	1.10	1.19	1.29	1.29
	12	Calculated Thickness	1.13	1.16	1.20	1.25	1.30	1.35	1.41
		Use { Thickness Class	25	26	26	27	27	28	28
		Thickness	1.10	1.19	1.19	1.29	1.29	1.39	1.39
	16	Calculated Thickness	1.22	1.25	1.29	1.33	1.38	1.43	1.48
		Use { Thickness Class	26	27	27	27	28	28	29
		Thickness	1.19	1.29	1.29	1.39	1.39	1.39	1.50
B	2½	Calculated Thickness	0.78*	0.81*	0.85*	0.90*	0.95*	1.01*	1.12
		Use { Thickness Class	21	21	22	22	23	24	25
		Thickness	0.81	0.81	0.87	0.87	0.94	1.02	1.10
	3½	Calculated Thickness	0.77*	0.81*	0.85*	0.89*	0.95*	1.03	1.12
		Use { Thickness Class	20	21	22	22	23	24	25
		Thickness	0.75	0.81	0.87	0.87	0.94	1.02	1.10
	5	Calculated Thickness	0.81*	0.85*	0.89*	0.93*	1.00	1.08	1.16
		Use { Thickness Class	21	22	22	23	24	25	26
		Thickness	.81	.87	.87	.94	1.02	1.10	1.19
	8	Calculated Thickness	0.88	0.92	0.97	1.02	1.08	1.15	1.22
		Use { Thickness Class	22	23	23	24	25	26	26
		Thickness	0.87	0.94	0.94	1.02	1.10	1.19	1.19
	12	Calculated Thickness	0.99	1.02	1.07	1.12	1.17	1.23	1.30
		Use { Thickness Class	24	24	25	25	26	26	27
		Thickness	1.02	1.02	1.10	1.10	1.19	1.19	1.29
	16	Calculated Thickness	1.06	1.10	1.14	1.19	1.24	1.29	1.35
		Use { Thickness Class	25	25	25	26	27	27	28
		Thickness	1.10	1.10	1.10	1.19	1.29	1.29	1.39
F	2½	Calculated Thickness	0.71*	0.74*	0.78*	0.83*	0.90	1.00	1.10
		Use { Thickness Class	20	20	21	21	22	24	25
		Thickness	0.75	0.75	0.81	0.81	0.87	1.02	1.10
	3½	Calculated Thickness	0.70*	0.74*	0.78*	0.83	0.92	1.01	1.11
		Use { Thickness Class	20	20	21	21	23	24	25
		Thickness	0.75	0.75	0.81	0.81	0.94	1.02	1.10
	5	Calculated Thickness	0.74*	0.77*	0.82	0.88	0.96	1.05	1.14
		Use { Thickness Class	20	20	21	22	23	24	25
		Thickness	0.75	0.75	0.81	0.87	0.94	1.02	1.10
	8	Calculated Thickness	0.81	0.85	0.90	0.96	1.02	1.10	1.18
		Use { Thickens Class	21	22	22	23	24	25	26
		Thickness	0.81	0.87	0.87	0.94	1.02	1.10	1.19
	12	Calculated Thickness	0.90	0.94	0.98	1.04	1.09	1.16	1.23
		Use { Thickness Class	22	23	24	24	25	26	26
		Thickness	0.87	0.94	1.02	1.02	1.10	1.19	1.19
	16	Calculated Thickness	0.96	1.00	1.04	1.09	1.15	1.21	1.28
		Use { Thickness Class	23	24	24	25	26	26	27
		Thickness	0.94	1.02	1.02	1.10	1.19	1.19	1.29

* Asterisk following total calculated thickness indicates that truck superload (Case 2) is the controlling factor When total calculated thickness is not followed by asterisk, surge pressure (Case 1) is the controlling factor. See Sec. 1–2.1.

TABLE 1–2 (*Continued*)
Schedule of Barrel Thickness for Water Pipe of 21/45 Iron Strength

5
4

Laying Condi-tion	Depth of Cover ft	Thickness Specifications	Internal Pressure—psi						
			50	100	150	200	250	300	350
			Barrel Thickness—in.						

Forty-two-Inch Water Pipe

Laying Condition	Depth	Thickness Specifications	50	100	150	200	250	300	350
A	2½	Calculated Thickness	0.99*	1.02*	1.07*	1.12*	1.17*	1.24*	1.31*
		Use Thickness Class	22	23	23	24	24	25	26
		Thickness	0.97	1.05	1.05	1.13	1.13	1.22	1.32
	3½	Calculated Thickness	0.99*	1.02*	1.07*	1.12*	1.17*	1.24*	1.30
		Use Thickness Class	22	23	23	24	24	25	26
		Thickness	0.97	1.05	1.05	1.13	1.13	1.22	1.32
	5	Calculated Thickness	1.04*	1.08*	1.12*	1.17*	1.22*	1.28*	1.37
		Use Thickness Class	23	23	24	24	25	26	26
		Thickness	1.05	1.05	1.13	1.13	1.22	1.32	1.32
	8	Calculated Thickness	1.13*	1.17*	1.21	1.27	1.33	1.40	1.48
		Use Thickness Class	24	24	25	26	26	27	27
		Thickness	1.13	1.13	1.22	1.32	1.32	1.43	1.43
	12	Calculated Thickness	1.26	1.30	1.35	1.40	1.46	1.52	1.59
		Use Thickness Class	25	26	26	27	27	28	28
		Thickness	1.22	1.32	1.32	1.43	1.43	1.54	1.54
	16	Calculated Thickness	1.37	1.41	1.46	1.51	1.56	1.62	1.68
		Use Thickness Class	26	27	27	28	28	29	29
		Thickness	1.32	1.43	1.43	1.54	1.54	1.66	1.66
B	2½	Calculated Thickness	0.85*	0.89*	0.94*	0.99*	1.05*	1.13	1.25
		Use Thickness Class	20	21	22	22	23	24	25
		Thickness	0.83	0.90	0.97	0.97	1.05	1.13	1.22
	3½	Calculated Thickness	0.85*	0.89*	0.94*	0.99*	1.05	1.15	1.27
		Use Thickness Class	20	21	22	22	23	24	26
		Thickness	0.83	0.90	0.97	0.97	1.05	1.13	1.32
	5	Calculated Thickness	0.89*	0.93*	0.98*	1.03*	1.11	1.21	1.31
		Use Thickness Class	21	21	22	23	24	25	26
		Thickness	0.90	0.90	0.97	1.05	1.13	1.22	1.32
	8	Calculated Thickness	0.97	1.02	1.07	1.13	1.21	1.29	1.37
		Use Thickness Class	22	23	23	24	25	26	26
		Thickness	0.97	1.05	1.05	1.13	1.22	1.32	1.32
	12	Calculated Thickness	1.09	1.13	1.18	1.24	1.30	1.38	1.45
		Use Thickness Class	24	24	25	25	26	27	27
		Thickness	1.13	1.13	1.22	1.22	1.32	1.43	1.43
	16	Calculated Thickness	1.18	1.22	1.27	1.32	1.38	1.45	1.52
		Use Thickness Class	25	25	26	26	27	27	28
		Thickness	1.22	1.22	1.32	1.32	1.43	1.43	1.54
F	2½	Calculated Thickness	0.77*	0.81*	0.86*	0.92*	1.00	1.12	1.24
		Use Thickness Class	20	20	20	21	22	24	25
		Thickness	0.83	0.83	0.83	0.90	0.97	1.13	1.22
	3½	Calculated Thickness	0.77*	0.81*	0.86*	0.92*	1.02	1.14	1.25
		Use Thickness Class	20	20	20	21	23	24	25
		Thickness	0.83	0.83	0.83	0.90	1.05	1.13	1.22
	5	Calculated Thickness	0.81*	0.85*	0.90	0.98	1.07	1.17	1.28
		Use Thickness Class	20	20	21	22	23	24	26
		Thickness	0.83	0.83	0.90	0.97	1.05	1.13	1.32
	8	Calculated Thickness	0.89	0.94	0.99	1.06	1.14	1.23	1.33
		Use Thickness Class	21	22	22	23	24	25	26
		Thickness	0.90	0.97	0.97	1.05	1.13	1.22	1.32
	12	Calculated Thickness	0.99	1.03	1.09	1.15	1.22	1.30	1.39
		Use Thickness Class	22	23	24	24	25	26	27
		Thickness	0.97	1.05	1.13	1.13	1.22	1.32	1.43
	16	Calculated Thickness	1.07	1.11	1.16	1.21	1.28	1.36	1.44
		Use Thickness Class	23	24	24	25	26	26	27
		Thickness	1.05	1.13	1.13	1.22	1.32	1.32	1.43

* Asterisk following total calculated thickness indicates that truck superload (Case 2) is the controlling factor. When total calculated thickness is not followed by asterisk, surge pressure (Case 1) is the controlling factor. See Sec. 1–2.1.

TABLE 1-2 (*Continued*)

Schedule of Barrel Thickness for Water Pipe of 21/45 Iron Strength

Laying Condition	Depth of Cover ft	Thickness Specifications	Internal Pressure—psi						
			50	100	150	200	250	300	350
			Barrel Thickness—in.						

Forty-eight-Inch Water Pipe

Laying Condition	Depth of Cover ft	Thickness Specifications	50	100	150	200	250	300	350
A	2½	Calculated Thickness	1.08*	1.12*	1.17*	1.23*	1.29*	1.37*	1.45*
		Use { Thickness Class	22	23	23	24	25	25	26
		{ Thickness	1.06	1.14	1.14	1.23	1.33	1.33	1.44
	3½	Calculated Thickness	1.09*	1.13*	1.18*	1.24*	1.30*	1.37*	1.46
		Use { Thickness Class	22	23	23	24	25	25	26
		{ Thickness	1.06	1.14	1.14	1.23	1.33	1.33	1.44
	5	Calculated Thickness	1.15*	1.19*	1.24*	1.30*	1.36*	1.43	1.53
		Use { Thickness Class	23	24	24	25	25	26	27
		{ Thickness	1.14	1.23	1.23	1.33	1.33	1.44	1.56
	8	Calculated Thickness	1.25*	1.29*	1.34	1.41	1.48	1.56	1.65
		Use { Thickness Class	24	25	25	26	26	27	28
		{ Thickness	1.23	1.33	1.33	1.44	1.44	1.56	1.68
	12	Calculated Thickness	1.41	1.45	1.51	1.57	1.63	1.70	1.77
		Use { Thickness Class	26	26	27	27	28	28	29
		{ Thickness	1.44	1.44	1.56	1.56	1.68	1.68	1.81
	16	Calculated Thickness	1.53	1.58	1.63	1.68	1.75	1.81	1.89
		Use { Thickness Class	27	27	28	28	29	29	30
		{ Thickness	1.56	1.56	1.68	1.68	1.81	1.81	1.95
B	2½	Calculated Thickness	0.92*	0.97*	1.02*	1.08*	1.16*	1.26	1.40
		Use { Thickness Class	20	21	22	22	23	24	26
		{ Thickness	0.91	0.98	1.06	1.06	1.14	1.23	1.44
	3½	Calculated Thickness	0.93*	0.97*	1.03*	1.09*	1.17	1.29	1.42
		Use { Thickness Class	20	21	22	22	23	25	26
		{ Thickness	0.91	0.98	1.06	1.06	1.14	1.33	1.44
	5	Calculated Thickness	0.98*	1.02*	1.08*	1.14	1.23	1.34	1.46
		Use { Thickness Class	21	22	22	23	24	25	26
		{ Thickness	0.98	1.06	1.06	1.14	1.23	1.33	1.44
	8	Calculated Thickness	1.06	1.12	1.18	1.25	1.34	1.43	1.53
		Use { Thickness Class	22	23	23	24	25	26	27
		{ Thickness	1.06	1.14	1.14	1.23	1.33	1.44	1.56
	12	Calculated Thickness	1.20	1.25	1.31	1.37	1.45	1.53	1.62
		Use { Thickness Class	24	24	25	25	26	27	28
		{ Thickness	1.23	1.23	1.33	1.33	1.44	1.56	1.68
	16	Calculated Thickness	1.30	1.35	1.41	1.47	1.54	1.61	1.70
		Use { Thickness Class	25	25	26	26	27	27	28
		{ Thickness	1.33	1.33	1.44	1.44	1.56	1.56	1.68
F	2½	Calculated Thickness	0.83*	0.88*	0.94*	1.01*	1.12	1.25	1.39
		Use { Thickness Class	20	20	20	21	23	24	26
		{ Thickness	0.91	0.91	0.91	0.98	1.14	1.23	1.44
	3½	Calculated Thickness	0.84*	0.89*	0.94*	1.02	1.14	1.27	1.40
		Use { Thickness Class	20	20	20	22	23	24	26
		{ Thickness	0.91	0.91	0.91	1.06	1.14	1.23	1.44
	5	Calculated Thickness	0.88*	0.93*	0.99	1.08	1.19	1.30	1.43
		Use { Thickness Class	20	20	21	22	24	25	26
		{ Thickness	0.91	0.91	0.98	1.06	1.23	1.33	1.44
	8	Calculated Thickness	0.97	1.03	1.09	1.17	1.27	1.37	1.48
		Use { Thickness Class	21	22	22	23	24	25	26
		{ Thickness	0.98	1.06	1.06	1.14	1.23	1.33	1.44
	12	Calculated Thickness	1.09	1.14	1.20	1.27	1.35	1.45	1.55
		Use { Thickness Class	22	23	24	24	25	26	27
		{ Thickness	1.06	1.14	1.23	1.23	1.33	1.44	1.56
	16	Calculated Thickness	1.17	1.23	1.28	1.35	1.43	1.51	1.60
		Use { Thickness Class	23	24	25	25	26	27	27
		{ Thickness	1.14	1.23	1.33	1.33	1.44	1.56	1.56

5
5

* Asterisk following total calculated thickness indicates that truck superload (Case 2) is the controlling factor. When total calculated thickness is not followed by asterisk, surge pressure (Case 1) is the controlling factor.　See Sec. 1-2.1.

TABLE 1–3

TABLE 1–3

Schedule of Barrel Thickness for Gas Pipe of 18/40 Iron Strength

Laying Condition	Depth of Cover ft	Thickness Specifications	Internal Pressure—psi					
			10	50	100	150	200	250
			Barrel Thicknesses—in.					

Four-Inch Gas Pipe

Laying Condition	Depth of Cover ft	Thickness Specifications	10	50	100	150	200	250
A	2½	Calculated Thickness Use {Thickness Class {Thickness	.23 22 .35	.23 22 .35	.23 22 .35	.24 22 .35	.24 22 .35	.24 22 .35
	3½	Calculated Thickness Use {Thickness Class {Thickness	.22 22 .35	.23 22 .35	.23 22 .35	.23 22 .35	.23 22 .35	.23 22 .35
	5	Calculated Thickness Use {Thickness Class {Thickness	.23 22 .35	.23 22 .35	.24 22 .35	.24 22 .35	.24 22 .35	.25 22 .35
	8	Calculated Thickness Use {Thickness Class {Thickness	.25 22 .35	.25 22 .35	.25 22 .35	.26 22 .35	.26 22 .35	.27 22 .35
	12	Calculated Thickness Use {Thickness Class {Thickness	.27 22 .35	.28 22 .35	.28 22 .35	.2ʹ 22 .35	.29 22 .35	.29 22 .35
	16	Calculated Thickness Use {Thickness Class {Thickness	.30 22 .35	.30 22 .35	.30 22 .35	.31 22 .35	.31 22 .35	.32 22 .35
B	2½	Calculated Thickness Use {Thickness Class {Thickness	.23 22 .35	.23 22 .35	.23 22 .35	.23 22 .35	.23 22 .35	.24 22 .35
	3½	Calculated Thickness Use {Thickness Class {Thickness	.22 22 .35	.23 22 .35	.23 22 .35	.23 22 .35	.24 22 .35	.24 22 .35
	5	Calculated Thickness Use {Thickness Class {Thickness	.23 22 .35	.23 22 .35	.23 22 .35	.23 22 .35	.24 22 .35	.24 22 .35
	8	Calculated Thickness Use {Thickness Class {Thickness	.24 22 .35	.24 22 .35	.24 22 .35	.25 22 .35	.25 22 .35	.26 22 .35
	12	Calculated Thickness Use {Thickness Class {Thickness	.28 22 .35	.29 22 .35	.29 22 .35	.29 22 .35	.30 22 .35	.30 22 .35
	16	Calculated Thickness Use {Thickness Class {Thickness	.29 22 .35	.29 22 .35	.29 22 .35	.30 22 .35	.30 22 .35	.31 22 .35
F	2½	Calculated Thickness Use {Thickness Class {Thickness	.22 22 .35	.22 22 .35	.22 22 .35	.22 22 .35	.23 22 .35	.23 22 .35
	3½	Calculated Thickness Use {Thickness Class {Thickness	.22 22 .35	.22 22 .35	.22 22 .35	.22 22 .35	.23 22 .35	.23 22 .35
	5	Calculated Thickness Use {Thickness Class {Thickness	.22 22 .35	.22 22 .35	.22 22 .35	.22 22 .35	.23 22 .35	.23 22 .35
	8	Calculated Thickness Use {Thickness Class {Thickness	.23 22 .35	.23 22 .35	.23 22 .35	.24 22 .35	.24 22 .35	.24 22 .35
	12	Calculated Thickness Use {Thickness Class {Thickness	.24 22 .35	.24 22 .35	.25 22 .35	.25 22 .35	.26 22 .35	.27 22 .35
	16	Calculated Thickness Use {Thickness Class {Thickness	.26 22 .35	.26 22 .35	.27 22 .35	.28 22 .35	.28 22 .35	.29 22 .35

5
6

TABLE 1–3 (*Continued*)

Schedule of Barrel Thickness for Gas Pipe of 18/40 Iron Strength

Laying Condi-tion	Depth of Cover ft	Thickness Specifications	Internal Pressure—*psi*					
			10	50	100	150	200	250
			Barrel Thicknesses—*in.*					

Six-Inch Gas Pipe

Laying Condi-tion	Depth of Cover ft	Thickness Specifications	10	50	100	150	200	250
A	2½	Calculated Thickness	.28	.29	.29	.30	.30	.32
		Use Thickness Class	22	22	22	22	22	22
		Thickness	.38	.38	.38	.38	.38	.38
	3½	Calculated Thickness	.27	.28	.28	.29	.30	.31
		Use Thickness Class	22	22	22	22	22	22
		Thickness	.38	.38	.38	.38	.38	.38
	5	Calculated Thickness	.28	.29	.29	.30	.31	.32
		Use Thickness Class	22	22	22	22	22	22
		Thickness	.38	.38	.38	.38	.38	.38
	8	Calculated Thickness	.31	.31	.32	.33	.33	.34
		Use Thickness Class	22	22	22	22	22	22
		Thickness	.38	.38	.38	.38	.38	.38
	12	Calculated Thickness	.34	.35	.35	.36	.37	.38
		Use Thickness Class	22	22	22	22	22	22
		Thickness	.38	.38	.38	.38	.38	.38
	16	Calculated Thickness	.38	.38	.39	.39	.40	.41
		Use Thickness Class	22	22	22	22	23	23
		Thickness	.38	.38	.38	.38	.41	.41
B	2½	Calculated Thickness	.27	.28	.28	.29	.30	.31
		Use Thickness Class	22	22	22	22	22	22
		Thickness	.38	.38	.38	.38	.38	.38
	3½	Calculated Thickness	.27	.27	.28	.28	.29	.30
		Use Thickness Class	22	22	22	22	22	22
		Thickness	.38	.38	.38	.38	.38	.38
	5	Calculated Thickness	.28	.28	.28	.29	.30	.31
		Use Thickness Class	22	22	22	22	22	22
		Thickness	.38	.38	.38	.38	.38	.38
	8	Calculated Thickness	.29	.30	.31	.31	.32	.33
		Use Thickness Class	22	22	22	22	22	22
		Thickness	.38	.38	.38	.38	.38	.38
	12	Calculated Thickness	.33	.34	.34	.35	.35	.36
		Use Thickness Class	22	22	22	22	22	22
		Thickness	.38	.38	.38	.38	.38	.38
	16	Calculated Thickness	.36	.36	.37	.38	.38	.39
		Use Thickness Class	22	22	22	22	22	22
		Thickness	.38	.38	.38	.38	.38	.38
F	2½	Calculated Thickness	.26	.26	.27	.27	.28	.29
		Use Thickness Class	22	22	22	22	22	22
		Thickness	.38	.38	.38	.38	.38	.38
	3½	Calculated Thickness	.26	.26	.26	.27	.28	.29
		Use Thickness Class	22	22	22	22	22	22
		Thickness	.38	.38	.38	.38	.38	.38
	5	Calculated Thickness	.26	.27	.27	.28	.28	.29
		Use Thickness Class	22	22	22	22	22	22
		Thickness	.38	.38	.38	.38	.38	.38
	8	Calculated Thickness	.28	28	.29	.29	.30	.31
		Use Thickness Class	22	22	22	22	22	22
		Thickness	.38	.38	.38	.38	.38	.38
	12	Calculated Thickness	.30	.31	.31	.32	.33	.34
		Use Thickness Class	22	22	22	22	22	22
		Thickness	.38	.38	.38	.38	.38	.38
	16	Calculated Thickness	.33	.33	.34	.35	.35	.36
		Use Thickness Class	22	22	22	22	22	22
		Thickness	.38	.38	.38	.38	.38	.38

5
7

TABLE 1–3 (*Continued*)

Schedule of Barrel Thickness for Gas Pipe of 18/40 Iron Strength

Laying Condition	Depth of Cover ft	Thickness Specifications	Internal Pressure—*psi*					
			10	50	100	150	200	250
			Barrel Thicknesses—*in.*					

Eight-Inch Gas Pipe

Laying Condition	Depth of Cover ft	Thickness Specifications	10	50	100	150	200	250
A	2½	Calculated Thickness	.33	.34	.35	.36	.37	.38
		Use Thickness Class	22	22	22	22	22	22
		Thickness	.41	.41	.41	.41	.41	.41
	3½	Calculated Thickness	.32	.33	.34	.34	.35	.36
		Use Thickness Class	22	22	22	22	22	22
		Thickness	.41	.41	.41	.41	.41	.41
	5	Calculated Thickness	.33	.34	.35	.36	.37	.38
		Use Thickness Class	22	22	22	22	22	22
		Thickness	.41	.41	.41	.41	.41	.41
	8	Calculated Thickness	.36	.37	.37	.38	.39	.40
		Use Thickness Class	22	22	22	22	22	22
		Thickness	.41	.41	.41	.41	.41	.41
	12	Calculated Thickness	.41	.42	.42	.43	.44	.45
		Use Thickness Class	22	22	22	23	23	23
		Thickness	.41	.41	.41	.44	.44	.44
	16	Calculated Thickness	.44	.45	.45	.46	.47	.48
		Use Thickness Class	23	23	23	24	24	24
		Thickness	.44	.44	.44	.48	.48	.48
B	2½	Calculated Thickness	.32	.32	.33	.34	.35	.36
		Use Thickness Class	22	22	22	22	22	22
		Thickness	.41	.41	.41	.41	.41	.41
	3½	Calculated Thickness	.31	.32	.33	.33	.34	.36
		Use Thickness Class	22	22	22	22	22	22
		Thickness	.41	.41	.41	.41	.41	.41
	5	Calculated Thickness	.32	.33	.34	.34	.35	.37
		Use Thickness Class	22	22	22	22	22	22
		Thickness	.41	.41	.41	.41	.41	.41
	8	Calculated Thickness	.35	.35	.36	.37	.38	.39
		Use Thickness Class	22	22	22	22	22	22
		Thickness	.41	.41	.41	.41	.41	.41
	12	Calculated Thickness	.39	.39	.40	.41	.42	.43
		Use Thickness Class	22	22	22	22	22	23
		Thickness	.41	.41	.41	.41	.41	.44
	16	Calculated Thickness	.42	.42	.43	.44	.45	.46
		Use Thickness Class	22	22	23	23	23	24
		Thickness	.41	.41	.44	.44	.44	.48
F	2½	Calculated Thickness	.29	.30	.31	.32	.33	.34
		Use Thickness Class	22	22	22	22	22	22
		Thickness	.41	.41	.41	.41	.41	.41
	3½	Calculated Thickness	.29	.29	.30	.30	.31	.33
		Use Thickness Class	22	22	22	22	22	22
		Thickness	.41	.41	.41	.41	.41	.41
	5	Calculated Thickness	.29	.30	.31	.32	.33	.34
		Use Thickness Class	22	22	22	22	22	22
		Thickness	.41	.41	.41	.41	.41	.41
	8	Calculated Thickness	.32	.33	.33	.34	.35	.36
		Use Thickness Class	22	22	22	22	22	22
		Thickness	.41	.41	.41	.41	.41	.41
	12	Calculated Thickness	.35	.36	.37	.37	.38	.39
		Use Thickness Class	22	22	22	22	22	22
		Thickness	.41	.41	.41	.41	.41	.41
	16	Calculated Thickness	.38	.38	.39	.40	.41	.42
		Use Thickness Class	22	22	22	22	22	22
		Thickness	.41	.41	.41	.41	.41	.41

TABLE 1-3 (*Continued*)

Schedule of Barrel Thickness for Gas Pipe of 18/40 Iron Strength

Laying Condition	Depth of Cover ft	Thickness Specifications	Internal Pressure—*psi*					
			10	50	100	150	200	250
			Barrel Thicknesses—*in.*					

Ten-Inch Gas Pipe

Laying Condition	Depth of Cover ft	Thickness Specifications	10	50	100	150	200	250
A	2½	Calculated Thickness	.39	.40	.41	.42	.43	.45
		Use {Thickness Class	22	22	22	22	22	22
		{Thickness	.44	.44	.44	.44	.44	.44
	3½	Calculated Thickness	.38	.39	.40	.41	.42	.44
		Use {Thickness Class	22	22	22	22	22	22
		{Thickness	.44	.44	.44	.44	.44	.44
	5	Calculated Thickness	.39	.40	.41	.42	.43	.44
		Use {Thickness Class	22	22	22	22	22	22
		{Thickness	.44	.44	.44	.44	.44	.44
	8	Calculated Thickness	.43	.43	.44	.46	.47	.48
		Use {Thickness Class	22	22	22	23	23	23
		{Thickness	.44	.44	.44	.48	.48	.48
	12	Calculated Thickness	.48	.49	.49	.50	.51	.52
		Use {Thickness Class	23	23	23	24	24	24
		{Thickness	.48	.48	.48	.52	.52	.52
	16	Calculated Thickness	.50	.51	.52	.53	.54	.56
		Use {Thickness Class	24	24	24	24	25	25
		{Thickness	.52	.52	.52	.52	.56	.56
B	2½	Calculated Thickness	.37	.38	.39	.40	.41	43
		Use {Thickness Class	22	22	22	22	22	22
		{Thickness	.44	.44	.44	.44	.44	.44
	3½	Calculated Thickness	.36	.37	.38	.39	.40	.42
		Use {Thickness Class	22	22	22	22	22	22
		{Thickness	.44	.44	.44	.44	.44	.44
	5	Calculated Thickness	.37	.38	.39	.40	.41	.43
		Use {Thickness Class	22	22	22	22	22	22
		{Thickness	.44	.44	.44	.44	.44	.44
	8	Calculated Thickness	.40	.41	.42	.43	.44	.46
		Use {Thickness Class	22	22	22	22	22	23
		{Thickness	.44	.44	.44	.44	.44	.48
	12	Calculated Thickness	.45	.46	.47	.48	.49	.50
		Use {Thickness Class	22	23	23	23	23	24
		{Thickness	.44	.48	.48	.48	.48	.52
	16	Calculated Thickness	.47	.48	.49	.50	.51	.53
		Use {Thickness Class	23	23	23	24	24	24
		{Thickness	.48	.48	.48	.52	.52	.52
F	2½	Calculated Thickness	.34	.35	.36	.37	.38	.40
		Use {Thickness Class	22	22	22	22	22	22
		{Thickness	.44	.44	.44	.44	.44	.44
	3½	Calculated Thickness	.33	.34	.35	.36	.37	.39
		Use {Thickness Class	22	22	22	22	22	22
		{Thickness	.44	.44	.44	.44	.44	.44
	5	Calculated Thickness	.34	.35	.36	.37	.38	.40
		Use {Thickness Class	22	22	22	22	22	22
		{Thickness	.44	.44	.44	.44	.44	.44
	8	Calculated Thickness	.36	.37	.38	.39	.41	.42
		Use {Thickness Class	22	22	22	22	22	22
		{Thickness	.44	.44	.44	.44	.44	.44
	12	Calculated Thickness	.41	.42	.43	.44	.45	.46
		Use {Thickness Class	22	22	22	22	22	23
		{Thickness	.44	.44	.44	.44	.44	.48
	16	Calculated Thickness	.43	.44	.45	.46	.47	.48
		Use {Thickness Class	22	22	22	23	23	23
		{Thickness	.44	.44	.44	.48	.48	.48

TABLE 1-3 (Continued)

Schedule of Barrel Thickness for Gas Pipe of 18/40 Iron Strength

Laying Condition	Depth of Cover ft	Thickness Specifications	Internal Pressure—psi					
			10	50	100	150	200	250
			Barrel Thicknesses—in.					
Twelve-Inch Gas Pipe								
A	2½	Calculated Thickness	.44	.45	.46	.47	.49	.51
		Use Thickness Class	22	22	22	22	22	23
		Thickness	.48	.48	.48	.48	.48	.52
	3½	Calculated Thickness	.43	.43	.45	.46	.48	.50
		Use Thickness Class	22	22	22	22	22	23
		Thickness	.48	.48	.48	.48	.48	.52
	5	Calculated Thickness	.44	.45	.46	.47	.48	.50
		Use Thickness Class	22	22	22	22	22	23
		Thickness	.48	.48	.48	.48	.48	.52
	8	Calculated Thickness	.48	.49	.50	.52	.53	.55
		Use Thickness Class	22	22	23	23	23	24
		Thickness	.48	.48	.52	.52	.52	.56
	12	Calculated Thickness	.53	.54	.55	.57	.58	.59
		Use Thickness Class	23.	24	24	24	25	25
		Thickness	.52	.56	.56	.56	.60	.60
	16	Calculated Thickness	.56	.56	.58	.59	.60	.62
		Use Thickness Class	24	24	25	25	25	25
		Thickness	.56	.56	.60	.60	.60	.60
B	2½	Calculated Thickness	.41	.42	.44	.45	.46	.48
		Use Thickness Class	22	22	22	22	22	22
		Thickness	.48	.48	.48	.48	.48	.48
	3½	Calculated Thickness	.40	.41	.42	.44	.45	.47
		Use Thickness Class	22	22	22	22	22	22
		Thickness	.48	.48	.48	.48	.48	.48
	5	Calculated Thickness	.41	.42	.43	.45	.46	.48
		Use Thickness Class	22	22	22	22	22	22
		Thickness	.48	.48	.48	.48	.48	.48
	8	Calculated Thickness	.45	.46	.47	.49	.50	.52
		Use Thickness Class	22	22	22	22	23	23
		Thickness	.48	.48	.48	.48	.52	.52
	12	Calculated Thickness	.50	.51	.52	.53	.55	.56
		Use Thickness Class	23	23	23	23	24	24
		Thickness	.52	.52	.52	.52	.56	.56
	16	Calculated Thickness	.52	.53	.54	.56	.57	.59
		Use Thickness Class	23	23	24	24	24	25
		Thickness	.52	.52	.56	.56	.56	.60
F	2½	Calculated Thickness	.38	.39	.40	.41	.43	.45
		Use Thickness Class	22	22	22	22	22	22
		Thickness	.48	.48	.48	.48	.48	.48
	3½	Calculated Thickness	.37	.38	.39	.40	.42	.44
		Use Thickness Class	22	22	22	22	22	22
		Thickness	.48	.48	.48	.48	.48	.48
	5	Calculated Thickness	.38	.39	.40	.41	.43	.45
		Use Thickness Class	22	22	22	22	22	22
		Thickness	.48	.48	.48	.48	.48	.48
	8	Calculated Thickness	.41	.42	.43	.45	.46	.48
		Use Thickness Class	22	22	22	22	22	22
		Thickness	.48	.48	.48	.48	.48	.48
	12	Calculated Thickness	.45	.46	.47	.48	.50	.52
		Use Thickness Class	22	22	22	22	23	23
		Thickness	.48	.48	.48	.48	.52	.52
	16	Calculated Thickness	.47	.48	.49	.50	.52	.53
		Use Thickness Class	22	22	22	23	23	23
		Thickness	.48	.48	.48	.52	52	.52

TABLE 1-3 (*Continued*)

Schedule of Barrel Thickness for Gas Pipe of 18/40 Iron Strength

Laying Condition	Depth of Cover ft	Thickness Specifications	Internal Pressure—*psi*					
			10	50	100	150	200	250
			Barrel Thicknesses—*in.*					

Sixteen-Inch Gas Pipe

Laying Condition	Depth of Cover ft	Thickness Specifications	10	50	100	150	200	250
A	2½	Calculated Thickness	.53	.54	.56	.58	.60	
		Use {Thickness Class	22	22	23	23	23	
		{Thickness	.54	.54	.58	.58	.58	
	3½	Calculated Thickness	.52	.54	.55	.57	.59	
		Use {Thickness Class	22	22	22	23	23	
		{Thickness	.54	.54	.54	.58	.58	
	5	Calculated Thickness	.54	.55	.57	.59	.61	
		Use {Thickness Class	22	22	23	23	24	
		{Thickness	.54	.54	.58	.58	.63	
	8	Calculated Thickness	.60	.61	.63	.65	.67	
		Use {Thickness Class	23	24	24	24	25	
		{Thickness	.58	.63	.63	.63	.68	
	12	Calculated Thickness	.65	.66	.67	.69	.71	
		Use {Thickness Class	24	25	25	25	26	
		{Thickness	.63	.68	.68	.68	.73	
	16	Calculated Thickness	.68	.69	.71	.73	.75	
		Use {Thickness Class	25	25	26	26	26	
		{Thickness	.68	.68	.73	.73	.73	
B	2½	Calculated Thickness	.49	.50	.52	.54	56	
		Use {Thickness Class	21	21	22	22	23	
		{Thickness	.50	.50	.54	.54	.58	
	3½	Calculated Thickness	.49	.50	.52	.53	.56	
		Use {Thickness Class	21	21	22	22	23	
		{Thickness	.50	.50	.54	.54	.58	
	5	Calculated Thickness	.51	.51	.53	.55	.57	
		Use {Thickness Class	21	21	22	22	23	
		{Thickness	.50	.50	.54	.54	.58	
	8	Calculated Thickness	.55	.56	.58	.60	.62	
		Use {Thickness Class	22	23	23	23	24	
		{Thickness	.54	.58	.58	.58	.63	
	12	Calculated Thickness	.60	.61	.62	.64	.66	
		Use {Thickness Class	23	24	24	24	25	
		{Thickness	.58	.63	.63	.63	.68	
	16	Calculated Thickness	.63	.64	.66	.67	.69	
		Use {Thickness Class	24	24	25	25	25	
		{Thickness	.63	.63	.68	.68	.68	
F	2½	Calculated Thickness	.46	.47	.48	.50	.52	
		Use {Thickness Class	21	21	21	21	22	
		{Thickness	.50	.50	.50	.50	.54	
	3½	Calculated Thickness	.45	.46	.47	.49	.52	
		Use {Thickness Class	21	21	21	21	22	
		{Thickness	.50	.50	.50	.50	.54	
	5	Calculated Thickness	.47	.48	.49	.51	.53	
		Use {Thickness Class	21	21	21	21	22	
		{Thickness	.50	.50	.50	.50	.54	
	8	Calculated Thickness	.51	.52	.53	.55	.57	
		Use {Thickness Class	21	22	22	22	23	
		{Thickness	.50	.54	.54	.54	.58	
	12	Calculated Thickness	.54	.55	.57	.59	.61	
		Use {Thickness Class	22	22	23	23	24	
		{Thickness	.54	.54	.58	.58	.63	
	16	Calculated Thickness	.57	.58	.59	.61	.63	
		Use {Thickness Class	23	23	23	24	24	
		{Thickness	.58	.58	.58	.63	.63	

6
1

TABLE 1-3 (*Continued*)

Schedule of Barrel Thickness for Gas Pipe of 18/40 Iron Strength

Laying Condition	Depth of Cover ft	Thickness Specifications	Internal Pressure—*psi*					
			10	50	100	150	200	250
			Barrel Thicknesses—*in.*					

Twenty-Inch Gas Pipe

Laying Condition	Depth of Cover ft	Thickness Specifications	10	50	100	150	200	250
A	2½	Calculated Thickness	.61	.62	.65	.67	.70	
		Use {Thickness Class	22	22	23	23	24	
		{Thickness	.62	.62	.67	.67	.72	
	3½	Calculated Thickness	.61	.62	.64	.66	.69	
		Use {Thickness Class	22	22	22	23	23	
		{Thickness	.62	.62	.62	.67	.67	
	5	Calculated Thickness	.63	.65	.67	.69	.72	
		Use {Thickness Class	22	23	23	23	24	
		{Thickness	.62	.67	.67	.67	.72	
	8	Calculated Thickness	.69	.70	.72	.75	.77	
		Use {Thickness Class	23	24	24	25	25	
		{Thickness	.67	.72	.72	.78	.78	
	12	Calculated Thickness	.74	.76	.78	.80	.83	
		Use {Thickness Class	24	25	25	25	26	
		{Thickness	.72	.78	.78	.78	.84	
	16	Calculated Thickness	.79	.80	.82	.84	.87	
		Use {Thickness Class	25	25	26	26	26	
		{Thickness	.78	.78	.84	.84	.84	
B	2½	Calculated Thickness	.56	.57	.59	.62	.65	
		Use {Thickness Class	21	21	21	22	23	
		{Thickness	.57	.57	.57	.62	.67	
	3½	Calculated Thickness	.55	.57	.59	.62	.64	
		Use {Thickness Class	21	21	21	22	22	
		{Thickness	.57	.57	.57	.62	.62	
	5	Calculated Thickness	.58	.59	.61	.64	.67	
		Use {Thickness Class	21	21	22	22	23	
		{Thickness	.57	.57	.62	.62	.67	
	8	Calculated Thickness	.63	.64	.66	.69	.71	
		Use {Thickness Class	22	22	23	23	24	
		{Thickness	.62	.62	.67	.67	.72	
	12	Calculated Thickness	.68	.69	.71	.73	.76	
		Use {Thickness Class	23	23	24	24	25	
		{Thickness	.67	.67	.72	.72	.78	
	16	Calculated Thickness	.71	.73	.75	.77	.80	
		Use {Thickness Class	24	24	25	25	25	
		{Thickness	.72	.72	.78	.78	.78	
F	2½	Calculated Thickness	.50	.51	.53	.57	.60	
		Use {Thickness Class	21	21	21	21	22	
		{Thickness	.57	.57	.57	.57	.62	
	3½	Calculated Thickness	.50	.51	.53	.57	.60	
		Use {Thickness Class	21	21	21	21	22	
		{Thickness	.57	.57	.57	.57	.62	
	5	Calculated Thickness	.52	.54	.56	.58	.62	
		Use {Thickness Class	21	21	21	21	22	
		{Thickness	.57	.57	.57	.57	.62	
	8	Calculated Thickness	.56	.58	.60	.62	.65	
		Use {Thickness Class	21	21	22	22	23	
		{Thickness	.57	.57	.62	.62	.67	
	12	Calculated Thickness	.61	.62	.64	.67	.70	
		Use {Thickness Class	22	22	22	23	24	
		{Thickness	.62	.62	.62	.67	.72	
	16	Calculated Thickness	.64	.66	.68	.70	.72	
		Use {Thickness Class	22	23	23	24	24	
		{Thickness	.62	.67	.67	.72	.72	

TABLE 1-3 (*Continued*)

Schedule of Barrel Thickness for Gas Pipe of 18/40 Iron Strength

Laying Condition	Depth of Cover ft	Thickness Specifications	Internal Pressure—*psi*					
			10	50	100	150	200	250
			Barrel Thicknesses—*in.*					

Twenty-four-Inch Gas Pipe

Laying Condition	Depth of Cover ft	Thickness Specifications	10	50	100	150	200	250
A	2½	Calculated Thickness	.67	.69	.72	.75	.78	
		Use {Thickness Class	22	22	23	23	24	
		{Thickness	.68	.68	.73	.73	.79	
	3½	Calculated Thickness	.67	.69	.71	.74	.78	
		Use {Thickness Class	22	22	23	23	24	
		{Thickness	.68	.68	.73	.73	.79	
	5	Calculated Thickness	.71	.72	.75	.78	.81	
		Use {Thickness Class	23	23	23	24	24	
		{Thickness	.73	.73	.73	.79	.79	
	8	Calculated Thickness	.77	.79	.81	.84	.87	
		Use {Thickness Class	24	24	24	25	25	
		{Thickness	.79	.79	.79	.85	.85	
	12	Calculated Thickness	.84	.86	.88	.91	.94	
		Use {Thickness Class	25	25	25	26	26	
		{Thickness	.85	.85	.85	.92	.92	
	16	Calculated Thickness	.90	.91	.94	.97	1.00	
		Use {Thickness Class	26	26	26	27	27	
		{Thickness	.92	.92	.92	.99	.99	
B	2½	Calculated Thickness	.61	.63	.65	.68	.72	
		Use {Thickness Class	21	21	21	22	23	
		{Thickness	.63	.63	.63	.68	.73	
	3½	Calculated Thickness	.59	.62	.65	.68	.71	
		Use {Thickness Class	21	21	21	22	23	
		{Thickness	.63	.63	.63	.68	.73	
	5	Calculated Thickness	.63	.65	.67	.71	.75	
		Use {Thickness Class	21	21	22	23	23	
		{Thickness	.63	.63	.68	.73	.73	
	8	Calculated Thickness	.69	.71	.73	.76	.79	
		Use {Thickness Class	22	23	23	24	24	
		{Thickness	.68	.73	.73	.79	.79	
	12	Calculated Thickness	.75	.77	.79	.82	.85	
		Use {Thickness Class	23	24	24	25	25	
		{Thickness	.73	.79	.79	.85	.85	
	16	Calculated Thickness	.80	.81	.84	.87	.90	
		Use {Thickness Class	24	24	25	25	26	
		{Thickness	.79	.79	.85	.85	.92	
F	2½	Calculated Thickness	.55	.56	.58	.63	.67	
		Use {Thickness Class	21	21	21	21	22	
		{Thickness	.63	.63	.63	.63	.68	
	3½	Calculated Thickness	.54	.56	.58	.62	.66	
		Use {Thickness Class	21	21	21	21	22	
		{Thickness	.63	.63	.63	.63	.68	
	5	Calculated Thickness	.57	.58	.62	.65	.68	
		Use {Thickness Class	21	21	21	21	22	
		{Thickness	.63	.63	.63	.63	.68	
	8	Calculated Thickness	.62	.64	.67	.70	.73	
		Use {Thickness Class	21	21	22	22	23	
		{Thickness	.63	.63	.68	.68	.73	
	12	Calculated Thickness	.67	.69	.72	.75	.78	
		Use {Thickness Class	22	22	23	23	24	
		{Thickness	.68	.68	.73	.73	.79	
	16	Calculated Thickness	.71	.73	.76	.79	.82	
		Use {Thickness Class	23	23	24	24	25	
		{Thickness	.73	.73	.79	.79	.85	

6
3

TABLE 1–3 (*Continued*)

Schedule of Barrel Thickness for Gas Pipe of 18/40 Iron Strength

Laying Condition	Depth of Cover ft	Thickness Specifications	Internal Pressure—*psi*					
			10	50	100	150	200	250
			Barrel Thicknesses—*in.*					

Thirty-Inch Gas Pipe

Laying Condition	Depth of Cover ft	Thickness Specifications	10	50	100	150	200	250
A	2½	Calculated Thickness	.80	.83	.86	.90		
		Use {Thickness Class	22	23	23	24		
		{Thickness	.79	.85	.85	.92		
	3½	Calculated Thickness	.79	.82	.85	.89		
		Use {Thickness Class	22	23	23	24		
		{Thickness	.79	.85	.85	.92		
	5	Calculated Thickness	.84	.86	.90	.93		
		Use {Thickness Class	23	23	24	24		
		{Thickness	.85	.85	.92	.92		
	8	Calculated Thickness	.92	.94	.97	1.01		
		Use {Thickness Class	24	24	25	25		
		{Thickness	.92	.92	.99	.99		
	12	Calculated Thickness	1.01	1.03	1.06	1.09		
		Use {Thickness Class	25	26	26	26		
		{Thickness	.99	1.07	1.07	1.07		
	16	Calculated Thickness	1.07	1.09	1.12	1.15		
		Use {Thickness Class	26	26	27	27		
		{Thickness	1.07	1.07	1.16	1.16		
B	2½	Calculated Thickness	.69	.74	.77	.81		
		Use {Thickness Class	21	21	22	22		
		{Thickness	.73	.73	.79	.79		
	3½	Calculated Thickness	.69	.73	.76	.80		
		Use {Thickness Class	21	21	22	22		
		{Thickness	.73	.73	.79	.79		
	5	Calculated Thickness	.75	.77	.80	.85		
		Use {Thickness Class	21	22	22	23		
		{Thickness	.73	.79	.79	.85		
	8	Calculated Thickness	.81	.83	.86	.90		
		Use {Thickness Class	22	23	23	24		
		{Thickness	.79	.85	.85	.92		
	12	Calculated Thickness	.88	.90	.93	.97		
		Use {Thickness Class	23	24	24	25		
		{Thickness	.85	.92	.92	.99		
	16	Calculated Thickness	.94	.96	.99	1.03		
		Use {Thickness Class	24	25	25	26		
		{Thickness	.92	.99	.99	1.07		
F	2½	Calculated Thickness	.63	.66	.70	.74		
		Use {Thickness Class	21	21	21	21		
		{Thickness	.73	.73	.73	.73		
	3½	Calculated Thickness	.63	.66	.70	.74		
		Use {Thickness Class	21	21	21	21		
		{Thickness	.73	.73	.73	.73		
	5	Calculated Thickness	.68	.70	.73	.77		
		Use {Thickness Class	21	21	21	22		
		{Thickness	.73	.73	.73	.79		
	8	Calculated Thickness	.73	.75	.79	.83		
		Use {Thickness Class	21	21	22	23		
		{Thickness	.73	.73	.79	.85		
	12	Calculated Thickness	.79	.81	.85	.88		
		Use {Thickness Class	22	22	23	23		
		{Thickness	.79	.79	.85	.85		
	16	Calculated Thickness	.84	.86	.90	.93		
		Use {Thickness Class	23	23	24	24		
		{Thickness	.85	.85	.92	.92		

TABLE 1–3 (*Continued*)

Schedule of Barrel Thickness for Gas Pipe of 18/40 Iron Strength

Laying Condition	Depth of Cover ft	Thickness Specifications	Internal Pressure—*psi*					
			10	50	100	150	200	250
			Barrel Thicknesses—*in.*					
Thirty-six-Inch Gas Pipe								
A	2½	Calculated Thickness	.91	.94	.98	1.02		
		Use { Thickness Class	23	23	24	24		
		{ Thickness	.94	.94	1.02	1.02		
	3½	Calculated Thickness	.90	.93	.97	1.02		
		Use { Thickness Class	22	23	23	24		
		{ Thickness	.87	.94	.94	1.02		
	5	Calculated Thickness	.95	.98	1.02	1.07		
		Use { Thickness Class	23	24	24	25		
		{ Thickness	.94	1.02	1.02	1.10		
	8	Calculated Thickness	1.04	1.07	1.11	1.15		
		Use { Thickness Class	24	25	25	26		
		{ Thickness	1.02	1.10	1.10	1.19		
	12	Calculated Thickness	1.15	1.17	1.21	1.25		
		Use { Thickness Class	26	26	26	27		
		{ Thickness	1.19	1.19	1.19	1.29		
	16	Calculated Thickness	1.23	1.26	1.29	1.33		
		Use { Thickness Class	26	27	27	27		
		{ Thickness	1.19	1.29	1.29	1.29		
B	2½	Calculated Thickness	.79	.82	.86	.91		
		Use { Thickness Class	21	21	22	23		
		{ Thickness	.81	.81	.87	.94		
	3½	Calculated Thickness	.79	.81	.85	.90		
		Use { Thickness Class	21	21	22	22		
		{ Thickness	.81	.81	.87	.87		
	5	Calculated Thickness	.83	.86	.90	.94		
		Use { Thickness Class	21	22	22	23		
		{ Thickness	.81	.87	.87	.94		
	8	Calculated Thickness	.90	.93	.96	1.01		
		Use { Thickness Class	22	23	23	24		
		{ Thickness	.87	.94	.94	1.02		
	12	Calculated Thickness	.99	1.01	1.05	1.10		
		Use { Thickness Class	24	24	24	25		
		{ Thickness	1.02	1.02	1.02	1.10		
	16	Calculated Thickness	1.06	1.08	1.13	1.17		
		Use { Thickness Class	25	25	25	26		
		{ Thickness	1.10	1.10	1.10	1.19		
F	2½	Calculated Thickness	.71	.73	.79	.84		
		Use { Thickness Class	21	21	21	22		
		{ Thickness	.81	.81	.81	.87		
	3½	Calculated Thickness	.71	.73	.78	.83		
		Use { Thickness Class	21	21	21	21		
		{ Thickness	.81	.81	.81	.81		
	5	Calculated Thickness	.75	.78	.82	.87		
		Use { Thickness Class	21	21	21	22		
		{ Thickness	.81	.81	.81	.87		
	8	Calculated Thickness	.81	.84	.88	.93		
		Use { Thickness Class	21	22	22	23		
		{ Thickness	.81	.87	.87	.94		
	12	Calculated Thickness	.89	.91	.95	1.00		
		Use { Thickness Class	22	23	23	24		
		{ Thickness	.87	.94	.94	1.02		
	16	Calculated Thickness	.95	.97	1.01	1.06		
		Use { Thickness Class	23	23	24	25		
		{ Thickness	.94	.94	1.02	1.10		

TABLE 1–3 (*Continued*)

Schedule of Barrel Thickness for Gas Pipe of 18/40 Iron Strength

Laying Condi-tion	Depth of Cover ft	Thickness Specifications	Internal Pressure—*psi*					
			10	50	100	150	200	250
			Barrel Thicknesses—*in.*					

Forty-two-Inch Gas Pipe

Laying Condi-tion	Depth of Cover ft	Thickness Specifications	10	50	100	150	200	250
A	2½	Calculated Thickness	*1.01*	*1.04*	*1.08*	*1.14*		
		Use {Thickness Class	23	23	23	24		
		Thickness	1.05	1.05	1.05	1.13		
	3½	Calculated Thickness	*1.01*	*1.04*	*1.08*	*1.14*		
		Use {Thickness Class	23	23	23	24		
		Thickness	1.05	1.05	1.05	1.13		
	5	Calculated Thickness	*1.07*	*1.09*	*1.14*	*1.19*		
		Use {Thickness Class	23	24	24	25		
		Thickness	1.05	1.13	1.13	1.22		
	8	Calculated Thickness	*1.16*	*1.19*	*1.24*	*1.29*		
		Use {Thickness Class	24	25	25	26		
		Thickness	1.13	1.22	1.22	1.32		
	12	Calculated Thickness	*1.29*	*1.32*	*1.36*	*1.41*		
		Use {Thickness Class	26	26	26	27		
		Thickness	1.32	1.32	1.32	1.43		
	16	Calculated Thickness	*1.39*	*1.42*	*1.46*	*1.51*		
		Use {Thickness Class	27	27	27	28		
		Thickness	1.43	1.43	1.43	1.54		
B	2½	Calculated Thickness	*.86*	*.89*	*.94*	*1.00*		
		Use {Thickness Class	21	21	22	22		
		Thickness	.90	.90	.97	.97		
	3½	Calculated Thickness	*.86*	*.89*	*.94*	*1.00*		
		Use {Thickness Class	21	21	22	22		
		Thickness	.90	.90	.97	.97		
	5	Calculated Thickness	*.91*	*.94*	*.98*	*1.04*		
		Use {Thickness Class	21	22	22	23		
		Thickness	.90	.97	.97	1 05		
	8	Calculated Thickness	*.99*	*1.02*	*1.07*	*1.12*		
		Use {Thickness Class	22	23	23	24		
		Thickness	.97	1.05	1.05	1.13		
	12	Calculated Thickness	*1.10*	*1.13*	*1.17*	*1.22*		
		Use {Thickness Class	24	24	24	25		
		Thickness	1.13	1.13	1.13	1.22		
	16	Calculated Thickness	*1.18*	*1.21*	*1.25*	*1.30*		
		Use {Thickness Class	25	25	25	26		
		Thickness	1.22	1.22	1.22	1.32		
F	2½	Calculated Thickness	*.77*	*.80*	*.86*	*.92*		
		Use {Thickness Class	21	21	21	21		
		Thickness	.90	.90	.90	.90		
	3½	Calculated Thickness	*.77*	*.80*	*.86*	*.92*		
		Use {Thickness Class	21	21	21	21		
		Thickness	.90	.90	.90	.90		
	5	Calculated Thickness	*.82*	*.85*	*.90*	*.96*		
		Use {Thickness Class	21	21	21	22		
		Thickness	.90	.90	.90	.97		
	8	Calculated Thickness	*.89*	*.92*	*.97*	*1.03*		
		Use {Thickness Class	21	21	22	23		
		Thickness	.90	.90	.97	1.05		
	12	Calculated Thickness	*.98*	*1.01*	*1.06*	*1.11*		
		Use {Thickness Class	22	23	23	24		
		Thickness	.97	1.05	1.05	1.13		
	16	Calculated Thickness	*1.06*	*1.08*	*1.13*	*1.18*		
		Use {Thickness Class	23	23	24	25		
		Thickness	1.05	1.05	1.13	1.22		

TABLE 1-3 (Continued)

Schedule of Barrel Thickness for Gas Pipe of 18/40 Iron Strength

Laying Condition	Depth of Cover ft	Thickness Specifications	Internal Pressure—psi					
			10	50	100	150	200	250
			Barrel Thicknesses—in.					

Forty-eight-Inch Gas Pipe

Laying Condition	Depth of Cover ft	Thickness Specifications	10	50	100	150	200	250
A	2½	Calculated Thickness	1.10	1.14	1.19	1.25		
		Use Thickness Class	23	23	24	24		
		Thickness	1.14	1.14	1.23	1.23		
	3½	Calculated Thickness	1.11	1.14	1.20	1.26		
		Use Thickness Class	23	23	24	24		
		Thickness	1.14	1.14	1.23	1.23		
	5	Calculated Thickness	1.18	1.22	1.27	1.32		
		Use Thickness Class	23	24	24	25		
		Thickness	1.14	1.23	1.23	1.33		
	8	Calculated Thickness	1.29	1.33	1.38	1.43		
		Use Thickness Class	25	25	25	26		
		Thickness	1.33	1.33	1.33	1.44		
	12	Calculated Thickness	1.43	1.47	1.52	1.57		
		Use Thickness Class	26	26	27	27		
		Thickness	1.44	1.44	1.56	1.56		
	16	Calculated Thickness	1.55	1.58	1.63	1.69		
		Use Thickness Class	27	27	28	28		
		Thickness	1.56	1.56	1.68	1.68		
B	2½	Calculated Thickness	.93	.97	1.03	1.09		
		Use Thickness Class	21	21	22	22		
		Thickness	.98	.98	1.06	1.06		
	3½	Calculated Thickness	.94	.98	1.03	1.10		
		Use Thickness Class	21	21	22	23		
		Thickness	.98	.98	1.06	1.14		
	5	Calculated Thickness	.99	1.03	1.09	1.15		
		Use Thickness Class	21	22	22	23		
		Thickness	.98	1.06	1.06	1.14		
		Calculated Thickness	1.09	1.12	1.17	1.24		
		Use Thickness Class	22	23	23	24		
		Thickness	1.06	1.14	1.14	1.23		
	12	Calculated Thickness	1.21	1.25	1.30	1.35		
		Use Thickness Class	24	24	25	25		
		Thickness	1.23	1.23	1.33	1.33		
	16	Calculated Thickness	1.30	1.34	1.38	1.44		
		Use Thickness Class	25	25	25	26		
		Thickness	1.33	1.33	1.33	1.44		
F	2½	Calculated Thickness	.84	.88	.93	1.00		
		Use Thickness Class	21	21	21	21		
		Thickness	.98	.98	.98	.98		
	3½	Calculated Thickness	.84	.88	.94	1.01		
		Use Thickness Class	21	21	21	21		
		Thickness	.98	.98	.98	.98		
	5	Calculated Thickness	.90	.93	.99	1.06		
		Use Thickness Class	21	21	21	22		
		Thickness	.98	.98	.98	1.06		
	8	Calculated Thickness	.97	1.01	1.07	1.13		
		Use Thickness Class	21	21	22	23		
		Thickness	.98	.98	1.06	1.14		
	12	Calculated Thickness	1.08	1.12	1.16	1.23		
		Use Thickness Class	22	23	23	24		
		Thickness	1.06	1.14	1.14	1.23		
	16	Calculated Thickness	1.15	1.20	1.25	1.31		
		Use Thickness Class	23	24	24	25		
		Thickness	1.14	1.23	1.23	1.33		

6
7

Sec. 1-2—General Procedure for Thickness Determination

Sec. 1-2.1—Scope

This section gives the general method for determining the thicknesses of cast-iron pressure pipe. Thickness nomograms (Fig. 1-1 through 1-5) are included for two commonly used iron strengths (18/40 and 21/45); thicknesses for other iron strengths may be computed by the method presented in Sec. 1-3.1.

The required thickness of cast-iron pipe is determined by considering trench load and internal pressure in combination, and calculations of net thickness are made for two cases, namely:

Case 1. Trench load (earth load but no truck superload) in combination with internal pressure (working pressure plus surge pressure) and with a 2.5 factor of safety applied to both trench load and internal pressure

Case 2. Trench load (earth load plus truck superload) in combination with internal pressure (working pressure but no surge pressure) and with a 2.5 factor of safety applied to both trench load and internal pressure.

The larger thickness thus determined is chosen as the net thickness (only Case 2 is used for gas pipe). To the net thickness is added a corrosion allowance to obtain the minimum manufacturing thickness and a casting tolerance to obtain the total calculated thickness. Finally, the thickness for specifying and ordering is selected from a table of standard class thicknesses.

An example of this method is shown in Sec. 1-2.3.

Sec. 1-2.2—Procedure for Thickness Determination

This section gives the procedure for determining total calculated thick-

nesses and standard class thicknesses for pipe. This procedure was used in calculating the values shown in Sec. 1-1, Tables 1-1, 1-2 and 1-3.

1-2.2.1—Determination of Net Thickness

The net thickness for the more usual conditions may be readily determined using Tables 1-4 and 1-5 and the nomograms in Fig. 1-1 through 1-5. The bases for these tables and nomograms are described in Sec. 1-3. The three most commonly used methods of laying pipe, called "laying conditions," are defined below:

Laying Condition	*Description*
A	Pipe laid on flat-bottom trench, backfill not tamped
B	Pipe laid on flat-bottom trench, backfill tamped
F	Pipe bedded in gravel or sand, backfill tamped

After the pipe size, working pressure, iron strength, laying condition, and depth of cover have been established, the procedure for determining net thickness is as follows:

a. From Table 1-4, select for both Case 1 and Case 2 the value of w, which is the ring test load equivalent of trench load including a 2.5 safety factor (see Sec. 1-3.1 for definition of ring test load equivalent).

b. From Table 1-5, select for both Case 1 and Case 2 the value of p, which is the internal pressure including a 2.5 safety factor.

c. Thickness nomograms are provided in Fig. 1-1 through 1-5 for iron strengths of 18/40 * and 21/45.* Se-

---

* The first figure designates the bursting tensile strength in 1,000 psi and the second figure designates the ring modulus of rupture in 1,000 psi.

45

lect the nomogram for the desired iron strength and range of loads and pressures. Locate the values of w and p for Case 1 on the vertical scales. Connect these values with a straightedge and read the required thickness for Case 1.

d. Repeat this procedure using the above values of w and p for Case 2 and read the required thickness for Case 2.

e. The larger thickness, as determined for Cases 1 and 2, is the net thickness for water pipe. (For gas pipe, only Case 2 is applicable.)

1–2.2.2—Addition of Allowances to Net Thickness

a. A corrosion allowance of 0.08 in. is added to the net thickness. The resulting thickness is the minimum manufacturing thickness. Where severe corrosion is anticipated, an analysis of the condition should be made.

b. A casting tolerance from Table 1–6 is added to the minimum manufacturing thickness and the resulting thickness is the total calculated thickness.

1–2.2.3—Selection of Standard Thickness

Refer to Table 1–7 and select the standard class thickness nearest to the total calculated thickness. When the calculated thickness is halfway between two standard thicknesses, the larger of these is selected. When the calculated thickness is less than the smallest standard thickness, the smallest standard thickness is selected.

Sec. 1–2.3—Example for Determining Thickness of 18-in. Water Pipe

Determine the thickness for 18-in. cast-iron pipe laid on flat-bottom trench with tamped backfill (laying condition B), under 5 ft of cover for a working pressure of 200 psi. Iron strength is 18/40.

1–2.3.1—Step 1—Determination of Net Thickness

a. From Table 1–4, the ring test load equivalents of trench load, including a safety factor of 2.5, are

For Case 1, $w = 2{,}786$ lb/lin ft
For Case 2, $w = 3{,}876$ lb/lin ft

b. From Table 1–5, the internal pressures, including a safety factor of 2.5, are

For Case 1, $p = 750$ psi
For Case 2, $p = 500$ psi

c. Using Fig. 1–1, locate $w = 2{,}786$ and $p = 750$ on the vertical scales. Connect these values with a straightedge and read the thickness for Case 1, which is 0.47 in.

d. Using Fig. 1–1, locate $w = 3{,}876$ and $p = 500$ and read the thickness for Case 2, which is 0.46 in.

e. The larger of these two thicknesses is the net thickness. The controlling one in this example is 0.47 in. (Case 1).

1–2.3.2—Step 2—Addition of Allowances to Net Thickness

The total calculated thickness, as shown in Table 1–1, is determined as follows:

Net thickness	0.47 in.
Corrosion allowance	0.08 in.
Minimum manufacturing thickness	0.55 in.
Casting tolerance (Table 1–6)	0.08 in.
Total calculated thickness	0.63 in.

1–2.3.3—Step 3—Selection of Standard Thickness

From Table 1–7, the total calculated thickness of 0.63 in. is exactly the same as Class 23. Therefore, Class 23 is the standard thickness class for the pipe in this example.

In ordering or specifying, the foregoing pipe is identified as 18-in. size, thickness Class 23, conforming to ANSI Standard A21.6 or A21.8 as applicable.

TABLE 1–4

Ring Test Load Equivalents (w) of Trench Loads—lb/lin ft*

Laying Condition	Pipe Size in.	Depth of Cover—ft							
		2½	3½	5	8	12	16	20	24
Case 1—Ring Test Load Equivalent of Earth Load (Use With Surge Pressure)									
A	3	396	565	817	1,324	2,000	2,674	3,343	4,026
	4	491	704	1,024	1,663	2,515	3,367	4,222	5,074
	6	672	974	1,428	2,337	3,548	4,756	5,487	5,754
	8	826	1,211	1,791	2,948	4,491	5,635	6,120	6,446
	10	974	1,448	2,157	3,576	5,376	6,198	6,770	7,163
	12	1,111	1,674	2,520	4,239	5,839	6,774	7,433	7,902
	14	1,235	1,887	2,865	4,822	6,304	7,356	8,115	8,659
	16	1,341	2,085	3,196	5,176	6,776	7,954	8,804	9,433
	18	1,446	2,265	3,513	5,506	7,250	8,548	9,509	10,224
	20	1,552	2,433	3,815	5,839	7,728	9,150	10,224	11,033
	24	1,770	2,730	4,372	6,509	8,693	10,378	11,678	12,680
	30	2,093	3,167	5,087	7,520	10,161	12,256	13,915	15,237
	36	2,437	3,626	5,713	8,537	11,643	14,167	16,220	17,882
	42	2,761	4,083	6,341	9,556	13,078	16,104	18,554	20,560
	48	3,102	4,550	6,991	10,580	14,646	18,061	20,928	23,354
	54	3,437	5,022	7,650	11,609	16,161	20,039	23,346	26,161
	60	3,778	5,493	8,313	12,643	17,815	22,028	25,776	28,998
B	3	355	508	734	1,189	1,797	2,402	3,004	3,617
	4	438	628	913	1,483	2,242	3,002	3,764	4,523
	6	585	848	1,244	2,036	3,091	4,144	4,780	5,014
	8	709	1,039	1,537	2,530	3,855	4,836	5,252	5,531
	10	824	1,224	1,823	3,024	4,546	5,241	5,724	6,057
	12	926	1,395	2,100	3,533	4,866	5,645	6,194	6,586
	14	1,007	1,539	2,337	3,933	5,142	6,000	6,618	7,062
	16	1,079	1,677	2,570	4,163	5,449	6,397	7,080	7,586
	18	1,147	1,797	2,786	4,367	5,750	6,779	7,541	8,109
	20	1,214	1,903	2,985	4,568	6,046	7,158	7,998	8,631
	24	1,339	2,066	3,307	4,924	6,577	7,852	8,836	9,593
	30	1,524	2,305	3,703	5,473	7,396	8,921	10,128	11,090
	36	1,709	2,543	4,006	5,986	8,165	9,935	11,373	12,540
	42	1,879	2,778	4,315	6,503	8,899	10,959	12,626	14,015
	48	2,074	3,042	4,674	7,074	9,792	12,076	13,993	15,615
	54	2,259	3,300	5,027	7,629	10,620	13,169	15,342	17,191
	60	2,455	3,569	5,401	8,215	11,575	14,312	16,747	18,841
F	3	261	374	540	875	1,322	1,767	2,210	2,661
	4	323	463	673	1,093	1,653	2,213	2,774	3,334
	6	434	629	923	1,510	2,292	3,073	3,545	3,718
	8	528	774	1,144	1,883	2,869	3,600	3,910	4,118
	10	612	910	1,355	2,247	3,378	3,895	4,254	4,502
	12	691	1,041	1,566	2,635	3,630	4,211	4,620	4,912
	14	755	1,154	1,753	2,949	3,856	4,500	4,964	5,297
	16	812	1,262	1,934	3,133	4,101	4,815	5,329	5,709
	18	861	1,350	2,093	3,281	4,320	5,093	5,666	6,092
	20	915	1,435	2,250	3,444	4,558	5,396	6,029	6,506
	24	1,018	1,570	2,514	3,742	4,999	5,968	6,715	7,291
	30	1,157	1,751	2,812	4,157	5,618	6,776	7,693	8,424
	36	1,310	1,949	3,070	4,588	6,257	7,613	8,716	9,610
	42	1,443	2,134	3,315	4,996	6,837	8,418	9,699	10,766
	48	1,586	2,326	3,573	5,408	7,485	9,231	10,697	11,937
	54	1,726	2,522	3,842	5,830	8,116	10,063	11,724	13,138
	60	1,881	2,735	4,139	6,295	8,869	10,967	12,832	14,436

* A safety factor of 2.5 is included. These ring test load equivalents are based on the earth loads in Table 1–8.

TABLE 1-4 (*Continued*)

Ring Test Load Equivalents (w) *of Trench Loads*—lb/lin ft*

Laying Condition	Pipe Size in.	Depth of Cover—ft							
		2¼	3¼	5	8	12	16	20	24

Case 2—Ring Test Load Equivalent of Earth Load Plus Truck Superload (Use Without Surge Pressure)†

Laying Condition	Pipe Size	2¼	3¼	5	8	12	16	20	24
	3	748	741	935	1,411	2,059	2,704	3,363	4,041
	4	1,137	1,057	1,200	1,780	2,602	3,426	4,261	5,104
	6	1,904	1,678	1,839	2,541	3,696	4,843	5,543	5,798
	8	2,528	2,267	2,437	3,270	4,696	5,752	6,198	6,506
	10	3,087	2,798	2,978	3,987	5,611	6,346	6,867	7,239
	12	3,635	3,317	3,517	4,767	6,104	6,950	7,548	7,991
	14	3,880	3,641	4,039	5,409	6,598	7,561	8,250	8,763
	16	4,183	3,996	4,478	5,880	7,128	8,220	8,980	9,567
	18	4,489	4,361	4,887	6,298	7,661	8,841	9,702	10,374
A	20	4,865	4,772	5,400	6,730	8,198	9,502	10,456	11,213
	24	5,383	5,250	6,043	7,513	9,250	10,761	11,930	12,876
	30	6,278	6,115	7,083	8,746	10,865	12,726	14,226	15,476
	36	7,180	7,054	8,083	9,911	12,465	14,724	16,587	18,167
	42	7,878	7,887	9,035	11,115	14,076	16,778	19,000	20,939
	48	8,596	8,728	10,035	12,350	15,761	18,824	21,432	23,743
	54	9,165	9,483	10,967	13,563	17,452	20,861	23,902	26,572
	60	9,789	10,237	11,904	14,783	19,193	22,909	26,393	29,437
	3	672	666	840	1,268	1,850	2,430	3,021	3,631
	4	1,014	942	1,070	1,587	2,320	3,054	3,798	4,550
	6	1,659	1,462	1,602	2,214	3,220	4,220	4,830	5,051
	8	2,170	1,946	2,091	2,806	4,030	4,937	5,319	5,584
	10	2,610	2,366	2,518	3,371	4,744	5,366	5,807	6,121
	12	3,029	2,765	2,931	3,973	5,087	5,792	6,290	6,659
	14	3,165	2,970	3,294	4,411	5,381	6,166	6,729	7,147
	16	3,364	3,213	3,601	4,729	5,733	6,610	7,222	7,694
B	18	3,560	3,459	3,876	4,995	6,076	7,012	7,695	8,228
	20	3,806	3,733	4,225	5.265	6.413	7,434	8,180	8,772
	24	4,072	3,972	4,572	5,684	6,998	8,141	9,026	9,742
	30	4,570	4,451	5,155	6,366	7,908	9,263	10,354	11,264
	36	5,035	4,947	5,668	6,950	8,741	10,325	11,631	12,739
	42	5,361	5,367	6,148	7,564	9,578	11,417	12,929	14,248
	48	5,747	5,836	6,709	8,257	10,538	12,586	14,330	15,875
	54	6,023	6,232	7,207	8,913	11,469	13,709	15,707	17,461
	60	6,360	6,651	7,734	9,604	12,470	14,884	17,148	19,126
	3	494	490	618	932	1,361	1,787	2,223	2,671
	4	747	694	789	1,170	1,710	2,251	2,800	3,354
	6	1,230	1,084	1,188	1,642	2,388	3,129	3,581	3,746
	8	1,615	1,449	1,557	2,089	3,000	3,675	3,960	4,157
	10	1,940	1,758	1,872	2,505	3,526	3,988	4,316	4,549
	12	2,260	2,062	2,187	2,964	3,795	4,320	4,692	4,968
	14	2,374	2,227	2,471	3,309	4,036	4,625	5,046	5,360
	16	2,532	2,418	2,711	3,559	4,315	4,975	5,436	5,791
F	18	2,675	2,598	2,912	3,752	4,565	5,268	5,781	6,181
	20	2,869	2,814	3,185	3,969	4,835	5,604	6,167	6,613
	24	3,095	3,019	3,475	4,320	5,319	6,188	6,860	7,404
	30	3,471	3,381	3,916	4,835	6,007	7,036	7,865	8,556
	36	3,859	3,791	4,343	5,326	6,698	7,912	8,914	9,763
	42	4,118	4,123	4,723	5,810	7,358	8,771	9,932	10,945
	48	4,393	4,461	5,129	6,312	8,055	9,621	10,954	12,136
	54	4,603	4,762	5,508	6,811	8,764	10,476	12,003	13,344
	60	4,874	5,097	5,927	7,360	9,556	11,405	13,140	14,655

*A safety factor of 2.5 is included. These ring test load equivalents are based on the earth loads and truck superloads in Table 1-8.

† Truck superload is based on two passing trucks with adjacent wheels 3 ft apart, having a 9,000-lb wheel load on unpaved road or flexible pavement and an impact factor of 1.50 (see Sec. 1-3.3).

TABLE 1-5

Internal Pressures (p)*

Pipe Size in.	Rated Working Pressure—*psi*							
	10	50	100	150	200	250	300	350
Case 1—Internal Pressure With Surge Pressure Allowances†								
3	—	425	550	675	800	925	1,050	1,175
4	—	425	550	675	800	925	1,050	1,175
6	—	425	550	675	800	925	1,050	1,175
8	—	425	550	675	800	925	1,050	1,175
10	—	425	550	675	800	925	1,050	1,175
12	—	400	525	650	775	900	1,025	1,150
14	—	400	525	650	775	900	1,025	1,150
16	—	375	500	625	750	875	1,000	1,125
18	—	375	500	625	750	875	1,000	1,125
20	—	350	475	600	725	850	975	1,100
24	—	338	463	588	713	838	963	1,088
30	—	325	450	575	700	825	950	1,075
36	—	313	438	563	688	813	938	1,063
42	—	300	425	550	675	800	925	1,050
48	—	300	425	550	675	800	925	1,050
54	—	300	425	550	675	800	925	1,050
60	—	300	425	550	675	800	925	1,050
Case 2—Internal Pressure Without Surge Pressure Allowances								
3–60 all	25	125	250	375	500	625	750	875

* Safety factor of 2.5 included.
† For surge pressure allowances, see Table 1–10.

TABLE 1-6

Allowances For Casting Tolerance

Pipe Size in.	Casting Tolerance in.	Pipe Size in.	Casting Tolerance in.
3–8	0.05	14–24	0.08
10–12	0.06	30–48	0.10

TABLE 1–7—*Standard Thickness Classes of Cast-Iron Pipe*

(See note on facing page)

Pipe Size in.	Thickness for Standard Thickness Class Number—*in.*										
	20	21	22	23	24	25	26	27	28	29	30
3			0.32*	0.35	0.38	0.41	0.44	0.48	0.52	0.56	0.60
4			0.35*	0.38	0.41	0.44	0.48	0.52	0.56	0.60	0.65
6		0.35*	0.38	0.41	0.44	0.48	0.52	0.56	0.60	0.65	0.70
8	0.35*	0.38	0.41	0.44	0.48	0.52	0.56	0.60	0.65	0.70	0.76
10	0.38*	0.41	0.44	0.48	0.52	0.56	0.60	0.65	0.70	0.76	0.82
12	0.41*	0.44	0.48	0.52	0.56	0.60	0.65	0.70	0.76	0.82	0.89
14	0.43	0.48*	0.51	0.55	0.59	0.64	0.69	0.75	0.81	0.87	0.94
16	0.46	0.50*	0.54	0.58	0.63	0.68	0.73	0.79	0.85	0.92	0.99
18	0.50	0.54*	0.58	0.63	0.68	0.73	0.79	0.85	0.92	0.99	1.07
20	0.53	0.57*	0.62	0.67	0.72	0.78	0.84	0.91	0.98	1.06	1.14
24	0.58	0.63*	0.68	0.73	0.79	0.85	0.92	0.99	1.07	1.16	1.25
30	0.68*	0.73	0.79	0.85	0.92	0.99	1.07	1.16	1.25	1.35	1.46
36	0.75*	0.81	0.87	0.94	1.02	1.10	1.19	1.29	1.39	1.50	1.62
42	0.83*	0.90	0.97	1.05	1.13	1.22	1.32	1.43	1.54	1.66	1.79
48	0.91*	0.98	1.06	1.14	1.23	1.33	1.44	1.56	1.68	1.81	1.95

7
3

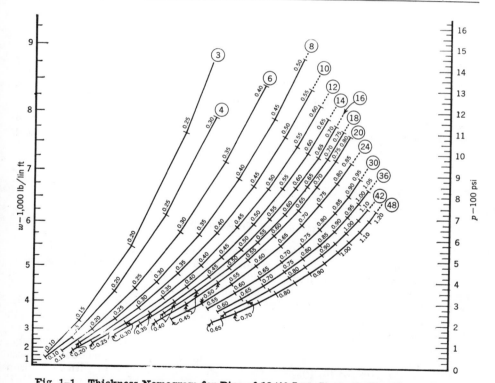

Fig. 1–1. Thickness Nomogram for Pipe of 18/40 Iron Strength, Low-Range Load

Thicknesses are net, and computations are made using nominal pipe diameter for inside diameter. S *equals 18,000,* R *equals 40,000. The encircled values at the end of each curve are for pipe size, in inches.*

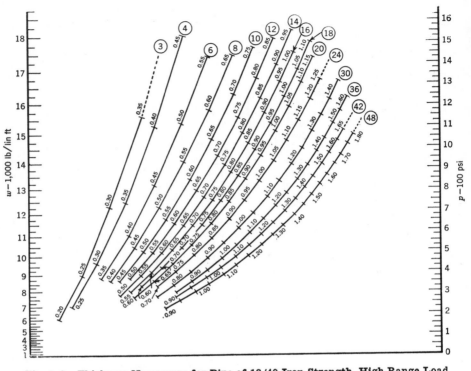

Fig. 1–2. Thickness Nomogram for Pipe of 18/40 Iron Strength, High-Range Load

Thicknesses are net, and computations are made using nominal pipe diameter for inside diameter. S equals 18,000, R equals 40,000. The encircled values at the end of each curve are for pipe size, in inches.

Explanatory Note to Table 1–7

Irrespective of calculated thickness, standard thicknesses are not made less than certain minimums, which are the smallest thickness classes shown in the table. They have been chosen by judgment based on experience with the shocks received by pipe in handling and transporting. ANSI Standards A21.6, A21.7, A21.8, and A21.9 are based on iron strength of 18/40. The recommended minimum nominal thicknesses for such pipe are shown in boldface type. Pipe of 21/45 iron strength for water service may have thicknesses less than those pipe of 18/40 iron strength. Pipe of such reduced thickness should be used only after consideration of possible adverse conditions of installation and environment. The recommended minimum nominal thicknesses for such pipe are shown with an asterisk. Pipe with 21/45 iron strength in Thickness Class 20 are sometimes used in sizes 14–24 in., when the calculated thicknesses permit and where provisions are made for special handling.

7
5

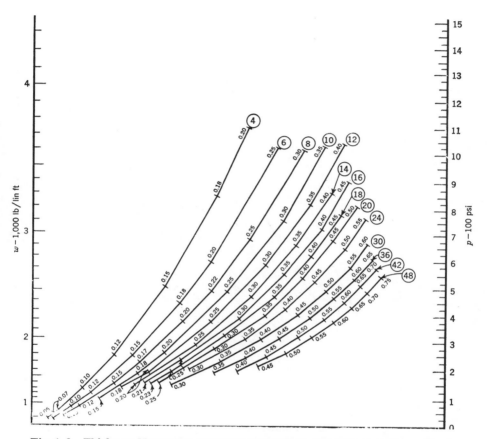

Fig. 1–3. Thickness Nomogram for Pipe of 21/45 Iron Strength, Low-Range Load

Thicknesses are net, and computations are made using nominal pipe diameter for inside diameter. S equals 21,000, R equals 45,000. The encircled values at the end of each curve are for pipe size, in inches.

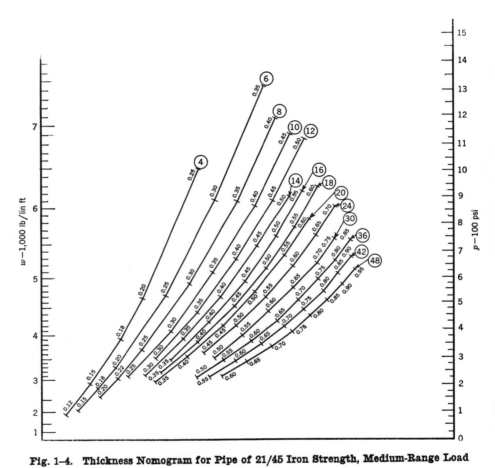

Fig. 1–4. Thickness Nomogram for Pipe of 21/45 Iron Strength, Medium-Range Load

Thicknesses are net, and computations are made using nominal pipe diameter for inside diameter. S equals 21,000, R equals 45,000. The encircled values at the end of each curve are for pipe size, in inches.

7
7

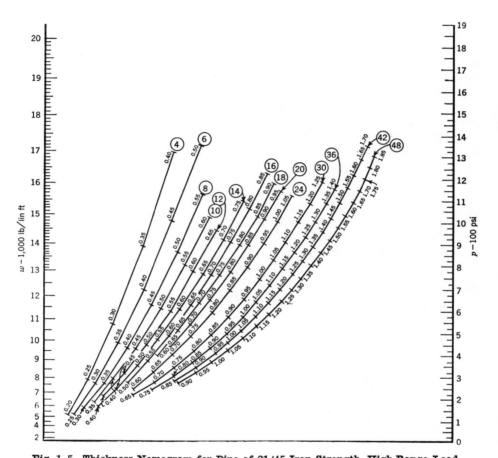

Fig. 1–5. Thickness Nomogram for Pipe of 21/45 Iron Strength, High-Range Load

Thicknesses are net, and computations are made using nominal pipe diameter for inside diameter. S equals 21,000, R equals 45,000. The encircled values at the end of each curve are for pipe size, in inches.

Sec. 1-3—Design Theory—Determination of Net Thickness, Earth Load, and Truck Superload

Sec. 1-3.1—Determination of Net Thickness

Tests made for ANSI Committee A21* showed that, when a pipe has both an external load applied in three-edge bearing (such as the laboratory ring test) and an internal pressure, the relation between external load and internal pressure at the point of breaking can be represented, with sufficient accuracy, by a parabola drawn as in Fig. 1-6.

The mathematical relationships expressed by the formula for the load-pressure curve are the basis for the whole system of pipe thickness calculations used in this standard.

The equation of the load-pressure parabola is

$$w = W \sqrt{\frac{P - p}{P}} \qquad (1)$$

in which:

> W = ring test crushing load with no internal pressure (lb/lin ft)
>
> P = bursting pressure with no external load (psi)
>
> p and w are any combination of internal pressure and external load that will just cause fracture.

The values of W and P are calculated as follows:

$$W = \frac{Rt^2}{0.0795(d + t)} \qquad (2)$$

$$P = \frac{2St}{d} \qquad (3)$$

---

* See Ref. 3, "Foreword," p. x.

in which:

> R = ring modulus of rupture (psi)
>
> S = bursting tensile strength (psi)
>
> t = net thickness (in.)
>
> d = nominal pipe size (in.).

The symbols S and R denote the strength of the iron in the pipe and are based on periodic full-length bursting tests and ring tests as specified in ANSI Standards A21.6 (AWWA C106), A21.7, A21.8 (AWWA C108), and A21.9.

The design values of w to be used in solving Eq 1, either by nomogram or by trial calculation, are determined as follows:

For Case 1,

$$w = \frac{2.5W_e}{L_f} \qquad (4)$$

For Case 2,

$$w = \frac{2.5(W_e + W_t)}{L_f} \qquad (5)$$

in which:

> w = ring test load equivalent of trench load (lb/lin ft), including a 2.5 safety factor
>
> W_e = earth load (lb/ft) (Table 1-8 and Sec. 1-3.2)
>
> W_t = truck superload (lb/ft) (Table 1-8 and Sec. 1-3.3)
>
> L_f = load factor dependent on laying condition (Table 1-9)
>
> 2.5 = safety factor.

The load factor L_f converts the trench load (earth load only or earth load plus truck superload) to an equivalent load in the laboratory ring test. Table 1–4 gives values of w computed from Eq 4 and 5 for standard depths of cover using the earth loads and truck superloads in Table 1–8 and the load factors in Table 1–9.

The design values of p to be used in solving Eq 1, either by nomogram or by trial calculation, are determined as follows:

For Case 1,

$$p = 2.5 \, (p_w + p_s) \qquad (6)$$

For Case 2,

$$p = 2.5 \, p_w \qquad (7)$$

in which:

p = internal pressure (psi), including a 2.5 safety factor

p_w = working pressure (psi)

p_s = allowance for surge pressure (psi) (Table 1–10)

2.5 = safety factor.

Table 1–5 gives values of p computed by Eq 6 and 7 for standard working pressures using the allowances for surge pressure in Table 1–10.

As described in Sec. 1–2.2, calculations for net thickness for water pipe are made for two conditions of combined loading: Case 1, which includes surge pressure but not truck superload; and Case 2, which includes truck superload but not surge pressure. The larger of the two thicknesses is used for design. Truck superload and surge pressure are transient and occasional loads, and it is considered extremely unlikely that they will occur simultaneously.

When Eq 2 and 3 are substituted in Eq 1, the result is a high-order equation which cannot be solved directly for the net thickness t by conventional mathematical procedure. It is necessary to resort to graphical solution or to successive approximation. The more convenient method for routine work is to use a nomogram as shown in Fig. 1–1 through 1–5. Such nomograms are prepared by the following steps:

a. Using the appropriate iron strength values of S and R, calculate W and P for a series of thicknesses for each pipe size by means of Eq 2 and 3.

b. For each thickness select two values of p and compute the corresponding values of w using Eq 1.

c. Set up parallel scales for w and p, using linear increments for the p scale and increments proportional to the square of the load for the w scale.

d. Using a straightedge to connect the corresponding values of w and p, locate and mark the intersection of the two lines determined by the two sets of values for each thickness.

e. Repeat for the full series of thicknesses to lay out the curve for each pipe size.

If it is desired to determine net thickness for iron strengths or loads and pressures not covered in the nomograms in Fig. 1–1 through 1–5, a trial calculation method may be used to solve Eq 1, as follows:

a. Based on the values of w and p corresponding to the design conditions, assume a trial value of t.

b. Using the known iron strength values S and R and the trial value of t, calculate W and P from Eq 2 and 3.

c. Using these values of W and P and the design value of p, solve Eq 1 for w.

d. Compare this calculated value of w to the design value of w and assume a smaller or larger value of t, as required, for the second trial calculation.

e. Continue until a change in assumed thickness of less than 0.01 in. results in a calculated w equal to or greater than the design w.

A graphical method may also be used for determining thicknesses for conditions not covered in the nomograms in Fig. 1–1 through 1–5. The parabolic graphical method is described in Sec. 1–12.3 of ANSI A21.1–1957 (AWWA H1–57).

Sec. 1–3.2—Earth Loads (W_e)

For computation of earth loads on cast-iron pipe the type of installation is identified as shown in Fig. 1–7. Ditch condition (Fig. 1–7a, b, c) denotes pipe laid in a relatively narrow trench and backfilled to the original ground surface. The trench width at the top of the pipe determines the load and the ditch may be widened above the top of the pipe for installation convenience (Fig. 1–7 b, c) without increasing the load on the pipe. Embankment condition includes two types of installation: positive projection condition (Fig. 1–7d) which denotes pipe laid on top of a subgrade and covered with fill, and negative projection condition (Fig. 1–7e) which denotes pipe laid in a trench in the subgrade and covered with fill, which extends substantially above the subgrade. Methods for calculating earth loads for these three installation conditions are given below:

1–3.2.1—*Ditch Condition*

The ditch condition is the most common method of installing cast-iron pipe and is the basis of the earth loads shown in Table 1–8. The load is obtained by selecting the lesser of the two loads computed by Eq 8 and 9.

$$W_e = C_d w B_d^2 \quad \text{(ditch condition)} \quad (8)$$

$$W_c = C_c w B_c^2 \quad \text{(positive projection condition)} \quad (9)$$

in which:

W_e = earth load, lb per linear ft
C_d = calculation coefficient, ditch condition
C_e = calculation coefficient, positive projection condition
w = soil density (120 lb/cu ft assumed in standard calculations)
B_d = width of trench at top of pipe (ft) (for standard calculations use nominal pipe diameter plus 2 ft)
B_e = outside diameter of pipe, ft.

This procedure was established by work done at Iowa State College,[*] which proved that for certain combinations of pipe size, trench width, and depth of cover, the load given by Eq 9 for positive projection condition should be used when it is the lesser of the two loads even though the pipe is laid in a trench.

The calculation coefficient C_d is obtained from Fig. 1–8 or from the following equation:

$$C_d = \frac{1 - e^{-2K\mu'\frac{H}{B_d}}}{2K\mu'} \quad (10)$$

in which:

K = ratio of active horizontal pressure at any point in the fill to the vertical pressure which causes the active horizontal pressure
μ' = coefficient of sliding friction between fill materials and sides of trench
$K\mu'$ = 0.130 for standard calculations

---

[*] SPANGLER, M. G. *Soil Engineering*. International Textbook Co., Scranton, Pa. (2nd ed., 1960). p. 416.

e = the base of natural logarithms (2.71828)

H = depth of cover to top of pipe (ft).

The calculation coefficient, C_c, is obtained from Fig. 1–9 or from the following equation:

$$C_c = \frac{e^{2K\mu \frac{H}{B_c}} - 1}{2K\mu} \qquad (11)$$

or

$$C_c = \frac{e^{2K\mu \frac{H_e}{B_c}} - 1}{2K\mu} + \left(\frac{H}{B_c} - \frac{H_e}{B_c}\right) e^{2K\mu \frac{H_e}{B_c}} \qquad (12)$$

in which:

μ = coefficient of internal friction in fill materials

$K\mu$ = 0.1924 for standard calculations

H_e = Height of equal settlement (ft) (vertical height from the top of conduit to the level at and above which the fill materials directly over the conduit settle equally with the adjacent fill materials).

Equation 11 is used when the height of fill, H, is equal to or less than the height of equal settlement, H_e. Equation 12 is used when the height of fill, H, is greater than the height of equal settlement, H_e.

The height of equal settlement, H_e, is obtained from the following equation:

$$e^{2K\mu \frac{H_e}{B_c}} - 2K\mu \frac{H_e}{B} = 2K\mu(r_{sd}p) + 1 \qquad (13)$$

in which:

r_{sd} = settlement ratio*
p = projection ratio*

For standard calculations the value of the product $r_{sd}p$ is taken to be 0.75. For this value, the height of equal settlement calculated by Eq 13 is, H_e = 1.75 B_c and Eq 12 reduces to:

$$C_c = 1.961 \frac{H}{B_c} - 0.934 \qquad (12a)$$

In most cases Eq 12a is used to calculate C_c becuase the depth of cover usually exceeds 1.75 times the pipe outside diameter.

Table 1–8 and Fig. 1–10 show earth loads computed by the above procedures for cast-iron pipe laid in trenches with widths, B_d, equal to the nominal pipe diameter plus 2 ft. Figure 1–11 shows a chart of earth loads computed by the above procedures for pipe installed in trenches where B_d is equal to the nominal pipe diameter plus 1 ft.

Figures 1–12 and 1–13 show charts computed by the above procedures for pipe laid in trenches with sides sloped 1:1 and 2:1, respectively.

For standard calculations, $K\mu$ = 0.1924 is used for C_c and $K\mu'$ = 0.130 is used for C_d in order to obtain conservative earth loads. Other values of $K\mu$ and $K\mu'$, which may be used in calculating earth loads for special soil conditions, are shown below with the corresponding soil type:

Soil Type	Value of $K\mu$ or $K\mu'$
Granular materials without cohesion	0.1924
Sand and gravel, maximum	0.165

*For a definition of these terms see: SPANGLER, M. G. *Soil Engineering*. International Textbook Co., Scranton, Pa (2nd ed., 1960). p. 403.

8
1

58

Saturated top soil, maximum	0.150
Clay, ordinary maximum	0.130
Saturated clay, maximum	0.110

1–3.2.2—Positive-Projection Embankment

The positive-projection condition may be encountered with pipe laid on top of a subgrade and covered with fill for highway or dam construction. The earth load is calculated from Eq 9 using Eq 12a for calculation of C_c when standard values of $r_{sd}p$ and $K\mu$ are used. For such standard values the load may also be read directly from Fig. 1–12 which was constructed from loads computed by Eq 9. The loads in Fig. 1–12 are based on soil weighing 110 lb/cu ft and may be adjusted to soil of 120 lb/cu ft by multiplying the graph load by 120/110. For calculation of pipe thicknesses, special load factors are used for positive projection embankment condition as shown in Table 1–9.

1–3.2.3—Negative-Projection Embankment Condition

The negative-projection embankment condition may be encountered in highway or dam construction when the pipe is laid in a relatively narrow trench cut in the subgrade, a more desirable method than installation directly on top of the subgrade. Loads for negative projection are calculated as follows.

$$W_e = C_n w B_d^2 \qquad (14)$$

in which:

C_n = calculation coefficient (negative projection condition).
Other terms are as defined for Eq 8 and 9.

The calculation coefficient, C_n, is read from Fig. 1–14 after the values of p' and H/B_d have been determined as follows:

$$p' = h/B_d$$
h = depth of cover in trench from top of pipe to subgrade (ft)
H = total height of fill from top of pipe to top of embankment (ft)
B_d = width of subgrade trench at top of pipe (ft)

Equation 14 is considered to give the maximum load that could occur on pipe laid in trenches under embankment conditions such as the case of embankment or additional fill added at some time, generally years, after the pipe was laid as a ditch conduit and backfill placed. In many embankment cases, Eq 14 may give loads which are too conservative.

In cases where pipe is laid in a subgrade trench with the embankment completed shortly thereafter, the load may be closer to that given by Eq 8 for ditch condition, and the embankment fill above the subgrade trench may be considered as equivalent to an increase of the trench width above the top of the pipe without significant effect on the earth load.

For other cases the load may lie somewhere between those given by Eq 8 and Eq 14. The correct load depends largely on relative soil compaction in the subgrade trench and overlying embankment and selection of the proper load for pipe design will be governed by engineering judgment based on the specific factors in each installation.

1–3.2.4—Sample Calculation of Earth Load

Determine earth load on 12-in. cast-iron pipe with 5 ft of cover. Pipe laid in a flat-bottom trench $(d + 2)$ ft wide.

Step 1. Calculate earth load for ditch condition using Eq 8.

$$C_d = \frac{1 - e^{-2K\mu'\frac{H}{B_d}}}{2K\mu'} \qquad (10)$$

$K\mu' = 0.130$
$H = 5$ ft

$$B_d = \frac{12}{12} + 2 = 3 \text{ ft}$$

$$C_d = \frac{1 - e^{-2(0.130)\left(\frac{5}{3}\right)}}{2(0.130)} = \frac{0.3516}{0.260} = 1.35$$

$$W_e = C_d w B_d^2 \qquad (8)$$
$$= 1.35(120)(3)^2$$
$$= 1,460 \text{ lb/ft}$$

Step 2. Calculate earth load for projection condition using Eq 9.

$$H_e = 1.75 \, B_c = 1.75 \left(\frac{13.20}{12}\right)$$
$$= 1.75(1.10) = 1.92 \text{ ft}$$

H is greater than H_e, therefore use Eq 12 or 12a to calculate C_c. Using Eq 12a:

$$C_c = 1.961 \frac{H}{B_c} - 0.934 \qquad (12a)$$
$$= 1.961 \left(\frac{5}{1.10}\right) - 0.934$$
$$= 7.98$$
$$W_e = C_c w B_c^2 \qquad (9)$$
$$= 7.98(120)(1.10)^2$$
$$= 1,159 \text{ lb/ft}$$

Step 3. Select lesser load from Step 1 or 2. The load for projection condition is the lesser. Therefore, the earth load, W_e, is 1,159 lb/ft. This load is shown in Table 1–8.

Sec. 1-3.3—Truck Superloads (W_t)

The procedures in this section may be used to compute truck superloads

for unpaved roads, flexible pavement or rigid pavement; one truck or two passing trucks; and any wheel load and impact factor. For unpaved road or flexible pavement Eq 15 is used. For rigid pavement Eq 16 is used.

$$W_t = CRPF \qquad (15)$$
$$W_t = KB_cPF \qquad (16)$$

in which:

W_t = Truck superload (lb/lin ft)
C = Surface load factor for unpaved road or flexible pavement (for one truck, see Table 1–11; for two trucks, see Table 1–12).
R = Reduction factor which takes account of the fact that the part of the pipe directly below the wheels receives the truck superload in its full intensity but is aided in carrying the load by adjacent parts of the pipe that receive little or no load from the truck. (See Table 1–13.)
P = Wheel load (lb)
F = Impact factor
K = Surface load factor for rigid pavement (see Table 1–14).
B_c = Outside diameter of pipe (ft) (see Table 1–15).

Equations 15 and 16 may be used in computing AASHO truck loading which is described in "Standard Specifications for Highway Bridges," American Assn. of State Highway Officials, 1961. The wheel loads and impact factors to be used in the equations are given in Art. 1.2.5 and

1.2.12 of the AASHO specification, as follows:

AASHO Truck	Gross Weight	Wheel Load, P
H-10	10 tons	8,000 lb
H-15	15 tons	12,000 lb
H-20	20 tons	16,000 lb

Depth of Cover	Impact Factor, F
0 ft to 1 ft, 0 in.	1.30
1 ft, 1 in., to 2 ft, 0 in.	1.20
2 ft, 1 in., to 2 ft, 11 in.	1.10
3 ft, 0 in., or more	1.00

The truck superload allowances given in Tables 1–4 and 1–8 are intended for standard conditions. They are based on two passing trucks with adjacent wheels 3 ft apart, 9,000-lb wheel load, unpaved road or flexible pavement, 1.50 impact factor. These loads in most cases equal or exceed the static load from a single AASHO H-20 truck with 16,000 lb on each rear wheel. These truck superloads are based on having the design depth of cover over the pipe. Consideration should be given to the loads that may be transmitted to the pipe if either truck superloads or heavy construction equipment is permitted to pass over the pipe at less than the design depth of cover.

TABLE 1-8

Earth Loads (We) and Truck Superloads (Wt)—lb/lin ft*

Depth of Cover—ft

Pipe Size in.	2½ W_e	2½ W_t	3½ W_e	3½ W_t	5 W_e	5 W_t	8 W_e	8 W_t	12 W_e	12 W_t	16 W_e	16 W_t	20 W_e	20 W_t	24 W_e	24 W_t
3	182	162	260	81	376	54	609	40	920	27	1,230	14	1,538	9	1,852	7
4	226	297	324	162	471	81	765	54	1,157	40	1,549	27	1,942	18	2,334	14
6	309	567	448	324	657	189	1,075	94	1,632	68	2,188	40	2,524	26	2,647	20
8	380	783	557	486	824	297	1,356	148	2,066	94	2,592	54	2,815	36	2,965	28
10	448	972	666	621	992	378	1,645	189	2,473	108	2,851	68	3,114	45	3,295	35
12	511	1,161	770	756	1,159	459	1,950	243	2,686	122	3,116	81	3,419	53	3,635	41
14	568	1,217	868	807	1,318	540	2,218	270	2,900	135	3,384	94	3,733	62	3,983	48
16	617	1,307	959	879	1,470	590	2,381	324	3,117	162	3,659	122	4,050	81	4,339	62
18	665	1,400	1,042	964	1,616	632	2,533	364	3,335	189	3,932	135	4,374	89	4,703	69
20	714	1,524	1,119	1,076	1,755	729	2,686	410	3,555	216	4,209	162	4,703	107	5,075	83
24	814	1,662	1,256	1,159	2,011	769	2,994	462	3,999	256	4,774	176	5,372	116	5,833	90
30	963	1,925	1,457	1,356	2,340	918	3,459	564	4,674	324	5,638	216	6,401	143	7,009	110
36	1,121	2,182	1,668	1,577	2,628	1,090	3,927	632	5,356	378	6,517	256	7,461	169	8,226	131
42	1,270	2,354	1,878	1,750	2,917	1,239	4,396	717	6,016	459	7,408	310	8,535	205	9,474	158
48	1,427	2,527	2,093	1,922	3,216	1,400	4,867	814	6,737	513	8,308	351	9,627	232	10,743	179
54	1,581	2,635	2,310	2,052	3,519	1,526	5,340	899	7,434	594	9,218	378	10,739	256	12,034	189
60	1,738	2,765	2,527	2,182	3,824	1,652	5,816	984	8,195	634	10,133	405	11,857	284	13,339	202

* Earth loads are based on the following conditions: trench width (B_d) equal to nominal pipe diameter plus 2 ft; unit weight of soil, 120 lb/cu ft; K_μ equal to 0.1924 and K_μ' equal to 0.130. Truck superloads are based on two passing trucks with adjacent wheels 3 ft apart, having a 9,000-lb wheel load on unpaved road or flexible pavement and a 1.50 impact factor.

TABLE 1–9

Load Factors for Cast-Iron Pipe in Ditch and Embankment Conditions

Pipe Size in.	$\frac{H*}{B_e}$	Laying Condition			Pipe Size in.	$\frac{H*}{B_e}$	Laying Condition		
		A	B	F			A	B	F
		Load Factor (L_f)					Load Factor (L_f)		
Ditch and Negative Projection Conditions†					Ditch and Negative Projection Conditions† *(Continued)*				
3		1.15	1.28	1.74					
4		1.15	1.29	1.75	54		1.15	1.75	2.29
6		1.15	1.32	1.78	60		1.15	1.77	2.31
8		1.15	1.34	1.80					
10		1.15	1.36	1.83	Positive-Projection Condition†				
12		1.15	1.38	1.85		0.5	1.50	2.16	
14		1.15	1.41	1.88		1.0	1.36	1.84	
16		1.15	1.43	1.90		1.5	1.29	1.74	
18		1.15	1.45	1.93		2.0	1.26	1.68	
20		1.15	1.47	1.95		3.0	1.26	1.64	
24		1.15	1.52	2.00		5.0	1.26	1.60	
30		1.15	1.58	2.08		10.0	1.26	1.58	
36		1.15	1.64	2.14					
42		1.15	1.69	2.20					
48		1.15	1.72	2.25					

* H is depth of cover to top of pipe, in feet; B_e is outside diameter of pipe, in feet (see Table 1–15).
† See Fig. 1–7 and See Sec. 1–3.2.

TABLE 1–10

Allowances for Surge Pressure

Pipe Size in.	Surge Pressure psi	Pipe Size in.	Surge Pressure psi
3–10	120	24	85
12–14	110	30	80
16–18	100	36	75
20	90	42–60	70

TABLE 1-11

Surface Load Factors (C) for One Truck on Unpaved Road or Flexible Pavement*

Pipe Size in.	Depth of Cover—ft												
	2	2½	3	3½	4	5	6	8	10	12	16	20	24
						Surface Load Factor							
3	0.028	0.020	0.014	0.011	0.009	0.006	0.004	0.002	0.0015	0.001	0.0006	0.0004	0.0002
4	0.034	0.024	0.017	0.013	0.011	0.007	0.005	0.003	0.002	0.0015	0.0008	0.0005	0.0003
6	0.048	0.034	0.025	0.020	0.015	0.010	0.007	0.004	0.003	0.002	0.001	0.0007	0.0004
8	0.062	0.044	0.033	0.026	0.020	0.013	0.009	0.006	0.0035	0.0025	0.0013	0.0008	0.0005
10	0.074	0.054	0.040	0.031	0.025	0.016	0.012	0.007	0.004	0.003	0.0016	0.001	0.0006
12	0.087	0.063	0.048	0.036	0.030	0.019	0.014	0.008	0.005	0.0035	0.002	0.0012	0.0007
14	0.099	0.072	0.055	0.042	0.034	0.022	0.016	0.010	0.006	0.004	0.0025	0.0015	0.0008
16	0.110	0.082	0.061	0.047	0.038	0.025	0.018	0.011	0.007	0.005	0.003	0.0017	0.001
18	0.122	0.090	0.068	0.052	0.042	0.028	0.020	0.012	0.008	0.0055	0.0035	0.002	0.0012
20	0.132	0.098	0.075	0.058	0.046	0.031	0.022	0.013	0.009	0.006	0.004	0.0025	0.0015
24	0.150	0.113	0.087	0.068	0.054	0.037	0.026	0.015	0.010	0.007	0.0045	0.003	0.0017
30	0.171	0.132	0.102	0.081	0.065	0.045	0.031	0.019	0.012	0.009	0.005	0.0035	0.002
36	0.188	0.148	0.117	0.093	0.076	0.052	0.037	0.022	0.015	0.010	0.006	0.004	0.0025
42	0.200	0.160	0.129	0.103	0.085	0.059	0.043	0.025	0.017	0.012	0.007	0.0045	0.003
48	0.210	0.170	0.139	0.113	0.093	0.066	0.048	0.029	0.019	0.013	0.008	0.005	0.0033
54	0.216	0.178	0.147	0.120	0.101	0.072	0.053	0.032	0.021	0.015	0.0085	0.0055	0.0036
60	0.222	0.184	0.153	0.126	0.107	0.077	0.057	0.035	0.023	0.016	0.009	0.006	0.004

* These factors are for a single concentrated wheel load centered over an effective pipe length of 3 ft. They were computed by the methods explained in Article 24.20 and Chap. 16 of M. G. Spangler's Soil Engineering (2nd ed., 1960; International Textbook Co., Scranton, Pa.).

87

TABLE 1-12

Surface Load Factors (C) for Two Passing Trucks on Unpaved Road or Flexible Pavement*

Pipe Size in.	Depth of Cover—ft												
	2	2½	3	3½	4	5	6	8	10	12	16	20	24
	Surface Load Factor												
3	0.019	0.012	0.008	0.006	0.005	0.004	0.0035	0.003	0.0025	0.002	0.001	0.0007	0.0005
4	0.032	0.022	0.016	0.012	0.009	0.006	0.005	0.004	0.0035	0.003	0.002	0.0013	0.0010
6	0.058	0.042	0.032	0.024	0.020	0.014	0.010	0.007	0.006	0.005	0.003	0.0019	0.0015
8	0.076	0.058	0.044	0.036	0.030	0.022	0.017	0.011	0.009	0.007	0.004	0.0027	0.0021
10	0.092	0.072	0.056	0.046	0.039	0.028	0.021	0.014	0.011	0.008	0.005	0.0033	0.0026
12	0.108	0.086	0.070	0.056	0.047	0.034	0.027	0.018	0.012	0.009	0.006	0.0039	0.0030
14	0.122	0.098	0.078	0.065	0.055	0.040	0.031	0.020	0.014	0.010	0.007	0.0046	0.0036
16	0.136	0.110	0.090	0.074	0.062	0.046	0.036	0.024	0.016	0.012	0.009	0.0060	0.0046
18	0.149	0.122	0.101	0.084	0.070	0.052	0.041	0.027	0.019	0.014	0.010	0.0066	0.0051
20	0.162	0.136	0.115	0.096	0.080	0.060	0.048	0.032	0.022	0.016	0.012	0.0079	0.0061
24	0.185	0.152	0.126	0.106	0.091	0.067	0.053	0.036	0.026	0.019	0.013	0.0086	0.0067
30	0.212	0.176	0.146	0.124	0.107	0.080	0.064	0.044	0.032	0.024	0.016	0.0106	0.0081
36	0.235	0.202	0.169	0.146	0.127	0.095	0.075	0.052	0.038	0.028	0.019	0.0125	0.0097
42	0.251	0.218	0.188	0.162	0.140	0.108	0.087	0.059	0.044	0.034	0.023	0.0152	0.0117
48	0.264	0.234	0.205	0.178	0.157	0.122	0.097	0.067	0.050	0.038	0.026	0.0172	0.0132
54	0.274	0.244	0.216	0.190	0.170	0.133	0.108	0.074	0.057	0.044	0.028	0.0190	0.0140
60	0.281	0.256	0.228	0.202	0.181	0.144	0.117	0.081	0.061	0.047	0.030	0.0210	0.0150

* The factors are for two trucks with 6-ft rear wheel spacing passing with inside rear wheels 3 ft apart. Effective pipe length is 3 ft, coinciding with the distance between the adjacent inside wheels. The factors were computed by the methods explained in Sec. 24.20 and Chap. 16 of M. G. Spangler's Soil Engineering (2nd ed., 1960; International Textbook Co., Scranton, Pa.)

AMERICAN NATIONAL STANDARD

TABLE 1-13

Reduction Factors (R)

Pipe Size in.	Depth of Cover—ft				Pipe Size in.	Depth of Cover—ft			
	2½–3½	4–7	8–10	>10		2½–3½	4–7	8–10	>10
	Reduction Factor					Reduction Factor			
3–12	1.00	1.00	1.00	1.00	20	0.83	0.90	0.95	1.00
14	0.92	1.00	1.00	1.00	24–30	0.81	0.85	0.95	1.00
16	0.88	0.95	1.00	1.00	36–60	0.80	0.85	0.90	1.00
18	0.85	0.90	1.00	1.00					

8
9

TABLE 1-14

Surface Load Factors (K) *For One Truck and Two Passing Trucks on Rigid Pavement**

Depth of Cover ft	One Truck				Two Passing Trucks			
	Pavement Thickness—in.				Pavement Thickness—in.			
	4	6	8	10	4	6	8	10
	Surface Load Factor							
2	0.0244	0.0149	0.0101	0.0076	0.0410	0.0263	0.0186	0.0142
2½	0.0213	0.0139	0.0097	0.0072	0.0364	0.0246	0.0177	0.0136
3	0.0186	0.0126	0.0090	0.0070	0.0333	0.0228	0.0167	0.0129
3½	0.0164	0.0114	0.0085	0.0066	0.0290	0.0206	0.0156	0.0122
4	0.0144	0.0102	0.0079	0.0061	0.0262	0.0187	0.0146	0.0117
5	0.0114	0.0084	0.0066	0.0054	0.0210	0.0156	0.0123	0.0102
6	0.0093	0.0071	0.0057	0.0047	0.0170	0.0133	0.0107	0.0088
8	0.0065	0.0052	0.0043	0.0036	0.0114	0.0097	0.0081	0.0069
10	0.0046	0.0039	0.0033	0.0029	0.0080	0.0070	0.0062	0.0055
12	0.0034	0.0030	0.0026	0.0023	0.0059	0.0054	0.0049	0.0045
16	0.0022	0.0019	0.0017	0.0016	0.0034	0.0032	0.0030	0.0028
20	0.0013	0.0011	0.0010	0.0009	0.0024	0.0023	0.0022	0.0021
24	0.0008	0.0007	0.0006	0.0005	0.0015	0.0014	0.0013	0.0012

* These factors were computed by the methods explained in "Vertical Pressure on Culverts under Wheel Loads on Concrete-Pavement Slabs", Bull. ST65, Portland Cement Assn., Chicago, Ill. In Bulletin ST65, K if expressed as c/L^2. These factors are based on a modulus of subgrade reaction of 100 lb/cu in. and a modulus os elasticity of 4,000,000 psi for the pavement concrete. The factors for two passing trucks are based on the inside rear wheels passing 3 ft apart.

66

TABLE 1–15

Outside Diameters of Cast-Iron Pipe

Pipe Size in.	Outside Diameter in.	Outside Diameter (B_e) ft	Pipe Size in.	Outside Diameter in.	Outside Diameter (B_e) ft
3	3.96	0.330	20	21.60	1.800
4	4.80	0.400	24	25.80	2.150
6	6.90	0.575	30	32.00	2.667
8	9.05	0.754	36	38.30	3.192
10	11.10	0.925	42	44.50	3.708
12	13.20	1.100	48	50.80	4.233
14	15.30	1.275	54	57.10	4.758
16	17.40	1.450	60	63.40	5.283
18	19.50	1.625			

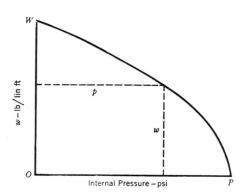

Fig. 1–6. Load-Pressure Curve

The parabola represents the relation between external load and internal pressure at the point of breaking.

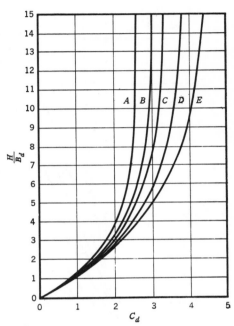

Fig. 1–8. Calculation Coefficients (C_d)
for Ditch Condition

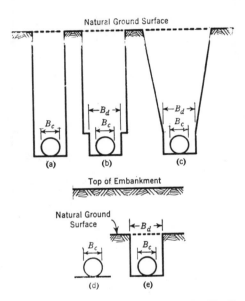

**Fig. 1–7. Installation Conditions for Earth
Load Calculations**

*Fig. 1–7(a)–(c) are for ditch conditions;
Fig. 1–7(d) and (e) are for positive
and negative projection embankment conditions, respectively.*

*Curve A is for C_d for $K\mu$ and $K\mu'$ of
0.1924, the minimum for granular materials without cohesion; Curve B, C_d for
$K\mu$ and $K\mu'$ of 0.165, the maximum for
sand and gravel; Curve C, C_d for $K\mu$ and
$K\mu'$ of 0.150, the maximum for saturated
topsoil; Curve D, C_d for $K\mu$ and $K\mu'$ of
0.130, the ordinary maximum for clay;
and Curve E, C_d for $K\mu$ and $K\mu'$ of
0.110, the maximum for saturated clay.*

9
2

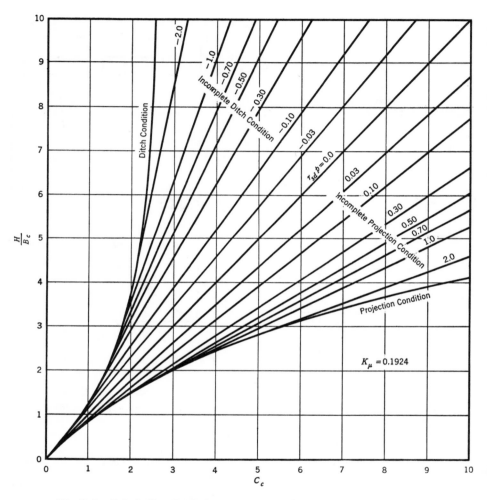

Fig. 1–9. Calculation Coefficients (C_c) for Positive-Projection Condition

The values of C_c may also be determined by Eq 11 or Eq 12, given in the text.

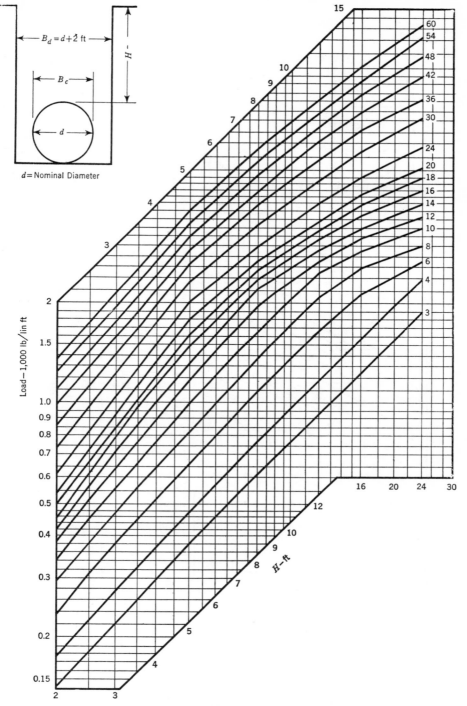

Fig. 1-10. Earth Loads on Pipe for Trench Width of $(d + 2)$ ft

Values associated with each curve are for pipe size, in inches. It is assumed that the unit weight of fill is 120 lb/cu ft, that $K\mu$ equals 0.1924 and $K\mu'$ equals 0.130, and that $r_{sd}p$ is 0.75 for all sizes. B_c is pipe OD. For 3- and 8-60-in. pipe, OD is as shown in Table 1-15. OD of 4- and 6-in. pipe is 5.00 and 7.10 in., respectively.

9
4

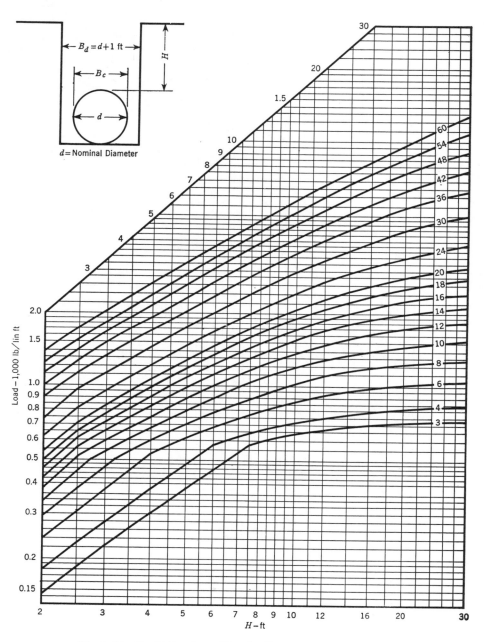

Fig. 1–11. Earth Loads on Pipe for Trench Width of $(d + 1)$ ft

Values associated with each curve are for pipe size, in inches. It is assumed that the unit weight of fill is 120 lb/cu ft, that $K\mu$ equals 0.1924 and $K\mu'$ equals 0.130. For wide ditches, $r_{sd}p$ is 0.75. B_c is pipe OD. For 3- and 8–60-in. pipe, OD is as shown in Table 1–15. OD of 4- and 6-in. pipe is 5.00 and 7.10 in., respectively.

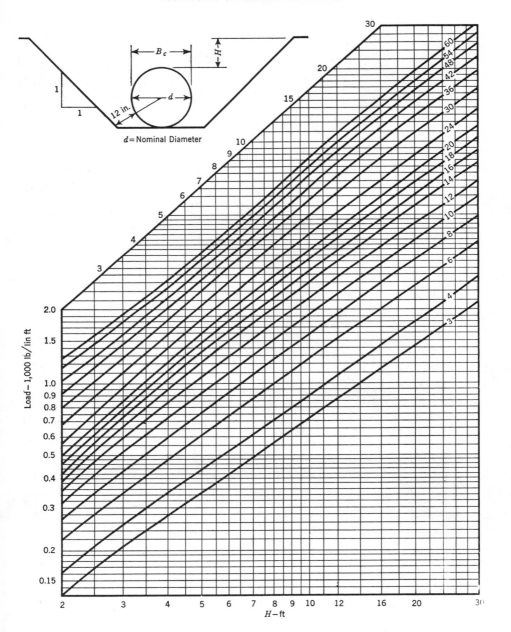

Fig. 1–12. Earth Loads on Pipe in Trench With 1:1 Side Slopes

Values associated with each curve are for pipe size, in inches. It is assumed that the unit weight of fill is 110 lb/cu ft; this load may be adjusted to soil of 120 lb/cu ft by multiplying the graph load by 120/110. For wide ditches, $r_{sd}p$ is 0.75. $K\mu$ equals 0.1924, $K\mu'$ equals 0.130. B_c is pipe OD. For 3- and 8–60-in. pipe, OD is as shown in Table 1–15. OD of 4- and 6-in. pipe is 5.00 and 7.10 in., respectively.

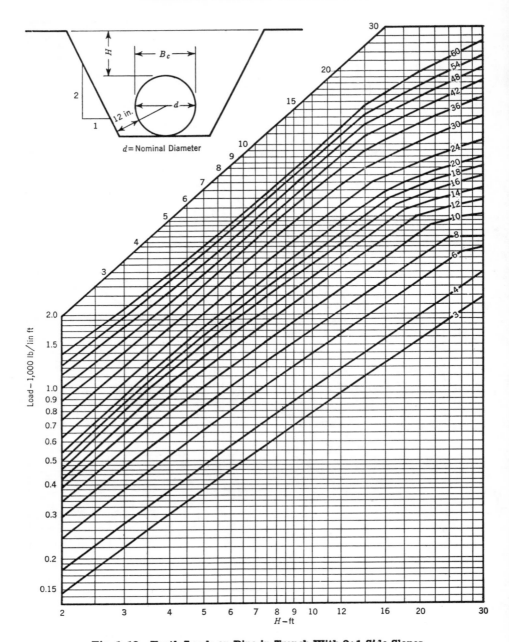

Fig. 1–13. Earth Loads on Pipe in Trench With 2:1 Side Slopes

Values associated with each curve are for pipe size, in inches. It is assumed that the unit weight of fill is 120 lb/cu ft. For wide ditches, $r_{sd}p$ equals 0.75. $K\mu$ equals 0.1924, $K\mu'$ equals 0.130. B_c is pipe OD. For 3- and 8–60-in. pipe, OD is as shown in Table 1–15. OD of 4- and 6-in. pipe is 5.00 and 7.10 in., respectively.

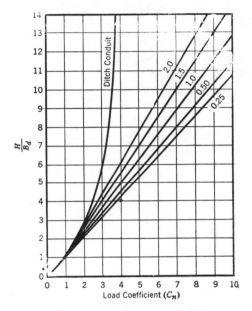

Fig. 1–14. Calculation Coefficients (C_n) for Negative-Projection Conditions

The values associated with the five curves to the right are for p'. Kµ' equals 0.130. r_{sd}p equals 0.0.

Sec. 1–4—Thickness Determination for Pipe on Piers or Piling Aboveground or Underground

Sec. 1–4.1—Scope

This section gives the procedures for determining the net thickness of cast-iron pipe supported at intervals rather than continuously. These procedures are applicable to pipe installed on piling bents and piers, with or without earth cover, as well as to pipe installed on bridges and other structures with hangers and other types of spaced supports.

Sec. 1–4.2—Pipe Installed Aboveground Without Earth Cover

a. Determine the ring test load equivalent of external load, including 2.5 safety factor, as follows:

$$w = \frac{2.5(W_p + W_w)}{L_f} \quad (17)$$

in which:

w = ring test load equivalent (lb/ft)

W_p = weight of pipe (lb/ft) (see Table 1–16)

W_w = weight of contained water (lb/ft) (see Table 1–16)

L_f = load factor (see Table 1–17)

b. Select the internal pressure for Case 1 from Table 1–5 which includes surge pressure and 2.5 safety factor.

c. Enter the above values of ring test load equivalent and internal pressure in the appropriate nomogram and read the net thickness. To obtain total calculated thickness, corrosion allowance and casting tolerance are added to this net thickness as described in Sec. 1–2.2.

Sec. 1–4.3—Pipe Installed Underground With Earth Cover

Thicknesses are computed for two cases as described in Sec. 1–2.2 and the larger of the two thicknesses is used for design.

Case 1:

a. Determine the ring test load equivalent from the following formula:

$$w = \frac{2.5W_e}{L_f} \quad (18)$$

in which:

w = ring test load equivalent (lb/ft)

W_e = earth load (see Table 1–8)

L_f = load factor (see Table 1–17)

b. Select the internal pressure to Case 1 from Table 1–5 which includes surge pressure and 2.5 safety factor.

c. Enter the above values of ring test load equivalent and internal pressure in the appropriate nomogram and read the net thickness.

Case 2:

a. Determine the ring test load equivalent from the following formula:

$$w = \frac{2.5 (W_e + W_t)}{L_f} \quad (19)$$

in which W_t is the truck superload in pounds per foot (see Table 1–8) and other factors are as defined for Eq 18.

b. Select the internal pressure for Case 2 from Table 1–5 which includes a safety factor of 2.5.

75

c. Enter the values of ring test load equivalent and internal pressure in the appropriate nomogram and read the net thickness.

The net thickness is selected from Case 1 or Case 2, whichever gives the greater computed thickness. To obtain total calculated thickness, corrosion allowance and casting tolerance are added to this net thickness as described in Sec. 1–2.2.

Sec. 1–4.4—Design Examples

a. Calculate the thickness of 24-in. cast-iron pipe, 18/40 iron strength, installed aboveground on piers spaced 18 ft apart on centers with 60-deg saddle support. Working pressure is 150 psi.

Load factor (Table 1–17),
$$L_f = 0.29 \times 1.55$$
$$= 0.45$$

Ring test load equivalent,
$$w = \frac{2.5(177 + 196)}{0.45}$$
$$= 2,071 \text{ lb/ft}$$

Internal pressure (Table 1–5),
$$p = 588 \text{ psi}$$

Using the above values of w and p in the nomogram, Fig. 1–1, the net thickness is determined to be 0.49 in. Adding 0.08 in. corrosion allowance and 0.08 in. casting tolerance, the total calculated thickness is determined to be 0.65 in.

b. Calculate the thickness of 16-in. cast-iron pipe, 18/40 iron strength, installed underground on piers spaced 18 ft apart on centers with 120-deg saddle support, with 5-ft cover. Working pressure is 150 psi.

Case 1:

Load factor (Table 1–17),
$$L_f = 0.24 \times 2.13$$
$$= 0.51$$

Ring test load equivalent,
$$w = \frac{2.5 \times 1470}{0.51}$$
$$= 7,206 \text{ lb/ft}$$

Internal pressure (Table 1–5),
$$p = 625 \text{ psi}$$

Using the above values of w and p in the nomogram, Fig. 1–1, the net thickness is determined to be 0.57 in.

Case 2:

Ring test load equivalent,
$$w = \frac{2.5(1,470 + 590)}{0.51}$$
$$= 10,980 \text{ lb/ft}$$

Internal pressure (Table 1–5),
$$p = 375 \text{ psi}$$

Using the above values of w and p in the nomogram, Fig. 1–2, the net thickness is determined to be 0.65 in. Case 2 gives the larger computed thickness, 0.65 in., which is selected as the net thickness. Adding 0.08 in. corrosion allowance and 0.08 in. casting tolerance, the total calculated thickness is determined to be 0.81 in.

Sec. 1–4.5—Calculation of Beam Stress and Deflection

For small-diameter pipe, generally 3 in. through 8 in., a check of the beam stress may be required. If the calculated beam stress exceeds 14,000 psi, the thickness is increased or the span between supports is decreased to limit the beam stress to 14,000 psi. In some types of installations, such as gravity flow sewers, beam deflection may be a significant factor in the design. Beam stress and deflection are calculated from the following two formulas:

$$f = \frac{15.28WDL^2}{D^4 - d^4} \qquad (20)$$

$$y = \frac{30fL^2}{DE} \qquad (21)$$

in which:

f = beam stress (14,000 psi maximum)

W = external load, lb per ft (for aboveground pipe $W = W_p + W_w$; for underground pipe, $W = W_e + W_t$)

L = distance on centers between supports (ft)

D = outside diameter of pipe (in.)

d = inside diameter (D-$2t$) (in.)

t = net thickness (in.)

y = deflection at mid-span (in.)

E = modulus of elasticity, 15,000,000 psi.

TABLE 1–16

Weights of Pipe and Contained Water for Design of Aboveground Pipe

Pipe Size *in.*	Weight—lb/ft		Pipe Size *in.*	Weight—lb/ft	
	Pipe (W_P)*	Water (W_w)†		Pipe (W_P)*	Water (W_w)†
3	12	3	18	114	110
4	16	6	20	135	136
6	26	12	24	177	196
8	37	22	30	257	307
10	49	34	36	339	442
12	63	49	42	439	601
14	78	67	48	545	785
16	94	88			

* Based on Class 22 pipe. Although the computed thickness may differ from that given for Class 22, the effect of the difference in pipe weight usually will not have a significant effect on the computed thickness and recalculation usually will not be necessary.

† Based on nominal pipe size.

TABLE 1-17

Load Factors for Pipe on Spaced Supports Aboveground and Underground

Pipe Size in.	Distance on Centers Between Supports—ft							
	6	8	9	10	12	16	18	20
	Load Factor for Flat Support							
3	0.19	0.14	0.13	0.11	0.10	0.07	0.06	0.06
4	0.22	0.17	0.15	0.13	0.11	0.08	0.07	0.07
6	0.31	0.23	0.21	0.19	0.16	0.12	0.10	0.09
8	0.40	0.30	0.27	0.24	0.20	0.15	0.13	0.12
10	0.50	0.38	0.33	0.30	0.25	0.19	0.17	0.15
12	0.60	0.45	0.40	0.36	0.30	0.23	0.20	0.18
14	0.67	0.50	0.45	0.40	0.33	0.25	0.22	0.20
16	0.73	0.55	0.49	0.44	0.36	0.27	0.24	0.22
18	0.78	0.59	0.52	0.47	0.39	0.29	0.26	0.23
20	0.81	0.61	0.54	0.49	0.40	0.30	0.27	0.24
24	0.87	0.65	0.58	0.52	0.43	0.33	0.29	0.26
30	0.93	0.70	0.62	0.56	0.46	0.35	0.31	0.28
36	0.96	0.72	0.64	0.58	0.48	0.36	0.32	0.29
42	0.98	0.73	0.65	0.59	0.49	0.37	0.33	0.29
48	0.99	0.74	0.66	0.60	0.50	0.37	0.33	0.30

Explanatory Note. 1. Load factor for saddle support is obtained by multiplying the load factor for flat support by the following modifiers:

Saddle Angle deg.	Modifier
30	1.25
45	1.40
60	1.55
90	1.87
120	2.13
180	2.35

2. Load factors for other distances between supports or for other saddle angles may be obtained by interpolation between tabulated values.

3. The load factor for flat support is equal to the load factor for laying condition C, A21.1–1957, multiplied by the ratio of 6-ft, the block spacing of laying condition C, to the distance between supports. The modifier for saddle support is obtained from Table 2 of Stresses in Pressure Pipelines and Protective Casing Pipes [M. G. Spangler, *J. Struct. Div. ASCE*, Vol. 82, No. ST5 (Sep. 1956)], by dividing K_b for 180 deg load and 0 deg support by K_b for 180 deg load and a support angle equal to the saddle angle.

4. The recommended minimum axial bearing length of supports, for underground pipe is 6 in. for 3–8-in. pipe, 12 in. for 10–24-in. pipe, and 18 in. for 30–48-in. pipe.

ANSI A21.6–1975
(AWWA C106–75)

Revision of
A21.6–1970
(AWWA C106–70)

AMERICAN NATIONAL STANDARD

for

CAST-IRON PIPE CENTRIFUGALLY CAST IN METAL MOLDS, FOR WATER OR OTHER LIQUIDS

ADMINISTRATIVE SECRETARIAT
AMERICAN WATER WORKS ASSOCIATION

CO-SECRETARIATS
AMERICAN GAS ASSOCIATION
NEW ENGLAND WATER WORKS ASSOCIATION

Revised edition approved by American National Standards Institute, Inc., May 28, 1975.

NOTICE

This Standard has been especially printed by the American Water Works Association for incorporation into this volume. It is current as of December 1, 1975. It should be noted, however, that all AWWA Standards are updated at least once in every five years. Therefore, before applying this Standard it is suggested that you confirm its currency with the American Water Works Association.

PUBLISHED BY

AMERICAN WATER WORKS ASSOCIATION
6666 West Quincy Avenue, Denver, Colorado 80235

Table of Contents

1
0
4

American National Standard

An American National Standard implies a consensus of those substantially concerned with its scope and provisions. An American National Standard is intended as a guide to aid the manufacturer, the consumer, and the general public. The existence of an American National Standard does not in any respect preclude anyone, whether he has approved the standard or not, from manufacturing, marketing, purchasing, or using products, processes, or procedures not conforming to the standard. American National Standards are subject to periodic review, and users are cautioned to obtain the latest editions. Producers of goods made in conformity with an American National Standard are encouraged to state on their own responsibility in advertising, promotion material, or on tags or labels that the goods are produced in conformity with particular American National Standards.

CAUTION NOTICE. This American National Standard may be revised or withdrawn at any time. The procedures of the American National Standards Institute require that action be taken to reaffirm, revise, or withdraw this standard no later than five (5) years from the date of publication. Purchasers of American National Standards may receive current information on all standards by calling or writing the American National Standards Institute, 1430 Broadway, New York, N. Y. 10018, (212) 868-1220.

Foreword

This foreword is provided for information only and is not a part of ANSI A21.6–1975 (AWWA C106–75).

I—History of Standard

On Sep. 10, 1902, NEWWA adopted a "Standard Specification for Cast-Iron Pipe and Special Castings," covering bell-and-spigot pit-cast pipe and fittings of ten thickness classes. The thickness classes were based on allowable internal pressures varying by increments of 50 ft of head.

On May 12, 1908, AWWA adopted a "Standard Specification for Cast-Iron Pipe and Special Castings," covering bell-and-spigot pit-cast pipe and fittings of eight classes, A through H, with allowable working pressures varying by increments of 100 ft of head from 100 to 800 ft. Dimensions and weights were given for pipe and fittings.

In 1926, ASA Sectional Committee (now American National Standards Committee) A21 on Cast-Iron Pipe and Fittings was organized under the sponsorship of A.G.A., ASTM, AWWA, and NEWWA and was assigned the following scope:

Unification of specifications for cast-iron pipe, including materials; dimensions; pressure ratings; methods of manufacture (including such new developments as centrifugal casting) insofar as they may be necessary to secure satisfactory specifications; elimination of unnecessary sizes and varieties; consideration of the possibility of developing a coordinated scheme of metallic pipe and fittings applicable to all common mediums; and methods of making up joints insofar as they are determining as to the dimensional design of cast-iron pipe.

The types of cast-iron pipe [are] to include bell-and-spigot pipe, flanged pipe, flanged and bell mouth fittings and wall castings, pipe elbows, tees, wyes, return bends, and other fittings not now included in standard lists; cast-iron pipe threaded for flanges or couplings. The standardization is not to include methods of installing pipe and similar matters, except as to the making up of joints in its relationship to the dimensional standardization of pipe and fittings, as noted above.

Sectional Committee A21 sponsored many tests of pipe and fittings; these included subjection of pipe to combined earth load and internal pressures (which form the basis of pipe thickness design), corrosion tests, measurement of hydraulic friction loss in fittings, and tests of bursting strengths of pipe and fittings. After exhaustive study of the test results and other research, the committee in 1939 issued A21.1, "American Standard Practice Manual for the Computation of Strength and Thickness of Cast Iron Pipe." The manual included nomograms and thickness tables for pit-cast pipe with 11/31* iron strength. As stated in the preface to that manual, however, the design method was applicable to pipe of any iron strength.

Discussions and interpretations of the method of design of cast-iron pipe

---

* The first figure designates the bursting tensile strength in units of 1 000 psi and the second figure designates the ring modulus of rupture in units of 1 000 psi.

were published in 1939 and presented to AWWA and A.G.A. As a result of these publications and because of the general acceptance of A21.1, a substantial volume of cast-iron pipe was designed by the new method and furnished to manufacturers' standards between 1939 and 1953. A standard (A21.2) for pit-cast pipe with 11/31 iron strength also was issued in 1939. Work on standards for centrifugally cast pipe with 18/40 iron strength was started after the design was completed in 1939, but, owing to the intervention of World War II and other causes, they were not formally issued until 1953.

In 1957, a revision of A21.1 was issued. In that revision, designated ASA A21.1-1957 (AWWA H1-57) the major change was the addition of a method for computing earth loads on pipe laid under embankments and of nomograms and thickness tables for centrifugally cast pipe with 18/40 iron strength.

In 1958, Sectional Committee A21 was reorganized. Subcommittees were established to study each group of standards in accordance with the review and revision policy of ASA (now ANSI). The subcommitteee on pipe (Subcommittee No. 1) was organized with the following assignment:

The scope of the committee actitivy shall include an examination of all present A21 standards for pipe to determine what is needed to bring these up to date. The examination shall include A21.1, A21.2, A21.3, A21.6, A21.7, A21.8, and A21.9, as well as any other matters pertaining to pipe standards.

As a result of the work of Subcommittee No. 1 on this assignment, revisions of the cast-iron pipe standards A21.6 (AWWA C106), A21.7, A21.8 (AWWA C108) and A21.9 were issued in 1962. A revision of A21.1-1957 (AWWA H1-57) was issued in 1967

and reaffirmed without revision in 1972. Revisions of the 1962 cast-iron pipe standards were issued in 1970. The major revisions of A21.6-1970 were: pipe with push-on joints were added; laying conditions C and D were deleted and condition F was added; and, the tables were revised to include the pipe lengths then being produced.

In 1974, Subcommittee 1 reviewed the 1970 edition and recommended minor editorial changes. Therefore, this edition is unchanged from the 1970 edition except for minor editorial changes and the updating of this foreword.

The tables and strength test requirements in this standard are for pipe with 18/40 iron strength. Advances in production technology have enabled the manufacturers to furnish pipe with greater strength. Pipe with 21/45 iron strength has been furnished for many years. Design details and standard thicknesses for pipe with 21/45 iron strength are shown in A21.1-1957 (AWWA C101-67), reaffirmed in 1972.

II—Acceptance Tests

Acceptance tests were established as routine control measures to ensure the design burst and ring strengths. The acceptance tests specified in this standard are the Talbot strip test and the hardness test. In establishing the acceptance values for the Talbot strip test, A21 Committee in the 1940s reviewed detailed data, including burst, ring, and Talbot strip tests on more than 400 pipe centrifugally cast in metal molds. Correlations of the data showed that the Talbot strip modulus of rupture and secant modulus of elasticity values specified in this standard represent acceptable pipe which meet the design burst and ring strengths.

The hardness test was specified as a means of assuring the ferritic matrix which is characteristic of the microstructure of pipe cast in metal molds. The specified acceptance values for the Talbot strip test provide additional control of the microstructure to ensure satisfactory machinability.

III—Options

This standard includes certain options which, if desired, must be specified in the invitation for bids and on the purchase order. Also, a number of items must be specified to describe completely the pipe required. The following summarizes these details and available options and lists the sections of the standard where they can be found:

1. Size, joint type, thickness or class, and laying lengths (Tables)
2. Special joints (Sec. 6–1)
3. Certification by manufacturer (Sec. 6–4)
4. Inspection by purchaser (Sec. 6–5)
5. Cement lining (Sec. 6–8.2) Experience has indicated that bituminous inside coating is not complete protection against loss in pipe capacity caused by tuberculation. Cement linings are recommended for most waters.
6. Special coatings and linings (Sec. 6–8.4)
7. Special marking on pipe (Sec. 6–10)
8. Written transcripts of foundry records (Sec. 6–15)
9 Special tests (Sec 6–16).

Committee Personnel

Subcommittee 1, Pipe, which reviewed this standard, had the following personnel at that time:

EDWARD C. SEARS, *Chairman*
WALTER AMORY, *Vice-Chairman*

User Members	*Producer Members*
ROBERT S. BRYANT	W. D. GOODE
FRANK E. DOLSON	CARL A. HENRIKSON
GEORGE F. KEENAN	THOMAS D. HOLMES
LEONARD ORLANDO JR.	SIDNEY P. TEAGUE
JOHN E. PERRY	

Standards Committee A21, Cast-Iron Pipe and Fittings, which reviewed and approved this standard, had the following personnel at the time of approval:

LLOYD W. WELLER, *Chairman*
CARL A. HENRIKSON, *Vice-Chairman*
JAMES B. RAMSEY, *Secretary*

Organization Represented	*Name of Representative*
American Gas Association	LEONARD ORLANDO JR.
American Society of Civil Engineers	KENNETH W. HENDERSON
American Society of Mechanical Engineers	JAMES S. VANICK
American Society for Testing and Materials	ALBERT H. SMITH JR.
American Water Works Association	ARNOLD M. TINKEY
	LLOYD W. WELLER
Cast Iron Pipe Research Association	CARL A. HENRIKSON
	EDWARD C. SEARS
	W. HARRY SMITH
Individual Producer	ALFRED F. CASE
Manufacturers' Standardization Society of the Valve and Fittings Industry	ABRAHAM FENSTER
New England Water Works Association	WALTER AMORY
Naval Facilities Engineering Command	STANLEY C. BAKER
Underwriters' Laboratories, Inc.	JOHN E. PERRY
Canadian Standards Association	W. F. SEMENCHUK*

---

* Liaison representative without vote.

American National Standard for

Cast-Iron Pipe Centrifugally Cast in Metal Molds, for Water or Other Liquids

Sec. 6–1—Scope

This standard covers 3-in. through 24-in. cast-iron pipe centrifugally cast in metal molds for water or other liquids. Characteristics of such pipe with push-on joints, mechanical joints and bell-and-spigot joints are given in the tables. This standard may be used for pipe with such other types of joints as may be agreed upon at the time of purchase. The thicknesses, weights, and strength test requirements shown in this standard are for pipe with 18/40 iron strength (18 000 psi minimum bursting tensile and 40 000 psi minimum ring modulus of rupture).

Sec. 6–2—Definitions

Under this standard, the following definitions shall apply:

6–2.1. *Purchaser*. The party entering into a contract or agreement to purchase pipe according to this standard.

6–2.2 *Manufacturer*. The party that produces the pipe.

6–2.3. *Inspector*. The representative of the purchaser, authorized to inspect in behalf of the purchaser to determine whether or not the pipe meet this standard.

6–2.4. *Cast iron*. The unqualified term *cast iron* shall apply to gray cast iron which is a cast ferrous material in which a major part of the carbon content occurs as free carbon in the form of flakes interspersed through the metal.

6–2.5. *Push-on joint*. The single rubber-gasket joint as described in ANSI A21.11 (AWWA C111) of latest revision.

6–2.6. *Mechanical joint*. The gasketed and bolted joint as detailed in ANSI A21.11 (AWWA C111) of latest revision.

6–2.7. *Bell-and-spigot joint*. The poured or caulked joint as detailed in Table 6.6.

Sec. 6–3—General Requirements

6–3.1. Pipe with push-on joints, mechanical joints, and bell-and-spigot joints shall conform to the applicable dimensions and weights shown in the tables in this standard and to the applicable requirements of ANSI A21.11 (AWWA C111) of latest revision. Pipe with other types of joints shall comply with the joint dimensions and weights agreed upon at the time of purchase, but in all other respects shall fulfill the requirements of this standard.

6–3.2. The nominal laying length of the pipe shall be as shown in the tables. A maximum of 10 per cent of the total number of pipe of each size specified in an order may be furnished by as much as 24 in. shorter than the nominal laying length, and an additional 10 per cent may be furnished by as much as 3 in. shorter than nominal laying length.

Sec. 6–4—Inspection and Certification by Manufacturer

6–4.1. The manufacturer shall establish the necessary quality control and inspection practice to assure compliance with this standard.

6–4.2. The manufacturer shall, if required on the purchase order, furnish a sworn statement that the inspection and all of the specified tests have been made and the results thereof comply with the requirements of this standard.

6–4.3. All pipe shall be clean and sound without defects which will impair their service. Repairing of defects by welding or other method shall not be allowed if such repairs will adversely affect the serviceability of the pipe or its capability to meet strength requirements of this standard.

Sec. 6–5—Inspection by Purchaser

6–5.1. If the purchaser desires to inspect pipe at the manufacturer's plant, the purchaser shall so specify on the purchase order, stating the conditions (such as time, and the extent of inspection) under which the inspection shall be made.

6–5.2. The inspector shall have free access to those parts of the manufacturer's plant that are necessary to assure compliance with this standard. The manufacturer shall make available for the inspector's use such gages as are necessary for inspection. The

manufacturer shall provide the inspector with assistance as necessary for the handling of pipe.

Sec. 6–6—Delivery and Acceptance

All pipe and accessories shall comply with this standard. Pipe and accessories not complying with this standard shall be replaced by the manufacturer at the agreed point of delivery. The manufacturer shall not be liable for shortages or damaged pipe after acceptance at the agreed point of delivery except as recorded on the delivery receipt or similar document by the carrier's agent.

Sec. 6–7—Tolerances or Permitted Variations

6–7.1. *Dimensions*. The spigot end, bell, and socket of the pipe and the accessories shall be gaged with suitable gages at sufficiently frequent intervals to ensure that the dimensions comply with the requirements of this standard. The smallest inside diameter of the sockets and the outside of the spigot ends shall be tested with circular gages. Other socket dimensions shall be gaged as appropriate.

6–7.2. *Thickness*. Minus thickness tolerances of pipe and bell shall not exceed those shown below:

Pipe Size in.	Minus Tolerance in.
3–8	0.05
10–12	0.06
14–24	0.08

NOTE: An additional tolerance of 0.02 in. shall be permitted over areas not exceeding 8 in. in any direction.

6–7.3. *Weight*. The weight of any single pipe shall not be less than the tabulated weight by more than 5 per cent for pipe 12 in. or smaller in diameter, nor by more than 4 per cent for pipe larger than 12 in. in diameter.

2

110

Sec. 6–8—Coatings and Linings

6–8.1. *Outside coating.* The outside coating for use under normal conditions shall be a bituminous coating approximately 1 mil thick. The coating shall be applied to the outside of all pipe, unless otherwise specified. The finished coating shall be continuous, smooth, neither brittle when cold nor sticky when exposed to the sun and shall be strongly adherent to the pipe.

6–8.2. *Cement-mortar linings.* Cement linings shall be in accordance with ANSI A21.4 (AWWA C104) of latest revision. If desired, cement linings shall be specified in the invitation for bids and on the purchase order.

6–8.3. *Inside coating.* Unless otherwise specified, the inside coating for pipe not cement lined shall be a bituminous material as thick as practicable (at least 1 mil) and conforming to all appropriate requirements for seal coat in ANSI A21.4 of latest revision.

6–8.4. *Special coatings and linings.* For special conditions, other types of coatings and linings may be available. Such special coatings and linings shall be specified in the invitation for bids and on the purchase order.

Sec. 6–9—Hydrostatic Test

Each pipe shall be subjected to a hydrostatic test of not less than 500 psi. This test may be made either before or after the outside coating and the inside coating have been applied, but shall be made before the application of cement lining or of a special lining.

The pipe shall be under the full test pressure for at least 10 s. Suitable controls and recording devices shall be provided so that the test pressure and duration may be adequately ascertained. Any pipe that leaks or does not withstand the test pressure shall be rejected.

In addition to the hydrostatic test before application of a cement lining or special lining, the pipe may be retested, at the manufacturer's option, after application of such lining.

Sec. 6–10—Marking Pipe

The weight, class or nominal thickness, and sampling period shall be shown on each pipe. The manufacturer's mark and the year in which the pipe was produced shall be cast or stamped on the pipe. When specified on the purchase order, initials not exceeding four in number shall be cast or stamped on the pipe. All required markings shall be clear and legible and all cast marks shall be on or near the bell. All letters and numerals on pipe sizes 8 in. and larger shall be not less than $\frac{1}{2}$ in. in height.

Sec. 6–11—Weighing Pipe

Each pipe shall be weighed before the application of any lining or coating other than the bituminous coating and the weight shall be shown on the outside or inside of the bell or spigot end.

Sec. 6–12—Acceptance Tests

The standard acceptance tests for the physical characteristics of the pipe shall be as follows:

6–12.1. *Talbot strip tests.* Talbot strip tests shall be used to determine the acceptability of 3–24 in. pipe for modulus of rupture and secant modulus of elasticity.

6–12.1.1. *Sampling.* At least one sample shall be taken during each period of approximately 3 hr. The sample for the first period shall be taken during the first hour, or if casting is direct from the melting unit, from the first ladle. Samples shall be

3

taken so that each size of pipe continously cast for 2 hr or longer and each source of iron continuously used for 2 hr or longer shall be fairly represented.

6–12.1.2. *Acceptance values.* The modulus of rupture as determined by the Talbot strip test shall be 40 000 psi minimum.

The secant modulus of elasticity value shall not exceed 300 times the actual value of the modulus of rupture. (For example: when the modulus of rupture is 40 000 psi, the secant modulus of elasticity shall not exceed 12 000 000 psi.)

6–12.1.3. *Test method.* Talbot strips (Fig. 6.1) shall be machined

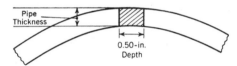

Fig. 6.1. Position From Which Talbot Strip is Cut

longitudinally from each pipe specimen selected for testing by this method. The Talbot strips may be cut from a part of the ring little stressed in the ring test—that is, near one of the elements marked *a* in the illustration of the ring test (Fig. 6.3). The strips in any case shall be in cross section as indicated in Fig. 6.1—that is, shall have for their width the thickness of the pipe and for their depth 0.50 in. Their length shall be at least $10\frac{1}{2}$ in. The strips shall be tested as beams on supports 10 in. apart with loads applied perpendicularly to the machined faces at two points $3\frac{1}{3}$ in. from the supports. The breaking load and the deflection shall be observed and recorded.

The strip shall be accurately calipered at the point of rupture and the modulus of rupture, *R*, shall be calculated by the usual beam formula,

which for this case reduces to the expression

$$R = \frac{10W}{td^2}$$

The secant modulus of elasticity, E_s, in pounds per square inch, shall be computed by the formula

$$E_s = \frac{21.3R}{dy}$$

In the above formulas, *R* is the modulus of rupture (psi); E_s, the secant modulus of elasticity (psi); *W*, the breaking load (lb); *d*, the depth (in.) of the strip (intended to be 0.50 in.); *t*, the width (in.) of the strip (pipe thickness); and *y*, the deflection (in.) of the strip at the center at breaking load.

Deflection measurements shall be that of the specimen and shall not include any compression of the supports or loading blocks, or backlash or distortion of the testing machine.

6–12.2. *Hardness tests.* Hardness tests shall be made on the outside surface of pipe. A sufficient number of pipe shall be tested to assure that the hardness does not exceed Rockwell B-95, or its equivalent. Pipe may be heat-treated to meet this requirement.

For the purpose of foundry records, hardness tests shall be made on a specimen from each Talbot strip selected for testing in Sec. 6–12.11. Rockwell B hardness determinations shall be made in accordance with ASTM E18-67 on the following surfaces at their approximate centers: (a) the outside pipe surface, (b) the inside pipe surface, and (c) either of the two cut surfaces. These three determinations shall be made at three locations (1-2-3) along the length of the specimen, as shown in Fig. 6.2. The three determinations for each surface shall be averaged and no average value shall exceed Rockwell B-95. No single

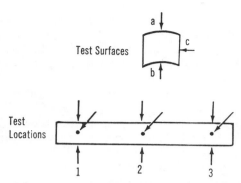

Fig. 6.2. Location of Rockwell Hardness Tests on Talbot Strip Specimen

determination on the Talbot strip shall exceed B-98.

Sec. 6–13—Ring Tests and Full-Length Bursting Tests

6–13.1. The manufacturer shall make bursting tests and ring tests in conjunction with strip tests so that he can certify the design values of the modulus of rupture (40 000 psi) and the bursting tensile strength of the iron in the pipe (18 000 psi). These tests shall be made in accordance with dimensions and methods given in Sec. 6–13.2 and Sec. 6–13.3. At least one pipe sample for the ring and bursting tests shall be selected from each of the following size groups from each calendar month's cast:

Group	Size in.
1	6, 8
2	10, 12
3	14, 16, 18
4	20, 24

When no pipe in a size group are manufactured during the calendar month, no tests on these sizes are required. Ring tests and bursting tests are not required on 3- and 4-in. pipe. At least three Talbot strips shall be tested from each pipe selected for bursting. Tests and records shall include the modulus of rupture of

each strip and ring and the modulus of elasticity and hardness of each strip.

6–13.2. *Ring test method.* The maximum length of any ring shall not exceed 12 in.; for pipe 14 in. and larger, the minimum length shall be $10\frac{1}{2}$ in.; for pipe 12 in. and smaller, the minimum length shall be one half the nominal diameter of the pipe. Each ring shall be tested by the three-edge bearing method as indicated in Fig. 6.3. The lower bearing for the

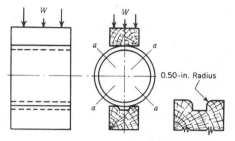

Fig. 6.3. Assembly for Ring Test

ring shall consist of two strips with vertical sides having their interior top edges rounded to a radius of approximately $\frac{1}{2}$ in. The strips shall be of hard wood or metal. If of metal, a piece of fabric or leather approximately $\frac{3}{16}$ in. thick shall be laid over them. They shall be straight and shall be securely fastened to a rigid block, with their interior vertical faces the following distances apart:

Pipe Size in.	Bearing Strip Spacing in.
6–12	$\frac{1}{2}$
14–24	1

The upper bearing shall be a hard wood block, straight and true from end to end. The upper and lower bearings shall extend the full length of the ring. The ring shall be placed symmetrically between the two bearings, and the center of application of the load shall

5

be so placed that the vertical deformation at the two ends of the ring shall be approximately equal. If the ring is not uniform in thickness, it shall be so placed that the thick and thin portions are near the ends of the horizontal diameter.

For purposes of Sec. 6–15, a record of the breaking load of each ring tested shall be made. The modulus of rupture is computed from the formula

$$R = 0.954 \frac{W(d + t)}{bt^2}$$

in which R is the modulus of rupture (psi); W, the breaking load (lb); d, the average inside diameter (in.) of the ring; t, the average thickness (in.) of metal along the line of fracture; and b, the length (in.) of the ring.

6–13.3. *Burst test method.* The bursting tensile strength shall be determined by testing full-length pipe (less the amount cut off for ring and strip test specimens) to destruction by hydraulic pressure. Bells may be removed to facilitate testing. A suitable means for holding the end thrust shall be used which will not subject the pipe to endwise tension or compression, or other parasitic stresses. A calibrated pressure gage shall be used for determining the bursting pressure. The gage shall be connected to the interior of the test pipe by a separate connection from that which supplies water for the test. The unit tensile strength in bursting shall be obtained by the use of the formula

$$S = \frac{Pd}{2t}$$

in which S is the bursting tensile strength (psi) of the iron; P, the internal pressure (psi) at bursting; d, the average inside diameter (in.) of the pipe; and t, the minimum average thickness (in.) of the pipe along the principal line of break.

Measurements of thickness shall be taken along the principal line of break at 1-ft intervals.

The minimum average thickness along the principal line of break shall be obtained by averaging the measurements at the thinnest section at a weight of two and at the adjacent sections on each side at a weight of one each; or, if the thinnest section is at the end of the break, by averaging the thinnest-section measurement at a weight of two and the measurements of the adjacent section and the next section at a weight of one each.

Sec. 6–14—Chemical Analyses

Analyses of the iron shall be made at sufficiently frequent intervals to determine compliance with the following limits:

Substance	Maximum Limit *per cent*
Phosphorus	0.90
Sulfur	0.12

Control of the other chemical constituents shall be maintained to meet the physical property requirements of this standard. Samples for chemical analyses shall be representative and shall be obtained from either acceptance test specimens or specimens cast for this purpose.

Sec. 6–15—Foundry Records

The results of the following tests shall be recorded and retained for one year and shall be available to the purchaser at the foundry. Written transcripts of the results of these tests shall be furnished when specified on the purchase order:

6–15.1. Talbot strip tests (see Sec. 6–12.1)

6–15.2. Hardness tests (see Sec. 6–12.2)

6–15.3. Ring tests and full-length bursting tests (see Sec. 6–13)

6–15.4. Chemical analyses (see Sec. 6–14).

Sec. 6–16—Additional Tests Required by Purchaser

When tests other than those provided in this standard are required by the purchaser, such tests shall be specified in the invitation for bids and on the purchase order.

Sec. 6–17—Defective Specimens and Retests

When any physical test specimen shows defective machining or lack of continuity of metal, it shall be discarded and replaced by another specimen. When any sound test specimen fails to meet the specified requirements, the pipe from which it was taken shall be rejected and a retest may be made on two additional sound specimens from pipe cast in the same sampling period as the specimen which failed. Both of the additional specimens shall meet the prescribed tests to qualify the pipe produced in that sampling period.

Sec. 6–18—Rejection of Pipe

When any routine chemical analysis fails to meet the requirements of Sec. 6–14 or when any physical acceptance test fails to meet the requirements of Sec. 6–12.1, 6–12.2, or 6–17, the pipe cast in the same sampling period shall be rejected except as subject to the provision of Sec. 6–19.

Sec. 6–19—Determining Rejection

The manufacturer may determine the amount of rejection by making similar additional tests of pipe of the same size as that rejected until the rejected lot is bracketed in order of manufacture by an acceptable test at each end of the interval in question. When pipe of one size is rejected from a sampling period, the acceptability of pipe of different sizes from that same period may be established by making the routine acceptance tests for these sizes.

115

TABLE 6.1

Selection Table for Push-On Joint Cast-Iron Pipe§

These thicknesses and weights are for pipe laid without blocks, on flat-bottom trench, with tamped backfill (laying condition B), under 5 ft of cover. For other conditions see tables 6.3 and 6.4 hereof and ANSI A21.1 (AWWA C101).

Size	Thickness	OD	Weight Based on 18-ft Laying Length		Weight Based on 20-ft Laying Length	
			Per Length†	Avg. per Foot‡	Per Length†	Avg. per Foot‡
in.			*lb*			

Working Pressure 50 psi—115 ft Head

Size	Thickness	OD	Per Length†	Avg. per Foot‡	Per Length†	Avg. per Foot‡
3*	0.32	3.96	215	12.0	240	11.9
4	0.35	4.80	290	16.1	320	16.0
6	0.38	6.90	460	25.6	510	25.6
8	0.41	9.05	665	36.9	735	36.8
10	0.44	11.10	880	49.0	975	48.7
12	0.48	13.20	1,140	63.4	1,260	63.1
14	0.48	15.30	1,335	74.1	1,470	73.6
16	0.54	17.40	1,700	94.5	1,880	94.0
18	0.54	19.50	1,920	106.7	2,120	106.1
20	0.57	21.60	2,250	124.9	2,485	124.2
24	0.63	25.80	2,975	165.3	3,285	164.4

Working Pressure 100 psi—231 ft Head

Size	Thickness	OD	Per Length†	Avg. per Foot‡	Per Length†	Avg. per Foot‡
3*	0.32	3.96	215	12.0	240	11.9
4	0.35	4.80	290	16.1	320	16.0
6	0.38	6.90	460	25.6	510	25.6
8	0.41	9.05	665	36.9	735	36.8
10	0.44	11.10	880	49.0	975	48.7
12	0.48	13.20	1,140	63.4	1,260	63.1
14	0.51	15.30	1,410	78.2	1,555	77.8
16	0.54	17.40	1,700	94.5	1,880	94.0
18	0.58	19.50	2,050	113.9	2,265	113.3
20	0.62	21.60	2,430	134.9	2,685	134.2
24	0.68	25.80	3,190	177.3	3,525	176.4

Working Pressure 150 psi—346 ft Head

Size	Thickness	OD	Per Length†	Avg. per Foot‡	Per Length†	Avg. per Foot‡
3*	0.32	3.96	215	12.0	240	11.9
4	0.35	4.80	290	16.1	320	16.0
6	0.38	6.90	460	25.6	510	25.6
8	0.41	9.05	665	36.9	735	36.8
10	0.44	11.10	880	49.0	975	48.7
12	0.48	13.20	1,140	63.4	1,260	63.1
14	0.51	15.30	1,410	78.2	1,555	77.8
16	0.54	17.40	1,700	94.5	1,880	94.0
18	0.58	19.50	2,050	113.9	2,265	113.3
20	0.62	21.60	2,430	134.9	2,685	134.2
24	0.73	25.80	3,410	189.4	3,765	188.4

Working Pressure 200 psi—462 ft Head

Size	Thickness	OD	Per Length†	Avg. per Foot‡	Per Length†	Avg. per Foot‡
3*	0.32	3.96	215	12.0	240	11.9
4	0.35	4.80	290	16.1	320	16.0
6	0.38	6.90	460	25.6	510	25.6
8	0.41	9.05	665	36.9	735	36.8
10	0.44	11.10	880	49.0	975	48.7
12	0.48	13.20	1,140	63.4	1,260	63.1
14	0.55	15.30	1,510	83.8	1,670	83.4
16	0.58	17.40	1,815	100.9	2,010	100.4
18	0.63	19.50	2,210	122.8	2,445	122.2
20	0.67	21.60	2,610	144.9	2,885	144.2
24	0.79	25.80	3,665	203.6	4,055	202.6

* Pipe of 3-in. size also available in 12-ft laying length. Weight per length† is 150 lb; average weight per foot‡ is 12.5 lb for all working pressures and heads.
† Including bell. Calculated weight of pipe rounded off to nearest 5 lb.
‡ Including bell. Average weight per foot based on calculated weight of pipe before rounding
§ Weights shown for push-on joint pipe are also applicable to bell-and-spigot joint pipe.

TABLE 6.1—(contd.)

Selection Table for Push-On Joint Cast-Iron Pipe§

These thicknesses and weights are for pipe laid without blocks, on flat-bottom trench, with tamped backfill (Laying Condition B), under 5 ft of cover. For other conditions see Tables 6.3 and 6.4 hereof and ANSI A21.1 (AWWA C101).

Size	Thickness	OD	Weight Based on 18-ft Laying Length		Weight Based on 20-ft Laying Length	
			Per Length†	Avg. per Foot‡	Per Length†	Avg. per Foot‡
	in.		*lb*			

Working Pressure 250 psi—577 ft Head

Size	Thickness	OD	Per Length†	Avg. per Foot‡	Per Length†	Avg. per Foot‡
3*	0.32	3.96	215	12.0	240	11.9
4	0.35	4.80	290	16.1	320	16.0
6	0.38	6.90	460	25.6	510	25.6
8	0.41	9.05	665	36.9	735	36.8
10	0.44	11.10	880	49.0	975	48.7
12	0.52	13.20	1,230	68.3	1,360	67.9
14	0.59	15.30	1,610	89.5	1,780	89.0
16	0.63	17.40	1,960	109.0	2,170	108.4
18	0.68	19.50	2,370	131.8	2,620	131.1
20	0.72	21.60	2,785	154.8	3,080	154.0
24	0.79	25.80	3,665	203.6	4,055	202.6

Working Pressure 300 psi—693 ft Head

Size	Thickness	OD	Per Length†	Avg. per Foot‡	Per Length†	Avg. per Foot‡
3*	0.32	3.96	215	12.0	240	11.9
4	0.35	4.80	290	16.1	320	16.0
6	0.38	6.90	460	25.6	510	25.6
8	0.41	9.05	665	36.9	735	36.8
10	0.48	11.10	955	53.0	1,055	52.7
12	0.52	13.20	1,230	68.3	1,360	67.9
14	0.59	15.30	1,610	89.5	1,780	89.0
16	0.68	17.40	2,100	116.8	2,325	116.2
18	0.73	19.50	2,530	140.6	2,800	140.0
20	0.78	21.60	3,000	166.6	3,315	165.8
24	0.85	25.80	3,920	217.8	4,335	216.8

Working Pressure 350 psi—808 ft Head

Size	Thickness	OD	Per Length†	Avg. per Foot‡	Per Length†	Avg. per Foot‡
3*	0.32	3.96	215	12.0	240	11.9
4	0.35	4.80	290	16.1	320	16.0
6	0.38	6.90	460	25.6	510	25.6
8	0.41	9.05	665	36.9	735	36.8
10	0.52	11.10	1,025	56.9	1,130	56.6
12	0.56	13.20	1,315	73.1	1,455	72.7
14	0.64	15.30	1,735	96.3	1,920	95.9
16	0.68	17.40	2,100	116.8	2,325	116.2
18	0.79	19.50	2,720	151.2	3,010	150.6
20	0.84	21.60	3,210	178.4	3,550	177.6
24	0.92	25.80	4,220	234.4	4,665	233.4

* Pipe of 3-in. size also available in 12-ft laying length. Weight per length† is 150 lb; average weight per foot‡ is 12.5 lb for all working pressures and heads.
† Including bell. Calculated weight of pipe rounded off to nearest 5 lb.
‡ Including bell. Average weight per foot based on calculated wieght of pipe before rounding.
§ Weights shown for push-on joint pipe are also applicable to bell-and-spigot joint pipe.

TABLE 6.2

Selection Table for Mechanical-Joint Cast-Iron Pipe

These thicknesses and weights are for pipe laid without blocks, on flat-bottom trench, with tamped backfill (Laying Condition B), under 5 ft of cover. For other conditions see Tables 6.3 and 6.5 hereof and ANSI A21.1 (AWWA C101).

Size	Thickness	OD	Weight Based on 18-ft Laying Length		Weight Based on 20-ft Laying Length	
			Per Length†	Avg. per Foot‡	Per Length†	Avg. per Foot‡
in.			*lb*			

Working Pressure 50 psi—115 ft Head

Size	Thickness	OD	Per Length†	Avg. per Foot‡	Per Length†	Avg. per Foot‡
3*	0.32	3.96	215	12.0	240	11.9
4	0.35	4.80	290	16.2	320	16.1
6	0.38	6.90	460	25.5	510	25.4
8	0.41	9.05	655	36.4	725	36.2
10	0.44	11.10	870	48.2	960	48.0
12	0.48	13.20	1,125	62.6	1,245	62.3
14	0.48	15.30	1,335	74.0	1,470	73.6
16	0.54	17.40	1,700	94.5	1,880	94.0
18	0.54	19.50	1,920	106.7	2,120	106.0
20	0.57	21.60	2,250	124.9	2,485	124.2
24	0.63	25.80	2,975	165.2	3,285	164.2

Working Pressure 100 psi—231 ft Head

Size	Thickness	OD	Per Length†	Avg. per Foot‡	Per Length†	Avg. per Foot‡
3*	0.32	3.96	215	12.0	240	11.9
4	0.35	4.80	290	16.2	320	16.1
6	0.38	6.90	460	25.5	510	25.4
8	0.41	9.05	655	36.4	725	36.2
10	0.44	11.10	870	48.2	960	48.0
12	0.48	13.20	1,125	62.6	1,245	62.3
14	0.51	15.30	1,410	78.2	1,555	77.8
16	0.54	17.40	1,700	94.5	1,880	94.0
18	0.58	19.50	2,050	113.9	2,265	113.2
20	0.62	21.60	2,430	134.9	2,685	134.2
24	0.68	25.80	3,190	177.2	3,525	176.2

Working Pressure 150 psi—346 ft Head

Size	Thickness	OD	Per Length†	Avg. per Foot‡	Per Length†	Avg. per Foot‡
3*	0.32	3.96	215	12.0	240	11.9
4	0.35	4.80	290	16.2	320	16.1
6	0.38	6.90	460	25.5	510	25.4
8	0.41	9.05	655	36.4	725	36.2
10	0.44	11.10	870	48.2	960	48.0
12	0.48	13.20	1,125	62.6	1,245	62.3
14	0.51	15.30	1,410	78.2	1,555	77.8
16	0.54	17.40	1,700	94.5	1,880	94.0
18	0.58	19.50	2,050	113.9	2,265	113.2
20	0.62	21.60	2,430	134.9	2,685	134.2
24	0.73	25.80	3,405	189.2	3,765	188.2

Working Pressure 200 psi—462 ft Head

Size	Thickness	OD	Per Length†	Avg. per Foot‡	Per Length†	Avg. per Foot‡
3*	0.32	3.96	215	12.0	240	11.9
4	0.35	4.80	290	16.2	320	16.1
6	0.38	6.90	460	25.5	510	25.4
8	0.41	9.05	655	36.4	725	36.2
10	0.44	11.10	870	48.2	960	48.0
12	0.48	13.20	1,125	62.6	1,245	62.3
14	0.55	15.30	1,510	83.8	1,670	83.4
16	0.58	17.40	1,815	100.9	2,005	100.3
18	0.63	19.50	2,210	122.8	2,445	122.2
20	0.67	21.60	2,610	144.9	2,885	144.2
24	0.79	25.80	3,665	203.5	4,050	202.6

* Pipe of 3-in. size also available in 12-ft laying length. Weight per length† is 150 lb; average weight per foot‡ is 12.3 lb for all working pressures and heads.
† Including bell. Calculated weight of pipe rounded off to nearest 5 lb.
‡ Including bell. Average weight per foot based on calculated weight of pipe before rounding.

TABLE 6.2—(contd.)

Selection Table for Mechanical-Joint Cast-Iron Pipe

These thicknesses and weights are for pipe laid without blocks, on flat-bottom trench, with tamped backfill (Laying Condition B), under 5 ft of cover. For other conditions see Tables 6.3 and 6.5 hereof and ANSI A21.1 (AWWA C101).

Size	Thickness	OD	Weight Based on 18-ft Laying Length		Weight Based on 20-ft Laying Length	
			Per Length†	Avg. per Foot‡	Per Length†	Avg. per Foot‡
	in.		*lb*			

Size	Thickness	OD	Per Length†	Avg. per Foot‡	Per Length†	Avg. per Foot‡
\multicolumn{7}{c}{Working Pressure 250 psi—577 ft Head}
3*	0.32	3.96	215	12.0	240	11.9
4	0.35	4.80	290	16.2	320	16.1
6	0.38	6.90	460	25.5	510	25.4
8	0.41	9.05	655	36.4	725	36.2
10	0.44	11.10	870	48.2	960	48.0
12	0.52	13.20	1,215	67.4	1,340	67.1
14	0.59	15.30	1,610	89.4	1,780	89.0
16	0.63	17.40	1,960	108.9	2,165	108.3
18	0.68	19.50	2,370	131.7	2,620	131.0
20	0.72	21.60	2,785	154.8	3,080	154.1
24	0.79	25.80	3,665	203.5	4,050	202.6
\multicolumn{7}{c}{Working Pressure 300 psi—693 ft Head}
3*	0.32	3.96	215	12.0	240	11.9
4	0.35	4.80	290	16.2	320	16.1
6	0.38	6.90	460	25.5	510	25.4
8	0.41	9.05	655	36.4	725	36.2
10	0.48	11.10	940	52.2	1,040	52.0
12	0.52	13.20	1,215	67.4	1,340	67.1
14	0.59	15.30	1,610	89.4	1,780	89.0
16	0.68	17.40	2,100	116.7	2,325	116.2
18	0.73	19.50	2,530	140.6	2,800	140.0
20	0.78	21.60	3,000	166.6	3,320	165.9
24	0.85	25.80	3,920	217.7	4,335	216.8
\multicolumn{7}{c}{Working Pressure 350 psi—808 ft Head}
3*	0.32	3.96	215	12.0	240	11.9
4	0.35	4.80	290	16.2	320	16.1
6	0.38	6.90	460	25.5	510	25.4
8	0.41	9.05	655	36.4	725	36.2
10	0.52	11.10	1,010	56.1	1,120	55.9
12	0.56	13.20	1,300	72.2	1,440	71.9
14	0.64	15.30	1,735	96.3	1,920	95.9
16	0.68	17.40	2,100	116.7	2,325	116.2
18	0.79	19.50	2,720	151.2	3,010	150.6
20	0.84	21.60	3,210	178.3	3,550	177.6
24	0.92	25.80	4,215	234.2	4,665	233.2

* Pipe of 3-in. size also available in 12-ft laying length. Weight per length† is 150 lb; average weight per foot‡ is 12.3 lb for all working pressures and heads.
† Including bell. Calculated weight of pipe rounded off to nearest 5 lb.
‡ Including bell. Average weight per foot based on calculated weight of pipe before rounding.

TABLE 6.3

Standard Thickness* Selection Table for Cast-Iron Pipe

Laying Condition A—flat-bottom trench, without blocks, untamped backfill
Laying Condition B—flat-bottom trench, without blocks, tamped backfill
Laying Condition F—pipe laid on gravel or sand bedding, backfill tamped

Size in.	Working Pressure psi	3½-ft Cover			5-ft Cover			8-ft Cover		
		Laying Condition			Laying Condition			Laying Condition		
		A	B	F	A	B	F	A	B	F
		Thickness*—in.								
3	50	0.32	0.32	0.32	0.32	0.32	0.32	0.32	0.32	0.32
	100	0.32	0.32	0.32	0.32	0.32	0.32	0.32	0.32	0.32
	150	0.32	0.32	0.32	0.32	0.32	0.32	0.32	0.32	0.32
	200	0.32	0.32	0.32	0.32	0.32	0.32	0.32	0.32	0.32
	250	0.32	0.32	0.32	0.32	0.32	0.32	0.32	0.32	0.32
	300	0.32	0.32	0.32	0.32	0.32	0.32	0.32	0.32	0.32
	350	0.32	0.32	0.32	0.32	0.32	0.32	0.32	0.32	0.32
4	50	0.35	0.35	0.35	0.35	0.35	0.35	0.35	0.35	0.35
	100	0.35	0.35	0.35	0.35	0.35	0.35	0.35	0.35	0.35
	150	0.35	0.35	0.35	0.35	0.35	0.35	0.35	0.35	0.35
	200	0.35	0.35	0.35	0.35	0.35	0.35	0.35	0.35	0.35
	250	0.35	0.35	0.35	0.35	0.35	0.35	0.35	0.35	0.35
	300	0.35	0.35	0.35	0.35	0.35	0.35	0.35	0.35	0.35
	350	0.35	0.35	0.35	0.35	0.35	0.35	0.35	0.35	0.35
6	50	0.38	0.38	0.38	0.38	0.38	0.38	0.38	0.38	0.38
	100	0.38	0.38	0.38	0.38	0.38	0.38	0.38	0.38	0.38
	150	0.38	0.38	0.38	0.38	0.38	0.38	0.38	0.38	0.38
	200	0.38	0.38	0.38	0.38	0.38	0.38	0.38	0.38	0.38
	250	0.38	0.38	0.38	0.38	0.38	0.38	0.38	0.38	0.38
	300	0.38	0.38	0.38	0.38	0.38	0.38	0.38	0.38	0.38
	350	0.38	0.38	0.38	0.38	0.38	0.38	0.38	0.38	0.38
8	50	0.41	0.41	0.41	0.41	0.41	0.41	0.41	0.41	0.41
	100	0.41	0.41	0.41	0.41	0.41	0.41	0.41	0.41	0.41
	150	0.41	0.41	0.41	0.41	0.41	0.41	0.41	0.41	0.41
	200	0.41	0.41	0.41	0.41	0.41	0.41	0.41	0.41	0.41
	250	0.41	0.41	0.41	0.41	0.41	0.41	0.44	0.41	0.41
	300	0.41	0.41	0.41	0.41	0.41	0.41	0.44	0.44	0.41
	350	0.41	0.41	0.41	0.52	0.41	0.41	0.48	0.44	0.44
10	50	0.44	0.44	0.44	0.44	0.44	0.44	0.44	0.44	0.44
	100	0.44	0.44	0.44	0.44	0.44	0.44	0.48	0.44	0.44
	150	0.44	0.44	0.44	0.44	0.44	0.44	0.48	0.44	0.44
	200	0.44	0.44	0.44	0.44	0.44	0.44	0.48	0.48	0.44
	250	0.44	0.44	0.44	0.48	0.44	0.44	0.52	0.48	0.48
	300	0.48	0.44	0.44	0.48	0.48	0.48	0.52	0.52	0.48
	350	0.48	0.48	0.48	0.52	0.52	0.48	0.56	0.52	0.52
12	50	0.48	0.48	0.48	0.48	0.48	0.48	0.52	0.48	0.48
	100	0.48	0.48	0.48	0.48	0.48	0.48	0.52	0.48	0.48
	150	0.48	0.48	0.48	0.48	0.48	0.48	0.52	0.52	0.48
	200	0.48	0.48	0.48	0.48	0.48	0.48	0.56	0.52	0.48
	250	0.52	0.48	0.48	0.52	0.52	0.48	0.56	0.56	0.52
	300	0.52	0.52	0.52	0.56	0.52	0.52	0.60	0.56	0.56
	350	0.56	0.56	0.52	0.56	0.56	0.56	0.60	0.60	0.60
14	50	0.51	0.48	0.48	0.51	0.48	0.48	0.59	0.55	0.51
	100	0.51	0.48	0.48	0.55	0.51	0.48	0.59	0.55	0.51
	150	0.55	0.51	0.48	0.55	0.51	0.48	0.64	0.59	0.55
	200	0.55	0.51	0.51	0.55	0.55	0.51	0.64	0.59	0.59
	250	0.59	0.55	0.55	0.59	0.59	0.55	0.64	0.64	0.59
	300	0.59	0.59	0.59	0.64	0.59	0.59	0.69	0.64	0.64
	350	0.64	0.64	0.64	0.64	0.64	0.64	0.75	0.69	0.64

* Thicknesses include allowances for foundry practice, corrosion, and either water hammer or truck load.

TABLE 6.3—(contd.)

Standard Thickness* Selection Table for Cast-Iron Pipe

Laying Condition A—flat-bottom trench, without blocks, untamped backfill
Laying Condition B—flat-bottom trench, without blocks, tamped backfill
Laying Condition F—pipe laid on gravel or sand bedding, backfill tamped

Size in.	Working Pressure psi	3½-ft Cover			5-ft Cover			8-ft Cover		
		Laying Condition			Laying Condition			Laying Condition		
		A	B	F	A	B	F	A	B	F
		Thickness*—in.								
16	50	0.54	0.50	0.50	0.58	0.54	0.50	0.63	0.58	0.54
	100	0.54	0.54	0.50	0.58	0.54	0.50	0.63	0.58	0.54
	150	0.58	0.54	0.50	0.58	0.54	0.54	0.68	0.63	0.58
	200	0.58	0.58	0.54	0.63	0.58	0.58	0.68	0.63	0.63
	250	0.63	0.58	0.58	0.63	0.63	0.58	0.73	0.68	0.63
	300	0.63	0.63	0.63	0.68	0.68	0.63	0.73	0.73	0.68
	350	0.68	0.68	0.68	0.73	0.68	0.68	0.79	0.73	0.73
18	50	0.58	0.54	0.54	0.58	0.54	0.54	0.68	0.63	0.58
	100	0.58	0.54	0.54	0.63	0.58	0.54	0.68	0.63	0.58
	150	0.63	0.58	0.54	0.63	0.58	0.58	0.73	0.68	0.63
	200	0.63	0.58	0.58	0.68	0.63	0.58	0.73	0.68	0.63
	250	0.68	0.63	0.63	0.68	0.68	0.63	0.79	0.73	0.68
	300	0.68	0.68	0.68	0.73	0.73	0.68	0.79	0.79	0.73
	350	0.79	0.73	0.73	0.79	0.79	0.73	0.85	0.85	0.79
20	50	0.62	0.57	0.57	0.67	0.57	0.57	0.72	0.67	0.62
	100	0.62	0.57	0.57	0.67	0.62	0.57	0.72	0.67	0.62
	150	0.67	0.62	0.57	0.67	0.62	0.62	0.78	0.72	0.67
	200	0.67	0.62	0.62	0.72	0.67	0.62	0.78	0.72	0.67
	250	0.72	0.67	0.67	0.78	0.72	0.67	0.84	0.78	0.72
	300	0.78	0.72	0.72	0.78	0.78	0.72	0.84	0.84	0.78
	350	0.84	0.78	0.78	0.84	0.84	0.78	0.91	0.84	0.84
24	50	0.68	0.63	0.63	0.73	0.63	0.63	0.79	0.73	0.68
	100	0.73	0.63	0.63	0.73	0.68	0.63	0.85	0.73	0.68
	150	0.73	0.68	0.63	0.79	0.73	0.68	0.85	0.79	0.73
	200	0.79	0.73	0.68	0.79	0.79	0.73	0.92	0.85	0.79
	250	0.79	0.79	0.73	0.85	0.79	0.79	0.92	0.85	0.85
	300	0.85	0.85	0.85	0.92	0.85	0.85	0.99	0.92	0.92
	350	0.92	0.92	0.92	0.99	0.92	0.92	1.07	0.99	0.92

* Thicknesses include allowances for foundry practice, corrosion, and either water hammer or truck load.

13

TABLE 6.4

Standard Dimensions and Weights of Push-On Joint Cast-Iron Pipe‖

Size in.	Thickness Class	Thickness	OD†	Weight of Barrel per Foot	Weight of Bell	Weight Based on 18-ft Laying Length		Weight Based on 20-ft Laying Length	
						Per Length‡	Avg. per Foot§	Per Length‡	Avg. per Foot§
		in.		*lb*					
3*	22	0.32	3.96	11.4	11	215	12.0	240	11.9
	23	0.35	3.96	12.4	11	235	13.0	260	12.9
	24	0.38	3.96	13.3	11	250	13.9	275	13.8
4	22	0.35	4.80	15.3	14	290	16.1	320	16.0
	23	0.38	4.80	16.5	14	310	17.3	345	17.2
	24	0.41	4.80	17.6	14	330	18.4	365	18.3
	25	0.44	4.80	18.8	14	350	19.5	390	19.5
6	22	0.38	6.90	24.3	25	460	25.6	510	25.6
	23	0.41	6.90	26.1	25	495	27.5	545	27.4
	24	0.44	6.90	27.9	25	525	29.3	585	29.2
	25	0.48	6.90	30.2	25	570	31.7	630	31.4
	26	0.52	6.90	32.5	25	610	33.9	675	33.8
8	22	0.41	9.05	34.7	41	665	36.9	735	36.8
	23	0.44	9.05	37.1	41	710	39.4	785	39.2
	24	0.48	9.05	40.3	41	765	42.6	845	42.4
	25	0.52	9.05	43.5	41	825	45.8	910	45.6
	26	0.56	9.05	46.6	41	880	48.9	975	48.6
	27	0.60	9.05	49.7	41	935	52.0	1,035	51.8
10	22	0.44	11.10	46.0	54	880	49.0	975	48.7
	23	0.48	11.10	50.0	54	955	53.0	1,055	52.7
	24	0.52	11.10	53.9	54	1,025	56.9	1,130	56.6
	25	0.56	11.10	57.9	54	1,095	60.9	1,210	60.6
	26	0.60	11.10	61.8	54	1,165	64.8	1,290	64.5
	27	0.65	11.10	66.6	54	1,255	69.6	1,385	69.3
12	22	0.48	13.20	59.8	66	1,140	63.4	1,260	63.1
	23	0.52	13.20	64.6	66	1,230	68.3	1,360	67.9
	24	0.56	13.20	69.4	66	1,315	73.1	1,455	72.7
	25	0.60	13.20	74.1	66	1,400	77.8	1,550	77.4
	26	0.65	13.20	80.0	66	1,505	83.7	1,665	83.3
	27	0.70	13.20	85.8	66	1,610	89.5	1,780	89.1
	28	0.76	13.20	92.7	66	1,735	96.4	1,920	96.0

* Pipe of 3-in. size also available in 12-ft laying length. Weight (lb) per length‡ for thickness class 22 is 150; for 23, 160; for 24, 170. Average weight (lb) per foot§ for thickness class 22 is 12.5; for 23, 13.5; for 24, 14.3.
‡ Including bell. Calculated weight of pipe rounded off to nearest 5 lb.
§ Including bell. Average weight per foot based on calculated weight of pipe before rounding.
† Tolerances of OD of spigot end: 3–12 in., ±0.06 in.; 14–24 in., +0.05 in., −0.08 in.
‖ Weights shown for push-on joint pipe are also applicable to bell-and-spigot joint pipe.

TABLE 6.4—(contd.)

Standard Dimensions and Weights of Push-On Joint Cast-Iron Pipe‖

Size in.	Thickness Class	Thickness	OD†	Weight of Barrel per Foot	Weight of Bell	Weight Based on 18-ft Laying Length		Weight Based on 20-ft Laying Length	
						Per Length‡	Avg. per Foot§	Per Length‡	Avg. per Foot§
		in.				*lb*			
14	21	0.48	15.30	69.7	78	1,335	74.1	1,470	73.6
	22	0.51	15.30	73.9	78	1,410	78.2	1,555	77.8
	23	0.55	15.30	79.5	78	1,510	83.8	1,670	83.4
	24	0.59	15.30	85.1	78	1,610	89.5	1,780	89.0
	25	0.64	15.30	92.0	78	1,735	96.3	1,920	95.9
	26	0.69	1.530	98.8	78	1,855	103.1	2,055	102.7
	27	0.75	15.30	107.0	78	2,005	111.3	2,220	110.9
	28	0 81	15.30	115.0	78	2,150	119.3	2,380	118.9
16	21	0.50	17.40	82.8	96	1,585	88.1	1,750	87.6
	22	0.54	17.40	89.2	96	1,700	94.5	1,880	94.0
	23	0.58	17.40	95.6	96	1,815	100.9	2,010	100.4
	24	0.63	17.40	103.6	96	1,960	109.0	2,170	108.4
	25	0.68	17.40	111.4	96	2,100	116.8	2,325	116.2
	26	0.73	17.40	119.3	96	2,245	124.6	2,480	124.1
	27	0.79	17.40	128.6	96	2,410	133.9	2,670	133.4
	28	0.85	17.40	137.9	96	2,580	143.2	2,855	142.7
18	21	0.54	19.50	100.4	114	1,920	106.7	2,120	106.1
	22	0.58	19.50	107.6	114	2,050	113.9	2,265	113.3
	23	0.63	19.50	116.5	114	2,210	122.8	2,445	122.2
	24	0.68	19.50	125.4	114	2,370	131.8	2,620	131.1
	25	0.73	19.50	134.3	114	2,530	140.6	2,800	140.0
	26	0.79	19.50	144.9	114	2,720	151.2	3,010	150.6
	27	0.85	19.50	155.4	114	2,910	161.7	3,220	161.1
	28	0.92	19.50	167.5	114	3,130	173.8	3,465	173.2
20	21	0.57	21.60	117.5	133	2,250	124.9	2,485	124.2
	22	0.62	21.60	127.5	133	2,430	134.9	2,685	134.2
	23	0.67	21.60	137.5	133	2,610	144.9	2,885	144.2
	24	0.72	21.60	147.4	133	2,785	154.8	3,080	154.0
	25	0.78	21.60	159.2	133	3,000	166.6	3,315	165.8
	26	0.84	21.60	170.9	133	3,210	178.4	3,550	177.6
	27	0.91	21.60	184.5	133	3,455	191.9	3,825	191.2
	28	0.98	21.60	198.1	133	3,700	205.5	4,095	204.8
24	21	0.63	25.80	155.4	179	2,975	165.3	3,285	164.4
	22	0.68	25.80	167.4	179	3,190	177.3	3,525	176.4
	23	0.73	25.80	179.4	179	3,410	189.4	3,765	188.4
	24	0.79	25.80	193.7	179	3,665	203.6	4,055	202.6
	25	0.85	25.80	207.9	179	3,920	217.8	4,335	216.8
	26	0.92	25.80	224.4	179	4,220	234.4	4,665	233.4
	27	0.99	25.80	240.8	179	4,515	250.7	4,995	249.8
	28	1.07	25.80	259.4	179	4,850	269.3	5,365	268.4

‡ Including bell. Calculated weight of pipe rounded off to nearest 5 lb.
§ Including bell. Average weight per foot based on calculated weight of pipe before rounding.
† Tolerances of OD of spigot end: 3–12 in., ±0.06 in.; 14–24 in., +0.05 in., −0.08 in.
‖ Weights shown for push-on joint pipe are also applicable to bell-and-spigot joint pipe.

TABLE 6.5

Standard Dimensions and Weights of Mechanical-Joint Cast-Iron Pipe

Size in.	Thickness Class	Thickness	OD†	Weight of Barrel per Foot	Weight of Bell	Weight Based on 18-ft Laying Length		Weight Based on 20-ft Laying Length	
						Per Length‡	Avg. per Foot§	Per Length‡	Avg. per Foot§
		in.				*lb*			
3*	22	0.32	3.96	11.4	11	215	12.0	240	11.9
	23	0.35	3.96	12.4	11	235	13.0	260	12.9
	24	0.38	3.96	13.3	11	250	13.9	275	13.8
4	22	0.35	4.80	15.3	16	290	16.2	320	16.1
	23	0.38	4.80	16.5	16	315	17.4	345	17.3
	24	0.41	4.80	17.6	16	335	18.5	370	18.4
	25	0.44	4.80	18.8	16	355	19.7	390	19.6
6	22	0.38	6.90	24.3	22	460	25.5	510	25.4
	23	0.41	6.90	26.1	22	490	27.3	545	27.2
	24	0.44	6.90	27.9	22	525	29.1	580	29.0
	25	0.48	6.90	30.2	22	565	31.4	625	31.3
	26	0.52	6.90	32.5	22	605	33.7	670	33.6
8	22	0.41	9.05	34.7	30	655	36.4	725	36.2
	23	0.44	9.05	37.1	30	700	38.8	770	38.6
	24	0.48	9.05	40.3	30	755	42.0	835	41.8
	25	0.52	9.05	43.5	30	815	45.2	900	45.0
	26	0.56	9.05	46.6	30	870	48.3	960	48.1
	27	0.60	9.05	49.7	30	925	51.4	1,025	51.2
10	22	0.44	11.10	46.0	40	870	48.2	960	48.0
	23	0.48	11.10	50.0	40	940	52.2	1,040	52.0
	24	0.52	11.10	53.9	40	1,010	56.1	1,120	55.9
	25	0.56	11.10	57.9	40	1,080	60.1	1,200	59.9
	26	0.60	11.10	61.8	40	1,150	64.0	1,275	63.8
	27	0.65	11.10	66.6	40	1,240	68.8	1,370	68.6
12	22	0.48	13.20	59.8	50	1,125	62.6	1,245	62.3
	23	0.52	13.20	64.6	50	1,215	67.4	1,340	67.1
	24	0.56	13.20	69.4	50	1,300	72.2	1,440	71.9
	25	0.60	13.20	74.1	50	1,385	76.9	1,530	76.6
	26	0.65	13.20	80.0	50	1,490	82.8	1,650	82.5
	27	0.70	13.20	85.8	50	1,595	88.6	1,765	88.3
	28	0.76	13.20	92.7	50	1,720	95.5	1,905	95.2

* Pipe of 3-in. size also available in 12-ft laying length. Weight (lb) per length‡ for thickness class 22 is 150; for 23, 160 ; for 24, 170. Average weight (lb) per foot§ for thickness class 22 is 12.3; for 23, 13.3; for 24, 14.2.
‡ Including bell. Calculated weight of pipe rounded off to nearest 5 lb.
§ Including bell. Average weight per foot based on calculated weight of pipe before rounding.
† Tolerances of OD of spigot end: 3–12 in., ±0.06 in.; 14–24 in., +0.05 in., −0.08 in.

TABLE 6.5—(contd.)

Standard Dimensions and Weights of Mechanical-Joint Cast-Iron Pipe

Size in.	Thickness Class	Thickness	OD†	Weight of Barrel per Foot	Weight of Bell	Weight Based on 18-ft Laying Length		Weight Based on 20-ft Laying Length	
						Per Length‡	Avg. per Foot§	Per Length‡	Avg. per Foot§
		in.				*lb*			
14	21	0.48	15.30	69.7	78	1.335	74.0	1,470	73.6
	22	0.51	15.30	73.9	78	1,410	78.2	1,555	77.8
	23	0.55	15.30	79.5	78	1,510	83.8	1,670	83.4
	24	0.59	15.30	85.1	78	1,610	89.4	1,780	89.0
	25	0.64	15.30	92.0	78	1,735	96.3	1,920	95.9
	26	0.69	15.30	98.8	78	1,855	103.1	2,055	102.7
	27	0.75	15.30	107.0	78	2,005	111.3	2,220	110.9
	28	0.81	15.30	115.0	78	2,150	119.3	2,380	118.9
16	21	0.50	17.40	82.8	95	1,585	88.1	1,750	87.6
	22	0.54	17.40	89.2	95	1,700	94.5	1,880	94.0
	23	0.58	17.40	95.6	95	1,815	100.9	2,005	100.3
	24	0.63	17.40	103.6	95	1,960	108.9	2,165	108.3
	25	0.68	17.40	111.4	95	2,100	116.7	2,325	116.2
	26	0.73	17.40	119.3	95	2,240	124.6	2,480	124.0
	27	0.79	17.40	128.6	95	2,410	133.9	2,665	133.3
	28	0.85	17.40	137.9	95	2,575	143.2	2,855	142.7
18	21	0.54	19.50	100.4	113	1,920	106.7	2,120	106.0
	22	0.58	19.50	107.6	113	2,050	113.9	2,265	113.2
	23	0.63	19.50	116.5	113	2,210	122.8	2,445	122.2
	24	0.68	19.50	125.4	113	2,370	131.7	2,620	131.0
	25	0.73	19.50	134.3	113	2,530	140.6	2,800	140.0
	26	0.79	19.50	144.9	113	2,720	151.2	3,010	150.6
	27	0.85	19.50	155.4	113	2,910	161.7	3,220	161.0
	28	0.92	19.50	167.5	113	3,130	173.8	3,465	173.2
20	21	0.57	21.60	117.5	134	2,250	124.9	2,485	124.2
	22	0.62	21.60	127.5	134	2,430	134.9	2,685	134.2
	23	0.67	21.60	137.5	134	2,610	144.9	2,885	144.2
	24	0.72	21.60	147.4	134	2,785	154.8	3,080	154.1
	25	0.78	21.60	159.2	134	3,000	166.6	3,320	165.9
	26	0.84	21.60	170.9	134	3,210	178.3	3,550	177.6
	27	0.91	21.60	184.5	134	3,455	191.9	3,825	191.2
	28	0.98	21.60	198.1	134	3,700	205.5	4,095	204.8
24	21	0.63	25.80	155.4	177	2,975	165.2	3,285	164.2
	22	0.68	25.80	167.4	177	3,190	177.2	3,525	176.2
	23	0.73	25.80	179.4	177	3,405	189.2	3,765	188.2
	24	0.79	25.80	193.7	177	3,665	203.5	4,050	202.6
	25	0.85	25.80	207.9	177	3,920	217.7	4,335	216.8
	26	0.92	25.80	224.4	177	4,215	234.2	4,665	233.2
	27	0.99	25.80	240.8	177	4,510	250.6	4,995	249.7
	28	1.07	25.80	259.4	177	4,845	269.2	5,365	268.2

† Including bell. Calcaulated weight of pipe rounded off to nearest 5 lb.
§ Including bell. Average weight per foot based on calculated weight of pipe before rounding.
‡ Tolerances of OD of spigot end: 3–12 in., +0.06 in.: 14–24 in., +0.05 in., −0.08 in.

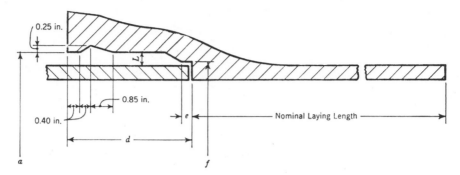

TABLE 6.6

Standard Bell-and-Spigot Joint Dimensions

For weights of bell-and-spigot joint pipe see Tables 6.1 and 6.4.

Size	Pipe OD†	Socket Diam. a†	Thickness of Joint L	Socket Depth d	Centering Shoulder	
					Depth e	ID f†
			in.			
3	3.96	4.76	0.40	3.30	0.30	4.10
4	4.80	5.60	0.40	3.30	0.30	4.94
6	6.90	7.70	0.40	3.88	0.38	7.06
8	9.05	9.85	0.40	4.38	0.38	9.21
10	11.10	11.90	0.40	4.38	0.38	11.28
12	13.20	14.00	0.40	4.38	0.38	13.38
14	15.30	16.10	0.40	4.50	0.50	15.52
16	17.40	18.40	0.50	4.50	0.50	17.62
18	19.50	20.50	0.50	4.50	0.50	19.72
20	21.60	22.60	0.50	4.50	0.50	21.82
24	25.80	26.80	0.50	4.50	0.50	26.02

† Tolerances for outside diameter of spigot ends, socket diameter a, and centering shoulder inside diameter shall be ±0.06 in. for sizes 3–12 in.; ±0.08 in. for sizes 14–24 in.

Dimensions of Pit Cast Pipe only
Not of current manufacture or standard

Dimensions of A. W. W. A. Standard Bell and Spigot Pipe
Classes A, B, C and D

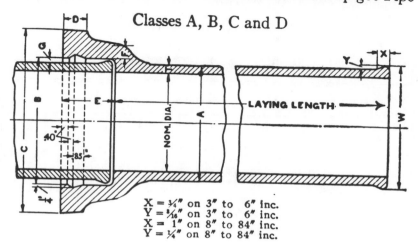

X = ¾" on 3" to 6" inc.
Y = ⅛₆" on 3" to 6" inc.
X = 1" on 8" to 84" inc.
Y = ¼" on 8" to 84" inc.

Table No. 1

Nominal Diameter Inches	Class	Dimensions, Inches								
		A	B	C	D	E	F	G	T	W
3	A	3.80	4.60	7.20	1.25	3.50	.65	.40	.39	4.18
	B	3.96	4.76	7.36	1.25	3.50	.65	.40	.42	4.34
	C	3.96	4.76	7.36	1.25	3.50	.65	.40	.45	4.34
	D	3.96	4.76	7.36	1.25	3.50	.65	.40	.48	4.34
4	A	4.80	5.60	8.20	1.50	3.50	.65	.40	.42	5.18
	B	5.00	5.80	8.40	1.50	3.50	.65	.40	.45	5.38
	C	5.00	5.80	8.40	1.50	3.50	.65	.40	.48	5.38
	D	5.00	5.80	8.40	1.50	3.50	.65	.40	.52	5.38
6	A	6.90	7.70	10.50	1.50	3.50	.70	.40	.44	7.28
	B	7.10	7.90	10.70	1.50	3.50	.70	.40	.48	7.48
	C	7.10	7.90	10.70	1.50	3.50	.70	.40	.51	7.48
	D	7.10	7.90	10.70	1.50	3.50	.70	.40	.55	7.48
8	A	9.05	9.85	12.85	1.50	4.00	.75	.40	.46	9.55
	B	9.05	9.85	12.85	1.50	4.00	.75	.40	.51	9.55
	C	9.30	10.10	13.10	1.50	4.00	.75	.40	.56	9.80
	D	9.30	10.10	13.10	1.50	4.00	.75	.40	.60	9.80

Dimensions continued on next page.

Dimensions of Pit Cast Pipe only
Not of current manufacture or standard

Dimensions of A. W. W. A. Standard Bell and Spigot Pipe
Classes A, B, C and D

Table No. 1 (continued)

Nominal Diameter Inches	Class	Dimensions, Inches								
		A	B	C	D	E	F	G	T	W
10	A	11.10	11.90	14.90	1.50	4.00	.75	.40	.50	11.60
	B	11.10	11.90	14.90	1.50	4.00	.75	.40	.57	11.60
	C	11.40	12.20	15.40	1.50	4.00	.80	.40	.62	11.90
	D	11.40	12.20	15.40	1.50	4.00	.80	.40	.68	11.90
12	A	13.20	14.00	17.20	1.50	4.00	.80	.40	.54	13.70
	B	13.20	14.00	17.20	1.50	4.00	.80	.40	.62	13.70
	C	13.50	14.30	17.70	1.50	4.00	.85	.40	.68	14.00
	D	13.50	14.30	17.70	1.50	4.00	.85	.40	.75	14.00
14	A	15.30	16.10	19.50	1.50	4.00	.85	.40	.57	15.80
	B	15.30	16.10	19.50	1.50	4.00	.85	.40	.66	15.80
	C	15.65	16.45	20.05	1.50	4.00	.90	.40	.74	16.15
	D	15.65	16.45	20.05	1.50	4.00	.90	.40	.82	16.15
16	A	17.40	18.40	22.00	1.75	4.00	.90	.50	.60	17.90
	B	17.40	18.40	22.00	1.75	4.00	.90	.50	.70	17.90
	C	17.80	18.80	22.60	1.75	4.00	1.00	.50	.80	18.30
	D	17.80	18.80	22.60	1.75	4.00	1.00	.50	.89	18.30
18	A	19.50	20.50	24.30	1.75	4.00	.95	.50	.64	20.00
	B	19.50	20.50	24.30	1.75	4.00	.95	.50	.75	20.00
	C	19.92	20.92	25.12	1.75	4.00	1.05	.50	.87	20.42
	D	19.92	20.92	25.12	1.75	4.00	1.05	.50	.96	20.42
20	A	21.60	22.60	26.60	1.75	4.00	1.00	.50	.67	22.10
	B	21.60	22.60	26.60	1.75	4.00	1.00	.50	.80	22.10
	C	22.06	23.06	27.66	1.75	4.00	1.15	.50	.92	22.56
	D	22.06	23.06	27.66	1.75	4.00	1.15	.50	1.03	22.56
24	A	25.80	26.80	31.00	2.00	4.00	1.05	.50	.76	26.30
	B	25.80	26.80	31.00	2.00	4.00	1.05	.50	.89	26.30
	C	26.32	27.32	32.32	2.00	4.00	1.25	.50	1.04	26.82
	D	26.32	27.32	32.32	2.00	4.00	1.25	.50	1.16	26.82
30	A	31.74	32.74	37.34	2.00	4.50	1.15	.50	.88	32.24
	B	32.00	33.00	37.60	2.00	4.50	1.15	.50	1.03	32.50
	C	32.40	33.40	38.60	2.00	4.50	1.32	.50	1.20	32.90
	D	32.74	33.74	39.74	2.00	4.50	1.50	.50	1.37	33.24
36	A	37.96	38.96	43.96	2.00	4.50	1.25	.50	.99	38.46
	B	38.30	39.30	44.90	2.00	4.50	1.40	.50	1.15	38.80
	C	38.70	39.70	45.90	2.00	4.50	1.60	.50	1.36	39.20
	D	39.16	40.16	46.96	2.00	4.50	1.80	.50	1.58	39.66

Dimensions continued on next page.

Dimensions of Pit Cast Pipe only
Not of current manufacture or standard

Dimensions of A. W. W. A. Standard
Bell and Spigot Pipe
Classes A, B, C and D
Table No. 1 (continued)

Nominal Diameter Inches	Class	Dimensions, Inches								
		A	B	C	D	E	F	G	T	W
42	A	44.20	45.20	50.80	2.00	5.00	1.40	.50	1.10	44.70
	B	44.50	45.50	51.50	2.00	5.00	1.50	.50	1.28	45.00
	C	45.10	46.10	52.90	2.00	5.00	1.75	.50	1.54	45.60
	D	45.58	46.58	54.18	2.00	5.00	1.95	.50	1.78	46.08
48	A	50.50	51.50	57.50	2.00	5.00	1.50	.50	1.26	51.00
	B	50.80	51.80	58.40	2.00	5.00	1.65	.50	1.42	51.30
	C	51.40	52.40	60.00	2.00	5.00	1.95	.50	1.71	51.90
	D	51.98	52.98	61.38	2.00	5.00	2.20	.50	1.96	52.48
54	A	56.66	57.66	64.06	2.25	5.50	1.60	.50	1.35	57.16
	B	57.10	58.10	65.30	2.25	5.50	1.80	.50	1.55	57.60
	C	57.80	58.80	66.80	2.25	5.50	2.15	.50	1.90	58.30
	D	58.40	59.40	68.20	2.25	5.50	2.45	.50	2.23	58.90
60	A	62.80	63.80	70.60	2.25	5.50	1.70	.50	1.39	63.30
	B	63.40	64.40	71.80	2.25	5.50	1.90	.50	1.67	63.90
	C	64.20	65.20	73.60	2.25	5.50	2.25	.50	2.00	64.70
	D	64.82	65.82	75.22	2.25	5.50	2.60	.50	2.38	65.32
72	A	75.34	76.59	84.19	2.25	5.50	1.87	.63	1.62	75.84
	B	76.00	77.25	85.65	2.25	5.50	2.20	.63	1.95	76.50
	C	76.88	78.13	87.33	2.25	5.50	2.64	.63	2.39	77.38
84	A	87.54	88.79	96.99	2.50	5.50	2.10	.63	1.72	88.04
	B	88.54	89.79	98.79	2.50	5.50	2.60	.63	2.22	89.04

Dimensions of Pit Cast Pipe only
Not of current manufacture or standard

Bell and Spigot Pipe
Classes E, F, G and H

Table No. 4

Nominal Diameter Inches	Class	Dimensions, Inches								
		A	B	C	D	E	F	G	T	W
6	E	7.22	8.02	11.52	1.50	4.00	.75	.40	.58	7.72
	F	7.22	8.02	11.52	1.50	4.00	.75	.40	.61	7.72
	G	7.38	8.18	11.88	1.50	4.00	.85	.40	.65	7.88
	H	7.38	8.18	11.88	1.50	4.00	.85	.40	.69	7.88
8	E	9.42	10.22	13.92	1.50	4.00	.85	.40	.66	9.92
	F	9.42	10.22	13.92	1.50	4.00	.85	.40	.71	9.92
	G	9.60	10.40	14.30	1.50	4.00	.95	.40	.75	10.10
	H	9.60	10.40	14.30	1.50	4.00	.95	.40	.80	10.10
10	E	11.60	12.40	16.30	1.75	4.50	.95	.40	.74	12.10
	F	11.60	12.40	16.30	1.75	4.50	.95	.40	.80	12.10
	G	11.84	12.64	16.74	1.75	4.50	1.05	.40	.86	12.34
	H	11.84	12.64	16.74	1.75	4.50	1.05	.40	.92	12.34
12	E	13.78	14.58	18.68	1.75	4.50	1.05	.40	.82	14.28
	F	13.78	14.58	18.68	1.75	4.50	1.05	.40	.89	14.28
	G	14.08	14.88	19.28	1.75	4.50	1.20	.40	.97	14.58
	H	14.08	14.88	19.28	1.75	4.50	1.20	.40	1.04	14.58
14	E	15.98	16.78	21.08	2.00	4.50	1.15	.40	.90	16.48
	F	15.98	16.78	21.08	2.00	4.50	1.15	.40	.99	16.48
	G	16.32	17.12	21.82	2.00	4.50	1.35	.40	1.07	16.82
	H	16.32	17.12	21.82	2.00	4.50	1.35	.40	1.16	16.82

Dimensions of Pit Cast Pipe only
Not of current manufacture or standard

Dimensions of A. W. W. A. Standard
Bell and Spigot Pipe
Classes E, F, G and H

Table No. 4 (continued)

Nominal Diameter Inches	Class	Dimensions, Inches								
		A	B	C	D	E	F	G	T	W
16	E	18.16	18.96	23.56	2.00	4.50	1.25	.40	.98	18.66
	F	18.16	18.96	23.56	2.00	4.50	1.25	.40	1.08	18.66
	G	18.54	19.34	24.44	2.00	4.50	1.45	.40	1.18	19.04
	H	18.54	19.34	24.44	2.00	4.50	1.45	.40	1.27	19.04
18	E	20.34	21.14	26.04	2.25	4.50	1.40	.40	1.07	20.84
	F	20.34	21.14	26.04	2.25	4.50	1.40	.40	1.17	20.84
	G	20.78	21.58	27.08	2.25	4.50	1.65	.40	1.28	21.28
	H	20.78	21.58	27.08	2.25	4.50	1.65	.40	1.39	21.28
20	E	22.54	23.34	28.44	2.25	4.50	1.50	.40	1.15	23.04
	F	22.54	23.34	28.44	2.25	4.50	1.50	.40	1.27	23.04
	G	23.02	23.82	29.52	2.25	4.50	1.75	.40	1.39	23.52
	H	23.02	23.82	29.52	2.25	4.50	1.75	:40	1.51	23.52
24	E	26.90	27.90	33.40	2.25	5.00	1.70	.50	1.31	27.40
	F	26.90	27.90	33.40	2.25	5.00	1.70	.50	1.45	27.40
	G	27.76	28.56	34.86	2.25	5.00	1.95	.50	1.75	28.26
	H	27.76	28.56	34.86	2.25	5.00	1.95	.50	1.88	28.26
30	E	33.10	34.10	40.60	2.25	5.00	1.80	.50	1.55	33.60
	F	33.46	34.46	41.46	2.25	5.00	2.00	.50	1.73	33.96
36	E	39.60	40.60	48.00	2.25	5.00	2.05	.50	1.80	40.10
	F	40.04	41.04	49.04	2.25	5.00	2.30	.50	2.02	40.54

STANDARD ANSI CLASS THICKNESSES OF 21/45 STRENGTH GRAY CAST IRON PIPE
American National Standards Institute Standard A21.1 (AWWA C101)

The tables below are reproduced from ANSI Standard A21.1 (AWWA C101) and are repeated here for convenience. They are computed for metal of 21,000 psi minimum burst tensile strength and 45,000 psi minimum modulus of rupture.

Laying Conditions

A—Flat bottom trench, untamped backfill.
B—Flat bottom trench, tamped backfill.
F—Bedded in gravel or sand, backfill tamped.

Nominal Inside Diameter Inches	Working Pressure psi	2½ Feet of Cover A	B	F	3½ Feet of Cover A	B	F	5 Feet of Cover A	B	F	8 Feet of Cover A	B	F
							ANSI Class Thickness*						
3	50	22	22	22	22	22	22	22	22	22	22	22	22
	100	22	22	22	22	22	22	22	22	22	22	22	22
	150	22	22	22	22	22	22	22	22	22	22	22	22
	200	22	22	22	22	22	22	22	22	22	22	22	22
	250	22	22	22	22	22	22	22	22	22	22	22	22
	300	22	22	22	22	22	22	22	22	22	22	22	22
	350	22	22	22	22	22	22	22	22	22	22	22	22
4	50	22	22	22	22	22	22	22	22	22	22	22	22
	100	22	22	22	22	22	22	22	22	22	22	22	22
	150	22	22	22	22	22	22	22	22	22	22	22	22
	200	22	22	22	22	22	22	22	22	22	22	22	22
	250	22	22	22	22	22	22	22	22	22	22	22	22
	300	22	22	22	22	22	22	22	22	22	22	22	22
	350	22	22	22	22	22	22	22	22	22	22	22	22
6	50	21	21	21	21	21	21	21	21	21	21	21	21
	100	21	21	21	21	21	21	21	21	21	21	21	21
	150	21	21	21	21	21	21	21	21	21	21	21	21
	200	21	21	21	21	21	21	21	21	21	21	21	21
	250	21	21	21	21	21	21	21	21	21	21	21	21
	300	21	21	21	21	21	21	21	21	21	21	21	21
	350	21	21	21	21	21	21	21	21	21	21	21	21
8	50	20	20	20	20	20	20	20	20	20	20	20	20
	100	20	20	20	20	20	20	20	20	20	21	20	20
	150	20	20	20	20	20	20	20	20	20	21	20	20
	200	20	20	20	20	20	20	20	20	20	21	21	20
	250	20	20	20	20	20	20	20	20	20	22	21	20
	300	21	20	20	20	20	20	21	20	20	22	22	21
	350	21	21	20	21	21	20	21	21	21	23	22	21
10	50	20	20	20	20	20	20	20	20	20	21	21	20
	100	20	20	20	20	20	20	20	20	20	22	21	20
	150	21	20	20	20	20	20	21	20	20	22	22	21
	200	21	20	20	21	20	20	21	21	20	23	22	21
	250	21	21	20	21	21	20	22	21	21	23	23	22
	300	22	21	21	22	21	21	22	22	21	23	23	22
	350	22	22	22	22	22	22	23	23	22	24	23	23
12	50	21	20	20	20	20	20	21	20	20	22	21	20
	100	21	20	20	21	20	20	21	20	20	22	22	21
	150	21	21	20	21	20	20	21	20	20	23	22	21
	200	22	21	20	21	21	20	22	21	20	23	22	22
	250	22	22	20	22	21	20	22	22	21	24	23	22
	300	22	22	21	22	22	21	23	22	22	24	23	23
	350	23	22	22	23	22	22	23	23	22	24	24	23

* Thickness class includes allowances for foundry practice, corrosion, and either water hammer or truck load, whichever is greater.

STANDARD ANSI CLASS THICKNESSES OF 21/45 STRENGTH GRAY CAST IRON PIPE

American National Standards Institute Manual A21.1 (AWWA H1)

Nominal Inside Diameter Inches	Working Pressure psi	2½ Feet of Cover Laying Condition			3½ Feet of Cover Laying Condition			5 Feet of Cover Laying Condition			8 Feet of Cover Laying Condition		
		A	B	F	A	B	F	A	B	F	A	B	F
		ANSI Class Thickness*											
14	50	21	21	21	21	21	21	21	21	21	23	22	21
	100	22	21	21	21	21	21	22	21	21	23	23	21
	150	22	21	21	22	21	21	22	21	21	24	23	22
	200	23	22	21	22	21	21	23	22	21	24	24	23
	250	23	22	21	23	22	21	23	23	22	25	24	23
	300	23	23	22	23	23	22	24	23	23	25	24	24
	350	24	23	23	24	24	23	24	24	24	25	25	24
16	50	22	21	21	21	21	21	22	21	21	23	22	21
	100	22	21	21	22	21	21	22	21	21	24	23	22
	150	22	21	21	22	21	21	23	22	21	24	23	22
	200	23	22	21	23	22	21	23	22	22	24	24	23
	250	23	22	22	23	22	22	23	23	22	25	24	23
	300	24	23	22	23	23	23	24	23	23	25	25	24
	350	24	23	23	24	24	23	25	24	24	26	25	25
18	50	22	21	21	21	21	21	22	21	21	23	22	21
	100	22	21	21	22	21	21	22	21	21	23	22	22
	150	22	21	21	22	21	21	23	22	21	24	23	22
	200	23	22	21	22	22	21	23	22	22	24	23	23
	250	23	22	21	23	22	22	23	23	22	25	24	23
	300	24	23	22	23	23	23	24	24	23	25	25	24
	350	24	23	23	24	24	24	25	24	24	26	25	25
20	50	21	21	21	21	21	21	22	21	21	23	22	21
	100	22	21	21	22	21	21	22	21	21	23	22	21
	150	22	21	21	22	21	21	23	22	21	24	23	22
	200	23	22	21	23	22	21	23	22	21	24	23	22
	250	23	22	21	23	22	22	24	23	22	25	24	23
	300	24	23	22	24	23	23	24	24	23	25	24	24
	350	24	24	23	24	24	24	25	24	24	26	25*	25
24	50	22	21	21	22	21	21	22	21	21	23	22	21
	100	22	21	21	22	21	21	23	21	21	24	23	21
	150	23	21	21	22	21	21	23	22	21	24	23	22
	200	23	22	21	23	22	21	24	22	22	25	24	23
	250	24	23	22	24	22	22	24	23	23	25	24	24
	300	24	23	23	24	23	23	25	24	24	26	25	24
	350	25	24	24	25	24	24	25	25	25	26	26	25

* Thickness class includes allowances for foundry practice, corrosion, and either water hammer or truck load, whichever is greater.

STANDARD THICKNESS CLASS FOR GRAY CAST IRON PIPE

Nominal Diameter Inches	Standard Class Thickness									
	20	21	22	23	24	25	26	27	28	29
3	..	..	.32	.35	.38	.41	.44	.48	.52	.56
4	..	.32	.35	.38	.41	.44	.48	.52	.56	.60
6	..	.35	.38	.41	.44	.48	.52	.56	.60	.65
8	.35	.38	.41	.44	.48	.52	.56	.60	.65	.70
10	.38	.41	.44	.48	.52	.56	.60	.65	.70	.76
12	.41	.44	.48	.52	.56	.60	.65	.70	.76	.82
14	.43	.48	.51	.55	.59	.64	.69	.75	.81	.87
16	.46	.50	.54	.58	.63	.68	.73	.79	.85	.92
18	.50	.54	.58	.63	.68	.73	.79	.85	.92	.99
20	.53	.57	.62	.67	.72	.78	.84	.91	.98	1.06
24	.58	.63	.68	.73	.79	.85	.92	.99	1.07	1.16

Federal Standards for Cast Iron Pipe

WW-P-421 Federal Specification for Pipe; Water Cast Iron (Bell and Spigot) adopted in July, 1931, contains tables showing pipe wall thicknesses for diameters 4 inches through 24 inches for Class 150 and Class 250 pipe. These thickness classes are now obsolete. Emergency Standard EWW-P-421 was used during World War II. The thickness classes in the original 1931 standard were revised in WW-P-421b (March, 1955) to include thickness classes which corresponded to those in ANSI Standards. At that time, ANSI Standards for cast iron pipe were made a part of the Federal Specification by reference. WW-P-421d (1977) also incorporates ANSI Standards for Gray and Ductile Cast Iron Pipe by reference. WW-P-421d does not list pipe wall thicknesses but rather specifies that the pipe shall be furnished for installation with five feet of cover, laying condition B, unless otherwise stated, and that pipe wall thicknesses shown in ANSI Standards A21.1 and A21.51 shall be used.

SECTION III

Ductile Iron Pipe

Introduction

Ductile iron was originally invented by a member company of CIPRA and the first ductile iron pipe was cast experimentally in 1948. Years of metallurgical, casting and quality control refinement followed and in 1955 ductile iron pipe was introduced into the .market place.

Its phenomenal strength and impact resistance, along with many other advantages, created a rapid increase in demand for this product as engineers and utility officials realized that it could be transported, handled and installed with virtually no damage to the pipe. In service, ductile iron pipe showed that expense of repair was practically eliminated. Its corrosion resistance exceeds that of gray cast iron, a pipe product with a reputation of centuries of service in the transmission and distribution of water and gas. Evidence of wide acceptance of ductile iron pipe is demonstrated by its adoption throughout the world as an accepted underground pressure pipe for the transportation of water, wastes, gas and many industrial materials.

Ductile iron pipe with dimensions that make it compatible with gray iron pipe and fittings is available with a wide variety of joints which equip it for specific types of service (see Section VI). As with gray cast iron pipe, protection from severe environmental factors is available (see Section VIII).

Ductile iron pipe provides versatility in design and while it may be designed in accordance with details of ANSI A21.50 (AWWA C150), it is not confined to this specific system. Engineers may choose other use conditions such as those recommended by the American Society of Civil Engineers (ASCE) and, observing the available strength and ductility of ductile iron pipe, follow the general procedure of ANSI A21.50.

Ductile iron pipe is manufactured in accordance with ANSI Standards A21.51 (AWWA C 151), A21.52 and Federal Specification WW-P-421d. The outstanding characteristics of ductile iron pipe result from its unique metallurgical properties effected during the manufacturing process.

Metallurgy of Ductile Iron Pipe

Structure

Ductile iron is usually defined as cast iron with primary graphite in the nodular or spheroidal form. This change in the graphite form is accomplished by adding an inoculant, usually magnesium, to molten iron of appropriate composition.

The matrix is predominantly ferritic for maximum impact resistance and ductility.

The chemical composition of ductile iron is similar to gray cast iron except for the inoculant addition. The chemistry is adjusted to meet the physical test requirements of the Standards.

**The high strength and ruggedness
of Ductile Iron Pipe are demonstrated**

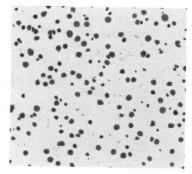

Comparison of Graphite Particles

The gray cast iron on the left shows flake graphitic particles. The ductile iron on the right has spheroidal graphite.

Properties

Ferritic ductile iron as compared to gray cast iron, will have about twice the strength as determined by tensile test, beam test, ring bending test, and bursting test. The tensile elongation and impact strength of ductile iron are many times that of gray cast iron.

Acceptance Test Requirements

The acceptance test requirements of the ANSI A21.51 (AWWA C151) and ANSI A21.52 Ductile Iron Pipe Standards are:

1. Strength and ductility properties of specimens machined from the pipe wall: ultimate strength, 60,000 psi minimum; yield strength, 42,000 psi minimum; elongation, 10 percent minimum.
2. Charpy V-Notch impact strengths on specimens cut from the pipe wall: 7 ft.-lb. minimum at 70°F; 3 ft.-lb. minimum at -40°F.

Quality Control

In addition to the acceptance tests, the manufacturers conduct other quality control tests to assure quality castings having the desired combination of properties.

A great deal of corrosion testing of ductile iron has been made since its commercial introduction. It has been found that the soil corrosion resistance of ductile iron pipe generally exceeds that of gray cast iron pipe.

Ductile Iron Pipe Design

The method of thickness design of ductile iron pipe presented in ANSI A21.50 (AWWA C150) is based on flexible-pipe principles. The principal characteristics that distinguish flexible pipe from more rigid types of pipe are as follows:

1. In a flexible pipe the bending stress from trench load is reduced by the lateral soil reaction that is developed as the pipe deflects under the trench load and pushes outward against the sidefill soil.

2. A flexible pipe, initially deflected by trench load, is partially rerounded by internal pressure, and the bending stress of trench load is thus reduced.

3. A flexible pipe is usually required to carry less earth load than a more rigid pipe because the flexible pipe, in deflecting under the earth load, transfers a significant part of the load to the sidefill soil columns.

These characteristics are expressed mathematically in equations from which may be calculated the earth loads on flexible pipe and the bending stresses and deflections of flexible pipe when subjected to: (1) external trench load and no internal pressure and (2) external trench load in combination with internal pressure. These equations are applicable to pipe made of various elastic metals, of which ductile iron is one.

During the development of the design procedures in this standard, a thorough investigation was carried out to establish conservative design criteria, metal stresses, and soil mechanics factors for use in applying the equations to the design of ductile iron pipe. This investigation included the following studies: ring-bending tests on a large number of ductile iron pipe; a review of the literature on the structural behavior of ductile iron pipe and other flexible pipe; and numerous calculations of pipe thicknesses, stresses, and deflections for various design criteria and loading conditions. From this investigation the design criteria and factors used in this standard were established as described below.

1. **Design criterion for external trench load.** The appropriate criterion for design of ductile iron pipe against external load is the bending stress developed at the pipe invert. This approach provides a uniform stress level in contrast to a design based on a uniform percentage deflection, which results in wide variations in stress and can result in undesirably high bending stresses. The standard provides a formula for supplementary calculation of deflection, if desired. Numerous calculations have shown, however, that the deflection of 3-in. through 54-in. pipe designed for bending stress according to the procedure in the standard will not exceed 3 percent of the outside diameter of the pipe, a deflection at

which the cement linings will not be damaged. Cement linings are not used for gas pipe.

2. **Design bending stress.** Results of a large number of ring-bending tests of ductile iron pipe of various sizes and thicknesses showed that a design bending stress of 48,000 psi is appropriate and conservative.

3. **Trench factors.** The design equation for bending stress contains the modulus of soil reaction and coefficients determined by the bedding angle, all of which are governed by the laying condition. The bedding angle is the angle subtended by the pipe surface that carries the load reaction at the bottom of the trench. Appropriate criteria are shown in Table 50.2 of the standard.

The design equation for bending stress, which incorporates the effect of lateral soil support, applies to the full range of sizes and thicknesses available in ductile iron pipe. For pipe with small diameter-thickness ratios, the effect of sidefill soil support is relatively small. For pipe with larger ratios of diameter to thickness the effect of lateral support in the equation is correspondingly larger.

4. **Separate stress design.** Because of partial rerounding by the internal pressure, the total stress in the pipe wall at the design pressure is less than the stress at zero pressure with external load only. Calculations showed that the larger of the two thicknesses obtained by designing separately for (1) external load with no internal pressure and (2) internal pressure with no external load is greater than the thickness calculated by the appropriate equation for combined external load and internal pressure. Therefore, the separate stress design detailed in the standard was selected.

5. **Design for internal pressure.** Results of a large number of wall tensile tests conducted in conjunction with full-length bursting tests of ductile iron pipe of various sizes and thicknesses showed that a design safety factor of 2 for internal pressure is appropriate and conservative. Internal pressures used for calculation of the thicknesses in ANSI A21.51 (AWWA C151) include an allowance of 100 psi for surges.

6. **Earth loads.** As stated previously, the deflection of a flexible pipe reduces the earth load to less than that imposed on a more rigid pipe in the same condition, because the sidefills carry a part of the trench load transferred to them by the deflection of the pipe. In the standard, the prism load is used for pipe laid in trenches of any width. This is conservative practice because it does not take into account friction forces on the trench walls.

In the calculations for earth loads, the unit weight of 120 lbs. per cubic foot for backfill soil is used. Soil weights generally vary from 110 to 130 lbs. per cubic foot. Experience has shown that 120 lbs. per cubic foot is commonly considered to be a conservative value for most installations.

7. **Allowance for truck superload.** In computing ductile iron pipe thicknesses, truck loads are added to earth loads. The truck superload allowance is a single AASHO H-20 truck with 16,000 lbs. on each rear wheel with an impact factor of 1.5. These truck superloads are based on having the design depth of cover over the pipe and are applied at all depths of cover. Consideration should be given to the loads that may be transmitted to the pipe if either truck superloads or heavy construction equipment is permitted to pass over the pipe at less than the design depth of cover.

8. **Beam load.** Beam stress may be calculated using standard engineering beam formulas, and the deflection may be calculated using a modulus of elasticity of 24,000,000 psi.

9. **Service allowance.** Comparative corrosion tests of cast iron and ductile iron pipe have proved that the corrosion resistance of ductile iron pipe is greater than that of gray cast iron pipe. ANSI design thickness is increased by 0.08 in. to provide an arbitrary service allowance. CIPRA studies have shown that most soils are not corrosive to ductile iron pipe. In those soils, the service allowance serves as an added safety factor. Some of the remaining soils are moderately aggressive and the service allowance serves as a protection against corrosion of the designed pipe wall thickness. In severely corrosive soils, special corrosion protection procedures must be employed (see Section VIII).

10. **Thickness tolerance.** The standard thickness includes a casting tolerance, and the standard weight is calculated from this thickness. The average thickness of a pipe is controlled by the minimum-weight limitation. The minimum thickness is limited by the casting tolerance.

11. **Thickness for tapping.** The number of threads for taps of various sizes in different pipe thicknesses are provided in ductile iron pipe Standards ANSI A21.51 (AWWA C151) and ANSI A21.52. Tests have demonstrated that standard ¾ and 1 inch corporation stops may be threaded directly into all classes of ductile iron pipe for pressures through 350 psi. Installation torque may be reduced significantly by using teflon tape as a thread lubricant.

American Society of Civil Engineers' trench classes B and C as well as other commonly used trench conditions are also used for ductile iron pipe. These trench conditions may be used in any design for underground ductile iron pipe.

ANSI A21.50–1976
(AWWA C150–76)

Revision of
A21.50–1971
(AWWA C150–71)

AMERICAN NATIONAL STANDARD

for the

THICKNESS DESIGN OF
DUCTILE-IRON PIPE

1
4
1

SECRETARIATS

AMERICAN GAS ASSOCIATION
AMERICAN WATER WORKS ASSOCIATION
NEW ENGLAND WATER WORKS ASSOCIATION

Revised edition approved by American National Standards Institute, Inc. Aug. 4, 1976

NOTICE

This Standard has been especially printed by the American Water Works Association for incorporation into this volume. It is current as of December 1, 1975. It should be noted, however, that all AWWA Standards are updated at least once in every five years. Therefore, before applying this Standard it is suggested that you confirm its currency with the American Water Works Association.

PUBLISHED BY

AMERICAN WATER WORKS ASSOCIATION
6666 West Quincy Avenue, Denver, Colorado 80235

American National Standard

An American National Standard implies a consensus of those substantially concerned with its scope and provisions. An American National Standard is intended as a guide to aid the manufacturer, the consumer, and the general public. The existence of an American National Standard does not in any respect preclude anyone, whether he has approved the standard or not, from manufacturing, marketing, purchasing, or using products, processes, or procedures not conforming to the standard. American National Standards are subject to periodic review, and users are cautioned to obtain the latest editions. Producers of goods made in conformity with an American National Standard are encouraged to state on their own responsibility in advertising and promotion material or on tags or labels that the goods are produced in conformity with particular American National Standards.

CAUTION NOTICE. This American National Standard may be revised or withdrawn at any time. The procedures of the American National Standards Institute require that action be taken to reaffirm, revise, or withdraw this standard no later than five (5) years from the date of publication. Purchasers of American National Standards may receive current information on all standards by calling or writing the American National Standards Institute, 1430 Broadway, New York, N.Y. 10018, (212) 868–1220.

Committee Personnel

Subcommittee 1, Pipe, which reviewed this standard, had the following personnel at that time:

EDWARD C. SEARS, *Chairman*
WALTER AMORY, *Vice-Chairman*

User Members	*Producer Members*
ROBERT S. BRYANT	ALFRED F. CASE
FRANK E. DOLSON	W. D. GOODE
GEORGE F. KEENAN	THOMAS D. HOLMES
LEONARD ORLANDO JR.	HAROLD KENNEDY JR.
JOHN E. PERRY	W. HARRY SMITH
	SIDNEY P. TEAGUE

4
3

Standards Committee A21, Cast-Iron Pipe and Fittings, which reviewed and approved this standard, had the following personnel at the time of approval:

LLOYD W. WELLER, *Chairman*
EDWARD C. SEARS, *Vice-Chairman*
JAMES B. RAMSEY, *Secretary*

Organization Represented	*Name of Representative*
American Gas Association	LEONARD ORLANDO JR.
American Society of Civil Engineers	KENNETH W. HENDERSON
American Society of Mechanical Engineers	JAMES S. VANICK
American Society for Testing and Materials	H. M. COBB*
American Water Works Association	ARNOLD M. TINKEY
	LLOYD W. WELLER
Cast Iron Pipe Research Association	THOMAS D. HOLMES
	HAROLD KENNEDY JR.
	EDWARD C. SEARS
	W. HARRY SMITH
Individual Producer	ALFRED F. CASE
Manufacturers' Standardization Society of the Valve and Fittings Industry	ABRAHAM FENSTER
New England Water Works Association	WALTER AMORY
Naval Facilities Engineering Command	STANLEY C. BAKER
Underwriters' Laboratories, Inc.	JOHN E. PERRY
Canadian Standards Association	W. F. SEMENCHUK†

---

* Alternate
† Liaison representative without vote

Table of Contents

Foreword

This foreword is for information only and is not a part of ANSI A21.50 (AWWA C150).

American National Standards Committee A21, Cast-Iron Pipe and Fittings, was organized in 1926 under the sponsorship of the American Gas Association, the American Society for Testing and Materials, the American Water Works Association, and the New England Water Works Association. Since 1972, the co-secretariats have been A.G.A., AWWA, and NEWWA, with AWWA serving as the administrative secretariat. The present scope of Committee A21 activity is

Standardization of specifications for cast-iron and ductile-iron pressure pipe for gas, water and other liquids, and fittings for use with such pipe. These specifications to include design, dimensions, materials, coatings, linings, joints, accessories and methods of inspection and test.

The work of Committee A21 is conducted by subcommittees. The directive to Subcommittee 1—Pipe is that

The scope of the subcommittee activity shall include the periodic review of all current A21 standards for pipe, the preparation of revisions and new standards when needed, as well as other matters pertaining to pipe standards.

The first edition of A21.50, the standard for thickness design of ductile-iron pipe, was issued in 1965 and a revision was issued in 1971. Subcommittee 1 reviewed the 1971 edition and submitted a proposed revision to American National Standards Committee A21 in 1975.

Major Revisions

The basic design procedures are unchanged from the 1965 and 1971 editions. Certain design parameters have been changed to the following.

1. Trench load, which includes earth load and truck load, is expressed as vertical pressure in pounds per square inch. Earth loads for all pipe sizes are based on the prism load concept. Truck load is based on a single AASHTO* H-20 truck with 16 000-lb wheel load and 1.5 impact factor. Truck load is included at all depths of cover. In addition to tabulated truck loads, equations are included for complete calculation of truck loads.

2. The design ring bending stress is 48 000 psi, which provides safety factors under trench loading of at least 1.5 based on ring yield strength, and at least 2.0 based on ultimate strength of the material.

3. The design pipe deflection is 3 per cent of the outside diameter. Tests have shown that 3 per cent deflection is permissible without causing any damage to cement linings of ductile-iron pipe.

4. The standard laying conditions have been expanded to include five types. Types 1, 2, and 5 replace A, B, and S, respectively, in A21.50–1971. Types 3 and 4 have been added to provide a wider selection of laying conditions.

5. The design for internal pressure is based on a 2.0 safety factor. The design pressure is obtained by adding 100-psi surge allowance to the working pressure and multiplying the sum by the 2.0 safety factor. (If anticipated surge pressures are greater than 100

*American Association of State Highway and Transportation Officials.

psi, the maximum anticipated pressure must be used.) The resulting design pressure is then applied to the minimum yield strength in tension of 42 000 psi.

6. The standard thickness classes have been renumbered. Class 1 becomes Class 51, Class 2 becomes Class 52, and so on. Class 50 has been added for 6–54-in. pipe and Class 51 has been expanded to include 3–12-in. pipe.

7. The tables have been modified to reflect the preceding changes. Also, a table has been added to show rated working pressure and maximum depth of cover for all the standard laying conditions and standard thickness classes.

American National Standard

for the

Thickness Design of Ductile-Iron Pipe

Sec. 50–1—Scope

This standard covers the thickness design of ductile-iron pipe complying with the requirements of ANSI A21.51 (AWWA, C151), "Ductile-Iron Pipe, Centrifugally Cast in Metal Molds or Sand-Lined Molds for Water or Other Liquids."

Section 50–2 outlines the design procedure and Sec. 50–3 gives a design example.

Section 50–4 explains the bases of design.

As opposed to using procedures in Sec. 50–2 or 50–4, the designer may reference Tables 12–14 directly.

Table 12 lists thicknesses for standard laying conditions and depths of cover up to 32 ft.

Table 13 lists thicknesses for 150–350 psi water working pressure.

The greater thickness from Table 12 or 13 for given trench load or internal pressure should be used.

Table 14 lists working pressures and maximum depths of cover for standard laying conditions and thickness classes.

Sec. 50–2—Procedure for Calculating Thickness

The thickness of ductile-iron pipe is determined by considering trench load and internal pressure separately.

50–2.1 *Step 1—Design for trench load.*

a. Determine trench load P_v. Table 1 gives trench load, including earth load, P_e, plus truck load P_t, for 2.5–32 ft cover.

b. Determine the standard laying condition from the descriptions in Table 2 and select the appropriate table for diameter–thickness ratios from Tables 7–11. Each table lists diameter–thickness ratios calculated for both bending stress and deflection over a range of trench loads.

c. For bending stress design, enter the column headed "Bending Stress Design" in the appropriate table of Tables 7–11 and locate the tabulated trench load, P_v, nearest to the calculated P_v from paragraph 50–2.1.a. (If the calculated P_v is halfway between two tabulated values, use the larger P_v value.)

Select the corresponding $\dfrac{D}{t}$ value for this P_v.

Divide the pipe's outside diameter D (Table 5) by the $\dfrac{D}{t}$ value to obtain the net thickness t.

d. For deflection design, enter the column headed "Deflection Design" in the appropriate table of Tables 7–11 and locate the tabulated trench load, P_v, nearest to the calculated P_v from paragraph 50–2.1.a. (If the calculated P_v is less than the minimum P_v listed in the table, design for trench load is not controlled by deflection and this determination need not be completed.) If the calculated P_v is halfway between two tabulated values, use the larger P_v value.

Select the corresponding $\dfrac{D}{t_1}$ value for this P_v.

Divide the pipe's outside diameter D (Table 5) by the $\dfrac{D}{t_1}$ value to obtain minimum manufacturing thickness t_1. Deduct 0.08-in. service allowance to obtain net thickness t.

e. Compare the net thicknesses from steps c and d and select the larger of the two. This will be the net thickness required for trench load.

50–2.2 *Step 2—Design for internal pressure.* Calculate the net thickness required for internal pressure using the equation for hoop stress:

$$t = \frac{P_i D}{2S}$$

in which t is net thickness in inches; P_i is the design internal pressure, which is equal to the safety factor of 2.0 times the sum of working pressure (P_w) in pounds per square inch, plus 100 psi surge allowance (P_s) for water pipe. That is, $P_i = 2.0\ (P_w + P_s)$.

If anticipated surge pressures are greater than 100 psi, then the maximum anticipated pressure must be used. D is outside diameter of pipe in inches, and S is minimum yield strength in tension (42 000 psi).

50–2.3 *Step 3—Selection of net thickness and addition of allowances.*

a. Select the net thickness t from step 1 or 2, whichever thickness is larger.

b. Add the service allowance of 0.08 in. to the net thickness t. The resulting thickness is the minimum manufacturing thickness t_1.

c. Add the casting tolerance from Table 3 to the minimum manufacturing thickness t_1. The resulting thickness is the total calculated thickness.

50–2.4 *Step 4—Selection of standard thickness and class.* Use the total calculated thickness from Sec. 50–2.3.c to select a standard class thickness from Table 5. Select the standard thickness nearest to the calculated thickness. When the calculated thickness is halfway between two standard thicknesses, select the larger of the two.

In specifying and ordering pipe, use the class number listed in Table 5 for this standard thickness.

Sec. 50–3—Design Example for Calculating Thickness

PROBLEM: Calculate the thickness for 30-in. ductile-iron pipe laid on a flat-bottom trench with backfill lightly consolidated to centerline of pipe, laying condition Type 2, under 10 ft of cover for a working pressure of 200 psi.

50–3.1 *Step 1—Design for trench load.*

a. Earth load (Table 1) P_e = 8.3 psi
   Truck load (Table 1) P_t = 0.7 psi
   Trench load, $P_v = P_e + P_t$ = 9.0 psi

b. Select Table 8 for diameter–thickness ratios for laying condition Type 2.

c. Entering P_v of 9.0 psi in Table 8, the bending stress design requires $\dfrac{D}{t}$ of 128. From Table 5, diameter D of 30-in. pipe is 32.00 in.

Net thickness t for bending stress

$$= \frac{D}{\dfrac{D}{t}} = \frac{32.00}{128} = 0.25 \text{ in.}$$

d. Also, from Table 8, the deflection design requires $\dfrac{D}{t_1}$ of 108.

Minimum thickness t_1 for deflection design

$$= \frac{D}{\dfrac{D}{t_1}} = \frac{32.00}{108} = 0.30 \text{ in.}$$

Deduct service allowance   −0.08 in.

Net thickness t for deflection control 0.22 in.

e. The larger net thickness is 0.25 in., obtained by the design for bending stress.

50–3.2 *Step 2—Design for internal pressure.*

P_i = 2.0 (Working pressure + 100 psi surge allowance)

(If anticipated surge pressures are greater than 100 psi, then the actual anticipated pressures must be used.)

P_i = 2.0 (200 + 100)
    = 600 psi

$$t = \frac{P_i D}{2S} = \frac{600 \times 32.00}{2 \times 42\,000} = 0.23 \text{ in.}$$

Net thickness t for internal pressure is 0.23 in.

50–3.3 *Step 3—Selection of net thickness and addition of allowances.* The larger of the thicknesses is given by the design for trench load, Step 1, and 0.25 in. is selected.

Net thickness	= 0.25 in.
Service allowance	= 0.08 in.
Minimum thickness	= 0.33 in.
Casting tolerance	= 0.07 in.
Total calculated thickness	= 0.40 in.

50–3.4 *Step 4—Selection of standard thickness and class.* The total calculated thickness of 0.40 in. is nearest to 0.39, Class 50, in Table 5. Therefore, Class 50 is selected for specifying and ordering.

Sec. 50–4—Design Method

50–4.1 The thickness of ductile-iron pipe is determined by considering trench load and internal pressure separately.

Calculations are made for the thicknesses required to resist the bending stress and the deflection caused by trench load. The larger of the two is selected as the thickness required to resist trench load. Calculations are then made for the thickness required to resist the hoop stress of internal pressure.

The larger of these is selected as the net design thickness. To this net thickness is added a service allowance to obtain the minimum manufacturing thickness and a casting tolerance to obtain the total calculated thickness.

The standard thickness and the thickness class for specifying and ordering are selected from a table of standard class thicknesses.

The reverse of the preceding procedure is used to determine the rated working pressure and maximum depth of cover for pipe of a given thickness class.

50–4.2 *Trench load,* P_v. Trench load is expressed as vertical pressure in pounds per square inch, and is equal to the sum of earth load P_e and truck load P_t.

50–4.3 *Earth load,* P_e. Earth load is computed by Eq 4 for the weight of the unit prism of soil with a height equal to the distance from the top of the pipe to the ground surface. The unit weight of backfill soil is taken to be 120 lb/cu ft. If the designer anticipates additional loads because of frost, the design load should be increased accordingly.

50–4.4 *Truck load*, P_t. The truck loads shown in Table 1 were computed by Eq 5 using the surface load factors in Table 6 and the reduction factors R from Table 4 for a single AASHTO H-20 truck on unpaved road or flexible pavement, 16 000-lb wheel load, and 1.5 impact factor. The surface load factors in Table 6 were calculated by Eq 6 for a single concentrated wheel load centered over an effective pipe length of 3 ft.

50–4.5 *Design for trench load.* Tables 7–11, the tables of diameter-thickness ratios used to design for trench load, were computed by Eq 2 and 3. Equation 2 is based on the bending stress at the bottom of the pipe. The design bending stress f is

48 000 psi, which provides at least a 1.5 safety factor based on minimum ring yield strength and a 2.0 safety factor based on ultimate strength. Equation 3 is based on the deflection of the pipe ring section. The design deflection Δx is 3 per cent of the outside diameter of the pipe, which is well below the deflection that might damage cement linings. Design values of the trench parameters E', K_b, and K_x are given in Table 2.

Tables similar to Tables 7–11 may be compiled for laying conditions other than those shown in this standard by calculating the trench loads P_v for a series of diameter–thickness ratios, $\dfrac{D}{t}$ and $\dfrac{D}{t_1}$, using Eq 2 and 3 with values of E', K_b, and K_x appropriate to the bedding and backfill conditions.

Design Equations

$$t = \frac{P_i D}{2S} \tag{1}$$

$$P_v = \frac{f}{3\left(\dfrac{D}{t}\right)\left(\dfrac{D}{t}-1\right)\left[K_b - \dfrac{K_x}{\dfrac{8E}{E'\left(\dfrac{D}{t}-1\right)^3}+0.732}\right]} \tag{2}$$

$$P_v = \frac{\Delta x}{\dfrac{D}{12K_x}}\left[\frac{8E}{\left(\dfrac{D}{t_1}-1\right)^3}+0.732E'\right] \tag{3}$$

$$P_e = \frac{wH}{144} = \frac{120H}{144} = \frac{H}{1.2} \tag{4}$$

$$P_t = \frac{C \cdot R \cdot P \cdot F}{12D} \tag{5}$$

$$C = \frac{1}{3} - \frac{2}{3\pi}\arcsin\left[H\sqrt{\frac{A^2+H^2+1.5^2}{(A^2+H^2)(H^2+1.5^2)}}\,\right] + \frac{AH}{\pi\sqrt{A^2+H^2+1.5^2}}\left[\frac{1}{A^2+H^2}+\frac{1}{H^2+1.5^2}\right] \tag{6}$$

4

Explanation of Symbols for Equations

A = Outside radius of pipe in feet = $\dfrac{D}{24}$

C = Surface load factor (Table 6)

D = Outside diameter in inches (Table 5)

E = Modulus of elasticity (24 × 10⁶ psi)

E' = Modulus of soil reaction in pounds per square inch (Table 2)

F = Impact factor—1.5

f = Design bending stress—48 000 psi

H = Depth of cover in feet

K_b = Bending moment coefficient (Table 2)

K_x = Deflection coefficient (Table 2)

P = Wheel load—16 000 lb

P_e = Earth load in pounds per square inch

P_i = Design internal pressure in pounds per square inch = 2.0 (working pressure + 100 psi surge allowance)

P_t = Truck load in pounds per square inch

P_v = Trench load in pounds per square inch = $P_e + P_t$

R = Reduction factor which takes account of the fact that the part of the pipe directly below the wheels is aided in carrying the truck load by adjacent parts of the pipe that receive little or no load from the wheels (Table 4)

S = Minimum yield strength in tension—42 000 psi

t = Net thickness in inches

t_1 = Minimum thickness in inches ($t + 0.08$)

w = Soil weight—120 pounds per cubic foot

Δx = Design deflection in inches $\left(\dfrac{\Delta x}{D} = 0.03 \right)$

NOTE: In Eq 6, angles are in radians.

TABLE 50.1

Earth Loads P_e, *Truck Loads* P_t, *and Trench Loads* P_v—psi

Depth of Cover ft	P_e	3-in. Pipe		4-in. Pipe		6-in. Pipe		8-in. Pipe	
		P_t	P_v	P_t	P_v	P_t	P_v	P_t	P_v
2.5	2.1	9.9	12.0	9.9	12.0	9.9	12.0	9.8	11.9
3	2.5	7.4	9.9	7.4	9.9	7.3	9.8	7.3	9.8
4	3.3	4.4	7.7	4.5	7.8	4.4	7.7	4.4	7.7
5	4.2	3.0	7.2	3.0	7.2	3.0	7.2	3.0	7.2
6	5.0	2.1	7.1	2.1	7.1	2.1	7.1	2.1	7.1
7	5.8	1.6	7.4	1.6	7.4	1.6	7.4	1.6	7.4
8	6.7	1.2	7.9	1.2	7.9	1.2	7.9	1.2	7.9
9	7.5	1.0	8.5	1.0	8.5	1.0	8.5	1.0	8.5
10	8.3	0.8	9.1	0.8	9.1	0.8	9.1	0.8	9.1
12	10.0	0.6	10.6	0.6	10.6	0.6	10.6	0.6	10.6
14	11.7	0.4	12.1	0.4	12.1	0.4	12.1	0.4	12.1
16	13.3	0.3	13.6	0.3	13.6	0.3	13.6	0.3	13.6
20	16.7	0.2	16.9	0.2	16.9	0.2	16.9	0.2	16.9
24	20.0	0.2	20.2	0.1	20.1	0.1	20.1	0.1	20.1
28	23.3	0.1	23.4	0.1	23.4	0.1	23.4	0.1	23.4
32	26.7	0.1	26.8	0.1	26.8	0.1	26.8	0.1	26.8

Depth of Cover ft	P_e	10-in. Pipe		12-in. Pipe		14-in. Pipe		16-in. Pipe	
		P_t	P_v	P_t	P_v	P_t	P_v	P_t	P_v
2.5	2.1	9.7	11.8	9.6	11.7	8.7	10.8	8.2	10.3
3	2.5	7.2	9.7	7.2	9.7	6.6	9.1	6.2	8.7
4	3.3	4.4	7.7	4.4	7.7	4.4	7.7	4.1	7.4
5	4.2	2.9	7.1	2.9	7.1	2.9	7.1	2.8	7.0
6	5.0	2.1	7.1	2.1	7.1	2.1	7.1	2.0	7.0
7	5.8	1.6	7.4	1.6	7.4	1.6	7.4	1.5	7.3
8	6.7	1.2	7.9	1.2	7.9	1.2	7.9	1.2	7.9
9	7.5	1.0	8.5	1.0	8.5	1.0	8.5	1.0	8.5
10	8.3	0.8	9.1	0.8	9.1	0.8	9.1	0.8	9.1
12	10.0	0.5	10.5	0.5	10.5	0.5	10.5	0.5	10.5
14	11.7	0.4	12.1	0.4	12.1	0.4	12.1	0.4	12.1
16	13.3	0.3	13.6	0.3	13.6	0.3	13.6	0.3	13.6
20	16.7	0.2	16.9	0.2	16.9	0.2	16.9	0.2	16.9
24	20.0	0.1	20.1	0.1	20.1	0.1	20.1	0.1	20.1
28	23.3	0.1	23.4	0.1	23.4	0.1	23.4	0.1	23.4
32	26.7	0.1	26.8	0.1	26.8	0.1	26.8	0.1	26.8

TABLE 50.1—(cont.)

Depth of Cover ft	P_e	18-in. Pipe		20-in. Pipe		24-in. Pipe		30-in. Pipe	
		P_t	P_v	P_t	P_v	P_t	P_v	P_t	P_v
2.5	2.1	7.8	9.9	7.5	9.6	7.1	9.2	6.7	8.8
3	2.5	5.9	8.4	5.7	8.2	5.4	7.9	5.2	7.7
4	3.3	3.9	7.2	3.9	7.2	3.6	6.9	3.5	6.8
5	4.2	2.6	6.8	2.6	6.8	2.4	6.6	2.4	6.6
6	5.0	1.9	6.9	1.9	6.9	1.7	6.7	1.7	6.7
7	5.8	1.4	7.2	1.4	7.2	1.3	7.1	1.3	7.1
8	6.7	1.2	7.9	1.1	7.8	1.1	7.8	1.1	7.8
9	7.5	1.0	8.5	0.9	8.4	0.9	8.4	0.9	8.4
10	8.3	0.8	9.1	0.7	9.0	0.7	9.0	0.7	9.0
12	10.0	0.5	10.5	0.5	10.5	0.5	10.5	0.5	10.5
14	11.7	0.4	12.1	0.4	12.1	0.4	12.1	0.4	12.1
16	13.3	0.3	13.6	0.3	13.6	0.3	13.6	0.3	13.6
20	16.7	0.2	16.9	0.2	16.9	0.2	16.9	0.2	16.9
24	20.0	0.1	20.1	0.1	20.1	0.1	20.1	0.1	20.1
28	23.3	0.1	23.4	0.1	23.4	0.1	23.4	0.1	23.4
32	26.7	0.1	26.8	0.1	26.8	0.1	26.8	0.1	26.8

Depth of Cover ft	P_e	36-in. Pipe		42-in. Pipe		48-in. Pipe		54-in. Pipe	
		P_t	P_v	P_t	P_v	P_t	P_v	P_t	P_v
2.5	2.1	6.2	8.3	5.8	7.9	5.4	7.5	5.0	7.1
3	2.5	4.9	7.4	4.6	7.1	4.4	6.9	4.1	6.6
4	3.3	3.4	6.7	3.3	6.6	3.1	6.4	3.0	6.3
5	4.2	2.3	6.5	2.3	6.5	2.2	6.4	2.1	6.3
6	5.0	1.7	6.7	1.7	6.7	1.6	6.6	1.6	6.6
7	5.8	1.3	7.1	1.3	7.1	1.2	7.0	1.2	7.0
8	6.7	1.1	7.8	1.0	7.7	1.0	7.7	1.0	7.7
9	7.5	0.8	8.3	0.8	8.3	0.8	8.3	0.8	8.3
10	8.3	0.7	9.0	0.7	9.0	0.7	9.0	0.7	9.0
12	10.0	0.5	10.5	0.5	10.5	0.5	10.5	0.5	10.5
14	11.7	0.4	12.1	0.4	12.1	0.4	12.1	0.4	12.1
16	13.3	0.3	13.6	0.3	13.6	0.3	13.6	0.3	13.6
20	16.7	0.2	16.9	0.2	16.9	0.2	16.9	0.2	16.9
24	20.0	0.1	20.1	0.1	20.1	0.1	20.1	0.1	20.1
28	23.3	0.1	23.4	0.1	23.4	0.1	23.4	0.1	23.4
32	26.7	0.1	26.8	0.1	26.8	0.1	26.8	0.1	26.8

153

TABLE 50.2

Design Values for Standard Laying Conditions

Laying Condition*	Description	E'	Bedding Angle deg	K_b	K_x
Type 1†	Flat-bottom trench.‡ Loose backfill.	150	30	0.235	0.108
Type 2	Flat-bottom trench. Backfill lightly consolidated to centerline of pipe.	300	45	0.210	0.105
Type 3	Pipe bedded in 4-in.-minimum loose soil.§ Backfill lightly consolidated to top of pipe.	400	60	0.189	0.103
Type 4	Pipe bedded in sand, gravel, or crushed stone to depth of ⅛ pipe diameter, 4-in. minimum. Backfill compacted to top of pipe. (Approx. 80 per cent Standard Proctor, AASHTO T-99)**	500	90	0.157	0.096
Type 5	Pipe bedded to its centerline in compacted granular material, 4-in. minimum under pipe. Compacted granular or select§ material to top of pipe. (Approx. 90 per cent Standard Proctor, AASHTO T-99)**	700	150	0.128	0.085

* See Fig. 1.
† For pipe 30 in. and larger, consideration should be given to the use of laying conditions other than Type 1.
‡ Flat-bottom is defined as "undisturbed earth."
§ Loose soil or select material is defined as "native soil excavated from the trench, free of rocks, foreign material, and frozen earth."
** AASHTO T-99, "Moisture Density Relations of Soils Using a 5.5 lb (2.5 kg) Rammer 12-in. (305-mm) Drop."

TABLE 50.3

Allowances for Casting Tolerance

Size in.	Casting Tolerance in.
3–8	0.05
10–12	0.06
14–42	0.07
48	0.08
54	0.09

TABLE 50.4

Reduction Factors R for Truck Load Calculations

Size in.	Depth of Cover—ft			
	<4	4–7	8–10	>10
	Reduction Factor			
3–12	1.00	1.00	1.00	1.00
14	0.92	1.00	1.00	1.00
16	0.88	0.95	1.00	1.00
18	0.85	0.90	1.00	1.00
20	0.83	0.90	0.95	1.00
24–30	0.81	0.85	0.95	1.00
36–54	0.80	0.85	0.90	1.00

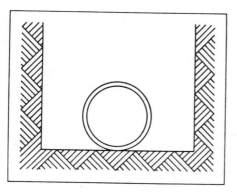

Type 1

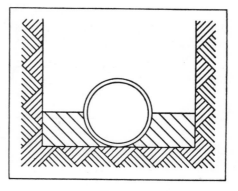

Type 2

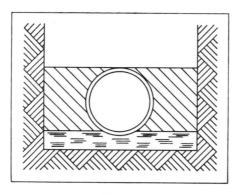

Type 3

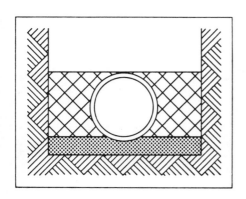

Type 4

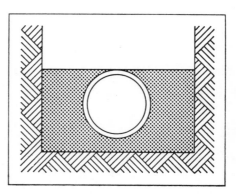

Type 5

Figure 1. Standard Pipe Laying Conditions

See Table 2

TABLE 50.5

Standard Thickness Classes of Ductile-Iron Pipe

Size in.	Outside Diameter—in.	Thickness Class						
		50	51	52	53	54	55	56
		Thickness—in.						
3	3.96	—	0.25	0.28	0.31	0.34	0.37	0.40
4	4.80	—	0.26	0.29	0.32	0.35	0.38	0.41
6	6.90	0.25	0.28	0.31	0.34	0.37	0.40	0.43
8	9.05	0.27	0.30	0.33	0.36	0.39	0.42	0.45
10	11.10	0.29	0.32	0.35	0.38	0.41	0.44	0.47
12	13.20	0.31	0.34	0.37	0.40	0.43	0.46	0.49
14	15.30	0.33	0.36	0.39	0.42	0.45	0.48	0.51
16	17.40	0.34	0.37	0.40	0.43	0.46	0.49	0.52
18	19.50	0.35	0.38	0.41	0.44	0.47	0.50	0.53
20	21.60	0.36	0.39	0.42	0.45	0.48	0.51	0.54
24	25.80	0.38	0.41	0.44	0.47	0.50	0.53	0.56
30	32.00	0.39	0.43	0.47	0.51	0.55	0.59	0.63
36	38.30	0.43	0.48	0.53	0.58	0.63	0.68	0.73
42	44.50	0.47	0.53	0.59	0.65	0.71	0.77	0.83
48	50.80	0.51	0.58	0.65	0.72	0.79	0.86	0.93
54	57.10	0.57	0.65	0.73	0.81	0.89	0.97	1.05

156

TABLE 50.6

Surface Load Factors for Single Truck on Unpaved Road

Depth of Cover ft	Pipe Size—in.							
	3	4	6	8	10	12	14	16
	Surface Load Factor—C							
2.5	0.0196	0.0238	0.0340	0.0443	0.0538	0.0634	0.0726	0.0814
3	0.0146	0.0177	0.0253	0.0330	0.0402	0.0475	0.0546	0.0614
4	0.0088	0.0107	0.0153	0.0201	0.0245	0.0290	0.0335	0.0379
5	0.0059	0.0071	0.0102	0.0134	0.0163	0.0194	0.0224	0.0254
6	0.0042	0.0050	0.0072	0.0095	0.0116	0.0138	0.0159	0.0181
7	0.0031	0.0038	0.0054	0.0071	0.0087	0.0103	0.0119	0.0135
8	0.0024	0.0029	0.0042	0.0055	0.0067	0.0079	0.0092	0.0104
9	0.0019	0.0023	0.0033	0.0043	0.0053	0.0063	0.0073	0.0083
10	0.0015	0.0019	0.0027	0.0035	0.0043	0.0051	0.0060	0.0068
12	0.0011	0.0013	0.0019	0.0025	0.0030	0.0036	0.0042	0.0047
14	0.0008	0.0010	0.0014	0.0018	0.0022	0.0027	0.0031	0.0035
16	0.0006	0.0007	0.0011	0.0014	0.0017	0.0020	0.0024	0.0027
20	0.0004	0.0005	0.0007	0.0009	0.0011	0.0013	0.0015	0.0017
24	0.0003	0.0003	0.0005	0.0006	0.0008	0.0009	0.0011	0.0012
28	0.0002	0.0002	0.0003	0.0005	0.0006	0.0007	0.0008	0.0009
32	0.0002	0.0002	0.0003	0.0003	0.0004	0.0005	0.0006	0.0007

Depth of Cover ft	Pipe Size—in.							
	18	20	24	30	36	42	48	54
	Surface Load Factor—C							
2.5	0.0899	0.0980	0.1130	0.1321	0.1479	0.1604	0.1705	0.1784
3	0.0681	0.0746	0.0867	0.1028	0.1169	0.1286	0.1384	0.1466
4	0.0422	0.0464	0.0545	0.0657	0.0761	0.0853	0.0936	0.1008
5	0.0283	0.0312	0.0369	0.0449	0.0525	0.0595	0.0661	0.0720
6	0.0202	0.0223	0.0264	0.0323	0.0381	0.0435	0.0486	0.0534
7	0.0151	0.0167	0.0198	0.0243	0.0288	0.0329	0.0370	0.0409
8	0.0117	0.0129	0.0154	0.0189	0.0224	0.0258	0.0290	0.0322
9	0.0093	0.0103	0.0122	0.0151	0.0179	0.0206	0.0233	0.0259
10	0.0076	0.0084	0.0100	0.0123	0.0147	0.0169	0.0191	0.0213
12	0.0053	0.0059	0.0070	0.0086	0.0103	0.0119	0.0135	0.0151
14	0.0039	0.0043	0.0052	0.0064	0.0076	0.0088	0.0100	0.0112
16	0.0030	0.0033	0.0040	0.0049	0.0059	0.0068	0.0077	0.0087
20	0.0019	0.0021	0.0025	0.0032	0.0038	0.0044	0.0050	0.0056
24	0.0013	0.0015	0.0018	0.0022	0.0026	0.0030	0.0035	0.0039
28	0.0010	0.0011	0.0013	0.0016	0.0019	0.0022	0.0026	0.0029
32	0.0008	0.0008	0.0010	0.0012	0.0015	0.0017	0.0020	0.0022

TABLE 50.7

Diameter–Thickness Ratios for Laying Condition Type 1*

Trench Load (P_v)—psi			Trench Load (P_v)—psi		
Bending Stress Design	Deflection Design	$\frac{D\dagger}{t}$ or $\frac{D}{t_1}$	Bending Stress Design	Deflection Design	$\frac{D\dagger}{t}$ or $\frac{D}{t_1}$
4.40	3.46	170	6.33	4.71	128
4.43	3.48	169	6.40	4.76	127
4.46	3.50	168	6.46	4.82	126
4.50	3.51	167			
4.54	3.53	166	6.53	4.87	125
			6.60	4.93	124
4.57	3.55	165	6.67	4.99	123
4.61	3.57	164	6.74	5.05	122
4.64	3.59	163	6.82	5.11	121
4.68	3.61	162			
4.72	3.63	161	6.89	5.18	120
			6.97	5.25	119
4.76	3.65	160	7.05	5.32	118
4.80	3.67	159	7.13	5.39	117
4.84	3.69	158	7.21	5.46	116
4.88	3.71	157			
4.92	3.74	156	7.29	5.54	115
			7.38	5.62	114
4.96	3.76	155	7.47	5.71	113
5.00	3.78	154	7.56	5.79	112
5.04	3.81	153	7.65	5.88	111
5.08	3.83	152			
5.13	3.86	151	7.75	5.97	110
			7.85	6.07	109
5.17	3.89	150	7.95	6.17	108
5.21	3.91	149	8.05	6.27	107
5.26	3.94	148	8.16	6.38	106
5.30	3.97	147			
5.35	4.00	146	8.27	6.49	105
			8.38	6.61	104
5.40	4.03	145	8.49	6.73	103
5.45	4.06	144	8.61	6.86	102
5.49	4.09	143	8.74	6.99	101
5.54	4.13	142			
5.59	4.16	141	8.86	7.12	100
			8.99	7.26	99
5.65	4.20	140	9.13	7.41	98
5.70	4.23	139	9.27	7.57	97
5.75	4.27	138	9.41	7.73	96
5.80	4.31	137			
5.86	4.35	136	9.56	7.89	95
			9.71	8.07	94
			9.87	8.25	93
5.91	4.39	135	10.03	8.44	92
5.97	4.43	134	10.20	8.64	91
6.03	4.47	133			
6.09	4.52	132	10.37	8.85	90
6.15	4.56	131	10.55	9.06	89
			10.74	9.29	88
6.21	4.61	130	10.93	9.53	87
6.27	4.66	129	11.13	9.78	86

1
5
8

TABLE 50.7—(cont.)

Trench Load (P_v)—psi			Trench Load (P_v)—psi		
Bending Stress Design	Deflection Design	$\frac{D\dagger}{t}$ or $\frac{D}{t_1}$	Bending Stress Design	Deflection Design	$\frac{D\dagger}{t}$ or $\frac{D}{t_1}$
11.34	10.04	85	21.91	26.54	58
11.55	10.31	84	22.63	27.85	57
11.78	10.60	83	23.38	29.26	56
12.01	10.90	82			
12.25	11.22	81	24.18	30.77	55
			25.02	32.39	54
12.50	11.56	80	25.92	34.15	53
12.76	11.91	79	26.86	36.05	52
13.03	12.28	78	27.87	38.10	51
13.31	12.67	77			
13.60	13.08	76	28.94	40.32	50
			30.07	42.73	49
13.91	13.51	75	31.28	45.35	48
14.23	13.97	74	32.57	48.20	47
14.56	14.45	73	33.95	51.31	46
14.91	14.96	72			
15.27	15.50	71	35.42	54.72	45
			37.00	58.44	44
15.65	16.07	70	38.69	62.53	43
16.05	16.68	69	40.50	67.03	42
16.46	17.32	68	42.46	71.99	41
16.89	18.00	67			
17.35	18.73	66	44.56	77.47	40
			46.84	83.54	39
17.83	19.50	65	49.30	90.28	38
18.33	20.32	64	51.96	97.80	37
18.85	21.19	63	54.86	106.20	36
19.40	22.12	62			
19.98	23.12	61	58.02	115.62	35
			61.46	126.21	34
			65.23	138.18	33
20.59	24.18	60	69.36	151.73	32
21.23	25.32	59	73.92	167.15	31

* $E' = 150$; $K_b = 0.235$; $K_x = 0.108$.

† The $\frac{D}{t}$ for the tabulated P_v nearest to the calculated P_v is selected; when the calculated P_v is halfway between two tabulated values, the smaller $\frac{D}{t}$ should be used.

1
5
9

TABLE 50.8

Diameter–Thickness Ratios for Laying Condition Type 2*

Trench Load (P_v)—psi			Trench Load (P_v)—psi		
Bending Stress Design	Deflection Design	$\frac{D\dagger}{t}$ or $\frac{D}{t_1}$	Bending Stress Design	Deflection Design	$\frac{D\dagger}{t}$ or $\frac{D}{t_1}$
6.29	6.18	170	8.99	7.46	128
6.34	6.19	169	9.07	7.51	127
6.39	6.21	168	9.16	7.57	126
6.44	6.23	167			
6.50	6.25	166	9.25	7.63	125
			9.33	7.69	124
6.55	6.26	165	9.42	7.75	123
6.60	6.28	164	9.51	7.81	122
6.66	6.30	163	9.60	7.87	121
6.71	6.32	162			
6.77	6.34	161	9.70	7.94	120
			9.79	8.01	119
6.82	6.37	160	9.89	8.08	118
6.88	6.39	159	9.99	8.16	117
6.94	6.41	158	10.09	8.23	116
6.99	6.43	157			
7.05	6.46	156	10.19	8.31	115
			10.29	8.40	114
7.11	6.48	155	10.40	8.48	113
7.17	6.50	154	10.51	8.57	112
7.23	6.53	153	10.62	8.66	111
7.29	6.56	152			
7.35	6.58	151	10.73	8.76	110
			10.84	8.86	109
7.42	6.61	150	10.96	8.96	108
7.48	6.64	149	11.08	9.07	107
7.54	6.67	148	11.21	9.18	106
7.61	6.70	147			
7.67	6.73	146	11.33	9.29	105
			11.46	9.41	104
7.74	6.76	145	11.59	9.54	103
7.80	6.79	144	11.73	9.67	102
7.87	6.83	143	11.87	9.80	101
7.94	6.86	142			
8.01	6.89	141	12.01	9.94	100
			12.16	10.09	99
8.08	6.93	140	12.31	10.24	98
8.15	6.97	139	12.46	10.40	97
8.22	7.01	138	12.62	10.56	96
8.29	7.05	137			
8.37	7.09	136	12.79	10.73	95
			12.96	10.91	94
			13.13	11.10	93
8.44	7.13	135	13.31	11.29	92
8.52	7.17	134	13.49	11.50	91
8.59	7.22	133			
8.67	7.26	132	13.68	11.71	90
8.75	7.31	131	13.88	11.94	89
			14.08	12.17	88
8.83	7.36	130	14.30	12.42	87
8.91	7.41	129	14.51	12.67	86

TABLE 50.8—(*cont.*)

Trench Load (P_v)—*psi*			Trench Load (P_v)—*psi*		
Bending Stress Design	Deflection Design	$\frac{D\dagger}{t}$ or $\frac{D}{t_1}$	Bending Stress Design	Deflection Design	$\frac{D\dagger}{t}$ or $\frac{D}{t_1}$
14.74	12.94	85	26.95	31.26	57
14.97	13.22	84	27.77	32.71	56
15.21	13.52	83			
15.46	13.83	82	28.64	34.26	55
15.72	14.16	81	29.56	35.93	54
			30.53	37.74	53
15.99	14.50	80	31.57	39.69	52
16.28	14.86	79	32.67	41.80	51
16.57	15.24	78			
16.87	15.64	77	33.84	44.09	50
17.19	16.06	76	35.08	46.56	49
			36.41	49.26	48
17.52	16.51	75	37.83	52.19	47
17.86	16.98	74	39.34	55.40	46
18.22	17.48	73			
18.59	18.00	72			
18.98	18.56	71	40.96	58.89	45
			42.70	62.73	44
19.39	19.14	70	44.57	66.93	43
19.82	19.77	69	46.57	71.56	42
20.27	20.43	68	48.73	76.66	41
20.73	21.13	67			
21.23	21.87	66	51.06	82.29	40
			53.57	88.54	39
21.74	22.67	65	56.30	95.48	38
22.28	23.51	64	59.25	103.21	37
22.85	24.41	63	62.46	111.85	36
23.45	25.37	62			
24.07	26.39	61	65.96	121.54	35
			69.79	132.44	34
24.74	27.49	60	73.98	144.74	33
25.43	28.66	59	78.57	158.68	32
26.17	29.91	58	83.64	174.54	31

* $E' = 300$; $K_b = 0.210$; $K_z = 0.105$.

† The $\frac{D}{t}$ for the tabulated P_v nearest to the calculated P_v is selected; when the calculated P_v is halfway between two tabulated values, the smaller $\frac{D}{t}$ should be used.

15

TABLE 50.9

Diameter–Thickness Ratios for Laying Condition Type 3*

Trench Load (P_v)—psi			Trench Load (P_v)—psi		
Bending Stress Design	Deflection Design	$\frac{D\dagger}{t}$ or $\frac{D}{t_1}$	Bending Stress Design	Deflection Design	$\frac{D\dagger}{t}$ or $\frac{D}{t_1}$
3.25	7.26	310	5.62	7.52	226
3.29	7.27	308	5.70	7.53	224
3.33	7.27	306	5.79	7.54	222
3.37	7.27	304			
3.41	7.28	302	5.87	7.55	220
			5.96	7.56	218
3.45	7.28	300	6.04	7.58	216
3.49	7.28	298	6.13	7.59	214
3.54	7.29	296	6.22	7.60	212
3.58	7.29	294			
3.63	7.30	292	6.32	7.62	210
			6.41	7.63	208
3.67	7.30	290	6.51	7.65	206
3.72	7.30	288	6.60	7.66	204
3.76	7.31	286	6.70	7.68	202
3.81	7.31	284			
3.86	7.32	282	6.80	7.70	200
			6.91	7.72	198
3.91	7.32	280	7.01	7.74	196
3.96	7.33	278	7.12	7.76	194
4.01	7.33	276	7.22	7.78	192
4.06	7.34	274			
4.11	7.34	272	7.33	7.80	190
			7.45	7.82	188
4.17	7.35	270	7.56	7.84	186
4.22	7.35	268	7.68	7.87	184
4.28	7.36	266	7.80	7.89	182
4.33	7.36	264			
4.39	7.37	262	7.92	7.92	180
			8.04	7.95	178
4.45	7.38	260	8.16	7.98	176
4.51	7.38	258	8.29	8.01	174
4.57	7.39	256	8.42	8.04	172
4.63	7.39	254			
4.69	7.40	252	8.55	8.07	170
			8.62	8.09	169
4.76	7.41	250	8.69	8.11	168
4.82	7.42	248	8.75	8.13	167
4.89	7.42	246	8.82	8.14	166
4.96	7.43	244			
5.03	7.44	242	8.89	8.16	165
			8.96	8.18	164
5.10	7.45	240	9.03	8.20	163
5.17	7.46	238	9.10	8.22	162
5.24	7.47	236	9.17	8.24	161
5.31	7.48	234			
5.39	7.48	232	9.25	8.27	160
			9.32	8.29	159
			9.39	8.31	158
5.47	7.49	230	9.47	8.33	157
5.54	7.51	228	9.54	8.36	156

TABLE 50.9—(cont.)

Trench Load (P_v)—psi			Trench Load (P_v)—psi		
Bending Stress Design	Deflection Design	$\dfrac{D\dagger}{t}$ or $\dfrac{D}{t_1}$	Bending Stress Design	Deflection Design	$\dfrac{D\dagger}{t}$ or $\dfrac{D}{t_1}$
9.62	8.38	155	13.71	10.51	112
9.69	8.41	154	13.83	10.61	111
9.77	8.43	153			
9.85	8.46	152	13.96	10.71	110
9.92	8.49	151	14.09	10.81	109
			14.22	10.91	108
10.00	8.52	150	14.36	11.02	107
10.08	8.54	149	14.50	11.13	106
10.16	8.57	148			
10.24	8.60	147			
10.33	8.64	146	14.64	11.25	105
			14.78	11.37	104
			14.93	11.50	103
10.41	8.67	145	15.08	11.63	102
10.49	8.70	144	15.23	11.77	101
10.58	8.73	143			
10.66	8.77	142			
10.75	8.81	141	15.39	11.91	100
			15.55	12.06	99
10.83	8.84	140	15.71	12.21	98
10.92	8.88	139	15.88	12.37	97
11.01	8.92	138	16.06	12.54	96
11.10	8.96	137			
11.19	9.00	136	16.23	12.72	95
			16.42	12.90	94
11.28	9.04	135	16.61	13.09	93
11.37	9.09	134	16.80	13.29	92
11.46	9.13	133	17.00	13.50	91
11.56	9.18	132			
11.65	9.23	131	17.21	13.72	90
			17.42	13.95	89
11.75	9.28	130	17.64	14.18	88
11.84	9.33	129	17.86	14.43	87
11.94	9.38	128	18.10	14.70	86
12.04	9.44	127			
12.14	9.49	126	18.34	14.97	85
			18.59	15.26	84
12.25	9.55	125	18.85	15.56	83
12.35	9.61	124	19.12	15.88	82
12.45	9.67	123	19.40	16.21	81
12.56	9.74	122			
12.67	9.80	121	19.68	16.56	80
			19.99	16.93	79
12.78	9.87	120	20.30	17.31	78
12.89	9.94	119	20.62	17.72	77
13.00	10.02	118	20.96	18.15	76
13.11	10.09	117			
13.23	10.17	116	21.31	18.61	75
			21.68	19.09	74
13.34	10.25	115	22.07	19.59	73
13.46	10.34	114	22.47	20.13	72
13.58	10.42	113	22.88	20.69	71

1
6
3

TABLE 50.9—(cont.)

Trench Load (P_v)—psi			Trench Load (P_v)—psi		
Bending Stress Design	Deflection Design	$\frac{D\dagger}{t}$ or $\frac{D}{t_1}$	Bending Stress Design	Deflection Design	$\frac{D\dagger}{t}$ or $\frac{D}{t_1}$
23.32	21.29	70	38.97	46.72	50
23.78	21.93	69	40.33	49.25	49
24.26	22.60	68	41.78	51.99	48
24.76	23.32	67	43.33	54.98	47
25.29	24.08	66	44.98	58.25	46
25.85	24.88	65	46.76	61.81	45
26.43	25.74	64	48.66	65.72	44
27.04	26.66	63	50.71	70.01	43
27.68	27.64	62	52.91	74.72	42
28.36	28.68	61	55.28	79.92	41
29.08	29.80	60	57.84	85.67	40
29.83	30.99	59	60.61	92.04	39
30.63	32.27	58	63.61	99.11	38
31.47	33.64	57	66.86	106.99	37
32.36	35.12	56	70.40	115.80	36
33.31	36.70	55	74.27	125.67	35
34.30	38.41	54	78.49	136.78	34
35.37	40.25	53	83.11	149.32	33
36.49	42.24	52	88.19	163.54	32
37.69	44.39	51	93.79	179.71	31

* $E' = 400$; $K_b = 0.189$; $K_x = 0.103$.

† The $\frac{D}{t}$ for the tabulated P_v nearest to the calculated P_v is selected; when the calculated P_v is halfway between two tabulated values, the smaller $\frac{D}{t}$ should be used.

TABLE 50.10

Diameter–Thickness Ratios for Laying Condition Type 4*

Trench Load (P_v)—psi			Trench Load (P_v)—psi		
Bending Stress Design	Deflection Design	$\dfrac{D\dagger}{t}$ or $\dfrac{D}{t_1}$	Bending Stress Design	Deflection Design	$\dfrac{D\dagger}{t}$ or $\dfrac{D}{t_1}$
5.93	9.70	310	9.95	9.97	226
6.00	9.70	308	10.08	9.98	224
6.07	9.71	306	10.21	9.99	222
6.14	9.71	304			
6.21	9.71	302	10.35	10.01	220
			10.49	10.02	218
6.29	9.72	300	10.62	10.03	216
6.36	9.72	298	10.77	10.05	214
6.43	9.73	296	10.91	10.06	212
6.51	9.73	294			
6.59	9.73	292	11.05	10.08	210
			11.20	10.09	208
6.67	9.74	290	11.35	10.11	206
6.74	9.74	288	11.50	10.13	204
6.83	9.75	286	11.66	10.15	202
6.91	9.75	284			
6.99	9.76	282	11.81	10.17	200
			11.97	10.19	198
7.08	9.76	280	12.13	10.21	196
7.16	9.77	278	12.29	10.23	194
7.25	9.77	276	12.45	10.25	192
7.34	9.78	274			
7.43	9.78	272	12.62	10.27	190
			12.79	10.30	188
7.52	9.79	270	12.96	10.32	186
7.61	9.79	268	13.13	10.35	184
7.71	9.80	266	13.30	10.37	182
7.80	9.81	264			
7.90	9.81	262	13.48	10.40	180
			13.66	10.43	178
8.00	9.82	260	13.84	10.46	176
8.10	9.83	258	14.02	10.50	174
8.20	9.83	256	14.20	10.53	172
8.31	9.84	254			
8.41	9.85	252	14.39	10.57	170
			14.48	10.59	169
8.52	9.86	250	14.57	10.60	168
8.63	9.86	248	14.67	10.62	167
8.74	9.87	246	14.76	10.64	166
8.85	9.88	244			
8.97	9.89	242	14.86	10.66	165
			14.96	10.69	164
9.08	9.90	240	15.05	10.71	163
9.20	9.91	238	15.15	10.73	162
9.32	9.92	236	15.25	10.75	161
9.44	9.93	234			
9.57	9.94	232	15.34	10.78	160
			15.44	10.80	159
			15.54	10.82	158
9.69	9.95	230	15.64	10.85	157
9.82	9.96	228	15.74	10.87	156

TABLE 50.10—(cont.)

Trench Load (P_v)—psi			Trench Load (P_v)—psi		
Bending Stress Design	Deflection Design	$\frac{D\dagger}{t}$ or $\frac{D}{t_1}$	Bending Stress Design	Deflection Design	$\frac{D\dagger}{t}$ or $\frac{D}{t_1}$
15.84	10.90	155	20.69	13.19	112
15.94	10.93	154	20.82	13.29	111
16.04	10.96	153			
16.14	10.98	152	20.96	13.39	110
16.24	11.01	151	21.10	13.50	109
			21.24	13.61	108
16.34	11.04	150	21.39	13.73	107
16.45	11.07	149	21.54	13.85	106
16.55	11.11	148			
16.65	11.14	147	21.69	13.98	105
16.76	11.17	146	21.84	14.11	104
			22.00	14.24	103
16.86	11.21	145	22.16	14.38	102
16.96	11.24	144	22.32	14.53	101
17.07	11.28	143			
17.18	11.31	142	22.49	14.68	100
17.28	11.35	141	22.66	14.84	99
			22.83	15.01	98
17.39	11.39	140	23.01	15.18	97
17.50	11.43	139	23.20	15.36	96
17.60	11.48	138			
17.71	11.52	137			
17.82	11.56	136	23.38	15.55	95
			23.58	15.75	94
17.93	11.61	135	23.78	15.95	93
18.04	11.66	134	23.99	16.17	92
18.15	11.71	133	24.20	16.39	91
18.26	11.76	132			
18.37	11.81	131	24.42	16.62	90
			24.64	16.87	89
18.49	11.86	130	24.88	17.12	88
18.60	11.92	129	25.12	17.39	87
18.72	11.97	128	25.37	17.67	86
18.83	12.03	127			
18.95	12.09	126	25.63	17.97	85
			25.90	18.28	84
19.06	12.15	125	26.18	18.60	83
19.18	12.22	124	26.47	18.94	82
19.30	12.28	123	26.77	19.30	81
19.42	12.35	122			
19.54	12.42	121	27.09	19.67	80
			27.42	20.07	79
19.66	12.50	120	27.76	20.48	78
19.78	12.57	119	28.11	20.92	77
19.91	12.65	118	28.49	21.38	76
20.04	12.73	117			
20.16	12.82	116	28.87	21.87	75
			29.28	22.38	74
20.29	12.91	115	29.70	22.93	73
20.42	13.00	114	30.15	23.50	72
20.55	13.09	113	30.62	24.11	71

1
6
6

TABLE 50.10—(cont.)

Trench Load (P_v)—psi			Trench Load (P_v)—psi		
Bending Stress Design	Deflection Design	$\frac{D\dagger}{t}$ or $\frac{D}{t_1}$	Bending Stress Design	Deflection Design	$\frac{D\dagger}{t}$ or $\frac{D}{t_1}$
31.11	24.75	70	49.11	52.03	50
31.62	25.43	69	50.70	54.74	49
32.16	26.16	68	52.41	57.69	48
32.72	26.92	67	54.23	60.90	47
33.32	27.74	66	56.18	64.40	46
33.95	28.60	65	58.27	68.23	45
34.61	29.53	64	60.52	72.42	44
35.30	30.51	63	62.93	77.02	43
36.04	31.56	62	65.54	82.08	42
36.81	32.68	61	68.35	87.66	41
37.63	33.88	60	71.39	93.82	40
38.50	35.16	59	74.67	100.65	39
39.42	36.53	58	78.24	108.24	38
40.39	38.00	57	82.11	116.70	37
41.42	39.58	56	86.33	126.15	36
42.51	41.28	55	90.93	136.74	35
43.67	43.12	54	95.97	148.66	34
44.91	45.09	53	101.49	162.12	33
46.22	47.22	52	107.56	177.37	32
47.62	49.53	51	114.25	194.72	31

1
6
7

* $E' = 500$; $K_b = 0.157$; $K_x = 0.096$.

† The $\frac{D}{t}$ for the tabulated P_v nearest to the calculated P_v is selected; when the calculated P_v is halfway between two tabulated values, the smaller $\frac{D}{t}$ should be used.

TABLE 50.11

Diameter–Thickness Ratios for Laying Condition Type 5*

Trench Load (P_v)—psi			Trench Load (P_v)—psi		
Bending Stress Design	Deflection Design	$\frac{D\dagger}{t}$ or $\frac{D}{t_1}$	Bending Stress Design	Deflection Design	$\frac{D\dagger}{t}$ or $\frac{D}{t_1}$
3.06	15.09	660	6.41	15.13	450
3.10	15.09	655	6.54	15.14	445
3.15	15.09	650	6.68	15.14	440
3.20	15.09	645			
3.25	15.09	640	6.83	15.14	435
			6.98	15.14	430
3.30	15.09	635	7.13	15.14	425
3.35	15.09	630	7.29	15.15	420
3.40	15.09	625	7.46	15.15	415
3.46	15.09	620			
3.51	15.09	615	7.63	15.15	410
			7.80	15.16	405
3.57	15.10	610	7.98	15.16	400
3.63	15.10	605	8.17	15.16	395
3.68	15.10	600	8.36	15.17	390
3.75	15.10	595			
3.81	15.10	590	8.56	15.17	385
			8.77	15.17	380
3.87	15.10	585	8.98	15.18	375
3.94	15.10	580	9.20	15.18	370
4.00	15.10	575	9.43	15.19	365
4.07	15.10	570			
4.14	15.10	565	9.66	15.19	360
			9.91	15.20	355
4.21	15.10	560	10.16	15.20	350
4.29	15.10	555	10.42	15.21	345
4.36	15.10	550	10.69	15.22	340
4.44	15.11	545			
4.52	15.11	540	10.97	15.22	335
			11.26	15.23	330
4.60	15.11	535	11.56	15.24	325
4.69	15.11	530	11.87	15.24	320
4.77	15.11	525	12.19	15.25	315
4.86	15.11	520			
4.95	15.11	515	12.52	15.26	310
			12.66	15.27	308
5.05	15.11	510	12.80	15.27	306
5.14	15.11	505	12.94	15.27	304
5.24	15.12	500	13.08	15.28	302
5.35	15.12	495			
5.45	15.12	490	13.23	15.28	300
			13.37	15.29	298
5.56	15.12	485	13.52	15.29	296
5.67	15.12	480	13.67	15.30	294
5.78	15.12	475	13.83	15.30	292
5.90	15.13	470			
6.02	15.13	465	13.98	15.30	290
			14.14	15.31	288
			14.30	15.31	286
6.15	15.13	460	14.46	15.32	284
6.28	15.13	455	14.62	15.33	282

TABLE 50.11—(cont.)

Trench Load (P_v)—psi			Trench Load (P_v)—psi		
Bending Stress Design	Deflection Design	$\frac{D\dagger}{t}$ or $\frac{D}{i_1}$	Bending Stress Design	Deflection Design	$\frac{D\dagger}{t}$ or $\frac{D}{i_1}$
14.79	15.33	280	24.24	15.86	194
14.96	15.34	278	24.50	15.88	192
15.13	15.34	276			
15.30	15.35	274	24.77	15.91	190
15.48	15 35	272	25.04	15.93	188
			25.31	15.96	186
15.65	15.36	270	25.59	15.99	184
15.83	15.37	268	25.86	16.02	182
16.02	15.37	266			
16.20	15.38	264			
16.39	15.39	262	26.13	16.06	180
			26.41	16.09	178
16.58	15.40	260	26.68	16.12	176
16.77	15.40	258	26.96	16.16	174
16.97	15.41	256	27.23	16.20	172
17.16	15.42	254			
17.36	15.43	252	27.51	16.24	170
			27.65	16.26	169
17.57	15.44	250	27.78	16.28	168
17.77	15.45	248	27.92	16.31	167
17.98	15.45	246	28.06	16.33	166
18.19	15.46	244			
18.40	15.47	242	28.19	16.35	165
			28.33	16.37	164
18.62	15.48	240	28.47	16.40	163
18.84	15.49	238	28.60	16.42	162
19.06	15.51	236	28.74	16.45	161
19.28	15.52	234			
19.51	15.53	232	28.87	16.48	160
			29.01	16.50	159
19.73	15.54	230	29.15	16.53	158
19.97	15.55	228	29.28	16.56	157
20.20	15.57	226	29.41	16.59	156
20.43	15.58	224			
20.67	15.59	222	29.55	16.62	155
			29.68	16.65	154
20.91	15.61	220	29.82	16.68	153
21.16	15.62	218	29.95	16.71	152
21.40	15.64	216	30.08	16.74	151
21.65	15.65	214			
21.90	15.67	212	30.21	16.78	150
			30.34	16.81	149
22.15	15.69	210	30.48	16.85	148
22.40	15.71	208	30.61	16.89	147
22.66	15.73	206	30.74	16.92	146
22.92	15.75	204			
23.18	15.77	202	30.87	16.96	145
			30.99	17.00	144
23.44	15.79	200	31.12	17.04	143
23.70	15.81	198	31.25	17.09	142
23.97	15.83	196	31.38	17.13	141

169

TABLE 50.11—(cont.)

Trench Load (P_v)—psi			Trench Load (P_v)—psi		
Bending Stress Design	Deflection Design	$\frac{D\dagger}{t}$ or $\frac{D}{t_1}$	Bending Stress Design	Deflection Design	$\frac{D\dagger}{t}$ or $\frac{D}{t_1}$
31.50	17.17	140	37.01	21.45	97
31.63	17.22	139	37.17	21.66	96
31.76	17.27	138			
31.88	17.32	137	37.34	21.87	95
32.01	17.37	136	37.52	22.09	94
			37.70	22.32	93
32.13	17.42	135	37.89	22.56	92
32.25	17.47	134	38.08	22.82	91
32.38	17.53	133			
32.50	17.58	132	38.28	23.08	90
32.62	17.64	131	38.49	23.36	89
			38.71	23.65	88
32.75	17.70	130	38.93	23.95	87
32.87	17.76	129	39.17	24.27	86
32.99	17.83	128			
33.11	17.89	127	39.41	24.60	85
33.23	17.96	126	39.67	24.95	84
			39.94	25.31	83
33.35	18.03	125	40.22	25.70	82
33.47	18.11	124	40.51	26.10	81
33.59	18.18	123			
33.71	18.26	122	40.82	26.52	80
33.83	18.34	121	41.14	26.97	79
			41.48	27.44	78
33.95	18.42	120	41.84	27.93	77
34.07	18.51	119	42.21	28.46	76
34.19	18.60	118			
34.31	18.69	117	42.60	29.01	75
34.43	18.78	116	43.02	29.59	74
			43.45	30.20	73
34.55	18.88	115	43.92	30.85	72
34.68	18.98	114	44.40	31.53	71
34.80	19.09	113			
34.92	19.20	112	44.91	32.26	70
35.05	19.31	111	45.46	33.03	69
			46.03	33.85	68
35.17	19.43	110	46.64	34.71	67
35.30	19.55	109	47.28	35.63	66
35.43	19.68	108			
35.56	19.81	107	47.96	36.61	65
35.69	19.95	106	48.68	37.65	64
			49.44	38.77	63
35.83	20.09	105	50.25	39.95	62
35.96	20.24	104	51.11	41.21	61
36.10	20.39	103			
36.25	20.55	102	52.02	42.57	60
36.39	20.72	101	52.99	44.01	59
			54.02	45.56	58
36.54	20.89	100	55.12	47.23	57
36.69	21.07	99	56.28	49.01	56
36.85	21.26	98			

TABLE 50.11—(cont.)

Trench Load (P_v)—psi			Trench Load (P_v)—psi		
Bending Stress Design	Deflection Design	$\frac{D\dagger}{t}$ or $\frac{D}{t_1}$	Bending Stress Design	Deflection Design	$\frac{D\dagger}{t}$ or $\frac{D}{t_1}$
57.53	50.93	55	84.50	97.01	42
58.86	53.00	54	87.85	103.31	41
60.28	55.23	53			
61.79	57.64	52	91.47	110.27	40
63.41	60.25	51	95.40	117.98	39
65.14	63.07	50	99.67	126.56	38
67.00	66.13	49	104.32	136.11	37
68.99	69.46	48	109.40	146.78	36
71.12	73.09	47			
73.41	77.04	46	114.94	158.75	35
			121.02	172.21	34
75.88	81.36	45	127.69	187.41	33
78.54	86.10	44	135.03	204.63	32
81.40	91.29	43	143.14	224.22	31

* $E' = 700$; $K_b = 0.128$; $K_x = 0.085$.

† The $\frac{D}{t}$ for the tabulated P_v nearest to the calculated P_v is selected; when the calculated P_v is halfway between two tabulated values, the smaller $\frac{D}{t}$ should be used.

TABLE 50.12

Thickness for Earth Load Plus Truck Load

Size in.	Depth of Cover ft	Laying Condition									
		Type 1		Type 2		Type 3		Type 4		Type 5	
		Total Calculated Thickness* in.	Use Class	Total Calculated Thickness* in.	Use Class	Total Calculated Thickness* in.	Use Class	Total Calculated Thickness* in.	Use Class	Total Calculated Thickness* in.	Use Class
3	2.5	0.18	51	0.17	51	0.16	51	0.15	51	0.14	51
	3	0.17	51	0.16	51	0.16	51	0.15	51	0.14	51
	4	0.17	51	0.16	51	0.15	51	0.14	51	0.14	51
	5	0.16	51	0.16	51	0.15	51	0.14	51	0.14	51
	6	0.16	51	0.16	51	0.15	51	0.14	51	0.14	51
	7	0.16	51	0.16	51	0.15	51	0.14	51	0.14	51
	8	0.17	51	0.16	51	0.15	51	0.15	51	0.14	51
	9	0.17	51	0.16	51	0.15	51	0.15	51	0.14	51
	10	0.17	51	0.16	51	0.15	51	0.15	51	0.14	51
	12	0.17	51	0.17	51	0.16	51	0.15	51	0.14	51
	14	0.18	51	0.17	51	0.16	51	0.15	51	0.14	51
	16	0.18	51	0.17	51	0.17	51	0.15	51	0.14	51
	20	0.19	51	0.18	51	0.17	51	0.16	51	0.15	51
	24	0.19	51	0.19	51	0.18	51	0.16	51	0.15	51
	28	0.20	51	0.19	51	0.19	51	0.17	51	0.15	51
	32	0.21	51	0.20	51	0.19	51	0.18	51	0.15	51
4	2.5	0.19	51	0.18	51	0.17	51	0.15	51	0.15	51
	3	0.18	51	0.17	51	0.16	51	0.15	51	0.14	51
	4	0.17	51	0.16	51	0.16	51	0.15	51	0.14	51
	5	0.17	51	0.16	51	0.15	51	0.15	51	0.14	51
	6	0.17	51	0.16	51	0.15	51	0.15	51	0.14	51
	7	0.17	51	0.16	51	0.16	51	0.15	51	0.14	51
	8	0.17	51	0.16	51	0.16	51	0.15	51	0.14	51
	9	0.18	51	0.17	51	0.16	51	0.15	51	0.14	51
	10	0.18	51	0.17	51	0.16	51	0.15	51	0.14	51
	12	0.18	51	0.17	51	0.16	51	0.15	51	0.14	51
	14	0.19	51	0.18	51	0.17	51	0.15	51	0.15	51
	16	0.19	51	0.18	51	0.17	51	0.16	51	0.15	51
	20	0.20	51	0.19	51	0.18	51	0.16	51	0.15	51
	24	0.21	51	0.20	51	0.19	51	0.17	51	0.15	51
	28	0.22	51	0.21	51	0.20	51	0.18	51	0.15	51
	32	0.22	51	0.21	51	0.21	51	0.19	51	0.16	51
6	2.5	0.21	50	0.20	50	0.18	50	0.16	50	0.15	50
	3	0.20	50	0.19	50	0.18	50	0.16	50	0.15	50
	4	0.19	50	0.18	50	0.17	50	0.16	50	0.15	50
	5	0.19	50	0.17	50	0.17	50	0.15	50	0.15	50
	6	0.19	50	0.17	50	0.17	50	0.15	50	0.15	50
	7	0.19	50	0.18	50	0.17	50	0.16	50	0.15	50
	8	0.19	50	0.18	50	0.17	50	0.16	50	0.15	50
	9	0.20	50	0.18	50	0.17	50	0.16	50	0.15	50
	10	0.20	50	0.18	50	0.17	50	0.16	50	0.15	50
	12	0.21	50	0.19	50	0.18	50	0.16	50	0.15	50
	14	0.21	50	0.20	50	0.18	50	0.17	50	0.15	50
	16	0.22	50	0.21	50	0.19	50	0.17	50	0.15	50
	20	0.23	50	0.22	50	0.21	50	0.18	50	0.16	50
	24	0.24	50	0.23	50	0.22	50	0.19	50	0.16	50
	28	0.25	50	0.24	50	0.23	50	0.20	50	0.16	50
	32	0.26	50	0.25	50	0.24	50	0.22	50	0.17	50

* Total calculated thickness includes service allowance and casting tolerance added to net thickness.

TABLE 50.12—(cont.)

Size in.	Depth of Cover ft	Laying Condition									
		Type 1		Type 2		Type 3		Type 4		Type 5	
		Total Calculated Thickness* in.	Use Class	Total Calculated Thickness* in.	Use Class	Total Calculated Thickness* in.	Use Class	Total Calculated Thickness* in.	Use Class	Total Calculated Thickness* in.	Use Class
8	2.5	0.24	50	0.22	50	0.20	50	0.18	50	0.16	50
	3	0.23	50	0.21	50	0.19	50	0.17	50	0.16	50
	4	0.21	50	0.19	50	0.18	50	0.16	50	0.15	50
	5	0.21	50	0.19	50	0.18	50	0.16	50	0.15	50
	6	0.21	50	0.19	50	0.18	50	0.16	50	0.15	50
	7	0.21	50	0.19	50	0.18	50	0.16	50	0.15	50
	8	0.21	50	0.19	50	0.18	50	0.16	50	0.15	50
	9	0.22	50	0.20	50	0.18	50	0.17	50	0.15	50
	10	0.22	50	0.20	50	0.19	50	0.17	50	0.15	50
	12	0.23	50	0.21	50	0.19	50	0.17	50	0.16	50
	14	0.24	50	0.22	50	0.20	50	0.18	50	0.16	50
	16	0.25	50	0.23	50	0.21	50	0.18	50	0.16	50
	20	0.27	50	0.25	50	0.23	50	0.19	50	0.17	50
	24	0.28	50	0.26	50	0.24	50	0.21	50	0.17	50
	28	0.29	51	0.28	50	0.26	50	0.23	50	0.18	50
	32	0.30	51	0.29	51	0.27	50	0.24	50	0.18	50
10	2.5	0.27	50	0.25	50	0.23	50	0.20	50	0.17	50
	3	0.26	50	0.23	50	0.21	50	0.19	50	0.17	50
	4	0.24	50	0.22	50	0.20	50	0.18	50	0.17	50
	5	0.23	50	0.21	50	0.20	50	0.18	50	0.17	50
	6	0.23	50	0.21	50	0.20	50	0.18	50	0.17	50
	7	0.24	50	0.21	50	0.20	50	0.18	50	0.17	50
	8	0.24	50	0.22	50	0.20	50	0.18	50	0.17	50
	9	0.25	50	0.22	50	0.20	50	0.18	50	0.17	50
	10	0.25	50	0.23	50	0.21	50	0.19	50	0.17	50
	12	0.26	50	0.24	50	0.22	50	0.19	50	0.17	50
	14	0.28	50	0.25	50	0.23	50	0.20	50	0.18	50
	16	0.29	50	0.26	50	0.24	50	0.20	50	0.18	50
	20	0.31	51	0.28	50	0.26	50	0.22	50	0.18	50
	24	0.32	51	0.30	50	0.28	50	0.24	50	0.19	50
	28	0.34	52	0.32	51	0.30	50	0.26	50	0.20	50
	32	0.35	52	0.33	51	0.32	51	0.28	50	0.20	50
12	2.5	0.30	50	0.27	50	0.24	50	0.21	50	0.18	50
	3	0.28	50	0.25	50	0.23	50	0.20	50	0.18	50
	4	0.26	50	0.23	50	0.21	50	0.19	50	0.17	50
	5	0.25	50	0.23	50	0.21	50	0.19	50	0.17	50
	6	0.25	50	0.23	50	0.21	50	0.19	50	0.17	50
	7	0.26	50	0.23	50	0.21	50	0.19	50	0.17	50
	8	0.26	50	0.23	50	0.21	50	0.19	50	0.17	50
	9	0.27	50	0.24	50	0.22	50	0.19	50	0.17	50
	10	0.27	50	0.24	50	0.22	50	0.20	50	0.18	50
	12	0.29	50	0.26	50	0.23	50	0.20	50	0.18	50
	14	0.30	50	0.27	50	0.24	50	0.21	50	0.18	50
	16	0.31	50	0.29	50	0.26	50	0.21	50	0.18	50
	20	0.34	51	0.31	50	0.28	50	0.23	50	0.19	50
	24	0.36	52	0.33	51	0.31	50	0.25	50	0.20	50
	28	0.38	52	0.35	51	0.33	51	0.28	50	0.21	50
	32	0.39	53	0.37	52	0.35	51	0.30	50	0.23	50

* Total calculated thickness includes service allowance and casting tolerance added to net thickness.

TABLE 50.12—(cont.)

Size in.	Depth of Cover ft	Type 1 Total Calculated Thickness* in.	Type 1 Use Class	Type 2 Total Calculated Thickness* in.	Type 2 Use Class	Type 3 Total Calculated Thickness* in.	Type 3 Use Class	Type 4 Total Calculated Thickness* in.	Type 4 Use Class	Type 5 Total Calculated Thickness* in.	Type 5 Use Class
14	2.5	0.32	50	0.29	50	0.26	50	0.22	50	0.20	50
	3	0.31	50	0.27	50	0.24	50	0.21	50	0.19	50
	4	0.29	50	0.26	50	0.23	50	0.21	50	0.19	50
	5	0.28	50	0.25	50	0.23	50	0.20	50	0.19	50
	6	0.28	50	0.25	50	0.23	50	0.20	50	0.19	50
	7	0.28	50	0.25	50	0.23	50	0.21	50	0.19	50
	8	0.29	50	0.26	50	0.23	50	0.21	50	0.19	50
	9	0.30	50	0.26	50	0.24	50	0.21	50	0.19	50
	10	0.31	50	0.27	50	0.24	50	0.21	50	0.19	50
	12	0.32	50	0.29	50	0.26	50	0.22	50	0.19	50
	14	0.34	50	0.30	50	0.27	50	0.23	50	0.20	50
	16	0.35	51	0.32	50	0.29	50	0.24	50	0.20	50
	20	0.38	52	0.35	51	0.32	50	0.26	50	0.21	50
	24	0.40	52	0.38	52	0.34	50	0.28	50	0.22	50
	28	0.42	53	0.40	52	0.37	51	0.31	50	0.24	50
	32	0.44	54	0.42	53	0.39	52	0.34	50	0.26	50
16	2.5	0.34	50	0.30	50	0.27	50	0.23	50	0.20	50
	3	0.32	50	0.28	50	0.25	50	0.22	50	0.20	50
	4	0.30	50	0.27	50	0.24	50	0.21	50	0.19	50
	5	0.30	50	0.26	50	0.24	50	0.21	50	0.19	50
	6	0.30	50	0.26	50	0.24	50	0.21	50	0.19	50
	7	0.30	50	0.26	50	0.24	50	0.21	50	0.19	50
	8	0.31	50	0.27	50	0.25	50	0.22	50	0.19	50
	9	0.32	50	0.28	50	0.25	50	0.22	50	0.20	50
	10	0.33	50	0.29	50	0.26	50	0.22	50	0.20	50
	12	0.35	50	0.31	50	0.27	50	0.23	50	0.20	50
	14	0.36	51	0.33	50	0.29	50	0.24	50	0.20	50
	16	0.38	51	0.34	50	0.30	50	0.25	50	0.21	50
	20	0.41	52	0.38	51	0.34	50	0.27	50	0.22	50
	24	0.44	53	0.41	52	0.37	51	0.30	50	0.24	50
	28	0.46	54	0.43	53	0.40	52	0.33	50	0.27	50
	32	0.48	55	0.46	54	0.43	53	0.36	51	0.29	50
18	2.5	0.36	50	0.32	50	0.28	50	0.24	50	0.20	50
	3	0.34	50	0.29	50	0.26	50	0.23	50	0.20	50
	4	0.32	50	0.28	50	0.25	50	0.22	50	0.20	50
	5	0.31	50	0.27	50	0.25	50	0.22	50	0.19	50
	6	0.31	50	0.27	50	0.25	50	0.22	50	0.19	50
	7	0.32	50	0.28	50	0.25	50	0.22	50	0.20	50
	8	0.33	50	0.29	50	0.26	50	0.22	50	0.20	50
	9	0.34	50	0.30	50	0.26	50	0.23	50	0.20	50
	10	0.35	50	0.30	50	0.27	50	0.23	50	0.20	50
	12	0.37	51	0.32	50	0.29	50	0.24	50	0.21	50
	14	0.39	51	0.35	50	0.30	50	0.25	50	0.21	50
	16	0.41	52	0.37	51	0.32	50	0.26	50	0.22	50
	20	0.44	53	0.40	52	0.36	50	0.29	50	0.23	50
	24	0.47	54	0.44	53	0.40	52	0.32	50	0.26	50
	28	0.50	55	0.46	54	0.43	53	0.36	50	0.29	50
	32	0.53	56	0.49	55	0.46	54	0.39	51	0.32	50

* Total calculated thickness includes service allowance and casting tolerance added to net thickness.

TABLE 50.12—(cont.)

Size in.	Depth of Cover ft	Type 1		Type 2		Type 3		Type 4		Type 5	
		Total Calculated Thickness* in.	Use Class	Total Calculated Thickness* in.	Use Class	Total Calculated Thickness* in.	Use Class	Total Calculated Thickness* in.	Use Class	Total Calculated Thickness* in.	Use Class
20	2.5	0.38	51	0.33	50	0.29	50	0.24	50	0.21	50
	3	0.35	50	0.31	50	0.27	50	0.23	50	0.20	50
	4	0.34	50	0.29	50	0.26	50	0.23	50	0.20	50
	5	0.33	50	0.28	50	0.26	50	0.23	50	0.20	50
	6	0.33	50	0.29	50	0.26	50	0.23	50	0.20	50
	7	0.34	50	0.29	50	0.26	50	0.23	50	0.20	50
	8	0.35	50	0.30	50	0.27	50	0.23	50	0.20	50
	9	0.36	50	0.31	50	0.28	50	0.24	50	0.21	50
	10	0.37	50	0.32	50	0.28	50	0.24	50	0.21	50
	12	0.39	51	0.34	50	0.30	50	0.25	50	0.21	50
	14	0.41	52	0.37	50	0.32	50	0.26	50	0.22	50
	16	0.43	52	0.39	51	0.34	50	0.27	50	0.22	50
	20	0.47	54	0.43	52	0.39	51	0.31	50	0.23	50
	24	0.50	55	0.47	54	0.42	52	0.34	50	0.28	50
	28	0.54	56	0.50	55	0.46	53	0.38	51	0.31	50
	32	0.57	—	0.53	56	0.49	54	0.42	52	0.34	50
24	2.5	0.42	51	0.35	50	0.31	50	0.26	50	0.22	50
	3	0.39	50	0.33	50	0.29	50	0.25	50	0.21	50
	4	0.37	50	0.31	50	0.28	50	0.24	50	0.21	50
	5	0.36	50	0.31	50	0.28	50	0.24	50	0.21	50
	6	0.36	50	0.31	50	0.28	50	0.24	50	0.21	50
	7	0.37	50	0.32	50	0.28	50	0.24	50	0.21	50
	8	0.39	50	0.33	50	0.29	50	0.25	50	0.21	50
	9	0.40	51	0.34	50	0.30	50	0.25	50	0.22	50
	10	0.41	51	0.35	50	0.31	50	0.26	50	0.22	50
	12	0.44	52	0.38	50	0.33	50	0.27	50	0.23	50
	14	0.46	53	0.41	51	0.35	50	0.28	50	0.23	50
	16	0.49	54	0.44	52	0.38	50	0.31	50	0.24	50
	20	0.54	55	0.49	54	0.43	52	0.36	50	0.25	50
	24	0.57	56	0.53	55	0.48	53	0.40	51	0.32	50
	28	0.61	—	0.57	56	0.52	55	0.43	52	0.36	50
	32	0.65	—	0.60	—	0.56	56	0.47	53	0.40	51
30	2.5	†	†	0.40	50	0.34	50	0.28	50	0.23	50
	3			0.37	50	0.32	50	0.27	50	0.23	50
	4			0.35	50	0.31	50	0.26	50	0.22	50
	5			0.35	50	0.31	50	0.26	50	0.22	50
	6			0.35	50	0.31	50	0.26	50	0.22	50
	7			0.36	50	0.31	50	0.26	50	0.23	50
	8			0.37	50	0.33	50	0.27	50	0.23	50
	9			0.39	50	0.34	50	0.28	50	0.23	50
	10			0.40	50	0.35	50	0.28	50	0.24	50
	12			0.44	51	0.37	50	0.30	50	0.24	50
	14			0.47	52	0.40	50	0.32	50	0.25	50
	16			0.51	53	0.43	51	0.37	50	0.26	50
	20			0.57	55	0.50	53	0.43	51	0.29	50
	24			0.62	56	0.56	54	0.48	52	0.37	50
	28			0.67	—	0.61	56	0.51	53	0.43	51
	32			0.71	—	0.66	—	0.55	54	0.47	52

* Total calculated thickness includes service allowance and casting tolerance added to net thickness.
† For pipe 30 in. and larger, consideration should be given to laying conditions other than Type 1.

TABLE 50.12—(cont.)

Size in.	Depth of Cover ft	Type 1 Total Calculated Thickness* in.	Type 1 Use Class	Type 2 Total Calculated Thickness* in.	Type 2 Use Class	Type 3 Total Calculated Thickness* in.	Type 3 Use Class	Type 4 Total Calculated Thickness* in.	Type 4 Use Class	Type 5 Total Calculated Thickness* in.	Type 5 Use Class
36	2.5	†	†	0.43	50	0.37	50	0.30	50	0.25	50
	3			0.40	50	0.35	50	0.29	50	0.24	50
	4			0.39	50	0.34	50	0.28	50	0.24	50
	5			0.38	50	0.34	50	0.28	50	0.24	50
	6			0.39	50	0.34	50	0.28	50	0.24	50
	7			0.40	50	0.35	50	0.29	50	0.24	50
	8			0.42	50	0.36	50	0.30	50	0.24	50
	9			0.43	50	0.37	50	0.30	50	0.25	50
	10			0.45	50	0.38	50	0.31	50	0.25	50
	12			0.49	51	0.42	50	0.33	50	0.26	50
	14			0.54	52	0.46	51	0.37	50	0.27	50
	16			0.58	53	0.49	51	0.42	50	0.28	50
	20			0.65	54	0.57	53	0.50	51	0.33	50
	24			0.71	56	0.63	54	0.55	52	0.43	50
	28			0.77	—	0.70	55	0.60	53	0.50	51
	32			0.82	—	0.76	—	0.64	54	0.55	52
42	2.5	†	†	0.46	50	0.40	50	0.32	50	0.26	50
	3			0.44	50	0.38	50	0.31	50	0.25	50
	4			0.42	50	0.37	50	0.30	50	0.25	50
	5			0.42	50	0.37	50	0.30	50	0.25	50
	6			0.42	50	0.37	50	0.30	50	0.25	50
	7			0.44	50	0.38	50	0.31	50	0.25	50
	8			0.46	50	0.39	50	0.32	50	0.26	50
	9			0.48	50	0.41	50	0.33	50	0.26	50
	10			0.50	51	0.42	50	0.33	50	0.27	50
	12			0.55	51	0.47	50	0.35	50	0.28	50
	14			0.60	52	0.52	51	0.42	50	0.29	50
	16			0.64	53	0.56	52	0.48	50	0.30	50
	20			0.73	54	0.64	53	0.57	52	0.37	50
	24			0.80	56	0.71	54	0.63	53	0.49	50
	28			0.87	—	0.79	55	0.69	54	0.57	52
	32			0.93	—	0.86	—	0.73	54	0.63	53
48	2.5	†	†	0.50	50	0.43	50	0.35	50	0.28	50
	3			0.48	50	0.42	50	0.34	50	0.28	50
	4			0.46	50	0.40	50	0.33	50	0.27	50
	5			0.46	50	0.40	50	0.33	50	0.27	50
	6			0.47	50	0.41	50	0.33	50	0.27	50
	7			0.48	50	0.42	50	0.34	50	0.28	50
	8			0.51	50	0.44	50	0.35	50	0.28	50
	9			0.53	50	0.45	50	0.36	50	0.29	50
	10			0.56	51	0.47	50	0.37	50	0.30	50
	12			0.61	51	0.53	50	0.39	50	0.31	50
	14			0.67	52	0.59	51	0.48	50	0.32	50
	16			0.72	53	0.64	52	0.55	51	0.33	50
	20			0.82	54	0.72	53	0.65	52	0.43	50
	24			0.91	56	0.80	54	0.72	53	0.56	51
	28			0.98	—	0.89	55	0.79	54	0.65	52
	32			1.05	—	0.97	—	0.84	55	0.72	53

* Total calculated thickness includes service allowance and casting tolerance added to net thickness.
† For pipe 30 in. and larger, consideration should be given to laying conditions other than Type 1.

TABLE 50.12—(cont.)

Size in.	Depth of Cover ft	Type 1 Total Calculated Thickness* in.	Type 1 Use Class	Type 2 Total Calculated Thickness* in.	Type 2 Use Class	Type 3 Total Calculated Thickness* in.	Type 3 Use Class	Type 4 Total Calculated Thickness* in.	Type 4 Use Class	Type 5 Total Calculated Thickness* in.	Type 5 Use Class
54	2.5	†	†	0.54	50	0.46	50	0.37	50	0.30	50
	3			0.52	50	0.45	50	0.37	50	0.30	50
	4			0.51	50	0.44	50	0.36	50	0.30	50
	5			0.51	50	0.44	50	0.36	50	0.30	50
	6			0.52	50	0.45	50	0.37	50	0.30	50
	7			0.53	50	0.46	50	0.37	50	0.30	50
	8			0.56	50	0.48	50	0.38	50	0.31	50
	9			0.59	50	0.50	50	0.39	50	0.32	50
	10			0.62	51	0.52	50	0.41	50	0.32	50
	12			0.68	51	0.60	50	0.43	50	0.34	50
	14			0.74	52	0.67	51	0.54	50	0.35	50
	16			0.80	53	0.72	52	0.62	51	0.36	50
	20			0.91	54	0.81	53	0.73	52	0.48	50
	24			1.01	56	0.89	54	0.81	53	0.63	51
	28			1.09	—	0.99	55	0.88	54	0.73	52
	32			1.17	—	1.08	—	0.94	55	0.81	53

* Total calculated thickness includes service allowance and casting tolerance added to net thickness.
† For pipe 30 in. and larger, consideration should be given to laying conditons other than Type 1.

TABLE 50.13

Thickness for Internal Pressure

Pipe Size in.	150 Total Calculated Thickness in.†	150 Use Class	200 Total Calculated Thickness in.†	200 Use Class	250 Total Calculated Thickness in.†	250 Use Class	300 Total Calculated Thickness in.†	300 Use Class	350 Total Calculated Thickness in.†	350 Use Class
3	0.15	51	0.16	51	0.16	51	0.17	51	0.17	51
4	0.16	51	0.16	51	0.17	51	0.18	51	0.18	51
6	0.17	50	0.18	50	0.19	50	0.20	50	0.20	50
8	0.18	50	0.19	50	0.21	50	0.22	50	0.23	50
10	0.21	50	0.22	50	0.23	50	0.25	50	0.26	50
12	0.22	50	0.23	50	0.25	50	0.27	50	0.28	50
14	0.24	50	0.26	50	0.28	50	0.30	50	0.31	50
16	0.25	50	0.27	50	0.30	50	0.32	50	0.34	50
18	0.27	50	0.29	50	0.31	50	0.34	50	0.36	50
20	0.28	50	0.30	50	0.33	50	0.36	50	0.38	51
24	0.30	50	0.33	50	0.37	50	0.40	51	0.43	52
30	0.34	50	0.38	50	0.42	51	0.45	52	0.49	53
36	0.38	50	0.42	50	0.47	51	0.51	52	0.56	53
42	0.41	50	0.47	50	0.52	51	0.57	52	0.63	53
48	0.46	50	0.52	50	0.58	51	0.64	52	0.70	53
54	0.51	50	0.58	50	0.65	51	0.71	52	0.78	53

Rated Water Working Pressure*—psi

* These pipe are adequate for the rated working pressure plus a surge allowance of 100 psi.
† Total calculated thickness includes service allowance and casting tolerance added to net thickness.

TABLE 50.14

Rated Working Pressure and Maximum Depth of Cover

Pipe Size in.	Thickness Class	Nominal Thickness in.	Rated Water Working Pressure psi*	Laying Condition				
				Type 1	Type 2	Type 3	Type 4	Type 5
				Maximum Depth of Cover—ft†				
3	51	0.25	350	98	100‡	100‡	100‡	100‡
	52	0.28	350	100‡	100‡	100‡	100‡	100‡
	53	0.31	350	100‡	100‡	100‡	100‡	100‡
	54	0.34	350	100‡	100‡	100‡	100‡	100‡
	55	0.37	350	100‡	100‡	100‡	100‡	100‡
	56	0.40	350	100‡	100‡	100‡	100‡	100‡
4	51	0.26	350	76	86	96	100‡	100‡
	52	0.29	350	100‡	100‡	100‡	100‡	100‡
	53	0.32	350	100‡	100‡	100‡	100‡	100‡
	54	0.35	350	100‡	100‡	100‡	100‡	100‡
	55	0.38	350	100‡	100‡	100‡	100‡	100‡
	56	0.41	350	100‡	100‡	100‡	100‡	100‡
6	50	0.25	350	32	38	44	56	75
	51	0.28	350	49	57	64	80	100‡
	52	0.31	350	67	77	86	100‡	100‡
	53	0.34	350	91	100‡	100‡	100‡	100‡
	54	0.37	350	100‡	100‡	100‡	100‡	100‡
	55	0.40	350	100‡	100‡	100‡	100‡	100‡
	56	0.43	350	100‡	100‡	100‡	100‡	100‡
8	50	0.27	350	25	30	36	46	64
	51	0.30	350	36	42	49	61	81
	52	0.33	350	47	54	62	77	99
	53	0.36	350	64	73	82	100‡	100‡
	54	0.39	350	80	91	100‡	100‡	100‡
	55	0.42	350	98	100‡	100‡	100‡	100‡
	56	0.45	350	100‡	100‡	100‡	100‡	100‡
10	50	0.29	350	19	24	29	38	55
	51	0.32	350	27	32	38	49	66
	52	0.35	350	35	41	47	59	79
	53	0.38	350	45	52	59	74	95
	54	0.41	350	57	65	74	91	100‡
	55	0.44	350	67	77	86	100‡	100‡
	56	0.47	350	81	92	100‡	100‡	100‡
12	50	0.31	350	17	22	27	36	52
	51	0.34	350	23	28	33	43	60
	52	0.37	350	30	35	41	53	71
	53	0.40	350	36	42	49	61	81
	54	0.43	350	45	52	59	74	95
	55	0.46	350	54	62	71	87	100‡
	56	0.49	350	64	73	83	100‡	100‡

* These pipe are adequate for the rated working pressure plus a surge allowance of 100 psi. Ductile-iron pipe for working pressures higher than 350 psi is available.
† An allowance for a single H-20 truck with 1.5 impact factor is included for all depths of cover.
‡ Calculated maximum depth of cover exceeds 100 ft.

TABLE 50.14—(cont.)

Pipe Size in.	Thickness Class	Nominal Thickness in.	Rated Water Working Pressure psi*	Laying Condition				
				Type 1	Type 2	Type 3	Type 4	Type 5
				Maximum Depth of Cover—ft†				
14	50	0.33	350	15	19	24	33	49
	51	0.36	350	19	23	28	38	55
	52	0.39	350	24	29	34	44	62
	53	0.42	350	30	35	41	53	71
	54	0.45	350	36	42	49	61	81
	55	0.48	350	43	50	57	71	92
	56	0.51	350	52	59	67	83	100‡
16	50	0.34	350	13	17	21	30	47
	51	0.37	350	16	21	25	34	51
	52	0.40	350	20	25	30	40	57
	53	0.43	350	25	30	36	46	64
	54	0.46	350	30	35	41	53	71
	55	0.49	350	35	41	47	59	79
	56	0.52	350	41	48	55	68	89
18	50	0.35	350	11	15	20	29	42
	51	0.38	350	14	19	23	32	49
	52	0.41	350	18	22	27	36	53
	53	0.44	350	22	26	31	41	58
	54	0.47	350	25	30	36	46	64
	55	0.50	350	30	35	41	53	71
	56	0.53	350	35	41	47	59	79
20	50	0.36	300	10	14	18	27	38
	51	0.39	350	13	17	21	30	44
	52	0.42	350	16	20	25	34	50
	53	0.45	350	19	23	28	38	54
	54	0.48	350	22	27	32	42	59
	55	0.51	350	26	31	37	47	65
	56	0.54	350	30	35	41	53	71
24	50	0.38	250	8	12	17	23	31
	51	0.41	300	10	15	19	27	36
	52	0.44	350	13	17	21	30	41
	53	0.47	350	15	19	24	33	47
	54	0.50	350	18	22	27	36	53
	55	0.53	350	20	25	30	40	57
	56	0.56	350	24	29	34	44	61
30	50	0.39	200	§	10	14	18	25
	51	0.43	250		12	16	21	29
	52	0.47	300		14	19	24	33
	53	0.51	350		17	21	29	38
	54	0.55	350		19	24	33	44
	55	0.59	350		22	27	36	51
	56	0.63	350		26	31	41	57

* These pipe are adequate for the rated working pressure plus a surge allowance of 100 psi. Ductile-iron pipe or working pressures higher than 350 psi is available.
† An allowance for a single H-20 truck with 1.5 impact factor is included for all depths of cover.
‡ Calculated maximum depth of cover exceeds 100 ft.
§ For pipe 30 in. and larger. consideration should be given to laying conditions other than Type 1.

33

TABLE 50.14—(cont.)

Pipe Size in.	Thickness Class	Nominal Thickness in.	Rated Water Working Pressure psi*	Laying Condition				
				Type 1	Type 2	Type 3	Type 4	Type 5
				Maximum Depth of Cover—ft†				
36	50	0.43	200	§	10	13	17	25
	51	0.48	250		12	16	20	28
	52	0.53	300		15	19	24	32
	53	0.58	350		17	21	28	37
	54	0.63	350		20	25	33	43
	55	0.68	350		23	28	37	50
	56	0.73	350		26	31	41	59
42	50	0.47	200	§	9	13	16	24
	51	0.53	250		12	15	19	27
	52	0.59	300		14	18	22	30
	53	0.65	350		17	22	27	35
	54	0.71	350		20	24	32	42
	55	0.77	350		23	28	38	48
	56	0.83	350		26	31	41	57
48	50	0.51	200	§	9	12	15	23
	51	0.58	250		12	14	18	26
	52	0.65	300		14	18	21	30
	53	0.72	350		17	21	25	34
	54	0.79	350		20	24	30	40
	55	0.86	350		23	28	37	47
	56	0.93	350		26	31	41	55
54	50	0.57	200	§	9	12	15	23
	51	0.65	250		12	14	18	25
	52	0.73	300		14	17	21	29
	53	0.81	350		17	21	25	34
	54	0.89	350		20	25	30	40
	55	0.97	350		23	28	37	47
	56	1.05	350		27	32	42	55

* These pipe are adequate for the rated working pressure plus a surge allowance of 100 psi. Ductile-iron pipe for working pressures higher than 350 psi is available.
† An allowance for a single H-20 truck with 1.5 impact factor is included for all depths of cover.
‡ Calculated maximum depth of cover exceeds 100 ft.
§ For pipe 30 in. and larger, consideration should be given to laying conditions other than Type 1.

ANSI A21.51–1976
(AWWA C151–76)

Revision of
A21.51–1971
(AWWA C151–71)

AMERICAN NATIONAL STANDARD

for

DUCTILE–IRON PIPE, CENTRIFUGALLY CAST IN METAL MOLDS OR SAND–LINED MOLDS, FOR WATER OR OTHER LIQUIDS

SECRETARIATS

AMERICAN GAS ASSOCIATION
AMERICAN WATER WORKS ASSOCIATION
NEW ENGLAND WATER WORKS ASSOCIATION

Revised edition approved by American National Standards Institute, Inc., Aug. 4, 1976

NOTICE

PUBLISHED BY

AMERICAN WATER WORKS ASSOCIATION
6666 West Quincy Avenue, Denver, Colorado 80235

i

1
8
2

Committee Personnel

Subcommittee 1, Pipe, which reviewed this standard, had the following personnel at that time:

EDWARD C. SEARS, *Chairman*
WALTER AMORY, *Vice-Chairman*

User Members	*Producer Members*
ROBERT S. BRYANT	ALFRED F. CASE
FRANK E. DOLSON	W. D. GOODE
GEORGE F. KEENAN	THOMAS D. HOLMES
LEONARD ORLANDO JR.	HAROLD KENNEDY JR.
JOHN E. PERRY	W. HARRY SMITH
	SIDNEY P. TEAGUE

Standards Committee A21, Cast-Iron Pipe and Fittings, which reviewed and approved this standard, had the following personnel at the time of approval:

LLOYD W. WELLER, *Chairman*
EDWARD C. SEARS, *Vice-Chairman*
JAMES B. RAMSEY, *Secretary*

Organization Represented	*Name of Representative*
American Gas Association	LEONARD ORLANDO JR.
American Society of Civil Engineers	KENNETH W. HENDERSON
American Society of Mechanical Engineers	JAMES S. VANICK
American Society for Testing and Materials	H. M. COBB *
American Water Works Association	ARNOLD M. TINKEY
	LLOYD W. WELLER
Cast Iron Pipe Research Association	THOMAS D. HOLMES
	HAROLD KENNEDY JR.
	EDWARD C. SEARS
	W. HARRY SMITH
Individual Producer	ALFRED F. CASE
Manufacturers' Standardization Society of the Valve and Fittings Industry	ABRAHAM FENSTER
New England Water Works Association	WALTER AMORY
Naval Facilities Engineering Command	STANLEY C. BAKER
Underwriters' Laboratories, Inc.	JOHN E. PERRY
Canadian Standards Association	W. F. SEMENCHUK †

* Alternate
† Liaison representative without vote

Table of Contents

Foreword

This foreword is for information only and is not a part of ANSI A21.51 (AWWA C151).

American National Standards Committee A21 on Cast-Iron Pipe and Fittings was organized in 1926 under the sponsorship of the American Gas Association, The American Society for Testing and Materials, the American Water Works Association, and the New England Water Works Association. Since 1972, the co-secretariats have been A.G.A., AWWA, and NEWWA, with AWWA serving as the administrative secretariat. The present scope of Committee A21 activity is

Standardization of specifications for cast-iron and ductile-iron pressure pipe for gas, water and other liquids, and fittings for use with such pipe. These specifications to include design, dimensions, materials, coatings, linings, joints, accessories and methods of inspection and test.

The work of Committee A21 is conducted by subcommittees. The directive to Subcommittee 1—Pipe is that

The scope of the subcommittee activity shall include the periodic review of all current A21 standards for pipe, the preparation of revisions and new standards when needed, as well as other matters pertaining to pipe standards.

The first edition of A21.51, the standard for ductile-iron pipe for water and other liquids, was issued in 1965, and a revision was issued in 1971. Subcommittee 1 reviewed the 1971 edition and submitted a proposed revision to American National Standards Committee A21 in 1975.

Major Revisions

1. An additional minus tolerance of 0.02 in. is permitted along the barrel of the pipe for a distance not to ex-

ceed 12 in. The weight tolerance permitted for any single pipe is 6 per cent for pipe 12 in. or smaller and 5 per cent for pipe larger than 12 in.

2. The tables include data for the five standard laying conditions covered in A21.50–1976. Types 1, 2, and 3 replace A, B, and S, respectively, in A21.51–1971. Types 3 and 4 have been added to provide a wider selection of laying conditions.

3. The pipe thickness selection tables have been revised to reflect the changes in the 1976 edition of A21.50, the standard for thickness design of ductile-iron pipe.

4. The standard thickness classes have been renumbered. Class 1 becomes Class 51; Class 2 becomes Class 52; and so on. Class 50 has been added for 6–54-in. pipe, and Class 51 has been added for 3–12-in. pipe.

5. Pipe weights have been adjusted for 3–24-in. push-on joint pipe to reflect lighter bell weights based on push-on joint bells, which are more compatible with the barrel thicknesses of ductile-iron pipe.

6. A table (Table 3) has been added to show rated working pressure and maximum depth of cover of the standard laying conditions and standard thickness classes.

Options

This standard includes certain options that, if desired, must be specified on the purchase order. Also, a number of items must be specified to describe completely the pipe required. The following summarizes the details and

v

available options and lists the sections of the standard where they can be found.

1. Size, joint type, thickness or class, and laying length (Tables).
2. a. Special joints (Sec. 51–3.1).
   b. Specifying ductile-iron gland, if required (Sec. 51–3.1).
3. Certification by manufacturer (Sec. 51–4.2).
4. Inspection by purchaser (Sec. 51–5).
5. a. No requirement for outside coating (Sec. 51–8.1).
   b. Cement lining (Sec. 51–8.2). Experience has indicated that bituminous inside coating is not complete protection against loss in pipe capacity caused by tuberculation. Cement linings are recommended for most waters.
   c. No requirement for inside coating (Sec. 51–8.3).
   d. Special coatings and linings (Sec. 51–8.4).
6. Special marking on pipe (Sec. 51–10).
7. Written transcripts of foundry records (Sec. 51–14).
8. Special tests (Sec. 51–15).

American National Standard for

Ductile-Iron Pipe, Centrifugally Cast in Metal Molds or Sand-Lined Molds, for Water or Other Liquids

1
8
7

Sec. 51–1—Scope

This standard covers 3-in. through 54-in. ductile-iron pipe centrifugally cast in metal molds or sand-lined molds for water or other liquids with push-on joints or mechanical joints. This standard may be used for pipe with such other types of joints as may be agreed upon at the time of purchase.

Sec. 51–2—Definitions

Under this standard, the following definitions shall apply:

51–2.1 *Purchaser.* The party entering into a contract or agreement to purchase pipe according to this standard.

51–2.2 *Manufacturer.* The party that produces the pipe.

51–2.3 *Inspector.* The representative of the purchaser, authorized to inspect on behalf of the purchaser to determine whether or not the pipe meets this standard.

51–2.4 *Ductile iron.* A cast ferrous material in which a major part of the carbon content occurs as free carbon in nodular or spheroidal form.

51–2.5 *Mechanical joint.* The gasketed and bolted joint as detailed in

ANSI A21.11 (AWWA C111) of latest revision.

51–2.6 *Push-on joint.* The single rubber-gasket joint as described in ANSI A21.11 (AWWA C111) of latest revision.

Sec. 51–3—General Requirements

51–3.1 Pipe with mechanical joints or push-on joints shall conform to the applicable dimensions and weights shown in the tables in this standard and to the applicable requirements of ANSI A21.11 (AWWA C111) of latest revision. Unless otherwise specified, the mechanical-joint glands shall be cast iron in accordance with ANSI A21.11 of latest revision and bolts shall conform to the requirements of the same standard. Pipe with other types of joints shall comply with the joint dimensions and weights agreed upon at the time of purchase but in all other respects shall fulfill the requirements of this standard.

51–3.2 The nominal laying length of the pipe shall be as shown in the tables. A maximum of 20 per cent of the total number of pipe of each size specified in an order may be furnished by as much as 24 in. shorter than the

1

nominal laying length, and an additional 10 per cent may be furnished by as much as 6 in. shorter than nominal laying length.

Sec. 51-4—Inspection and Certification by Manufacturer

51-4.1 The manufacturer shall establish the necessary quality-control and inspection practice to ensure compliance with this standard.

51-4.2 The manufacturer shall, if required on the purchase order, furnish a sworn statement that the inspection and all of the specified tests have been made and the results thereof comply with the requirements of this standard.

51-4.3 All pipe shall be clean and sound without defects that could impair service. Repairing of defects by welding or other methods shall not be allowed if such repairs could adversely affect the serviceability of the pipe or its capability to meet strength requirements of this standard.

Sec. 51-5—Inspection by Purchaser

51-5.1 If the purchaser desires to inspect pipe at the manufacturer's plant, the purchaser shall so specify on the purchase order, stating the conditions (such as time and the extent of inspection) under which the inspection shall be made.

51-5.2 The inspector shall have free access to those parts of the manufacturer's plant that are necessary to ensure compliance with this standard. The manufacturer shall make available for the inspector's use such gages as are necessary for inspection. The manufacturer shall provide the inspector with assistance as necessary for handling of pipe.

Sec. 51-6—Delivery and Acceptance

All pipe and accessories shall comply with this standard. Pipe and accessories not complying with this standard shall be replaced by the manufacturer at the agreed point of delivery. The manufacturer shall not be liable for shortages or damaged pipe after acceptance at the agreed point of delivery, except as recorded on the delivery receipt or similar document by the carrier's agent.

Sec. 51-7—Tolerances or Permitted Variations

51-7.1 *Dimensions.* The spigot end, bell, and socket of the pipe and the accessories shall be gaged with suitable gages at sufficiently frequent intervals to ensure that the dimensions comply with the requirements of this standard. The smallest inside diameter of the sockets and the outside of the spigot ends shall be tested with circular gages. Other socket dimensions shall be gaged as may be appropriate.

51-7.2 *Thickness.* Minus thickness tolerances of pipe and bell shall not exceed the following:

Size in.	Minus Tolerance in.
3–8	0.05
10–12	0.06
14–42	0.07
48	0.08
54	0.09

An additional minus tolerance of 0.02 in. shall be permitted along the barrel of the pipe for a distance not to exceed 12 in.

51-7.3 *Weight.* The weight of any single pipe shall not be less than the tabulated weight by more than 6 per cent for pipe 12 in. or smaller in diameter, or by more than 5 per cent for pipe larger than 12 in. in diameter.

Sec. 51-8—Coatings and Linings

51-8.1 *Outside coating.* The out-

1
8
8

2

side coating for use under normal conditions shall be a bituminous coating approximately 1 mil thick. The coating shall be applied to the outside of all pipe, unless otherwise specified. The finished coating shall be continuous, smooth, neither brittle when cold nor sticky when exposed to the sun, and shall be strongly adherent to the pipe.

51–8.2 *Cement–mortar linings.* Cement linings shall be in accordance with ANSI A21.4 (AWWA C104) of latest revision. If desired by the purchaser, cement linings shall be specified in the invitation for bids and on the purchase order.

51–8.3 *Inside coating.* Unless otherwise specified, the inside coating for pipe that is not cement-lined shall be a bituminous material as thick as practicable (at least 1 mil) which conforms to all appropriate requirements for seal coat in ANSI A21.4 of latest revision.

51–8.4 *Special coatings and linings.* For special conditions, other types of coatings and linings may be available. Such special coatings and linings shall be specified in the invitation for bids and on the purchase order.

Sec. 51–9—Hydrostatic Test

Each pipe shall be subjected to a hydrostatic test of not less than 500 psi. This test may be made either before or after the outside coating and inside coating have been applied, but shall be made before the application of cement lining or of a special lining.

The pipe shall be under the full test pressure for at least 10 s. Suitable controls and recording devices shall be provided so that the test pressure and duration may be adequately ascertained. Any pipe that leaks or does not withstand the test pressure shall be rejected.

In addition to the hydrostatic test before application of a cement lining or special lining, the pipe may be retested, at the manufacturer's option, after application of such lining.

Sec. 51–10—Marking Pipe

The weight, class or nominal thickness, and casting period shall be shown on each pipe. The manufacturer's mark, the year in which the pipe was produced, and the letters "DI" or "DUCTILE" shall be cast or stamped on the pipe. When specified on the purchase order, initials not exceeding four in number shall be cast or stamped on the pipe. All required markings shall be clear and legible, and all cast marks shall be on or near the bell. All letters and numerals on pipe sizes 14 in. and larger shall be not less than ½ in. in height.

Sec. 51–11—Weighing Pipe

Each pipe shall be weighed before the application of any lining or coating other than the bituminous coating and the weight shown on the outside or inside of the bell or spigot end.

Sec. 51–12—Acceptance Tests

The standard acceptance tests for the physical characteristics of the pipe shall be as follows:

51–12.1 *Tensile test.* A tensile test specimen shall be cut longitudinally from the midsection of the pipe wall. This specimen shall be machined and tested in accordance with Fig. 1 and ASTM E8-69, "Tension Testing of Metallic Materials." The yield strength shall be determined by the 0.2 per cent offset, halt-of-pointer, or extension-under-load method. If check tests are to be made, the 0.2 per cent offset

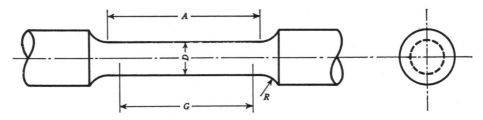

Fig. 1. Tensile-Test Specimen

The tensile-test specimen dimensions are given in the following table:

| Dimension | Standard Specimen 0.500 in. round | Small-Size Specimens Proportional to Standard | | | |
		0.350 in. round	0.250 in. round	0.175 in. round	0.125 in. round
		Dimensions—*in.*			
*T**	0.71 and greater	0.50–0.70	0.35–0.49	0.25–0.34	0.18–0.24
G	2.000 ± 0.005	1.400 ± 0.005	1.000 ± 0.005	0.700 ± 0.005	0.500 ± 0.005
D	0.500 ± 0.010	0.350 ± 0.007	0.250 ± 0.005	0.175 ± 0.005	0.125 ± 0.005
R	$\frac{3}{8}$ (min.)	$\frac{1}{4}$ (min.)	$\frac{3}{16}$ (min.)	$\frac{3}{32}$ (min.)	$\frac{3}{32}$ (min.)
A	$2\frac{1}{4}$ (min.)	$1\frac{3}{4}$ (min.)	$1\frac{1}{4}$ (min.)	$\frac{3}{4}$ (min.)	$\frac{5}{8}$ (min.)

* Thickness of the section from the wall of the pipe from which the tensile specimen is to be machined.
    Note. 1. The reduced section *A* may have a gradual taper from the ends toward the center with the ends not more than 0.005 in. larger in diameter than the center on the standard specimen and not more than 0.003 in. larger in diameter than the center on the small size specimens.
    Note. 2. If desired, on the small size specimens the length of the reduced section may be increased to accommodate an extensometer. However, reference marks for the measurement of elongation should nevertheless be spaced at the indicated gage length *G*.
    Note. 3. The gage length and fillets shall be as shown, but the ends may be of any form to fit the holders of the testing machine in such a way that the load shall be axial. If the ends are to be held in grips it is desirable, if possible, to make the length of the grip section great enough to allow the specimen to extend into the grips a distance equal to two thirds or more of the length of the grips.

method shall be used. All specimens shall be tested at room temperature [70F ± 10 (21C ± 6)].

51–12.1.1 *Acceptance values.* The acceptance values for test specimens shall be as follows:

Grade of iron: 60–42–10

1. Minimum tensile strength: 60 000 psi.
2. Minimum yield strength: 42 000 psi.
3. Minimum elongation: 10 per cent.

51–12.2 *Impact test.* Tests shall be made in accordance with ASTM E23–72 "Notched Charpy Tests," ex-cept that specimens shall be 0.500 in. by full thickness of pipe wall. The notched impact test specimen shall be in accordance with Fig. 2. If the pipe wall thickness exceeds 0.40 in., the impact specimen may be machined to a nominal thickness of 0.40 in. In all tests, impact values are to be corrected to 0.40-in. wall thickness by calculations as follows:

Impact value (corrected)

$$= \frac{0.40}{t} \times \text{impact value (actual)}$$

in which *t* is the thickness of the specimen in inches (wall thickness of pipe).

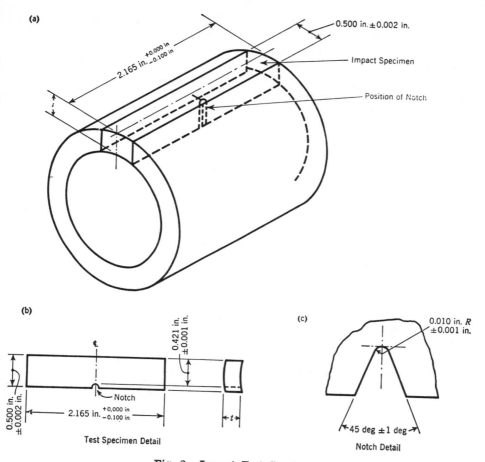

Fig. 2. Impact Test Specimen

In Diagrams (a) and (b) the symbol t is for the pipe-wall thickness.

The Charpy test machine anvil shall not be moved to compensate for the variation of cross section dimensions of the test specimen.

51–12.2.1 *Acceptance value.* The corrected acceptance value for notched impact test specimens shall be a minimum of 7 ft-lb for tests conducted at 70 F ± 10 (21C ± 6).

51–12.3 *Sampling.* At least one tensile and impact sample shall be taken during each casting period of approximately 3 hr. Samples shall be selected to represent extremes of pipe diameters and thicknesses properly.

Sec. 51–13—Additional Control Tests by Manufacturer

Low-temperature impact tests shall be made from at least one third of the test pipe specified in Sec. 51–12.3 to ensure compliance with a minimum corrected value of 3 ft-lb for tests conducted at −40 F (−40C). Test specimens shall be prepared and tested in accordance with Sec. 51–12.2.

In addition, the manufacturer shall conduct such other control tests as necessary to assure continuing compliance with this standard.

Sec. 51-14—Foundry Records

The results of the acceptance tests (Sec. 51–12) and low-temperature impact tests (Sec. 51–13) shall be recorded and retained for one year and shall be available to the purchaser at the foundry. Written transcripts shall be furnished, if specified on the purchase order.

Sec. 51-15—Additional Tests Required by Purchaser

When tests other than those required in this standard are required by the purchaser, such tests shall be specified in the invitation for bids and on the purchase order.

Sec. 51-16—Defective Specimens and Retests

When any physical-test specimen shows defective machining or lack of continuity of metal, it shall be discarded and replaced by another specimen. When any sound test specimen fails to meet the specified requirements, the pipe from which it was taken shall be rejected, and a retest may be made on two additional sound specimens from pipe cast in the same period as the specimen that failed. Both of the additional specimens shall meet the prescribed tests to qualify the pipe produced in that period.

Sec. 51-17—Rejection of Pipe

If the results of any physical acceptance test fail to meet the requirements of Sec. 51–12 or Sec. 51–16, the pipe cast in the same period shall be rejected, except as provided in Sec. 51–18.

Sec. 51-18—Determining Rejection

The manufacturer may determine the amount of rejection by making similar additional tests of pipe of the same size until the rejected lot is bracketed, in order of manufacture, by an acceptable test at each end of the interval in question. When pipe of one size is rejected from a casting period, the acceptability of pipe of different sizes from that same period may be established by making the acceptance tests for these sizes as specified in Sec. 51–12.

DUCTILE—IRON PIPE

TABLE 51.1

Standard Thickness for Earth Load Plus Truck Load*

Size in.	Depth of Cover ft.	Type 1†		Type 2†		Type 3†		Type 4†		Type 5†	
		Thickness in.	Thickness Class	Thickness in.	Thickness Class	Thickness in.	Thickness Class	Thickness in.	Thickness Class	Thickness in.	Thickness Class
3	2.5	0.25	51	0.25	51	0.25	51	0.25	51	0.25	51
	3	0.25	51	0.25	51	0.25	51	0.25	51	0.25	51
	4	0.25	51	0.25	51	0.25	51	0.25	51	0.25	51
	5	0.25	51	0.25	51	0.25	51	0.25	51	0.25	51
	6	0.25	51	0.25	51	0.25	51	0.25	51	0.25	51
	7	0.25	51	0.25	51	0.25	51	0.25	51	0.25	51
	8	0.25	51	0.25	51	0.25	51	0.25	51	0.25	51
	9	0.25	51	0.25	51	0.25	51	0.25	51	0.25	51
	10	0.25	51	0.25	51	0.25	51	0.25	51	0.25	51
	12	0.25	51	0.25	51	0.25	51	0.25	51	0.25	51
	14	0.25	51	0.25	51	0.25	51	0.25	51	0.25	51
	16	0.25	51	0.25	51	0.25	51	0.25	51	0.25	51
	20	0.25	51	0.25	51	0.25	51	0.25	51	0.25	51
	24	0.25	51	0.25	51	0.25	51	0.25	51	0.25	51
	28	0.25	51	0.25	51	0.25	51	0.25	51	0.25	51
	32	0.25	51	0.25	51	0.25	51	0.25	51	0.25	51
4	2.5	0.26	51	0.26	51	0.26	51	0.26	51	0.26	51
	3	0.26	51	0.26	51	0.26	51	0.26	51	0.26	51
	4	0.26	51	0.26	51	0.26	51	0.26	51	0.26	51
	5	0.26	51	0.26	51	0.26	51	0.26	51	0.26	51
	6	0.26	51	0.26	51	0.26	51	0.26	51	0.26	51
	7	0.26	51	0.26	51	0.26	51	0.26	51	0.26	51
	8	0.26	51	0.26	51	0.26	51	0.26	51	0.26	51
	9	0.26	51	0.26	51	0.26	51	0.26	51	0.26	51
	10	0.26	51	0.26	51	0.26	51	0.26	51	0.26	51
	12	0.26	51	0.26	51	0.26	51	0.26	51	0.26	51
	14	0.26	51	0.26	51	0.26	51	0.26	51	0.26	51
	16	0.26	51	0.26	51	0.26	51	0.26	51	0.26	51
	20	0.26	51	0.26	51	0.26	51	0.26	51	0.26	51
	24	0.26	51	0.26	51	0.26	51	0.26	51	0.26	51
	28	0.26	51	0.26	51	0.26	51	0.26	51	0.26	51
	32	0.26	51	0.26	51	0.26	51	0.26	51	0.26	51
6	2.5	0.25	50	0.25	50	0.25	50	0.25	50	0.25	50
	3	0.25	50	0.25	50	0.25	50	0.25	50	0.25	50
	4	0.25	50	0.25	50	0.25	50	0.25	50	0.25	50
	5	0.25	50	0.25	50	0.25	50	0.25	50	0.25	50
	6	0.25	50	0.25	50	0.25	50	0.25	50	0.25	50
	7	0.25	50	0.25	50	0.25	50	0.25	50	0.25	50
	8	0.25	50	0.25	50	0.25	50	0.25	50	0.25	50
	9	0.25	50	0.25	50	0.25	50	0.25	50	0.25	50
	10	0.25	50	0.25	50	0.25	50	0.25	50	0.25	50
	12	0.25	50	0.25	50	0.25	50	0.25	50	0.25	50
	14	0.25	50	0.25	50	0.25	50	0.25	50	0.25	50
	16	0.25	50	0.25	50	0.25	50	0.25	50	0.25	50
	20	0.25	50	0.25	50	0.25	50	0.25	50	0.25	50
	24	0.25	50	0.25	50	0.25	50	0.25	50	0.25	50
	28	0.25	50	0.25	50	0.25	50	0.25	50	0.25	50
	32	0.25	50	0.25	50	0.25	50	0.25	50	0.25	50
8	2.5	0.27	50	0.27	50	0.27	50	0.27	50	0.27	50
	3	0.27	50	0.27	50	0.27	50	0.27	50	0.27	50
	4	0.27	50	0.27	50	0.27	50	0.27	50	0.27	50
	5	0.27	50	0.27	50	0.27	50	0.27	50	0.27	50
	6	0.27	50	0.27	50	0.27	50	0.27	50	0.27	50
	7	0.27	50	0.27	50	0.27	50	0.27	50	0.27	50
	8	0.27	50	0.27	50	0.27	50	0.27	50	0.27	50
	9	0.27	50	0.27	50	0.27	50	0.27	50	0.27	50
	10	0.27	50	0.27	50	0.27	50	0.27	50	0.27	50
	12	0.27	50	0.27	50	0.27	50	0.27	50	0.27	50
	14	0.27	50	0.27	50	0.27	50	0.27	50	0.27	50
	16	0.27	50	0.27	50	0.27	50	0.27	50	0.27	50
	20	0.27	50	0.27	50	0.27	50	0.27	50	0.27	50
	24	0.27	50	0.27	50	0.27	50	0.27	50	0.27	50
	28	0.30	51	0.27	50	0.27	50	0.27	50	0.27	50
	32	0.30	51	0.30	51	0.27	50	0.27	50	0.27	50

* Truckloads used in computing this table are based on a single H-20 truck with 16 000-lb wheel load and 1.5 impact factor.
† See corresponding illustrations in Fig. 3 of types of laying conditions.

TABLE 51.1—(cont.)

Size in.	Depth of Cover ft.	Type 1† Thickness in.	Type 1† Thickness Class	Type 2† Thickness in.	Type 2† Thickness Class	Type 3† Thickness in.	Type 3† Thickness Class	Type 4† Thickness in.	Type 4† Thickness Class	Type 5† Thickness in.	Type 5† Thickness Class
10	2.5	0.29	50	0.29	50	0.29	50	0.29	50	0.29	50
	3	0.29	50	0.29	50	0.29	50	0.29	50	0.29	50
	4	0.29	50	0.29	50	0.29	50	0.29	50	0.29	50
	5	0.29	50	0.29	50	0.29	50	0.29	50	0.29	50
	6	0.29	50	0.29	50	0.29	50	0.29	50	0.29	50
	7	0.29	50	0.29	50	0.29	50	0.29	50	0.29	50
	8	0.29	50	0.29	50	0.29	50	0.29	50	0.29	50
	9	0.29	50	0.29	50	0.29	50	0.29	50	0.29	50
	10	0.29	50	0.29	50	0.29	50	0.29	50	0.29	50
	12	0.29	50	0.29	50	0.29	50	0.29	50	0.29	50
	14	0.29	50	0.29	50	0.29	50	0.29	50	0.29	50
	16	0.29	50	0.29	50	0.29	50	0.29	50	0.29	50
	20	0.32	51	0.29	50	0.29	50	0.29	50	0.29	50
	24	0.32	51	0.29	50	0.29	50	0.29	50	0.29	50
	28	0.35	52	0.32	51	0.29	50	0.29	50	0.29	50
	32	0.35	52	0.32	51	0.32	51	0.29	50	0.29	50
12	2.5	0.31	50	0.31	50	0.31	50	0.31	50	0.31	50
	3	0.31	50	0.31	50	0.31	50	0.31	50	0.31	50
	4	0.31	50	0.31	50	0.31	50	0.31	50	0.31	50
	5	0.31	50	0.31	50	0.31	50	0.31	50	0.31	50
	6	0.31	50	0.31	50	0.31	50	0.31	50	0.31	50
	7	0.31	50	0.31	50	0.31	50	0.31	50	0.31	50
	8	0.31	50	0.31	50	0.31	50	0.31	50	0.31	50
	9	0.31	50	0.31	50	0.31	50	0.31	50	0.31	50
	10	0.31	50	0.31	50	0.31	50	0.31	50	0.31	50
	12	0.31	50	0.31	50	0.31	50	0.31	50	0.31	50
	14	0.31	50	0.31	50	0.31	50	0.31	50	0.31	50
	16	0.31	50	0.31	50	0.31	50	0.31	50	0.31	50
	20	0.34	51	0.31	50	0.31	50	0.31	50	0.31	50
	24	0.37	52	0.34	51	0.31	50	0.31	50	0.31	50
	28	0.37	52	0.34	51	0.34	51	0.31	50	0.31	50
	32	0.40	53	0.37	52	0.34	51	0.31	50	0.31	50
14	2.5	0.33	50	0.33	50	0.33	50	0.33	50	0.33	50
	3	0.33	50	0.33	50	0.33	50	0.33	50	0.33	50
	4	0.33	50	0.33	50	0.33	50	0.33	50	0.33	50
	5	0.33	50	0.33	50	0.33	50	0.33	50	0.33	50
	6	0.33	50	0.33	50	0.33	50	0.33	50	0.33	50
	7	0.33	50	0.33	50	0.33	50	0.33	50	0.33	50
	8	0.33	50	0.33	50	0.33	50	0.33	50	0.33	50
	9	0.33	50	0.33	50	0.33	50	0.33	50	0.33	50
	10	0.33	50	0.33	50	0.33	50	0.33	50	0.33	50
	12	0.33	50	0.33	50	0.33	50	0.33	50	0.33	50
	14	0.33	50	0.33	50	0.33	50	0.33	50	0.33	50
	16	0.36	51	0.33	50	0.33	50	0.33	50	0.33	50
	20	0.39	52	0.36	51	0.33	50	0.33	50	0.33	50
	24	0.39	52	0.39	52	0.33	50	0.33	50	0.33	50
	28	0.42	53	0.39	52	0.36	51	0.33	50	0.33	50
	32	0.45	54	0.42	53	0.39	52	0.33	50	0.33	50
16	2.5	0.34	50	0.34	50	0.34	50	0.34	50	0.34	50
	3	0.34	50	0.34	50	0.34	50	0.34	50	0.34	50
	4	0.34	50	0.34	50	0.34	50	0.34	50	0.34	50
	5	0.34	50	0.34	50	0.34	50	0.34	50	0.34	50
	6	0.34	50	0.34	50	0.34	50	0.34	50	0.34	50
	7	0.34	50	0.34	50	0.34	50	0.34	50	0.34	50
	8	0.34	50	0.34	50	0.34	50	0.34	50	0.34	50
	9	0.34	50	0.34	50	0.34	50	0.34	50	0.34	50
	10	0.34	50	0.34	50	0.34	50	0.34	50	0.34	50
	12	0.34	50	0.34	50	0.34	50	0.34	50	0.34	50
	14	0.37	51	0.34	50	0.34	50	0.34	50	0.34	50
	16	0.37	51	0.34	50	0.34	50	0.34	50	0.34	50
	20	0.40	52	0.37	51	0.34	50	0.34	50	0.34	50
	24	0.43	53	0.40	52	0.37	51	0.34	50	0.34	50
	28	0.46	54	0.43	53	0.40	52	0.34	50	0.34	50
	32	0.49	55	0.46	54	0.43	53	0.37	51	0.34	50

† See corresponding illustrations in Fig. 3 of types of laying conditions.

TABLE 51.1—(cont.)

Size in.	Depth of Cover ft.	Type 1† Thickness in.	Type 1† Thickness Class	Type 2† Thickness in.	Type 2† Thickness Class	Type 3† Thickness in.	Type 3† Thickness Class	Type 4† Thickness in.	Type 4† Thickness Class	Type 5† Thickness in.	Type 5† Thickness Class
18	2.5	0.35	50	0.35	50	0.35	50	0.35	50	0.35	50
	3	0.35	50	0.35	50	0.35	50	0.35	50	0.35	50
	4	0.35	50	0.35	50	0.35	50	0.35	50	0.35	50
	5	0.35	50	0.35	50	0.35	50	0.35	50	0.35	50
	6	0.35	50	0.35	50	0.35	50	0.35	50	0.35	50
	7	0.35	50	0.35	50	0.35	50	0.35	50	0.35	50
	8	0.35	50	0.35	50	0.35	50	0.35	50	0.35	50
	9	0.35	50	0.35	50	0.35	50	0.35	50	0.35	50
	10	0.35	50	0.35	50	0.35	50	0.35	50	0.35	50
	12	0.38	51	0.35	50	0.35	50	0.35	50	0.35	50
	14	0.38	51	0.35	50	0.35	50	0.35	50	0.35	50
	16	0.41	52	0.38	51	0.35	50	0.35	50	0.35	50
	20	0.44	53	0.41	52	0.35	50	0.35	50	0.35	50
	24	0.47	54	0.44	53	0.41	52	0.35	50	0.35	50
	28	0.50	55	0.47	54	0.44	53	0.35	50	0.35	50
	32	0.53	56	0.50	55	0.47	54	0.38	51	0.35	50
20	2.5	0.39	51	0.36	50	0.36	50	0.36	50	0.36	50
	3	0.36	50	0.36	50	0.36	50	0.36	50	0.36	50
	4	0.36	50	0.36	50	0.36	50	0.36	50	0.36	50
	5	0.36	50	0.36	50	0.36	50	0.36	50	0.36	50
	6	0.36	50	0.36	50	0.36	50	0.36	50	0.36	50
	7	0.36	50	0.36	50	0.36	50	0.36	50	0.36	50
	8	0.36	50	0.36	50	0.36	50	0.36	50	0.36	50
	9	0.36	50	0.36	50	0.36	50	0.36	50	0.36	50
	10	0.36	50	0.36	50	0.36	50	0.36	50	0.36	50
	12	0.39	51	0.36	50	0.36	50	0.36	50	0.36	50
	14	0.42	52	0.36	50	0.36	50	0.36	50	0.36	50
	16	0.42	52	0.39	51	0.36	50	0.36	50	0.36	50
	20	0.48	54	0.42	52	0.39	51	0.36	50	0.36	50
	24	0.51	55	0.48	54	0.42	52	0.36	50	0.36	50
	28	0.54	56	0.51	55	0.45	53	0.39	51	0.36	50
	32	—	—	0.54	56	0.48	54	0.42	52	0.36	50
24	2.5	0.41	51	0.38	50	0.38	50	0.38	50	0.38	50
	3	0.38	50	0.38	50	0.38	50	0.38	50	0.38	50
	4	0.38	50	0.38	50	0.38	50	0.38	50	0.38	50
	5	0.38	50	0.38	50	0.38	50	0.38	50	0.38	50
	6	0.38	50	0.38	50	0.38	50	0.38	50	0.38	50
	7	0.38	50	0.38	50	0.38	50	0.38	50	0.38	50
	8	0.38	50	0.38	50	0.38	50	0.38	50	0.38	50
	9	0.41	51	0.38	50	0.38	50	0.38	50	0.38	50
	10	0.41	51	0.38	50	0.38	50	0.38	50	0.38	50
	12	0.44	52	0.38	50	0.38	50	0.38	50	0.38	50
	14	0.47	53	0.41	51	0.38	50	0.38	50	0.38	50
	16	0.50	54	0.44	52	0.38	50	0.38	50	0.38	50
	20	0.53	55	0.50	54	0.44	52	0.38	50	0.38	50
	24	0.56	56	0.53	55	0.47	53	0.41	51	0.38	50
	28	—	—	0.56	56	0.53	55	0.44	52	0.38	50
	32	—	—	—	—	0.56	56	0.47	53	0.41	51
30	2.5	‡	‡	0.39	50	0.39	50	0.39	50	0.39	50
	3			0.39	50	0.39	50	0.39	50	0.39	50
	4			0.39	50	0.39	50	0.39	50	0.39	50
	5			0.39	50	0.39	50	0.39	50	0.39	50
	6			0.39	50	0.39	50	0.39	50	0.39	50
	7			0.39	50	0.39	50	0.39	50	0.39	50
	8			0.39	50	0.39	50	0.39	50	0.39	50
	9			0.39	50	0.39	50	0.39	50	0.39	50
	10			0.39	50	0.39	50	0.39	50	0.39	50
	12			0.43	51	0.39	50	0.39	50	0.39	50
	14			0.47	52	0.39	50	0.39	50	0.39	50
	16			0.51	53	0.43	51	0.39	50	0.39	50
	20			0.59	55	0.51	53	0.43	51	0.39	50
	24			0.63	56	0.55	54	0.47	52	0.39	50
	28			—	—	0.63	56	0.51	53	0.43	51
	32			—	—	—	—	0.55	54	0.47	52

† See corresponding illustrations in Fig. 3 of types of laying conditions.
‡ For pipe 30 in. and larger, consideration should be given to the use of laying conditions other than Type 1.

TABLE 51.1—(cont.)

Size in.	Depth of Cover ft.	Type 1† Thickness in.	Type 1† Thickness Class	Type 2† Thickness in.	Type 2† Thickness Class	Type 3† Thickness in.	Type 3† Thickness Class	Type 4† Thickness in.	Type 4† Thickness Class	Type 5† Thickness in.	Type 5† Thickness Class
36	2.5	‡	‡	0.43	50	0.43	50	0.43	50	0.43	50
	3			0.43	50	0.43	50	0.43	50	0.43	50
	4			0.43	50	0.43	50	0.43	50	0.43	50
	5			0.43	50	0.43	50	0.43	50	0.43	50
	6			0.43	50	0.43	50	0.43	50	0.43	50
	7			0.43	50	0.43	50	0.43	50	0.43	50
	8			0.43	50	0.43	50	0.43	50	0.43	50
	9			0.43	50	0.43	50	0.43	50	0.43	50
	10			0.43	50	0.43	50	0.43	50	0.43	50
	12			0.48	51	0.43	50	0.43	50	0.43	50
	14			0.53	52	0.48	51	0.43	50	0.43	50
	16			0.58	53	0.48	51	0.43	50	0.43	50
	20			0.63	54	0.58	53	0.48	51	0.43	50
	24			0.73	56	0.63	54	0.53	52	0.43	50
	28			—	—	0.68	55	0.58	53	0.48	51
	32			—	—	—	—	0.63	54	0.53	52
42	2.5	‡	‡	0.47	50	0.47	50	0.47	50	0.47	50
	3			0.47	50	0.47	50	0.47	50	0.47	50
	4			0.47	50	0.47	50	0.47	50	0.47	50
	5			0.47	50	0.47	50	0.47	50	0.47	50
	6			0.47	50	0.47	50	0.47	50	0.47	50
	7			0.47	50	0.47	50	0.47	50	0.47	50
	8			0.47	50	0.47	50	0.47	50	0.47	50
	9			0.47	50	0.47	50	0.47	50	0.47	50
	10			0.53	51	0.47	50	0.47	50	0.47	50
	12			0.53	51	0.47	50	0.47	50	0.47	50
	14			0.59	52	0.53	51	0.47	50	0.47	50
	16			0.65	53	0.59	52	0.47	50	0.47	50
	20			0.71	54	0.65	53	0.59	52	0.47	50
	24			0.83	56	0.71	54	0.65	53	0.47	50
	28			—	—	0.77	55	0.71	54	0.59	52
	32			—	—	—	—	0.71	54	0.65	53
48	2.5	‡	‡	0.51	50	0.51	50	0.51	50	0.51	50
	3			0.51	50	0.51	50	0.51	50	0.51	50
	4			0.51	50	0.51	50	0.51	50	0.51	50
	5			0.51	50	0.51	50	0.51	50	0.51	50
	6			0.51	50	0.51	50	0.51	50	0.51	50
	7			0.51	50	0.51	50	0.51	50	0.51	50
	8			0.51	50	0.51	50	0.51	50	0.51	50
	9			0.51	50	0.51	50	0.51	50	0.51	50
	10			0.58	51	0.51	50	0.51	50	0.51	50
	12			0.58	51	0.51	50	0.51	50	0.51	50
	14			0.65	52	0.58	51	0.51	50	0.51	50
	16			0.72	53	0.65	52	0.58	51	0.51	50
	20			0.79	54	0.72	53	0.65	52	0.51	50
	24			0.93	56	0.79	54	0.72	53	0.58	51
	28			—	—	0.86	55	0.79	54	0.65	52
	32			—	—	—	—	0.86	55	0.72	53
54	2.5	‡	‡	0.57	50	0.57	50	0.57	50	0.57	50
	3			0.57	50	0.57	50	0.57	50	0.57	50
	4			0.57	50	0.57	50	0.57	50	0.57	50
	5			0.57	50	0.57	50	0.57	50	0.57	50
	6			0.57	50	0.57	50	0.57	50	0.57	50
	7			0.57	50	0.57	50	0.57	50	0.57	50
	8			0.57	50	0.57	50	0.57	50	0.57	50
	9			0.57	50	0.57	50	0.57	50	0.57	50
	10			0.65	51	0.57	50	0.57	50	0.57	50
	12			0.65	51	0.57	50	0.57	50	0.57	50
	14			0.73	52	0.65	51	0.57	50	0.57	50
	16			0.81	53	0.73	52	0.65	51	0.57	50
	20			0.89	54	0.81	53	0.73	52	0.57	50
	24			1.05	56	0.89	54	0.81	53	0.65	51
	28			—	—	0.97	55	0.89	54	0.73	52
	32			—	—	—	—	0.97	55	0.81	53

Laying Condition

† See corresponding illustrations in Fig. 3 of types of laying conditions.
‡ For pipe 30 in. and larger, consideration should be given to the use of laying conditions other than Type 1

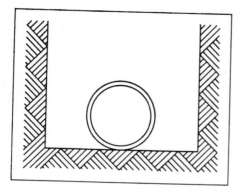
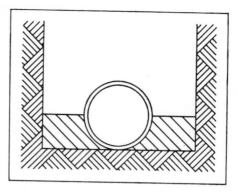

Type 1 * Flat-bottom trench.† Loose back-fill.

Type 2 Flat-bottom trench.† Backfill lightly consolidated to centerline of pipe.

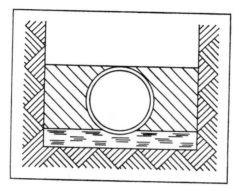

Type 3 Pipe bedded in 4-in.-minimum loose soil.‡ Backfill lightly consolidated to top of pipe.

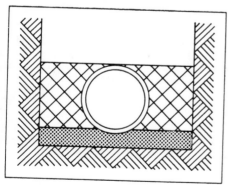
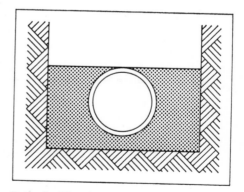

Type 4 Pipe bedded in sand, gravel, or crushed stone to depth of ⅛ pipe diameter, 4-in. minimum. Backfill compacted to top of pipe. (Approx. 80 per cent Standard Proctor, AASHTO § T-99)

Type 5 Pipe bedded to its centerline in compacted granular material, 4-in. minimum under pipe. Compacted granular or select ‡ material to top of pipe. (Approx. 90 per cent Standard Proctor, AASHTO § T-99)

Fig. 3. Standard Laying Conditions

* For pipe 30 in. and larger, consideration should be given to the use of laying conditions other than Type 1.
† "Flat-bottom" is defined as undisturbed earth.
‡ "Loose soil" or "select material" is defined as native soil excavated from the trench, free of rocks, foreign materials, and frozen earth.
§ American Association of State Highway and Transportation Officials, 341 National Press Bldg., Washington, D.C. 20004.

TABLE 51.2

Standard Thickness for Internal Pressure

Pipe Size in.	Rated Water Working Pressure*—psi									
	150		200		250		300		350	
	Thickness in.	Thickness Class	Thickness in.	Thickness Class	Thickness in.	Thickness Class	Thickness in.	Thickness Class	Thickness in.	Thickness Class
3	0.25	51	0.25	51	0.25	51	0.25	51	0.25	51
4	0.26	51	0.26	51	0.26	51	0.26	51	0.26	51
6	0.25	50	0.25	50	0.25	50	0.25	50	0.25	50
8	0.27	50	0.27	50	0.27	50	0.27	50	0.27	50
10	0.29	50	0.29	50	0.29	50	0.29	50	0.29	50
12	0.31	50	0.31	50	0.31	50	0.31	50	0.31	50
14	0.33	50	0.33	50	0.33	50	0.33	50	0.33	50
16	0.34	50	0.34	50	0.34	50	0.34	50	0.34	50
18	0.35	50	0.35	50	0.35	50	0.35	50	0.35	50
20	0.36	50	0.36	50	0.36	50	0.36	50	0.39	51
24	0.38	50	0.38	50	0.38	50	0.41	51	0.44	52
30	0.39	50	0.39	50	0.43	51	0.47	52	0.51	53
36	0.43	50	0.43	50	0.48	51	0.53	52	0.58	53
42	0.47	50	0.47	50	0.53	51	0.59	52	0.65	53
48	0.51	50	0.51	50	0.58	51	0.65	52	0.72	53
54	0.57	50	0.57	50	0.65	51	0.73	52	0.81	53

* These pipe are adequate for the rated working pressure plus a surge allowance of 100 psi.

TABLE 51.3

Rated Working Pressure and Maximum Depth of Cover

Pipe Size in.	Thickness Class	Nominal Thickness in.	Rated Water Working Pressure* psi	Laying Condition				
				Type 1	Type 2	Type 3	Type 4	Type 5
				Maximum Depth of Cover—ft†				
3	51	0.25	350	98	100‡	100‡	100‡	100‡
	52	0.28	350	100‡	100‡	100‡	100‡	100‡
	53	0.31	350	100‡	100‡	100‡	100‡	100‡
	54	0.34	350	100‡	100‡	100‡	100‡	100‡
	55	0.37	350	100‡	100‡	100‡	100‡	100‡
	56	0.40	350	100‡	100‡	100‡	100‡	100‡
4	51	0.26	350	76	86	96	100‡	100‡
	52	0.29	350	100‡	100‡	100‡	100‡	100‡
	53	0.32	350	100‡	100‡	100‡	100‡	100‡
	54	0.35	350	100‡	100‡	100‡	100‡	100‡
	55	0.38	350	100‡	100‡	100‡	100‡	100‡
	56	0.41	350	100‡	100‡	100‡	100‡	100‡
6	50	0.25	350	32	38	44	56	75
	51	0.28	350	49	57	64	80	100‡
	52	0.31	350	67	77	86	100‡	100‡
	53	0.34	350	91	100‡	100‡	100‡	100‡
	54	0.37	350	100‡	100‡	100‡	100‡	100‡
	55	0.40	350	100‡	100‡	100‡	100‡	100‡
	56	0.43	350	100‡	100‡	100‡	100‡	100‡
8	50	0.27	350	25	30	36	46	64
	51	0.30	350	36	42	49	61	81
	52	0.33	350	47	54	62	77	99
	53	0.36	350	64	73	82	100‡	100‡
	54	0.39	350	80	91	100‡	100‡	100‡
	55	0.42	350	98	100‡	100‡	100‡	100‡
	56	0.45	350	100‡	100‡	100‡	100‡	100‡
10	50	0.29	350	19	24	29	38	55
	51	0.32	350	27	32	38	49	66
	52	0.35	350	35	41	47	59	79
	53	0.38	350	45	52	59	74	95
	54	0.41	350	57	65	74	91	100‡
	55	0.44	350	67	77	86	100‡	100‡
	56	0.47	350	81	92	100‡	100‡	100‡
12	50	0.31	350	17	22	27	36	52
	51	0.34	350	23	28	33	43	60
	52	0.37	350	30	35	41	53	71
	53	0.40	350	36	42	49	61	81
	54	0.43	350	45	52	59	74	95
	55	0.46	350	54	62	71	87	100‡
	56	0.49	350	64	73	83	100‡	100‡
14	50	0.33	350	15	19	24	33	49
	51	0.36	350	19	23	28	38	55
	52	0.39	350	24	29	34	44	62
	53	0.42	350	30	35	41	53	71
	54	0.45	350	36	42	49	61	81
	55	0.48	350	43	50	57	71	92
	56	0.51	350	52	59	67	83	100‡
16	50	0.34	350	13	17	21	30	47
	51	0.37	350	16	21	25	34	51
	52	0.40	350	20	25	30	40	57
	53	0.43	350	25	30	36	46	64
	54	0.46	350	30	35	41	53	71
	55	0.49	350	35	41	47	59	79
	56	0.52	350	41	48	55	68	89
18	50	0.35	350	11	15	20	29	42
	51	0.38	350	14	19	23	32	49
	52	0.41	350	18	22	27	36	53
	53	0.44	350	22	26	31	41	58
	54	0.47	350	25	30	36	46	64
	55	0.50	350	30	35	41	53	71
	56	0.53	350	35	41	47	59	79

* These pipe are adequate for the rated working pressure plus a surge allowance of 100 psi. Ductile-iron pipe for working pressures higher than 350 psi is available.
† An allowance for a single H-20 truck with 1.5 impact factor is included for all depths of cover.
‡ Calculated maximum depth of cover exceeds 100 ft.

TABLE 51.3—(cont.)

Pipe Size in.	Thickness Class	Nominal Thickness in.	Rated Water Working Pressure* psi	Laying Condition				
				Type 1	Type 2	Type 3	Type 4	Type 5
				Maximum Depth of Cover—ft†				
20	50	0.36	300	10	14	18	27	38
	51	0.39	350	13	17	21	30	44
	52	0.42	350	16	20	25	34	50
	53	0.45	350	19	23	28	38	54
	54	0.48	350	22	27	32	42	59
	55	0.51	350	26	31	37	47	65
	56	0.54	350	30	35	41	53	71
24	50	0.38	250	8	12	17	23	31
	51	0.41	300	10	15	19	27	36
	52	0.44	350	13	17	21	30	41
	53	0.47	350	15	19	24	33	47
	54	0.50	350	18	22	27	36	53
	55	0.53	350	20	25	30	40	57
	56	0.56	350	24	29	34	44	61
30	50	0.39	200	§	10	14	18	25
	51	0.43	250		12	16	21	29
	52	0.47	300		14	19	24	33
	53	0.51	350		17	21	29	38
	54	0.55	350		19	24	33	44
	55	0.59	350		22	27	36	51
	56	0.63	350		26	31	41	57
36	50	0.43	200	§	10	13	17	25
	51	0.48	250		12	16	20	28
	52	0.53	300		15	19	24	32
	53	0.58	350		17	21	28	37
	54	0.63	350		20	25	33	43
	55	0.68	350		23	28	37	50
	56	0.73	350		26	31	41	59
42	50	0.47	200	§	9	13	16	24
	51	0.53	250		12	15	19	27
	52	0.59	300		14	18	22	30
	53	0.65	350		17	22	27	35
	54	0.71	350		20	24	32	42
	55	0.77	350		23	28	38	48
	56	0.83	350		26	31	41	57
48	50	0.51	200	§	9	12	15	23
	51	0.58	250		12	14	18	26
	52	0.65	300		14	18	21	30
	53	0.72	350		17	21	25	34
	54	0.79	350		20	24	30	40
	55	0.86	350		23	28	37	47
	56	0.93	350		26	31	41	55
54	50	0.57	200	§	9	12	15	23
	51	0.65	250		12	14	18	25
	52	0.73	300		14	17	21	29
	53	0.81	350		17	21	25	34
	54	0.89	350		20	25	30	40
	55	0.97	350		23	28	37	47
	56	1.05	350		27	32	42	55

* These pipe are adequate for the rated working pressure plus a surge allowance of 100 psi. Ductile-iron pipe for working pressures higher than 350 psi is available.

† An allowance for a single H-20 truck with 1.5 impact factor is included for all depths of cover.

§ For pipe 30 in. and larger, consideration should be given to the use of laying conditions other than Type 1.

14

TABLE 51.4

Standard Dimensions and Weights of Push-On-Joint Ductile-Iron Pipe

Size in.	Thickness Class	Thickness in.	OD* in.	Wt. of Barrel Per Ft lb	Wt. of Bell lb	18-Ft Laying Length		20-Ft Laying Length	
						Wt. Per Lgth.† lb	Avg. Wt. Per Ft‡ lb	Wt. Per Lgth.† lb	Avg. Wt. Per Ft‡ lb
3§	51	0.25	3.96	8.9	9	170		185	
	52	0.28	3.96	9.9	9	185	9.4	205	9.4
	53	0.31	3.96	10.9	9	205	10.4	225	10.4
	54	0.34	3.96	11.8	9	220	11.4	245	11.4
	55	0.37	3.96	12.8	9	240	12.3	265	12.2
	56	0.40	3.96	13.7	9	255	13.3	285	13.2
							14.2		14.2
4	51	0.26	4.80	11.3	11	215		235	
	52	0.29	4.80	12.6	11	240	11.9	265	11.8
	53	0.32	4.80	13.8	11	260	13.2	285	13.2
	54	0.35	4.80	15.0	11	280	14.4	310	14.4
	55	0.38	4.80	16.1	11	300	15.6	335	15.6
	56	0.41	4.80	17.3	11	320	16.7	355	16.6
							17.9		17.8
6	50	0.25	6.90	16.0	18	305		340	
	51	0.28	6.90	17.8	18	340	17.0	375	16.9
	52	0.31	6.90	19.6	18	370	18.8	410	18.7
	53	0.34	6.90	21.4	18	405	20.6	445	20.5
	54	0.37	6.90	23.2	18	435	22.4	480	22.3
	55	0.40	6.90	25.0	18	470	24.2	520	24.1
	56	0.43	6.90	26.7	18	500	26.0	550	25.9
							27.7		27.6
8	50	0.27	9.05	22.8	26	435		480	
	51	0.30	9.05	25.2	26	480	24.2	530	24.1
	52	0.33	9.05	27.7	26	525	26.6	580	26.5
	53	0.36	9.05	30.1	26	570	29.1	630	29.0
	54	0.39	9.05	32.5	26	610	31.5	675	31.4
	55	0.42	9.05	34.8	26	650	33.9	720	33.8
	56	0.45	9.05	37.2	26	695	36.2	770	36.1
							38.6		38.5
10	50	0.29	11.10	30.1	34	575		635	
	51	0.32	11.10	33.2	34	630	32.0	700	31.8
	52	0.35	11.10	36.2	34	685	35.1	760	34.9
	53	0.38	11.10	39.2	34	740	38.1	820	37.9
	54	0.41	11.10	42.1	34	790	41.1	875	40.9
	55	0.44	11.10	45.1	34	845	44.0	935	43.8
	56	0.47	11.10	48.0	34	900	47.0	995	46.8
							49.9		49.7

* Tolerance of OD of spigot end: 3–12 in., ±0.06 in.; 14–24 in., +0.05 in., −0.08 in.; 30–54 in., +0.08 in., −0.06 in.
† Including bell; calculated weight of pipe rounded off to nearest 5 lb.
‡ Including bell; average weight, per foot, based on calculated weight of pipe before rounding.
§ Pipe of 3-in. size also available in 12-ft laying length with following weights:

Thickness Class	Thickness in.	Wt. Per Lgth.† lb	Avg. Wt. Per Ft‡ lb
51	0.25	115	
52	0.28	130	9.6
53	0.31	140	10.6
54	0.34	150	11.6
55	0.37	165	12.6
56	0.40	175	13.6
			14.4

TABLE 51.4—(cont.)

Size in.	Thick-ness Class	Thick-ness in.	OD* in.	Wt. of Barrel Per Ft lb	Wt. of Bell lb	18-Ft Laying Length		20-Ft Laying Length	
						Wt. Per Lgth.† lb	Avg. Wt. Per Ft‡ lb	Wt. Per Lgth.† lb	Avg. Wt. Per Ft‡ lb
12	50	0.31	13.20	38.4	43	735	40.8	810	40.6
	51	0.34	13.20	42.0	43	800	44.4	885	44.2
	52	0.37	13.20	45.6	43	865	48.0	955	47.8
	53	0.40	13.20	49.2	43	930	51.6	1 025	51.4
	54	0.43	13.20	52.8	43	995	55.2	1 100	55.0
	55	0.46	13.20	56.3	43	1 055	58.7	1 170	58.4
	56	0.49	13.20	59.9	43	1 120	62.3	1 240	62.0
14	50	0.33	15.30	47.5	63	920	51.0	1 015	50.6
	51	0.36	15.30	51.7	63	995	55.2	1 095	54.8
	52	0.39	15.30	55.9	63	1 070	59.4	1 180	59.0
	53	0.42	15.30	60.1	63	1 145	63.6	1 265	63.2
	54	0.45	15.30	64.2	63	1 220	67.7	1 345	67.4
	55	0.48	15.30	68.4	63	1 295	71.9	1 430	71.6
	56	0.51	15.30	72.5	63	1 370	76.0	1 515	75.6
16	50	0.34	17.40	55.8	76	1 080	60.0	1 190	59.6
	51	0.37	17.40	60.6	76	1 165	64.8	1 290	64.4
	52	0.40	17.40	65.4	76	1 255	69.6	1 385	69.2
	53	0.43	17.40	70.1	76	1 340	74.3	1 480	73.9
	54	0.46	17.40	74.9	76	1 425	79.1	1 575	78.7
	55	0.49	17.40	79.7	76	1 510	83.9	1 670	83.5
	56	0.52	17.40	84.4	76	1 595	88.6	1 765	88.2
18	50	0.35	19.50	64.4	87	1 245	69.2	1 375	68.8
	51	0.38	19.50	69.8	87	1 345	74.6	1 485	74.2
	52	0.41	19.50	75.2	87	1 440	80.0	1 590	79.6
	53	0.44	19.50	80.6	87	1 540	85.4	1 700	85.0
	54	0.47	19.50	86.0	87	1 635	90.8	1 805	90.4
	55	0.50	19.50	91.3	87	1 730	96.1	1 915	95.6
	56	0.53	19.50	96.7	87	1 830	101.5	2 020	101.0
20	50	0.36	21.60	73.5	97	1 420	78.9	1 565	78.4
	51	0.39	21.60	79.5	97	1 530	84.9	1 685	84.4
	52	0.42	21.60	85.5	97	1 635	90.9	1 805	90.4
	53	0.45	21.60	91.5	97	1 745	96.9	1 925	96.4
	54	0.48	21.60	97.5	97	1 850	102.9	2 045	102.4
	55	0.51	21.60	103.4	97	1 950	108.8	2 165	108.2
	56	0.54	21.60	109.3	97	2 065	114.7	2 285	114.2

* Tolerance of OD of spigot end: 3–12 in., ±0.06 in.; 14–24 in., +0.05 in., −0.08 in.; 30–54 in., +0.08 in., −0.06 in.
† Including bell; calculated weight of pipe rounded off to nearest 5 lb.
‡ Including bell; average weight per foot, based on calculated weight of pipe before rounding.

DUCTILE–IRON PIPE

TABLE 51.4—(cont.)

Size in.	Thickness Class	Thickness in.	OD* in.	Wt. of Barrel Per Ft lb	Wt. of Bell lb	18-Ft Laying Length		20-Ft Laying Length	
						Wt. Per Lgth.† lb	Avg. Wt. Per Ft‡ lb	Wt. Per Lgth.† lb	Avg. Wt. Per Ft‡ lb
24	50	0.38	25.80	92.9	120	1 790	99.6	1 980	98.9
	51	0.41	25.80	100.1	120	1 929	106.8	2 120	106.1
	52	0.44	25.80	107.3	120	2 050	114.0	2 265	113.3
	53	0.47	25.80	114.4	120	2 180	121.1	2 410	120.4
	54	0.50	25.80	121.6	120	2 310	128.3	2 550	127.6
	55	0.53	25.80	128.8	120	2 440	135.5	2 695	134.8
	56	0.56	25.80	135.9	120	2 565	142.6	2 840	141.9
30	50	0.39	32.00	118.5	**	2 350	130.5	2 535	126.6
	51	0.43	32.00	130.5		2 565	142.5	2 775	138.6
	52	0.47	32.00	142.5		2 780	154.5	3 015	150.6
	53	0.51	32.00	154.4		2 995	166.4	3 250	162.6
	54	0.55	32.00	166.3		3 210	178.3	3 490	174.4
	55	0.59	32.00	178.2		3 425	190.2	3 725	186.4
	56	0.63	32.00	190.0		3 635	202.0	3 965	198.2
36	50	0.43	38.30	156.5	††	3 110	172.7	3 345	167.3
	51	0.48	38.30	174.5		3 435	190.7	3 705	185.3
	52	0.53	38.30	192.4		3 755	208.6	4 065	203.2
	53	0.58	38.30	210.3		4 075	226.5	4 420	221.1
	54	0.63	38.30	228.1		4 400	244.3	4 780	238.9
	55	0.68	38.30	245.9		4 720	262.1	5 135	256.7
	56	0.73	38.30	263.7		5 040	279.9	5 490	274.5
42	50	0.47	44.50	198.9	261			4 240	212.0
	51	0.53	44.50	224.0	261			4 740	237.0
	52	0.59	44.50	249.1	261			5 245	262.2
	53	0.65	44.50	274.0	261			5 740	287.0
	54	0.71	44.50	298.9	261			6 240	312.0
	55	0.77	44.50	323.7	261			6 735	336.8
	56	0.83	44.50	348.4	261			7 230	361.4
48	50	0.51	50.80	246.6	316			5 250	262.4
	51	0.58	50.80	280.0	316			5 915	295.8
	52	0.65	50.80	313.4	316			6 585	329.2
	53	0.72	50.80	346.6	316			7 250	362.4
	54	0.79	50.80	379.8	316			7 910	395.6
	55	0.86	50.80	412.9	316			8 575	428.7
	56	0.93	50.80	445.9	316			9 235	461.7
54	50	0.57	57.10	309.8	370			6 565	328.3
	51	0.65	57.10	352.7	370			7 425	371.2
	52	0.73	57.10	395.6	370			8 280	414.1
	53	0.81	57.10	438.3	370			9 135	456.8
	54	0.89	57.10	480.9	370			9 990	499.4
	55	0.97	57.10	523.4	370			10 840	541.9
	56	1.05	57.10	565.8	370			11 685	584.3

* Tolerances of OD of spigot end: 3–12 in., ±0.06 in.; 14–24 in., +0.05 in., −0.08 in.; 30–54 in., +0.08 in., −0.06 in.
† Including bell; calculated weight of pipe rounded off to nearest 5 lb.
‡ Including bell; average weight per foot, based on calculated weight of pipe before rounding.
** Weight of 30-in. bell is 216 lb for 18-ft pipe and 163 lb for 20-ft pipe.
†† Weight of 36-in. bell is 292 lb for 18-ft pipe and 216 lb for 20-ft pipe.

203

17

TABLE 51.5

Standard Dimensions and Weights of Mechanical-Joint Ductile-Iron Pipe

Size in.	Thick-ness Class	Thick-ness in.	OD* in.	Wt. of Barrel Per Ft lb	Wt. of Bell lb	18-ft Laying Length		20-ft Laying Length	
						Wt. Per Lgth.† lb	Avg. Wt. Per Ft‡ lb	Wt. Per Lgth.† lb	Avg. Wt. Per Ft‡ lb
3§	51	0.25	3.96	8.9	11	170	9.5	190	9.4
	52	0.28	3.96	9.9	11	190	10.5	210	10.4
	53	0.31	3.96	10.9	11	205	11.5	230	11.4
	54	0.34	3.96	11.8	11	225	12.4	245	12.4
	55	0.37	3.96	12.8	11	240	13.4	265	13.4
	56	0.40	3.96	13.7	11	260	14.3	285	14.2
4	51	0.26	4.80	11.3	16	220	12.2	240	12.1
	52	0.29	4.80	12.6	16	245	13.5	270	13.4
	53	0.32	4.80	13.8	16	265	14.7	290	14.6
	54	0.35	4.80	15.0	16	285	15.9	315	15.8
	55	0.38	4.80	16.1	16	305	17.0	340	16.9
	56	0.41	4.80	17.3	16	325	18.2	360	18.1
6	50	0.25	6.90	16.0	22	310	17.2	340	17.1
	51	0.28	6.90	17.8	22	340	19.0	380	18.9
	52	0.31	6.90	19.6	22	375	20.8	415	20.7
	53	0.34	6.90	21.4	22	405	22.6	450	22.5
	54	0.37	6.90	23.2	22	440	24.4	485	24.3
	55	0.40	6.90	25.0	22	470	26.2	520	26.1
	56	0.43	6.90	26.7	22	505	27.9	555	27.8
8	50	0.27	9.05	22.8	29	440	24.4	485	24.2
	51	0.30	9.05	25.2	29	485	26.8	535	26.6
	52	0.33	9.05	27.7	29	530	29.3	585	29.2
	53	0.36	9.05	30.1	29	570	31.7	630	31.6
	54	0.39	9.05	32.5	29	615	34.1	680	34.0
	55	0.42	9.05	34.8	29	655	36.4	725	36.2
	56	0.45	9.05	37.2	29	700	38.8	775	38.6
10	50	0.29	11.10	30.1	39	580	32.3	640	32.0
	51	0.32	11.10	33.2	39	635	35.4	705	35.2
	52	0.35	11.10	36.2	39	690	38.4	765	38.2
	53	0.38	11.10	39.2	39	745	41.4	825	41.2
	54	0.41	11.10	42.1	39	795	44.3	880	44.0
	55	0.44	11.10	45.1	39	850	47.3	940	47.0
	56	0.47	11.10	48.0	39	905	50.2	1 000	50.0
12	50	0.31	13.20	38.4	49	740	41.1	815	40.8
	51	0.34	13.20	42.0	49	805	44.7	890	44.4
	52	0.37	13.20	45.6	49	870	48.3	960	48.0
	53	0.40	13.20	49.2	49	935	51.9	1 035	51.6
	54	0.43	13.20	52.8	49	1 000	55.5	1 105	55.2
	55	0.46	13.20	56.3	49	1 060	59.0	1 175	58.8
	56	0.49	13.20	59.9	49	1 125	62.6	1 245	62.3
14	50	0.33	15.30	47.5	76	930	51.7	1 025	51.3
	51	0.36	15.30	51.7	76	1 005	55.9	1 110	55.5
	52	0.39	15.30	55.9	76	1 080	60.1	1 195	59.7
	53	0.42	15.30	60.1	76	1 160	64.3	1 280	63.9
	54	0.45	15.30	64.2	76	1 230	68.4	1 360	68.0
	55	0.48	15.30	68.4	76	1 305	72.6	1 445	72.2
	56	0.51	15.30	72.5	76	1 380	76.7	1 525	76.3

* Tolerances of OD of spigot end: 3–12 in., ±0.06 in.; 14–24 in., +0.05 in., −0.08 in.; 30–48 in., +0.08 in., −0.06 in.
† Including bell; calculated weight of pipe rounded off to nearest 5 lb.
‡ Including bell; average weight, per foot, based on calculated weight of pipe before rounding.
§ Pipe of 3-in. size also available in 12-ft laying length with following weights:

Thickness Class	Thickness in.	Wt. Per Lgth.† lb	Avg. Wt. Per Ft‡ lb
51	0.25	120	9.8
52	0.28	130	10.8
53	0.31	140	11.8
54	0.34	155	12.7
55	0.37	165	13.7
56	0.40	175	14.6

204

TABLE 51.5—(cont.)

Size in.	Thickness Class	Thickness in.	OD* in.	Wt. of Barrel Per Ft lb	Wt. of Bell†† lb	18-ft Laying Length		20-ft Laying Length	
						Wt. Per Lgth.† lb	Avg. Wt. Per Ft‡ lb	Wt. Per Lgth.† lb	Avg. Wt. Per Ft‡ lb
16	50	0.34	17.40	55.8	93	1 095	61.0	1 210	60.4
	51	0.37	17.40	60.6	93	1 185	65.8	1 305	65.2
	52	0.40	17.40	65.4	93	1 270	70.6	1 400	70.0
	53	0.43	17.40	70.1	93	1 355	75.3	1 495	74.8
	54	0.46	17.40	74.9	93	1 440	80.1	1 590	79.6
	55	0.49	17.40	79.7	93	1 530	84.9	1 685	84.4
	56	0.52	17.40	84.4	93	1 610	89.6	1 780	89.0
18	50	0.35	19.50	64.4	111	1 270	70.6	1 400	70.0
	51	0.38	19.50	69.8	111	1 365	76.0	1 505	75.4
	52	0.41	19.50	75.2	111	1 465	81.4	1 615	80.8
	53	0.44	19.50	80.6	111	1 560	86.8	1 725	86.2
	54	0.47	19.50	86.0	111	1 660	92.2	1 830	91.6
	55	0.50	19.50	91.3	111	1 755	97.5	1 935	96.8
	56	0.53	19.50	96.7	111	1 850	102.9	2 045	102.2
20	50	0.36	21.60	73.5	131	1 455	80.8	1 600	80.0
	51	0.39	21.60	79.5	131	1 560	86.8	1 720	86.0
	52	0.42	21.60	85.5	131	1 670	92.8	1 840	92.0
	53	0.45	21.60	91.5	131	1 780	98.8	1 960	98.0
	54	0.48	21.60	97.5	131	1 885	104.8	2 080	104.0
	55	0.51	21.60	103.4	131	1 990	110.7	2 200	110.0
	56	0.54	21.60	109.3	131	2 100	116.6	2 315	115.8
24	50	0.38	25.80	92.9	174	1 845	102.6	2 030	101.6
	51	0.41	25.80	100.1	174	1 975	109.8	2 175	108.8
	52	0.44	25.80	107.3	174	2 105	117.0	2 320	116.0
	53	0.47	25.80	114.4	174	2 235	124.1	2 460	123.1
	54	0.50	25.80	121.6	174	2 365	131.3	2 605	130.3
	55	0.53	25.80	128.8	174	2 490	138.5	2 750	137.5
	56	0.56	25.80	135.9	174	2 620	145.6	2 890	144.6
30	50	0.39	32.00	118.5	216	2 350	130.5	2 585	129.3
	51	0.43	32.00	130.5	216	2 565	142.5	2 825	141.3
	52	0.47	32.00	142.5	216	2 780	154.5	3 065	153.3
	53	0.51	32.00	154.4	216	2 995	166.4	3 305	165.2
	54	0.55	32.00	166.3	216	3 210	178.3	3 540	177.1
	55	0.59	32.00	178.2	216	3 425	190.2	3 780	189.0
	56	0.63	32.00	190.0	216	3 635	202.0	4 015	200.8
36	50	0.43	38.30	156.5	310	3 125	173.7	3 440	172.0
	51	0.48	38.30	174.5	310	3 450	191.7	3 800	190.0
	52	0.53	38.30	192.4	310	3 775	209.6	4 160	207.9
	53	0.58	38.30	210.3	310	4 095	227.5	4 515	225.8
	54	0.63	38.30	228.1	310	4 415	245.3	4 870	243.6
	55	0.68	38.30	245.9	310	4 735	263.1	5 230	261.4
	56	0.73	38.30	263.7	310	5 055	280.9	5 585	279.2
42	50	0.47	44.50	198.9	405			4 385	219.2
	51	0.53	44.50	224.0	405			4 885	244.2
	52	0.59	44.50	249.1	405			5 385	269.4
	53	0.65	44.50	274.0	405			5 885	294.2
	54	0.71	44.50	298.9	405			6 385	319.2
	55	0.77	44.50	323.7	405			6 880	344.0
	56	0.83	44.50	348.4	405			7 375	368.6
48	50	0.51	50.80	246.6	505			5 435	271.8
	51	0.58	50.80	280.0	505			6 105	305.2
	52	0.65	50.80	313.4	505			6 775	338.6
	53	0.72	50.80	346.6	505			7 435	371.8
	54	0.79	50.80	379.8	505			8 100	405.0
	55	0.86	50.80	412.9	505			8 765	438.2
	56	0.93	50.80	445.9	505			9 425	471.2

* Tolerances of OD of spigot end: 3–12 in., ±0.06 in.; 14–24 in., +0.05 in., −0.08 in.; 30–48 in., +0.08 in., −0.06 in.

†† The mechanical joint bell for 30–48 in. sizes of ductile-iron pipe have thicknesses different from those shown in ANSI A21.11 (AWWA C111), which are based on gray-iron pipe. These reduced thicknesses provide a lighter weight bell, which is compatible with the wall thicknesses of ductile-iron pipe. The internal socket dimensions, bolt circle and bolt holes of the redesigned bell remain identical to those specified in A21.11 (AWWA C111) to assure interchangeability of the joint.

† Including bell; calculated weight of pipe rounded off to nearest 5 lb.

‡ Including bell; average weight per foot, based on calculated weight of pipe before rounding.

Appendix

This appendix is for information and is not a part of ANSI A21.51 (AWWA C151).

TABLE A.1

Pipe Thicknesses Required for Different Tap Sizes as per ANSI B2.1 for Standard Taper Pipe Threads With Two, Three, and Four Full Threads

Pipe Size in.	No. of Threads	Tap Size—in.									
		$\frac{1}{2}$	$\frac{3}{4}$	1	$1\frac{1}{4}$	$1\frac{1}{2}$	2	$2\frac{1}{2}$	3	$3\frac{1}{2}$	4
		Pipe Thickness—in.									
3	2	0.18	0.21	0.28							
3	3	0.26	0.29	0.37							
3	4	0.33	0.36	0.46							
4	2	0.17	0.19	0.26	0.31						
4	3	0.25	0.27	0.35	0.40						
4	4	0.32	0.34	0.44	0.49						
6	2	0.17	0.18	0.23	0.27	0.30					
6	3	0.25	0.26	0.32	0.36	0.39					
6	4	0.32	0.33	0.41	0.45	0.48					
8	2	0.16	0.17	0.22	0.24	0.27	0.33				
8	3	0.24	0.25	0.31	0.33	0.36	0.42				
8	4	0.31	0.32	0.40	0.42	0.45	0.51				
10	2	0.15	0.17	0.21	0.23	0.25	0.30	0.44			
10	3	0.23	0.25	0.30	0.32	0.34	0.39	0.56			
10	4	0.30	0.32	0.39	0.41	0.43	0.48	0.69			
12	2	0.15	0.16	0.20	0.22	0.24	0.28	0.40	0.48		
12	3	0.23	0.24	0.29	0.31	0.33	0.37	0.52	0.60		
12	4	0.30	0.31	0.38	0.40	0.42	0.46	0.65	0.73		
14	2	0.15	0.16	0.20	0.22	0.23	0.26	0.38	0.45	0.51	0.58
14	3	0.23	0.24	0.29	0.31	0.32	0.35	0.50	0.58	0.64	0.70
14	4	0.30	0.31	0.38	0.40	0.41	0.44	0.63	0.70	0.76	0.83
16	2	0.15	0.16	0.20	0.21	0.22	0.25	0.37	0.43	0.48	0.54
16	3	0.23	0.24	0.29	0.30	0.31	0.34	0.50	0.56	0.60	0.66
16	4	0.30	0.31	0.38	0.39	0.40	0.43	0.62	0.68	0.73	0.79
18	2	0.15	0.15	0.19	0.21	0.22	0.24	0.35	0.41	0.46	0.51
18	3	0.23	0.23	0.28	0.30	0.31	0.33	0.48	0.54	0.58	0.64
18	4	0.30	0.30	0.37	0.39	0.40	0.42	0.60	0.66	0.71	0.76
20	2	0.15	0.15	0.19	0.20	0.21	0.23	0.34	0.39	0.44	0.49
20	3	0.23	0.23	0.28	0.29	0.30	0.32	0.46	0.52	0.56	0.62
20	4	0.30	0.30	0.37	0.38	0.39	0.41	0.59	0.64	0.69	0.74
24	2	0.14	0.15	0.19	0.20	0.21	0.22	0.32	0.37	0.40	0.45
24	3	0.22	0.23	0.28	0.29	0.30	0.31	0.44	0.50	0.52	0.58
24	4	0.29	0.30	0.37	0.38	0.39	0.40	0.57	0.62	0.65	0.70
30	2	0.14	0.15	0.19	0.19	0.20	0.21	0.31	0.34	0.37	0.41
30	3	0.22	0.23	0.28	0.28	0.29	0.30	0.44	0.46	0.50	0.54
30	4	0.29	0.30	0.37	0.37	0.38	0.39	0.56	0.59	0.62	0.66
36	2	0.14	0.14	0.18	0.19	0.20	0.21	0.30	0.33	0.35	0.38
36	3	0.22	0.22	0.27	0.28	0.29	0.30	0.42	0.46	0.48	0.50
36	4	0.29	0.29	0.36	0.37	0.38	0.39	0.55	0.58	0.60	0.63
42	2	0.14	0.14	0.18	0.19	0.19	0.20	0.29	0.32	0.34	0.36
42	3	0.22	0.22	0.27	0.28	0.28	0.29	0.42	0.44	0.46	0.48
42	4	0.29	0.29	0.36	0.37	0.37	0.38	0.54	0.57	0.59	0.61
48	2	0.14	0.14	0.18	0.18	0.19	0.20	0.29	0.31	0.32	0.35
48	3	0.22	0.22	0.27	0.27	0.28	0.29	0.42	0.44	0.44	0.48
48	4	0.29	0.29	0.36	0.36	0.37	0.38	0.54	0.56	0.57	0.60
54	2	0.14	0.14	0.17	0.18	0.18	0.19	0.28	0.30	0.32	0.34
54	3	0.21	0.21	0.26	0.27	0.27	0.28	0.41	0.43	0.44	0.47
54	4	0.29	0.29	0.35	0.35	0.36	0.37	0.53	0.55	0.57	0.59

TABLE A.2

Pipe Thicknesses Required for Different Tap Sizes as per AWWA C800 for Standard Corporation Stop Threads* With Two, Three, and Four Full Threads

Pipe Size in.	No. of Threads	Tap Size—in.						
		½	⅝	¾	1	1¼	1½	2
		Pipe Thickness—in.						
3	2	0.21	0.24	0.25	0.33			
3	3	0.29	0.32	0.33	0.41			
3	4	0.36	0.39	0.40	0.49			
4	2	0.19	0.22	0.23	0.30	0.36		
4	3	0.27	0.30	0.31	0.38	0.45		
4	4	0.34	0.37	0.38	0.46	0.54		
6	2	0.18	0.20	0.20	0.26	0.30	0.35	
6	3	0.26	0.28	0.28	0.34	0.39	0.44	
6	4	0.33	0.35	0.35	0.42	0.48	0.53	
8	2	0.17	0.18	0.19	0.24	0.27	0.31	0.39
8	3	0.25	0.26	0.27	0.32	0.36	0.40	0.48
8	4	0.32	0.33	0.34	0.40	0.45	0.49	0.57
10	2	0.17	0.17	0.18	0.23	0.25	0.28	0.35
10	3	0.25	0.25	0.26	0.31	0.34	0.37	0.44
10	4	0.32	0.32	0.33	0.39	0.43	0.46	0.53
12	2	0.16	0.17	0.17	0.22	0.24	0.26	0.32
12	3	0.24	0.25	0.25	0.30	0.33	0.35	0.41
12	4	0.31	0.32	0.32	0.38	0.42	0.44	0.50
14	2	0.16	0.17	0.17	0.21	0.23	0.25	0.30
14	3	0.24	0.25	0.25	0.29	0.32	0.34	0.39
14	4	0.31	0.32	0.32	0.37	0.41	0.43	0.48
16	2	0.16	0.16	0.17	0.21	0.22	0.24	0.28
16	3	0.24	0.24	0.25	0.29	0.31	0.33	0.37
16	4	0.31	0.31	0.32	0.37	0.40	0.42	0.46
18	2	0.15	0.16	0.16	0.20	0.21	0.23	0.27
18	3	0.23	0.24	0.24	0.28	0.30	0.32	0.36
18	4	0.30	0.31	0.31	0.36	0.39	0.41	0.45
20	2	0.15	0.16	0.16	0.20	0.21	0.23	0.26
20	3	0.23	0.24	0.24	0.28	0.30	0.32	0.35
20	4	0.30	0.31	0.31	0.36	0.39	0.41	0.44
24	2	0.15	0.15	0.16	0.19	0.21	0.22	0.24
24	3	0.23	0.23	0.24	0.27	0.30	0.31	0.33
24	4	0.30	0.30	0.31	0.35	0.39	0.40	0.42
30	2	0.15	0.15	0.16	0.19	0.20	0.21	0.23
30	3	0.23	0.23	0.24	0.27	0.29	0.30	0.32
30	4	0.30	0.30	0.31	0.35	0.38	0.39	0.41
36	2	0.14	0.15	0.15	0.19	0.20	0.20	0.22
36	3	0.22	0.23	0.23	0.27	0.29	0.29	0.31
36	4	0.29	0.30	0.30	0.35	0.38	0.38	0.40
42	2	0.14	0.14	0.15	0.18	0.19	0.20	0.21
42	3	0.22	0.22	0.23	0.26	0.28	0.29	0.30
42	4	0.29	0.29	0.30	0.34	0.37	0.38	0.39
48	2	0.14	0.14	0.15	0.18	0.18	0.19	0.20
48	3	0.22	0.22	0.23	0.26	0.27	0.28	0.29
48	4	0.29	0.29	0.30	0.34	0.36	0.37	0.38
54	2	0.14	0.14	0.14	0.17	0.18	0.19	0.20
54	3	0.22	0.22	0.22	0.25	0.27	0.28	0.29
54	4	0.29	0.29	0.29	0.34	0.36	0.36	0.38

* This thread is commonly known to the trade as the Mueller thread.

ANSI A21.52-1976
REVISION OF
A21.52-1971

AMERICAN NATIONAL STANDARD
for
DUCTILE-IRON PIPE, CENTRIFUGALLY CAST, IN METAL MOLDS OR SAND-LINED MOLDS, FOR GAS

SECRETARIATS

AMERICAN GAS ASSOCIATION

AMERICAN WATER WORKS ASSOCIATION

NEW ENGLAND WATER WORKS ASSOCIATION

Revised Edition Approved By

American National Standards Institute June 25, 1976

FOREWORD

This foreword is for information and is not a part of ANSI A21.52.

American National Standards Committee A21 on Cast-Iron Pipe and Fittings was organized in 1926 under the sponsorship of the American Gas Association, the American Society for Testing and Materials, the American Water Works Association and the New England Water Works Association. The present scope of Committee A21 activity is: Since 1972 the co-secretariats have been AGA, AWWA and NEWWA. AWWA is the administrative secretariat.

Standardization of specifications for cast-iron and ductile-iron pressure pipe for gas, water and other liquids, and fittings for use with such pipe. These specifications to include design, dimensions, materials, coatings, linings, joints, accessories, and methods of inspection and test.

The work of Committee A21 is carried out by subcommittees. The directive to Subcommittee 1—Pipe is that:

The scope of the Subcommittee activity shall include the periodic review of all current A21 standards for pipe, the preparation of revisions or new standards when needed, as well as other matters pertaining to pipe standards.

The first edition of A21.52, the standard for ductile-iron pipe for gas, was issued in 1965 and a revision was issued in 1971. Subcommittee 1 reviewed this standard and submitted a proposed revision to American National Standards Committee A21 in 1975.

Revisions

1. An additional minus tolerance of 0.02 in. is permitted along the barrel of the pipe for a distance not to exceed 12-in. The weight tolerance permitted for any single pipe is 6 percent for 12-in. or smaller pipe and 5 percent for pipe larger than 12-in.

2. The standard laying conditions have been renamed. A becomes Type 1 and B becomes Type 2.

3. The standard thickness classes have been renumbered. Class 1 becomes Class 51, Class 2 becomes Class 52, etc.

4. Some thicknesses have been changed to reflect the effects of the following changes in A21.50: (a) change to AASHO H-20 truck loads; (b) change to prism earth loads for all pipe sizes; (c) change to 45° bedding angle for Laying Condition Type 2. The design stresses and design safety factors are unchanged from the 1971 edition.

Options

This standard includes certain options which, if desired, must be specified on the purchase order. Also, a number of items must be specified to describe completely the pipe required. The following summarizes these details and available options, and lists the sections of the standard where they can be found.

1. Size, thickness or class and laying length-Tables

2. a. Special Joints—Sec. 52-3.1
    b. Specify ductile-iron gland if required—Sec. 52-3.1

3. Certification by manufacturer—Sec. 52-4.2

4. Inspection by purchaser—Sec. 52-5

5. Coating—Sec. 52-8.1

6. Special Coating—Sec. 52-8.2

7. Special marking on pipe—Sec. 52-11

8. Written transcripts of foundry records—Sec. 52-14

9. Special tests—Sec. 52-15

AMERICAN NATIONAL STANDARD

for

DUCTILE-IRON PIPE CENTRIFUGALLY CAST,

IN METAL MOLDS OR SAND-LINED MOLDS

FOR GAS

Section 52-1—Scope

This standard covers 3 inch through 24 inch Mechanical Joint ductile-iron pipe, centrifugally cast, in metal molds or sand-lined molds for gas. This standard may be used for pipe with such other types of joints as may be agreed upon at the time of purchase.

Section 52-2—Definitions

Under this standard, the following definitions shall apply:

52-2.1 *Purchaser.* The party entering into a contract or agreement to purchase pipe according to this standard.

52-2.2 *Manufacturer.* The party that produces the pipe.

52-2.3 *Inspector.* The representative of the purchaser, authorized to inspect in behalf of the purchaser, to determine whether or not the pipe meet this standard.

52-2.4 *Ductile Iron.* A cast ferrous material in which a major part of the carbon content occurs as free carbon in nodular or spheroidal form.

52-2.5 *Mechanical-Joint.* The gasketed and bolted joint as detailed in ANSI A21.11 of latest revision.

Section 52-3—General Requirements

52-3.1 Pipe with mechanical joints shall conform to the applicable dimensions and weights shown in the tables in this standard and to the applicable requirements of ANSI A21.11 of latest revision. Unless otherwise specified, the mechanical-joint glands shall be cast iron in accordance with ANSI A21.11 of latest revision and bolts shall conform to the requirements of the same standard. Pipe with other types of joints shall comply with the joint dimensions and weights agreed upon at the time of purchase, but in all other respects shall fulfill the requirements of this standard.

52-3.2　The nominal laying length of the pipe shall be as shown in the tables. A maximum of 20 percent of the total number of pipe of each size specified in an order may be furnished as much as 24 inches shorter than the nominal laying length and an additional 10 percent may be furnished as much as 6 inches shorter than nominal laying length.

Section 52-4—Inspection and Certification by Manufacturer

52-4.1　The manufacturer shall establish the necessary quality control and inspection practice to assure compliance with this standard.

52-4.2　The manufacturer shall, if required on the purchase order, furnish a sworn statement that the inspection and all of the specified tests have been made and the results thereof comply with the requirements of this standard.

52-4.3　All pipe shall be clean and sound without defects which will impair their service. Repairing of defects by welding or other method shall not be allowed if such repairs will adversely affect the serviceability of the pipe or its capability to meet strength requirements of this standard.

Section 52-5—Inspection by Purchaser

52-5.1　When the purchaser desires to inspect pipe at the manufacturer's plant, the purchaser shall so specify on the purchase order, stating the conditions (such as time, and the extent of inspection) under which the inspection shall be made.

52-5.2　The inspector shall have free access to those parts of the manufacturer's plant which are necessary to assure compliance with this standard. The manufacturer shall make available for the inspector's use such gages as are necessary for inspection. The manufacturer shall provide the inspector with assistance as necessary for handling of pipe.

Section 52-6—Delivery and Acceptance

All pipe and accessories shall comply with this standard. Pipe and accessories not complying with this standard shall be replaced by the manufacturer at the agreed point of delivery. The manufacturer shall not be liable for shortages or damaged pipe after acceptance at the agreed point of delivery except as recorded on the delivery receipt or similar document by the carrier's agent.

Section 52-7—Tolerances or Permitted Variations

52-7.1 *Dimensions.* The spigot end, bell and socket of the pipe and the accessories shall be gaged with suitable gages at sufficiently frequent intervals to assure that the dimensions comply with the requirements of this standard.

The smallest inside diameter of the sockets and the outside of the spigot ends shall be tested with circular gages. Other socket dimensions shall be gaged as appropriate.

52-7.2 *Thickness.* Minus thickness tolerances of pipe and bell shall not exceed those shown below:

Size, Inch	Minus Tolerance, Inch
3-8	0.05
10-12	0.06
14-24	0.07

An additional minus tolerance of .02 in. shall be permitted along the barrel of the pipe for a distance not to exceed 12 in.

52-7.3 *Weight.* The weight of any single pipe shall not be less than the tabulated weight by more than 6 percent for pipe 12 inches or smaller in diameter, nor by more than 5 percent for pipe larger than 12 inches in diameter.

Section 52-8—Coating

52-8.1 *Outside Coating.* Unless otherwise specified on the purchase order, pipe may be coated on the outside at the manufacturer's option with a bituminous coating approximately one mil thick. The finished coating shall be continuous, smooth, neither brittle when cold nor sticky when exposed to the sun and shall be strongly adherent to the pipe.

52-8.2 *Special Coating.* For special conditions, other types of coatings may be available. Such special coatings shall be specified in the invitation for bids and on the purchase order.

Section 52-9—Hydrostatic Test

Each pipe shall be subjected to a hydrostatic test of not less than 500 psi. The pipe shall be under the full test pressure for at least 10 seconds. Suitable controls and recording devices shall be provided so that the test pressure and duration may be adequately ascertained. Any pipe that leaks or does not withstand the test pressure shall be rejected.

Section 52-10—Air Test

In addition to the hydrostatic test, each length of pipe and its socket sealing surface shall be subjected to an air test prior to coating. These tests may be made simultaneously or separately at the manufacturer's option. The air test pressure shall be a minimum of 50 psi. The joint sealing surface test shall be made in such a manner as to assure no leakage when the joint is properly assembled. The pipe shall be under the test pressure a sufficient length of time to permit careful inspection for leakage. While under this air pressure, the pipe shall be immersed in water or covered with a soapy water solution. Any length of pipe that leaks shall be rejected.

Section 52-11—Marking Pipe

The weight, class or nominal thickness, and casting period shall be shown on each pipe. The manufacturer's mark, the year in which the pipe was produced and the letters "DI" or "DUCTILE" shall be cast or stamped on the pipe. When specified on the purchase order initials not exceeding four in number shall be cast or stamped on the pipe. All required markings shall be clear and legible and all cast marks shall be on or near the bell.

All letters and numerals on pipe sizes 8-in. and larger shall be not less than ½-in. in height.

Section 52-12—Acceptance Tests

The standard acceptance tests for the physical characteristics of the pipe shall be as follows:

52-12.1 *Tensile Test.* A tensile test specimen shall be cut longitudinally from the midsection of the pipe wall. This specimen shall be machined and tested in accordance with Figure 1 and ASTM (American Society for Testing and Materials) designation E8-69 Tension Testing of Metallic Materials. The yield strength shall be determined by the 0.2% offset, halt of pointer, or extension under load methods. If check tests are to be made, the 0.2% offset method shall be used. All specimens shall be tested at room temperature $(70 \pm 10F)$.

52-12.1.1 *Acceptance Values.* The acceptance values for test specimens shall be as follows:

Grade of Iron	Tensile Strength, Minimum psi	Yield Strength, Minimum psi	Per cent Elongation, Minimum
60-42-10	60,000	42,000	10

52-12.2 *Impact Test.* Tests shall be made in accordance with ASTM designation E23-66 Notched Charpy Tests, except that specimens shall be 0.500 inch and full thickness of pipe wall. The notched impact test specimen shall be in accordance with Figure 2. In cases when the pipe wall thickness exceeds 0.40 inch, the impact specimen may be machined to a nominal thickness of 0.40 inch.

In all cases, impact values are to be corrected to 0.40 inch wall thickness by calculation as follows:

$$\text{Impact value corrected} = \frac{0.40 \text{ inch}}{t} \times \text{Impact value actual}$$

where t = thickness of the specimen in inches (wall thickness of pipe).

Charpy test machine anvil shall not be moved to compensate for the variation of cross section dimensions of the test specimen.

52-12.2.1 *Acceptance Value.* The corrected acceptance value for notched impact test specimens shall be a minimum of 7 foot pounds for tests conducted at 70 ± 10F.

52-12.3 *Sampling.* At least one tensile and impact sample shall be taken during each casting period of approximately three hours. Samples shall be selected to properly represent extremes of pipe diameters and thicknesses.

Section 52-13—Additional Control Tests by Manufacturer

Low temperature impact tests shall be made from at least one-third of the test pipe specified in Section 52-12.3 to assure compliance with a minimum corrected value of 3 foot pounds for tests conducted at −40F. Test specimens shall be prepared and tested in accordance with Section 52-12.2.

In addition, the manufacturer shall conduct such other control tests as necessary to assure continuing compliance with this standard.

Section 52-14—Foundry Records

The results of the following tests shall be recorded and retained for one year and shall be available to the purchaser at the foundry. Written transcripts of the results of the following tests shall be furnished when specified on the purchase order:

Acceptance tests, Section 52-12

Low-temperature impact tests, Section 52-13.

Section 52-15—Additional Tests Required by Purchaser

When tests other than those provided in this standard are required by the purchaser, such tests shall be specified in the invitation for bids and on the purchase order.

Section 52-16—Defective Specimens and Retests

When any physical test specimen shows defective machining or lack of continuity of metal, it shall be discarded and replaced by another specimen. When any sound test specimen fails to meet the specified requirements, the pipe from which it was taken shall be rejected and a retest may be made on two additional sound specimens from pipe cast in the same period as the specimen which failed. Both of the additional specimens shall meet the prescribed tests to qualify the pipe produced in that period.

Section 52-17—Rejection of Pipe

When any physical acceptance test fails to meet the requirements of Section 52-12 or Section 52-16, the pipe cast in the same period shall be rejected except as subject to the provision of Section 52-18.

Section 52-18—Determining Rejection

The manufacturer may determine the amount of rejection by making similar additional tests of pipe, of the same size, until the rejected lot is bracketed, in order of manufacture, by an acceptable test at each end of the interval in question. When pipe of one size is rejected from a casting period, the acceptability of pipe of different sizes from that same period may be established by making the acceptance tests for these sizes specified in Section 52-12.

DIMENSIONS

| | Std. Specimen 0.500" round | Small Size Specimens Proportional to Standard | | | |
		0.350" round	0.250" round	0.175" round	0.125" round
G–Gage Length	2.000±0.005"	1.400±0.005"	1.000±0.005"	0.700±0.005"	0.500±0.005"
D–Diameter (Note 1)	0.500±0.010"	0.350±0.007"	0.250±0.005"	0.175±0.005"	0.125±0.005"
R–Radius of Fillet	$\frac{3}{8}$", Min.	$\frac{1}{4}$", Min.	$\frac{3}{16}$", Min.	$\frac{3}{32}$", Min.	$\frac{3}{32}$", Min.
A–Length of Reduced Section (Note 2)	$2\frac{1}{4}$", Min.	$1\frac{3}{4}$", Min.	$1\frac{1}{4}$", Min.	$\frac{3}{4}$", Min.	$\frac{5}{8}$", Min.
Thickness of the section from the wall of the pipe from which the tensile specimen is to be machined.	.71" & Greater	.50" thru .70"	.35" thru .49"	.25" thru .34"	.18" thru .24"

NOTE 1. The reduced section may have a gradual taper from the ends toward the center with the ends not more than 0.005 in. larger in diameter than the center on the standard specimen and not more than 0.003 in. larger in diameter than the center on the small size specimens.

NOTE 2. If desired, on the small size specimens the length of the reduced section may be increased to accommodate an extensometer. However, reference marks for the measurement of elongation should nevertheless be spaced at the indicated gage length.

NOTE 3. The gage length and fillets shall be as shown, but the ends may be of any form to fit the holders of the testing machine in such a way that the load shall be axial. If the ends are to be held in grips it is desirable, if possible, to make the length of the grip section great enough to allow the specimen to extend into the grips a distance equal to two thirds or more of the length of the grips.

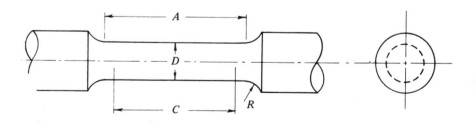

Figure 1. Tensile Test Specimen

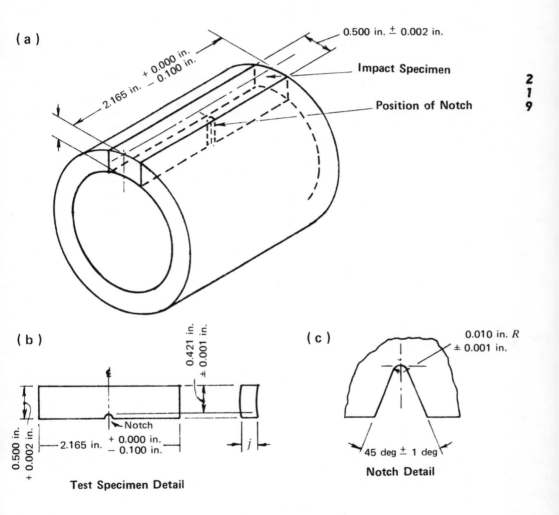

(a)

0.500 in. ± 0.002 in.

2.165 in. + 0.000 in. − 0.100 in.

Impact Specimen

Position of Notch

**2
1
9**

(b)

0.421 in. ± 0.001 in.

Notch

0.500 in. + 0.002 in.

2.165 in. + 0.000 in. − 0.100 in.

j

Test Specimen Detail

(c)

0.010 in. *R* ± 0.001 in.

45 deg ± 1 deg

Notch Detail

Figure 2. Impact Test Specimen

TABLE 1

Selection Table for Mechanical-Joint Ductile Iron Pipe

The thicknesses in this table are equal to or in excess of those required to withstand 250 psi working pressure.

All pipe in this table for the depths of cover indicated are adequate for trench loads including truck superloads under laying condition Type 2 (flat-bottom trench, backfill lightly consolidated to centerline of pipe).

For other depths of cover and laying condition Type 1 see Table 2. The basis of design is ANSI A21.50 except that for gas pipe the bending stress, f, is 36,000 psi and the safety factor for internal pressure is 2.5. A surge allowance is not required for gas pipe.

Thread engagement in taps for service connections and bag holes may require consideration in selecting pipe thicknesses. See Appendix.

Size, in.	Thickness Class	Thick- ness in.	OD	Weight Based on 18-ft. Laying Length Per Length†	Avg. Per Foot*	Weight Based on 20-ft. Laying Length Per Length†	Avg. Per Foot*
					lb.		
5-ft. Cover							
3**	52	0.28	3.96	190	10.5	210	10.4
4	52	0.29	4.80	245	13.5	270	13.4
6	52	0.31	6.90	375	20.8	415	20.7
8	52	0.33	9.05	530	29.3	585	29.2
10	52	0.35	11.10	690	38.4	765	38.2
12	52	0.37	13.20	870	48.3	960	48.0
14	51	0.36	15.30	1005	55.9	1110	55.5
16	51	0.37	17.40	1185	65.8	1305	65.2
18	51	0.38	19.50	1365	76.0	1505	75.4
20	51	0.39	21.60	1560	86.8	1720	86.0
24	51	0.41	25.80	1975	109.8	2175	108.8
8-ft. Cover							
3**	52	0.28	3.96	190	10.5	210	10.4
4	52	0.29	4.80	245	13.5	270	13.4
6	52	0.31	6.90	375	20.8	415	20.7
8	52	0.33	9.05	530	29.3	585	29.2
10	52	0.35	11.10	690	38.4	765	38.2
12	52	0.37	13.20	870	48.3	960	48.0
14	51	0.36	15.30	1005	55.9	1110	55.5
16	51	0.37	17.40	1185	65.8	1305	65.2
18	51	0.38	19.50	1365	76.0	1505	75.4
20	51	0.39	21.60	1560	86.8	1720	86.0
24	51	0.41	25.80	1975	109.8	2175	108.8

† Including bell. Calculated weights of pipe rounded off to nearest 5 lb.
* Including bell. Average weight, per foot, based on calculated weight before rounding.
**Pipe of 3-in. size also available in 12-ft. laying length. Weight, per length† is 130 lb. and average weight, per foot*, is 10.8 lb. for all standard depths of cover.

TABLE 1 (Continued)

Selection Table for Mechanical-Joint Ductile Iron-Pipe

The thicknesses in this table are equal to or in excess of those required to withstand 250 psi working pressure.

All pipe in this table for the depths of cover indicated are adequate for trench loads including truck superloads under laying condition Type 2 (flat-bottom trench, backfill lightly consolidated to centerline of pipe).

For other depths of cover and laying condition Type 1 see Table 2. The basis of design is ANSI A21.50 except that for gas pipe the bending stress, f, is 36,000 psi and the safety factor for internal pressure is 2.5. A surge allowance is not required for gas pipe.

Thread engagement in taps for service connections and bag holes may require consideration in selecting pipe thicknesses. See Appendix.

Size, in.	Thickness Class	Thick-ness in.	OD	Weight Based on 18-ft. Laying Length		Weight Based on 20-ft. Laying Length	
				Per Length†	Avg. Per Foot*	Per Length†	Avg. Per Foot*
					lb.		
12-ft. Cover							
3**	52	0.28	3.96	190	10.5	210	10.4
4	52	0.29	4.80	245	13.5	270	13.4
6	52	0.31	6.90	375	20.8	415	20.7
8	52	0.33	9.05	530	29.3	585	29.2
10	52	0.35	11.10	690	38.4	765	38.2
12	52	0.37	13.20	870	48.3	960	48.0
14	51	0.36	15.30	1005	55.9	1110	55.5
16	51	0.37	17.40	1185	65.8	1305	65.2
18	51	0.38	19.50	1365	76.0	1505	75.4
20	51	0.39	21.60	1560	86.8	1720	86.0
24	52	0.44	25.80	2105	117.0	2320	116.0
16-ft. Cover							
3**	52	0.28	3.96	190	10.5	210	10.4
4	52	0.29	4.80	245	13.5	270	13.4
6	52	0.31	6.90	375	20.8	415	20.7
8	52	0.33	9.05	530	29.3	585	29.2
10	52	0.35	11.10	690	38.4	765	38.2
12	52	0.37	13.20	870	48.3	960	48.0
14	51	0.36	15.30	1005	55.9	1110	55.5
16	52	0.40	17.40	1270	70.6	1400	70.0
18	52	0.41	19.50	1465	81.4	1615	80.8
20	53	0.45	21.60	1780	98.8	1960	98.0
24	54	0.50	25.80	2365	131.3	2605	130.3

† Including bell. Calculated weights of pipe rounded off to nearest 5 lb.
* Including bell. Average weight, per foot, based on calculated weight before rounding.
**Pipe of 3-in. size also available in 12-ft. laying length. Weight, per length† is 130 lb. and average weight, per foot*, is 10.8 lb. for all standard depths of cover.

TABLE 2

Standard Thickness Selection Table for Ductile-Iron Pipe

Laying Condition Type 1—flat-bottom trench, loose backfill.

Laying Condition Type 2—flat-bottom trench, backfill lightly consolidated to center-line of pipe.

The thicknesses in this table are equal to or in excess of those required to withstand 250 psi working pressure.

All thicknesses shown in this table for the depths of cover indicated are adequate for trench loads including truck superloads.

The basis of design is ANSI A21.50 except that for gas pipe the bending stress, f, is 36,000 psi and the safety factor for internal pressure is 2.5. A surge allowance is not required for gas pipe.

Thread engagement in taps for service connections and bag holes may require consideration in selecting pipe thicknesses. See Appendix.

Size, in.	Laying Condition	Depth of Cover-ft.							
		2½	3½	5	8	12	16	20	24
3	1	0.28	0.28	0.28	0.28	0.28	0.28	0.28	0.28
	2	0.28	0.28	0.28	0.28	0.28	0.28	0.28	0.28
4	1	0.29	0.29	0.29	0.29	0.29	0.29	0.29	0.29
	2	0.29	0.29	0.29	0.29	0.29	0.29	0.29	0.29
6	1	0.31	0.31	0.31	0.31	0.31	0.31	0.31	0.31
	2	0.31	0.31	0.31	0.31	0.31	0.31	0.31	0.31
8	1	0.33	0.33	0.33	0.33	0.33	0.33	0.33	0.33
	2	0.33	0.33	0.33	0.33	0.33	0.33	0.33	0.33
10	1	0.35	0.35	0.35	0.35	0.35	0.35	0.35	0.35
	2	0.35	0.35	0.35	0.35	0.35	0.35	0.35	0.35
12	1	0.37	0.37	0.37	0.37	0.37	0.37	0.37	0.40
	2	0.37	0.37	0.37	0.37	0.37	0.37	0.37	0.37
14	1	0.36	0.36	0.36	0.36	0.36	0.39	0.42	0.45
	2	0.36	0.36	0.36	0.36	0.36	0.36	0.39	0.42
16	1	0.37	0.37	0.37	0.37	0.37	0.43	0.46	0.49
	2	0.37	0.37	0.37	0.37	0.37	0.40	0.43	0.46
18	1	0.41	0.38	0.38	0.38	0.41	0.44	0.50	0.53
	2	0.38	0.38	0.38	0.38	0.38	0.41	0.47	0.50
20	1	0.42	0.39	0.39	0.39	0.45	0.48	0.54	
	2	0.39	0.39	0.39	0.39	0.39	0.45	0.48	0.54
24	1	0.47	0.41	0.41	0.44	0.50	0.56		
	2	0.41	0.41	0.41	0.41	0.44	0.50	0.56	

TABLE 3

Standard Dimensions and Weights of Mechanical-Joint Ductile-Iron Pipe

Size, in.	Thick- ness Class	Thick- ness in.	OD	Weight of Barrel per Foot	Weight of Bell	Weight Based on 18-ft. Laying Length		Weight Based on 20-ft. Laying Length	
						Per Length†	Avg. per Foot* lb.	Per Length†	Avg. per Foot*
3**	52	0.28	3.96	9.9	11	190	10.5	210	10.4
	53	0.31	3.96	10.9	11	205	11.5	230	11.4
	54	0.34	3.96	11.8	11	225	12.4	245	12.4
	55	0.37	3.96	12.8	11	240	13.4	265	13.4
	56	0.40	3.96	13.7	11	260	14.3	285	14.2
4	52	0.29	4.80	12.6	16	245	13.5	270	13.4
	53	0.32	4.80	13.8	16	265	14.7	290	14.6
	54	0.35	4.80	15.0	16	285	15.9	315	15.8
	55	0.38	4.80	16.1	16	305	17.0	340	16.9
	56	0.41	4.80	17.3	16	325	18.2	360	18.1
6	52	0.31	6.90	19.6	22	375	20.8	415	20.7
	53	0.34	6.90	21.4	22	405	22.6	450	22.5
	54	0.37	6.90	23.2	22	440	24.4	485	24.3
	55	0.40	6.90	25.0	22	470	26.2	520	26.1
	56	0.43	6.90	26.7	22	505	27.9	555	27.8
8	52	0.33	9.05	27.7	29	530	29.3	585	29.2
	53	0.36	9.05	30.1	29	570	31.7	630	31.6
	54	0.39	9.05	32.5	29	615	34.1	680	34.0
	55	0.42	9.05	34.8	29	655	36.4	725	36.2
	56	0.45	9.05	37.2	29	700	38.8	775	38.6
10	52	0.35	11.10	36.2	39	690	38.4	765	38.2
	53	0.38	11.10	39.2	39	745	41.4	825	41.2
	54	0.41	11.10	42.1	39	795	44.3	880	44.0
	55	0.44	11.10	45.1	39	850	47.3	940	47.0
	56	0.47	11.10	48.0	39	905	50.2	1000	50.0

† Including bell. Calculated weight of pipe rounded off to nearest 5 lb.
* Including bell. Average weight, per foot, based on calculated weight of pipe before rounding.
**Pipe of 3-in. size also available in 12-ft. laying length with following weights:

Thickness Class	Thickness, in.	Weight (lb.)	
		Per Length†	Avg. per Foot*
52	0.28	130	10.8
53	0.31	140	11.8
54	0.34	155	12.7
55	0.37	165	13.7
56	0.40	175	14.6

TABLE 3 (Continued)

Standard Dimensions and Weights of Mechanical-Joint Ductile-Iron Pipe

Size, in.	Thickness Class	Thickness in.	OD	Weight of Barrel per Foot	Weight of Bell	Weight Based on 18-ft. Laying Length		Weight Based on 20-ft. Laying Length	
						Per Length†	Avg. per Foot*	Per Length†	Avg. per Foot*
						lb.			
12	52	0.37	13.20	45.6	49	870	48.3	960	48.0
	53	0.40	13.20	49.2	49	935	51.9	1035	51.6
	54	0.43	13.20	52.8	49	1000	55.5	1105	55.2
	55	0.46	13.20	56.3	49	1060	59.0	1175	58.8
	56	0.49	13.20	59.9	49	1125	62.6	1245	62.3
14	51	0.36	15.30	51.7	76	1005	55.9	1110	55.5
	52	0.39	15.30	55.9	76	1080	60.1	1195	59.7
	53	0.42	15.30	60.1	76	1160	64.3	1280	63.9
	54	0.45	15.30	64.2	76	1230	68.4	1360	68.0
	55	0.48	15.30	68.4	76	1305	72.6	1445	72.2
	56	0.51	15.30	72.5	76	1380	76.7	1525	76.3
16	51	0.37	17.40	60.6	93	1185	65.8	1305	65.2
	52	0.40	17.40	65.4	93	1270	70.6	1400	70.0
	53	0.43	17.40	70.1	93	1355	75.3	1495	74.8
	54	0.46	17.40	74.9	93	1440	80.1	1590	79.6
	55	0.49	17.40	79.7	93	1530	84.9	1685	84.4
	56	0.52	17.40	84.4	93	1610	89.6	1780	89.0
18	51	0.38	19.50	69.8	111	1365	76.0	1505	75.4
	52	0.41	19.50	75.2	111	1465	81.4	1615	80.8
	53	0.44	19.50	80.6	111	1560	86.8	1725	86.2
	54	0.47	19.50	86.0	111	1660	92.2	1830	91.6
	55	0.50	19.50	91.3	111	1755	97.5	1935	96.8
	56	0.53	19.50	96.7	111	1850	102.9	2045	102.2
20	51	0.39	21.60	79.5	131	1560	86.8	1720	86.0
	52	0.42	21.60	85.5	131	1670	92.8	1840	92.0
	53	0.45	21.60	91.5	131	1780	98.8	1960	98.0
	54	0.48	21.60	97.5	131	1885	104.8	2080	104.0
	55	0.51	21.60	103.4	131	1990	110.7	2200	110.0
	56	0.54	21.60	109.3	131	2100	116.6	2315	115.8
24	51	0.41	25.80	100.1	174	1975	109.8	2175	108.8
	52	0.44	25.80	107.3	174	2105	117.0	2320	116.0
	53	0.47	25.80	114.4	174	2235	124.1	2460	123.1
	54	0.50	25.80	121.6	174	2365	131.3	2605	130.3
	55	0.53	25.80	128.8	174	2490	138.5	2750	137.5
	56	0.56	25.80	135.9	174	2620	145.6	2890	144.6

† Including bell. Calculated weight of pipe rounded off to nearest 5 lb.
* Including bell. Average weight, per foot, based on calculated weight of pipe before rounding.

APPENDIX

This appendix is for information and is not a part of A21.52 American National Standard for Ductile-Iron Pipe, Centrifugally Cast, in Metal Molds or Sand-Lined Molds for Gas.

PIPE THICKNESS REQUIRED FOR DIFFERENT TAP SIZES USING AMERICAN NATIONAL STANDARD B2.1 STANDARD TAPER PIPE THREADS WITH 2, 3 AND 4 FULL THREADS.

Pipe Size	No. of Threads	Tap Sizes									
		½"	¾"	1"	1¼"	1½"	2"	2½"	3"	3½"	4"
3"	2	.18	.21	.28							
	3	.26	.29	.37							
	4	.33	.36	.46							
4"	2	.17	.19	.26	.31						
	3	.25	.27	.35	.40						
	4	.32	.34	.44	.49						
6"	2	.17	.18	.23	.27	.30					
	3	.25	.26	.32	.36	.39					
	4	.32	.33	.41	.45	.48					
8"	2	.16	.17	.22	.24	.27	.33				
	3	.24	.25	.31	.33	.36	.42				
	4	.31	.32	.40	.42	.45	.51				
10"	2	.15	.17	.21	.23	.25	.30	.44			
	3	.23	.25	.30	.32	.34	.39	.56			
	4	.30	.32	.39	.41	.43	.48	.69			
12"	2	.15	.16	.20	.22	.24	.28	.40	.48		
	3	.23	.24	.29	.31	.33	.37	.52	.60		
	4	.30	.31	.38	.40	.42	.46	.65	.73		
14"	2	.15	.16	.20	.22	.23	.26	.38	.45	.51	.58
	3	.23	.24	.29	.31	.32	.35	.50	.58	.64	.70
	4	.30	.31	.38	.40	.41	.44	.63	.70	.76	.83
16"	2	.15	.16	.20	.21	.22	.25	.37	.43	.48	.54
	3	.23	.24	.29	.30	.31	.34	.50	.56	.60	.66
	4	.30	.31	.38	.39	.40	.43	.62	.68	.73	.79
18"	2	.15	.15	.19	.21	.22	.24	.35	.41	.46	.51
	3	.23	.23	.28	.30	.31	.33	.48	.54	.58	.64
	4	.30	.30.	37	.39	.40	.42	.60	.66	.71	.76

Pipe Size	No. of Threads	Tap Sizes									
		½″	¾″	1″	1¼″	1½″	2″	2½″	3″	3½″	4″
20″	2	.15	.15	.19	.20	.21	.23	.34	.39	.44	.49
	3	.23	.23	.28	.29	.30	.32	.46	.52	.56	.62
	4	.30	.30	.37	.38	.39	.41	.59	.64	.69	.74
24″	2	.14	.15	.19	.20	.21	.22	.32	.37	.40	.45
	3	.22	.23	.28	.29	.30	.31	.44	.50	.52	.58
	4	.29	.30	.37	.38	.39	.40	.57	.62	.65	.70
30″	2	.14	.15	19.	.19	.20	.21	.31	.34	.37	.41
	3	.22	.23	.28	.28	.29	.30	.44	.46	.50	.54
	4	.29	.30	.37	.37	.38	.39	.56	.59	.62	.66
36″	2	.14	.14	.18	.19	.20	.21	.30	.33	.35	.38
	3	.22	.22	.27	.28	.29	.30	.42	.46	.48	.50
	4	.29	.29	.36	.37	.38	.39	.55	.58	.60	.63
42″	2	.14	.14	.18	.19	.19	.20	.29	.32	.34	.36
	3	.22	.22	.27	.28	.28	.29	.42	.44	.46	.48
	4	.29	.29	.36	.37	.37	.38	.54	.57	.59	.61
48″	2	.14	.14	.18	.18	.19	.20	.29	.31	.32	.35
	3	.22	.22	.27	.27	.28	.29	.42	.44	.44	.48
	4	.29	.29	.36	.36	.37	.38	.54	.56	.57	.60

SECTION IV

Ductile Iron Pipe for Gravity Flow Service

Because of its high strength and ability to carry great earth loads, ductile iron pipe has found wide acceptance in service as gravity flow waste lines and culvert pipes.

In sewer service, ductile iron pipe with standard push-on joints has virtually eliminated infiltration as well as exfiltration. Its resistance to impact, convenient pipe lengths and easy joint assembly have caused engineers and those responsible for construction to become aware of its many advantages and its use in sewer service has increased rapidly in recent years.

Ductile iron pipe joints are ideally suited for gravity flow pipelines as exhibited by the following test results:

1,000 psi internal pressure

430 psi external pressure

14 psi negative air pressure

No leakage—No infiltration

An advantage of ductile iron pipe in sewer service is the fact that its inside diameter is greater than nominal. This results in greater flow capacity for a given pipe size and thus considerable savings may be effected.

Ductile iron pipe is available with standard shop linings or cement-mortar linings for normal domestic sewage. Special linings are available for more aggressive wastes.

Still another valuable feature of ductile iron pipe in sewer service is its ability to withstand great depths of earth cover under nominal laying conditions (see ASTM A746 Table 12).

Ductile iron gravity flow pipelines should be designed and specified in accordance with ASTM Standard Specification A746–75.

Exceptional ring and beam strengths make ductile iron pipe an ideal structure for culvert pipes. ASTM Standard Specification A716 is a useful reference when specifying culvert piping.

ASTM Designation: A 716 – 75 *

Reprinted by permission of the
AMERICAN SOCIETY FOR TESTING AND MATERIALS
1916 Race St., Philadelphia, Pa., 19103
Reprinted from the Annual Book of ASTM Standards, Copyright ASTM
If not listed in the current combined Index, will appear in the next edition

Standard Specification for
DUCTILE IRON CULVERT PIPE[1]

This Standard is issued under the fixed designation A 716; the number immediately following the designation indicates the year of original adoption or, in the case of revision, the year of last revision. A number in parentheses indicates the year of last reapproval.

1. Scope

1.1 This specification covers 14 to 54-in. (0.356 to 1.372-m) ductile-iron culvert pipe centrifugally cast in metal molds or sand-lined molds.

2. Applicable Documents

2.1 *ASTM Standards.*
A 377, Specification for Cast Iron and Ductile Iron Pressure Pipe[2]
E 8, Tension Testing of Metallic Material[3]
E 23, Notch and Bar Impact Testing of Metallic Materials[3]
2.2 *ANSI Standards:*
A21.50, Thickness Design of Ductile Iron Pipe[4]
A21.51, Ductile-Iron Pipe Centrifugally Cast in Metal Molds or Sand-Lined Molds for Water or Other Liquids[4]

3. General Requirements

3.1 The pipe shall be manufactured of ductile iron that meets the requirements of ANSI A21.51.

3.2 The pipe shall be provided with suitable joints, such as push-on or other types of joints that prevent lateral displacement. Plain-end pipe for use with suitable couplings may be furnished.

3.3 Unless otherwise specified, pipe shall have a nominal length of 18 or 20 ft (5.5 or 6.1 m) and may be shorter than nominal in accordance with ANSI A21.51.

4. Inspection and Certification by Manufacturer

4.1 The manufacturer shall establish the necessary quality-control and inspection practice to assure compliance with this standard specification.

4.2 The manufacturer shall, if required on the purchase order, furnish a sworn statement that the inspection and all of the specified tests have been made and the results thereof comply with the requirements of this standard specification.

4.3 All pipes shall be clean and sound without defects that will impair their service. Repairing of defects by welding or another method shall not be allowed if such repairs will affect the serviceability of the pipe or its capability to meet strength requirements of this standard specification.

5. Inspection by Purchaser

5.1 If the purchaser desires to inspect pipe at the manufacturer's plant, the purchaser shall so specify on the purchase order, stating the conditions (such as time and the extent of inspection) under which the inspection shall be made.

5.2 The inspector shall have free access to those parts of the manufacturer's plant that are necessary to assure compliance with this standard specification. The manufacturer shall make available for the inspector's use such gages as are necessary for inspection. The manufacturer shall provide the inspector

---

[1] This specification is under the jurisdiction of ASTM Committee A-4 on Iron Castings.
Current edition approved Aug. 29, 1975. Published October 1975.
[2] *Annual Book of ASTM Standards,* Part 2.
[3] *Annual Book of ASTM Standards,* Part 10.
[4] Available from the American National Standards Institute, 1430 Broadway, New York, N. Y. 10018.

This standard has been included in this handbook by arrangement with the American Society for Testing and Materials.
* This standard is being revised at the time of this printing. Before employing this standard it is suggested that you confirm its currency with ASTM.

with assistance as necessary for handling of pipe.

6. Delivery and Acceptance

6.1 All pipe and accessories shall comply with this standard specification. Pipe and accessories not complying with this standard specification shall be replaced by the manufacturer at the agreed point of delivery. The manufacturer shall not be liable for shortages or damaged pipe after acceptance at the agreed point of delivery, except as recorded on the delivery receipt or similar document by the carrier's agent.

7. Tolerances or Permitted Variations

7.1 *Dimensions*—The spigot end, bell, and socket of the pipe and the accessories shall be gaged with suitable gages at sufficiently frequent intervals to assure that the dimensions comply with the requirements of this standard specification. The smallest inside diameter of the sockets and the outside of the spigot ends shall be tested with circular gages. Other socket dimensions shall be gaged as appropriate.

7.2 *Thickness*—Minus thickness tolerances of pipe and bell shall not exceed those shown below:

Nominal Size, in.	Minus Tolerance, in. (mm)
14 to 42	0.07 (1.8)
48	0.08 (2.0)
54	0.09 (2.3)

7.3 *Weight*—The weight of any single pipe shall be not less than the tabulated weight by more than 4 %.

8. Coating and Lining

8.1 All pipe shall be coated inside and outside with a bituminous material approximately 1 mil (0.025 mm) thick. The finished coating shall be continuous, smooth, neither brittle when cold nor sticky when exposed to the sun, and shall be strongly adherent to the pipe.

9. Acceptance Tests

9.1 The standard acceptance tests for the physical characteristics of the pipe shall be as follows:

9.1.1 *Tension Test*—A tension test specimen shall be cut longitudinally from the midsection of the pipe wall. This specimen shall be machined and tested in accordance with Fig. 1 and Specification E 8. The yield strength shall be determined by the 0.2 % offset, halt-of-pointer, or extension-underload methods. If check tests are to be made, the 0.2 % offset method shall be used. All specimens shall be tested at room temperature (70 ± 10°F (21.1 ± 5.5°C)).

9.1.2 *Acceptance Values*—The acceptance values for test specimens shall be as follows:

Grade of Ductile Iron:	60-42-10
Minimum tensile strength, psi (kPa):	60 000 (414)
Minimum yield strength, psi (kPa):	42 000 (290)
Minimum elongation, %:	10

9.1.3 *Impact Test*—Tests shall be made in accordance with Specification E 23, except that specimens shall be 0.50 in. (12.7 mm) and full thickness of pipe wall. The notched impact test specimen shall be in accordance with Fig. 2. If the pipe wall thickness exceeds 0.40 in. (10.2 mm), the impact specimen may be machined to a nominal thickness of 0.40 in. (10.2 mm). In all tests, impact values are to be corrected to 0.40-in. (10.2-mm) wall thickness by calculations as follows:

Impact value (corrected) = $(0.40/t) \times$ impact value (actual)

where:

t = thickness of the specimen, in. (wall thickness of pipe).

The Charpy test machine anvil shall not be moved to compensate for the variation of cross-section dimensions of the test specimen.

9.1.4 *Acceptance Value*—The corrected acceptance value for notched impact test specimens shall be a minimum of 7 lbf·ft (9 J) for tests conducted at 70 ± 10°F (21.1 ± 5.5°C).

9.2 *Sampling*—At least one tension and impact sample shall be taken during each casting period of approximately 3 h. Samples shall be selected to represent extremes of pipe diameters and thicknesses properly.

10. Additional Control Tests by Manufacturer

10.1 Low-temperature impact tests shall

be made from at least one third of the test pipe specified in 9.2 to assure compliance with a minimum corrected value of 3 lbf·ft (4 J) for test conducted at −40°F (−40°C). Test specimens shall be prepared and tested in accordance with 9.1.3.

10.2 In addition, the manufacturer shall conduct such other control tests as necessary to assure continuing compliance with this standard specification.

11. Foundry Records

11.1 The results of the acceptance tests (Section 9) and low-temperature impact tests (Section 10) shall be recorded and retained for 1 year, and shall be available to the purchaser at the foundry. Written transcripts shall be furnished, if specified on the purchase order.

12. Marking Pipe

12.1 The weight, class or nominal thickness, and casting period shall be shown on each pipe. The manufacturer's mark, the year in which the pipe was produced, and the letters "DI" or "DUCTILE" shall be cast or stamped on the pipe. When specified on the purchase order, initials not exceeding four in number shall be cast or stamped on the pipe. All required markings shall be clear and legible and all cast marks shall be on or near the bell. All letters and numerals shall be not less than ½ in. (12.7 mm) in height.

13. Weighing the Pipe

13.1 Each pipe shall be weighed and the weight shown on the outside or inside of the bell or spigot end.

14. Additional Tests Required by Purchaser

14.1 When tests other than those required in this standard specification are required by the purchaser, such tests shall be specified in the invitation for bids and on the purchase order.

15. Defective Specimens and Retests

15.1 When any physical-test specimen shows defective machining or lack of continuity of metal, it shall be discarded and replaced by another specimen. When any sound test specimen fails to meet the specified requirements, the pipe from which it was taken shall be rejected, and a retest may be made on two additional sound specimens from pipe cast in the same period as the specimen that failed. Both of the additional specimens shall meet the prescribed tests to qualify the pipe produced in that period.

16. Rejection of Pipe

16.1 If the results of any physical acceptance test fail to meet the requirements of Section 9 or Section 15, the pipe cast in the same period shall be rejected, except as provided in Section 17.

17. Determining Rejection

17.1 The manufacturer may determine the amount of rejection by making similar additional tests of pipe, of the same size as the rejected pipe, until the rejected lot is bracketed, in order of manufacture, by an acceptable test at each end of the interval in question. When pipe of one size is rejected from a casting period, the acceptability of pipe of different sizes from that same period may be established by making the acceptance tests for these sizes as specified in Section 9.

TABLE 1 Standard Wall Thickness[a] and Weights of Push-On Joint Ductile-Iron Culvert Pipe

Nominal Diameter,		Nominal Thickness for Depths of Cover, 2 to 30 ft,		Thickness Class	18-ft (5.5-m) Laying Length Weight per Length,		20-ft (6 m) Laying Length Weight per Length,	
in.	(mm)	in.	(mm)		lb	(kg)	lb	(kg)
14	(356)	0.36	(9.1)	1	1005	(456)	1110	(503)
16	(406)	0.37	(9.4)	1	1185	(538)	1305	(592)
18	(457)	0.38	(9.7)	1	1370	(621)	1510	(685)
20	(508)	0.39	(9.9)	1	1560	(708)	1720	(780)
24	(610)	0.41	(10.4)	1	1970	(894)	2170	(984)
30	(762)	0.43	(10.9)	1	2565	(1163)	2775	(1259)
36	(914)	0.48	(12.2)	1	3435	(1558)	3705	(1681)
42	(1067)	0.53	(13.5)	1	—		4740	(2150)
48	(1219)	0.58	(12.2)	1	—		5915	(2683)
54	(1372)	0.65	(16.5)	1	—		7425	(3368)

[a] Nominal thickness based on "S" trench condition of ANSI A21.50 with maximum ring deflection of 5 % and ring stress of 36 000 psi (248 kPa). Wall thickness of pipe to serve at greater depths of cover may be calculated in accordance with ANSI A21.50.

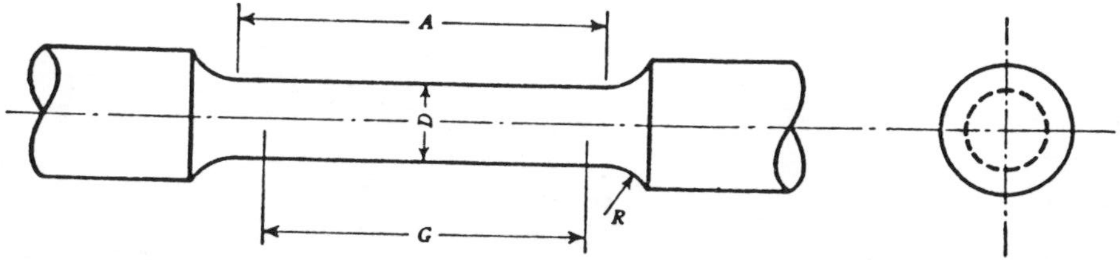

NOTE 1—The reduced section (A) may have a gradual taper from the ends toward the center with the ends not more than 0.005 in. (0.13 mm) larger in diameter than the center on the standard specimen and not more than 0.003 in. (0.08 mm) larger in diameter than the center on the small size specimens.

NOTE 2—If desired, on the small size specimens the length of the reduced section may be increased to accommodate an extensometer. However, reference marks for the measurement of elongation should nevertheless be spaced at the indicated gage length (G).

NOTE 3—The gage length and fillets shall be as shown, but the ends may be of any form to fit the holders of the testing machine in such a way that the load shall be axial. If the ends are to be held in grips it is desirable, if possible to make the length of the grip section great enough to allow the specimen to extend into the grips a distance equal to two thirds or more of the length of the grips.

Small-Size Specimens Proportional to Standard

Dimension	Standard Specimen 0.50-in. (12.7-mm) Round	0.350-in. (8.89-mm) Round	0.250-in. (6.35-mm) Round	0.175-in. (4.45-mm) Round	0.125-in. (3.18-mm) Round
			Dimensions, in. (mm)		
G	2.000 ± 0.005 (50.80 ± 0.13)	1.400 ± 0.005 (35.56 ± 0.13)	1.000 ± 0.005 (25.40 ± 0.13)	0.700 ± 0.005 (12.78 ± 0.13)	0.500 ± 0.005 (12.70 ± 0.13)
D	0.500 ± 0.010 (12.70 ± 0.25)	0.350 ± 0.007 (8.89 ± 0.18)	0.250 ± 0.005 (6.35 ± 0.13)	0.175 ± 0.005 (4.44 ± 0.13)	0.125 ± 0.005 (3.18 ± 0.13)
R, min	3/8 (9.5)	¼ (6.4)	3/16 (4.8)	3/32 (2.4)	3/32 (2.4)
A, min	2¼ (57.2)	1¾ (44.4)	1¼ (31.8)	¾ (19)	5/8 (15.9)
T[a]	0.71 and greater (18.0)	0.50 to 0.70 (12.2 to 17.8)	0.35 to 0.49 (8.9 to 12.4)	0.25 to 0.34 (6.4 to 8.6)	0.18 to 0.24 (4.6 to 6.1)

[a] Thickness of the section from the wall of the pipe from which the tension specimen is to be machined.

FIG. 1 Tension-Test Specimen.

231

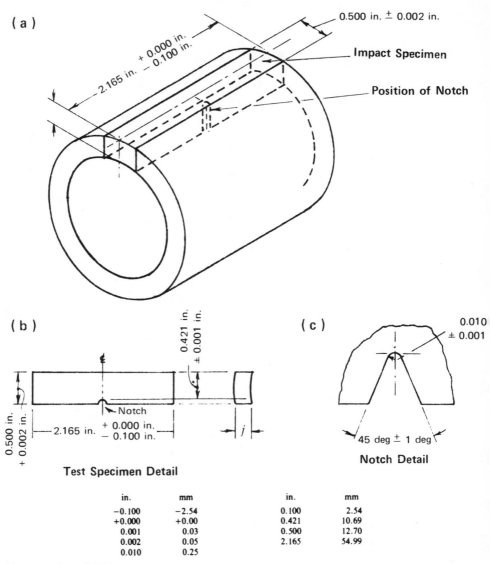

Test Specimen Detail

Notch Detail

in.	mm	in.	mm
−0.100	−2.54	0.100	2.54
+0.000	+0.00	0.421	10.69
0.001	0.03	0.500	12.70
0.002	0.05	2.165	54.99
0.010	0.25		

NOTE—*t* = pipe-wall thickness.

FIG. 2 Impact Test Specimen.

Standard Specification for
DUCTILE IRON GRAVITY SEWER PIPE[1]

This Standard is issued under the fixed designation A 746; the number immediately following the designation indicates the year of original adoption or, in the case of revision, the year of last revision. A number in parentheses indicates the year of last reapproval.

1. Scope

1.1 This specification covers 4 to 54-in. ductile iron gravity sewer pipe centrifugally cast in metal molds or sand-lined molds with push-on joints. This specification may be used for pipe with other types of joints, as may be agreed upon at the time of purchase.

1.2 This specification covers trench load design procedures for both cement-lined pipe and flexible-lined pipe. Maximum depth of cover tables are included for both types of linings.

2. Applicable Documents

2.1 *ASTM Standards:*

A 377 Specification for Cast Iron and Ductile Iron Pressure Pipe[2]

E 8 Tension Testing of Metallic Materials[3]

E 23 Notched Bar Impact Testing of Metallic Materials[4]

2.2 *ANSI Standards:*[5]

A 21.4 Cement-Mortar Lining for Cast-Iron and Ductile-Iron Pipe and Fittings for Water

A 21.11 Rubber-Gasket Joints for Cast-Iron and Ductile-Iron Pressure Pipe and Fittings

A 21.50 Thickness Design of Ductile-Iron Pipe

2.3 *ASCE Standards:*[6]

Manuals and Reports on Engineering Practice, No. 37, (WPCF Manual of Practice No. 9). "Design and Construction of Sanitary and Storm Sewers."

3. Symbols

P_r = trench load, psi = $P_e + P_t$

f = design bending stress, 48 000 psi (331 MPa)

$\Delta\chi$ = design deflection, in. ($\Delta\chi/D = 0.03$) or ($\Delta\chi/D = 0.05$ (for flexible linings))

t = net thickness, in.

t_1 = minimum thickness, in. ($t + 0.08$)

D = outside diameter, in. (Table 5)

E = modulus of elasticity, 24×10^6 psi (165 GPa)

E' = modulus of soil reaction, psi (Table 2)

K_b = bending moment coefficient (Table 2)

K_x = deflection coefficient (Table 2)

P_e = earth load, psi

H = depth of cover, ft

w = soil weight, 120 lb/ft³ (54 kg/m³)

P_t = truck load, psi

C = surface load factor (Table 6)

R = reduction factor that takes into account the fact that the part of the pipe directly below the wheels is aided in carrying the truck load by adjacent parts of the pipe that receive little or no load from the wheels (Table 4)

P = wheel load, 16 000 lbf (71.2 kN)

F = impact factor, 1.5

A = outside radius of pipe, ft = $D/24$

4. General Requirements

4.1 The pipe shall be ductile iron in accordance with Section 9.

4.2 Push-on joints shall comply with all ap-

[1] This specification is under the jurisdiction of ASTM Committee A-4 on Iron Castings.
Current edition approved April 29, 1977. Published June 1977.
[2] *Annual Book of ASTM Standards,* Part 2.
[3] *Annual Book of ASTM Standards,* Parts 6, 7, and 10.
[4] *Annual Book of ASTM Standards,* Part 10.
[5] Available from the American National Standards Institute, 1430 Broadway, New York, N. Y. 10018.
[6] Available from the American Society of Civil Engineers, 345 East 47th St., New York, N. Y. 10017.

plicable requirements of ANSI A21.11. Pipe with other types of joints shall comply with the joint dimensions and weights agreed upon at the time of purchase, but in all other respects shall fulfill the requirements of this specification.

4.3 Unless otherwise specified, pipe shall have a nominal length of 18 or 20 ft (5.5 or 6.1 m). A maximum of 20 % of the total number of pipe of each size specified in an order may be furnished as much as 24 in. (610 mm) shorter than the nominal laying length, and an additional 10 % may be furnished as much as 6 in. (152 mm) shorter than the nominal laying length.

5. Tolerances or Permitted Variations

5.1 *Dimensions* — The plain end, bell, and socket of the pipe shall be gaged with suitable gages at sufficiently frequent intervals to assure that the dimensions comply with the requirements of this specification.

5.2 *Thickness* — Minus thickness tolerances of pipe and bell shall not exceed those shown in Table 3.

NOTE — An additional minus tolerance of 0.02 in. (0.51 mm) shall be permitted along the barrel of the pipe for a distance not to exceed 12 in. (305 mm)

5.3 *Weight* — The weight of any single pipe shall not be less than the tabulated weight by more than 6 % for pipe 12 in. or smaller in diameter, nor by more than 5 % for pipe larger than 12 in. in diameter.

6. Coating and Lining

6.1 *Outside Coating* — The outside coating for use under normal conditions shall be a bituminous coating approximately 1 mil (0.025 mm) thick. The coating shall be applied to the outside of all pipe, unless otherwise specified. The finished coating shall be continuous, smooth, neither brittle when cold nor sticky when exposed to the sun, and shall be strongly adherent to the pipe.

6.2 *Cement-Mortar Linings* — If desired, cement linings shall be specified in the invitation for bids and on the purchase order. Cement linings shall be in accordance with ANSI A21.4.

6.3 *Bituminous Lining* — Unless otherwise specified, the lining for pipe not cement-lined shall be a bituminous material, minimum 1 mil (0.025 mm) thick, and conforming to all appropriate requirements for seal coat in ANSI A21.4.

6.4 Unless otherwise specified, the manufacturer at his option may furnish either cement-mortar-lined or bituminous-lined pipe.

6.5 *Special Linings* — For severely aggressive wastes, other types of linings may be available. Such special linings shall be specified in the invitation for bids and on the purchase order.

7. Pipe Design

7.1 This section covers the design of ductile iron pipe for trench loads.

7.2 *Determining the Total Calculated Thickness and Standard Thickness:*

7.2.1 Determine the trench load, P_v. Table 1 gives the trench load, including the earth load, P_e, plus the truck load, P_t, for 2.5 to 32 ft (0.76 to 9.75 m) of cover.

7.2.2 Determine the standard trench class from the descriptions in Table 2 and select the appropriate table of diameter-thickness ratios from Tables 7 to 11. Enter the calculated trench load, P_v, in the column headed "Bending Stress Design" and read the required ratio D/t. Divide the pipe outside diameter, D (Table 5) by the ratio D/t to obtain the net thickness, t, required by the design for bending stress.

7.2.3 Enter the calculated trench load, P_v, in the appropriate column headed "Deflection Design" and read the required D/t_1. Divide the outside diameter, D, by the ratio D/t_1 to obtain the minimum thickness, t_1, required by the design for deflection. Deduct the 0.08-in. (2.0-mm) service allowance to obtain the corresponding net thickness, t. (If the calculated P_v is less than the minimum P_v listed in the table, the design for the trench load is not controlled by deflection and this determination need not be completed.)

7.2.4 Compare the net thicknesses from 7.2.2 and 7.2.3 and select the larger of the two. This will be the net thickness, t.

7.2.5 Add the service allowance of 0.08 in. (2.0 mm) to the net thickness, t. The resulting thickness is the minimum manufacturing thickness, t_1.

7.2.6 Add the casting tolerance from Table 3 to the minimum manufacturing thickness, t_1,

The resulting thickness is the total calculated thickness.

7.2.7 In specifying and ordering pipe, use the total calculated thickness to select one of the standard class thicknesses in Table 5. Select the standard thickness nearest to the calculated thickness. When the calculated thickness is halfway between two standard thicknesses, select the larger of the two.

7.3 *Design Example* — Calculate the thickness for 30-in. cement-lined ductile iron pipe laid on a flat-bottom trench with backfill lightly tamped to the centerline of pipe, Laying Condition Type 2, under 10 ft (3 m) of cover.

7.3.1

Earth load, Table 1, P_e = 8.3 psi
Truck load, Table 1, P_t = 0.7 psi
Trench load, $P_r = P_e + P_t$ = 9.0 psi

7.3.2 Entering P_r of 9.0 psi in Table 8, the bending stress design requires D/t of 128.

Net thickness, t, for bending stress

$$= D/(D/t) = 32.00/128 = 0.25$$

7.3.3 Reentering P_r of 9.0 psi in Table 8, the deflection design requires D/t_1 of 108.

Minimum thickness t_1 for deflection design
$$= D/(D/t_1) = 32.00/108$$
= 0.30 in.
Deduct service allowance = −0.08 in.
Net thickness t for deflection control = 0.22 in.

7.3.4 The larger net thickness is 0.25 in., obtained by the design for bending stress.

7.3.5 Net thickness = 0.25 in.
       Service allowance = 0.08 in.
       Minimum thickness = 0.33 in.
7.3.6 Casting tolerance = 0.07 in.
       Total calculated thickness = 0.40 in.

7.3.7 The total calculated thickness of 0.40 in. is nearest to 0.39, Class 50, in Table 5. Therefore, Class 50 is selected for specifying and ordering.

7.4 *Design Method:*

7.4.1 Calculations are made for the thicknesses required to resist the bending stress and the deflection due to trench load. The larger of the two is selected as the thickness required to resist trench load.

7.4.2 To this net thickness is added a service allowance to obtain the minimum manufacturing thickness and a casting tolerance to obtain the total calculated thickness.

7.4.3 The thickness for specifying and ordering is selected from a table of standard class thicknesses.

7.4.4 The reverse of the above procedure is used to determine the maximum depth of cover for pipe of a given thickness class.

7.4.5 *Trench Load, P_r* — Trench load is expressed as vertical pressure, psi, and is equal to the sum of earth load, P_e, and truck load, P_t.

7.4.6 *Earth Load, P_e* — Earth load is computed by Eq 3 for the weight of the unit prism of soil with a height equal to the distance from the top of the pipe to the ground surface. The unit weight of backfill soil is taken to be 120 lb/ft³ (54 kg/m³).

7.4.7 *Truck Load, P_t* — The truck loads shown in Table 1 were computed by Eq 4 using the surface load factors in Table 6 for a single AASHO H-20 truck on an unpaved road or flexible pavement, 16 000-lbf (71.2 kN) wheel load and 1.5 impact factor. The surface load factors in Table 6 were calculated by Eq 5 for a single concentrated wheel load centered over an effective pipe length of 3 ft (0.91 m).

7.4.8 *Design for Trench Load* — Tables 7 through 11, the tables of diameter-thickness ratios used to design for trench load, were computed by Eqs 1 and 2. Equation 1 is based on the bending stress at the bottom of the pipe. The design bending stress, f, is 48 000 psi (331 MPa) which provides at least 1.5 safety factor based on minimum yield strength and 2.0 safety factor based on ultimate strength. Equation 2 is based on the deflection of the pipe ring section. The design deflection Δ_x is 3 % of the outside diameter of the pipe for cement-lined pipe and 5 % for pipe with flexible linings. Design values of the trench parameters, E', K_b, and K_x are given in Table 2.

7.4.9 Tables similar to Tables 7 through 11 may be compiled for laying conditions other than those shown in this specification by calculating the trench loads, P_r, for a series of diameter-thickness ratios, D/t and D/t_1, using Eqs 1 and 2 with values of E', K_b, and K_x appropriate to the bedding and backfill conditions.

2
3
5

3

2
3
6

7.5 Design Equations:

$$P_v = \frac{f}{3\left(\dfrac{D}{t}\right)\left(\dfrac{D}{t}-1\right)\left[K_b - \dfrac{K_x}{\dfrac{8E}{E'\left(\dfrac{D}{t}-1\right)^3} + 0.732}\right]} \quad (1)$$

$$P_r = \frac{\Delta\chi/D}{12\,K_x}\left[\frac{8E}{\left(\dfrac{D}{t_1}-1\right)^3} + 0.732\,E'\right] \quad (2)$$

$$P_e = \frac{wH}{144} = \frac{120\,H}{144} = \frac{H}{1.2} \quad (3)$$

$$P_t = \frac{CRPF}{12D} \quad (4)$$

$$C = \frac{1}{3} - \frac{2}{3\pi}\arcsin\left[H\sqrt{\frac{A^2 + H^2 + 1.5^2}{(A^2 + H^2)(H^2 + 1.5^2)}}\right]$$

$$+ \frac{AH}{\pi\sqrt{A^2 + H^2 + 1.5^2}}\left[\frac{1}{A^2 + H^2} + \frac{1}{H^2 + 1.5^2}\right] \quad (5)$$

NOTE — In Eq 5, angles are in radians.

8. Hydrostatic Test

8.1 Each pipe shall be subjected to a hydrostatic test of not less than 500 psi (3.43 MPa). This test may be made either before or after the outside coating and the bituminous lining have been applied, but shall be made before the application of cement lining or of a special lining.

8.2 The pipe shall be under the full test pressure for at least 10 s. Suitable controls and recording devices shall be provided so that the test pressure and duration may be adequately ascertained. Any pipe that leaks or does not withstand the test pressure shall be rejected.

8.3 In addition to the hydrostatic test before application of a cement lining or special lining, the pipe may be retested, at the manufacturer's option, after application of such lining.

9. Acceptance Tests

9.1 The standard acceptance tests for the physical characteristics of the pipe shall be as follows:

9.2 *Tension Test* — A tension test specimen shall be cut longitudinally from the midsection of the pipe wall. This specimen shall be ma-chined and tested in accordance with Fig. 1 and Method E 8. The yield strength shall be determined by the 0.2 % offset, halt-of-pointer, or extension-under-load methods. If check tests are to be made, the 0.2 % offset method shall be used. All specimens shall be tested at room temperature 70 ± 10°F (21.1 ± 5.5°C).

9.2.1 *Acceptable Values* — The acceptance values for test specimens shall be as follows:

Grade of Iron:	60–42–10
Minimum tensile strength, psi (MPa)	60 000 (414)
Minimum yield strength, psi (MPa):	42 000 (290)
Minimum elongation, %:	10

9.3 *Impact Test* — Tests shall be made in accordance with Method E 23, except that specimens shall be 0.500 in. (12.70 mm) and full thickness of pipe wall. The notched impact test specimen shall be in accordance with Fig. 2. If the pipe wall thickness exceeds 0.40 in. (10.2 mm), the impact specimen may be machined to a nominal thickness of 0.40 in. In all tests, impact values are to be corrected to 0.40-in. wall thickness by calculations as follows:

Impact value (corrected)
$$= (0.40/t) \times \text{impact value (actual)}$$

where:

t = thickness of the specimen, in. (wall thickness of pipe).

The Charpy test machine anvil shall not be moved to compensate for the variation of cross-sectional dimensions of the test specimen.

9.3.1 *Acceptance Value* — The corrected acceptance value for notched impact test specimens shall be a minimum of 7 ft·lbf (9 J) for tests conducted at 70 ± 10°F (21.1 ± 5.5°C).

9.4 *Sampling* — At least one tension and impact sample shall be taken during each casting period of approximately 3 h. Samples shall be selected to represent extremes of pipe diameters and thicknesses properly.

10. Additional Control Tests by Manufacturer

10.1 Low-temperature impact tests shall be made from at least one third of the test pipe specified in 9.4 to assure compliance with a minimum corrected value of 3 ft·lbf (4 J) for tests conducted at −40°F (−40°C). Test specimens shall be prepared and tested in accordance with 9.3.

10.2 In addition, the manufacturer shall conduct such other control tests as necessary to assure continuing compliance with this specification.

11. Additional Tests Required by Purchaser

11.1 When tests other than those required in this specification are required by the purchaser, such tests shall be specified in the invitation for bids and on the purchase order.

12. Inspection and Certification by Manufacturer

12.1 The manufacturer shall establish the necessary quality-control and inspection practice to ensure compliance with this specification.

12.2 The manufacturer shall, if required on the purchase order, furnish a sworn statement that the inspection and all of the specified tests have been made and the results thereof comply with the requirements of this specification.

12.3 All pipes shall be clean and sound without defects that will impair their service. Repairing of defects by welding or other method shall not be allowed if such repairs will adversely affect the serviceability of the pipe or its capability to meet strength requirements of this specification.

13. Inspection by Purchaser

13.1 If the purchaser desires to inspect pipe at the manufacturer's plant, the purchaser shall so specify on the purchase order, stating the conditions (such as time and the extent of inspection) under which the inspection shall be made.

13.2 The inspector shall have free access to those parts of the manufacturer's plant that are necessary to assure compliance with this specification. The manufacturer shall make available for the inspector's use such gages as are necessary for inspection. The manufacturer shall provide the inspector with assistance as necessary for handling of pipe.

14. Delivery and Acceptance

14.1 All pipe and accessories shall comply with this specification. Pipe and accessories not complying with this specification shall be replaced by the manufacturer at the agreed

point of delivery. The manufacturer shall not be liable for shortages or damaged pipe after acceptance at the agreed point of delivery, except as recorded on the delivery receipt or similar document by the carrier's agent.

15. Foundry Records

15.1 The results of the acceptance tests (Section 9) and low-temperature impact tests (Section 10) shall be recorded and retained for 1 year, and shall be available to the purchaser at the foundry. Written transcripts shall be furnished, if specified on the purchase order.

16. Defective Specimens and Retests

16.1 When any physical test specimen shows defective machining or lack of continuity of metal, it shall be discarded and replaced by another specimen. When any sound test specimen fails to meet the specified requirements, the pipe from which it was taken shall be rejected, and a retest may be made on two additional sound specimens from pipe cast in the same period as the specimen that failed. Both of the additional specimens shall meet the prescribed tests to qualify the pipe produced in that period.

17. Rejection of Pipe

17.1 If the results of any physical acceptance test fail to meet the requirements of Section 9 or 16, the pipe cast in the same period shall be rejected, except as provided in Section 18.

18. Determining Rejection

18.1 The manufacturer may determine the amount of rejection by making similar additional tests of pipe, of the same size as the rejected pipe, until the rejected lot is bracketed, in order of manufacture, by an acceptable test at each end of the interval in question. When pipe of one size is rejected from a casting period, the acceptability of pipe of different sizes from that same period may be established by making the acceptance tests for these sizes as specified in Section 9.

19. Marking Pipe

19.1 The weight, class or nominal thickness, and casting period shall be shown on each pipe. The manufacturer's mark, the year in which the pipe was produced, and the letters "DI" or "DUCTILE" shall be cast or stamped on the pipe. Markings shall be clear and legible and all cast marks shall be on or near the bell. All letters and numerals on pipe sizes 14 in. and larger shall not be less than $^1/_2$

in. (12.7 mm) in height.

20. Weighing Pipe

20.1 Each pipe shall be weighed before the application of any lining or coating other than the bituminous coating and the weight shown on the outside or inside of the bell or plain end.

2
3
8

TABLE 1 Earth Loads (P_e), Truck Loads (P_t), and Trench Loads (P_v), psi

Depth of Cover, ft (m)	P_e	4-in. Pipe P_t	P_v	6-in. Pipe P_t	P_v	8-in. Pipe P_t	P_v	10-in. Pipe P_t	P_v	12-in. Pipe P_t	P_v	14-in. Pipe P_t	P_v	16-in. Pipe P_t	P_v
2.5(0.76)	2.1	9.9	12.0	9.9	12.0	9.8	11.9	9.7	11.8	9.6	11.7	8.7	10.8	8.2	10.3
3 (0.91)	2.5	7.4	9.9	7.3	9.8	7.3	9.8	7.2	9.7	7.2	9.7	6.6	9.1	6.2	8.7
4 (1.21)	3.3	4.5	7.8	4.4	7.7	4.4	7.7	4.4	7.7	4.4	7.7	4.4	7.7	4.1	7.4
5 (1.52)	4.2	3.0	7.2	3.0	7.2	3.0	7.2	2.9	7.1	2.9	7.1	2.9	7.1	2.8	7.0
6 (1.82)	5.0	2.1	7.1	2.1	7.1	2.1	7.1	2.1	7.1	2.1	7.1	2.1	7.1	2.0	7.0
7 (2.13)	5.8	1.6	7.4	1.6	7.4	1.6	7.4	1.6	7.4	1.6	7.4	1.6	7.4	1.5	7.3
8 (2.43)	6.7	1.2	7.9	1.2	7.9	1.2	7.9	1.2	7.9	1.2	7.9	1.2	7.9	1.2	7.9
9 (2.74)	7.5	1.0	8.5	1.0	8.5	1.0	8.5	1.0	8.5	1.0	8.5	1.0	8.5	1.0	8.5
10 (3.04)	8.3	0.8	9.1	0.8	9.1	0.8	9.1	0.8	9.1	0.8	9.1	0.8	9.1	0.8	9.1
12 (3.65)	10.0	0.6	10.6	0.6	10.6	0.6	10.6	0.5	10.5	0.5	10.5	0.5	10.5	0.5	10.5
14 (4.26)	11.7	0.4	12.1	0.4	12.1	0.4	12.1	0.4	12.1	0.4	12.1	0.4	12.1	0.4	12.1
16 (4.87)	13.3	0.3	13.6	0.3	13.6	0.3	13.6	0.3	13.6	0.3	13.6	0.3	13.6	0.3	13.6
20 (6.09)	16.7	0.2	16.9	0.2	16.9	0.2	16.9	0.2	16.9	0.2	16.9	0.2	16.9	0.2	16.9
24 (7.31)	20.0	0.1	20.1	0.1	20.1	0.1	20.1	0.1	20.1	0.1	20.1	0.1	20.1	0.1	20.1
28 (8.53)	23.3	0.1	23.4	0.1	23.4	0.1	23.4	0.1	23.4	0.1	23.4	0.1	23.4	0.1	23.4
32 (9.75)	26.7	0.1	26.8	0.1	26.8	0.1	26.8	0.1	26.8	0.1	26.8	0.1	26.8	0.1	26.8

Depth of Cover, ft (m)	P_e	18-in. Pipe P_t	P_v	20-in. Pipe P_t	P_v	24-in. Pipe P_t	P_v	30-in. Pipe P_t	P_v	36-in. Pipe P_t	P_v	42-in. Pipe P_t	P_v	48-in. Pipe P_t	P_v	54-in. Pipe P_t	P_v
2.5(0.76)	2.1	7.8	9.9	7.5	9.6	7.1	9.2	6.7	8.8	6.2	8.3	5.8	7.9	5.4	7.5	5.0	7.1
3 (0.91)	2.5	5.9	8.4	5.7	8.2	5.4	7.9	5.2	7.7	4.9	7.4	4.6	7.1	4.4	6.9	4.1	6.6
4 (1.21)	3.3	3.9	7.2	3.9	7.2	3.6	6.9	3.5	6.8	3.4	6.7	3.3	6.6	3.1	6.4	3.0	6.3
5 (1.52)	4.2	2.6	6.8	2.6	6.8	2.4	6.6	2.4	6.6	2.3	6.5	2.3	6.5	2.2	6.4	2.1	6.3
6 (1.82)	5.0	1.9	6.9	1.9	6.9	1.7	6.7	1.7	6.7	1.7	6.7	1.7	6.7	1.6	6.6	1.6	6.6
7 (2.13)	5.8	1.4	7.2	1.4	7.2	1.3	7.1	1.3	7.1	1.3	7.1	1.3	7.1	1.2	7.0	1.2	7.0
8 (2.43)	6.7	1.2	7.9	1.1	7.8	1.1	7.8	1.1	7.8	1.1	7.8	1.0	7.7	1.0	7.7	1.0	7.7
9 (2.74)	7.5	1.0	8.5	0.9	8.4	0.9	8.4	0.9	8.4	0.8	8.3	0.8	8.3	0.8	8.3	0.8	8.3
10 (3.04)	8.3	0.8	9.1	0.7	9.0	0.7	9.0	0.7	9.0	0.7	9.0	0.7	9.0	0.7	9.0	0.7	9.0
12 (3.65)	10.0	0.5	10.5	0.5	10.5	0.5	10.5	0.5	10.5	0.5	10.5	0.5	10.5	0.5	10.5	0.5	10.5
14 (4.26)	11.7	0.4	12.1	0.4	12.1	0.4	12.1	0.4	12.1	0.4	12.1	0.4	12.1	0.4	12.1	0.4	12.1
16 (4.87)	13.3	0.3	13.6	0.3	13.6	0.3	13.6	0.3	13.6	0.3	13.6	0.3	13.6	0.3	13.6	0.3	13.6
20 (6.09)	16.7	0.2	16.9	0.2	16.9	0.2	16.9	0.2	16.9	0.2	16.9	0.2	16.9	0.2	16.9	0.2	16.9
24 (7.31)	20.0	0.1	20.1	0.1	20.1	0.1	20.1	0.1	20.1	0.1	20.1	0.1	20.1	0.1	20.1	0.1	20.1
28 (8.53)	23.3	0.1	23.4	0.1	23.4	0.1	23.4	0.1	23.4	0.1	23.4	0.1	23.4	0.1	23.4	0.1	23.4
32 (9.75)	26.7	0.1	26.8	0.1	26.8	0.1	26.8	0.1	26.8	0.1	26.8	0.1	26.8	0.1	26.8	0.1	26.8

TABLE 2 Design Values for Standard Laying Conditions

Laying Condition	Description	E'	Bedding Angle, deg	K_b	K_x
\n\nType 1[A]	Flat-bottom trench.[B] Backfill not tamped.	150	30	0.235	0.108
\n\nType 2	Flat-bottom trench.[B] Backfill lightly tamped[C] to centerline of pipe.	300	45	0.210	0.105
\n\nType 3	Pipe bedded in 4-in. (102 mm) min loose soil.[D] Backfill lightly tamped[C] to top of pipe.	400	60	0.189	0.103
\n\nType 4	Pipe bedded in sand, gravel, or crushed stone to depth of $^1/_8$ pipe diameter, 4-in. (102 mm) min. Backfill lightly compacted[C] to top of pipe.	500	90	0.157	0.096
\n\nType 5	Pipe bedded in compacted granular material to centerline of pipe. Carefully compacted[C] granular or select[D] material to top of pipe.	700	150	0.128	0.085

[A] Laying condition Type 1 is limited to 16-in. and smaller pipe.

[B] Flat-bottom is defined as undisturbed earth.

[C] These laying conditions can be expected to develop the following backfill densities (Standard Proctor): Types 2 and 3, approximately 70 %; Type 4, approximately 75 %; Type 5, approximately 85 %.

[D] Loose soil or select material is defined as native soil excavated from the trench, free of rocks, foreign materials, and frozen earth.

TABLE 3 Allowances for Casting Tolerance

Size, in.	Casting Tolerance, in. (mm)
4–8	0.05 (1.3)
10–12	0.06 (1.5)
14–42	0.07 (1.8)
48	0.08 (2.0)
54	0.09 (2.3)

TABLE 4 Reduction Factors (R) for Truck Load Calculations

Size, in.	Depth of Cover, ft (m)			
	<4 (1.21)	4 to 7 (1.21 to 2.13)	8 to 10 (2.43 to 3.04)	>10 (3.04)
	Reduction Factor			
4 to 12	1.00	1.00	1.00	1.00
14	0.92	1.00	1.00	1.00
16	0.88	0.95	1.00	1.00
18	0.85	0.90	1.00	1.00
20	0.83	0.90	0.95	1.00
24 to 30	0.81	0.85	0.95	1.00
36 to 54	0.80	0.85	0.90	1.00

TABLE 5 Standard Outside Diameters and Thickness Classes

Size, in.	Outside diameter, in. (mm)	Thickness Class		
		50	51	52
		Thickness, in. (mm)		
4	4.80 (121.9)	. . .	0.26 (6.6)	0.29 (7.4)
6	6.90 (175.2)	0.25 (6.4)	0.28 (7.1)	0.31 (7.9)
8	9.05 (229.9)	0.27 (6.9)	0.30 (7.6)	0.33 (8.4)
10	11.10 (281.9)	0.29 (7.4)	0.32 (8.1)	0.35 (8.9)
12	13.20 (335.2)	0.31 (7.9)	0.34 (8.6)	0.37 (9.4)
14	15.30 (388.6)	0.33 (8.4)	0.36 (9.1)	0.39 (9.9)
16	17.40 (441.9)	0.34 (8.6)	0.37 (9.4)	0.40 (10.1)
18	19.50 (495.3)	0.35 (8.9)	0.38 (9.7)	0.41 (10.4)
20	21.60 (548.6)	0.36 (9.1)	0.39 (9.9)	0.42 (10.7)
24	25.80 (655.3)	0.38 (9.7)	0.41 (10.4)	0.44 (11.1)
30	32.00 (812.8)	0.39 (9.9)	0.43 (10.9)	0.47 (11.9)
36	38.30 (972.8)	0.43 (10.9)	0.48 (12.2)	0.53 (13.5)
42	44.50 (1130.3)	0.47 (11.9)	0.53 (13.5)	0.59 (15.1)
48	50.80 (1290.3)	0.51 (12.9)	0.58 (14.7)	0.65 (16.5)
54	57.10 (1450.3)	0.57 (14.5)	0.65 (16.5)	0.73 (18.5)

2
4
1

TABLE 6 Surface Load Factors for Single Truck on Unpaved Road

Depth of Cover, ft (m)	Pipe Size														
	4-in.	6-in.	8-in.	10-in.	12-in.	14-in.	16-in.	18-in.	20-in.	24-in.	30-in.	36-in.	42-in.	48-in.	54-in.
	Surface Load Factor, C														
2.5(0.76)	0.0238	0.0340	0.0443	0.0538	0.0634	0.0726	0.0814	0.0899	0.0980	0.1130	0.1321	0.1479	0.1604	0.1705	0.1784
3 (0.91)	0.0177	0.0253	0.0330	0.0402	0.0475	0.0546	0.0614	0.0681	0.0746	0.0867	0.1028	0.1169	0.1286	0.1384	0.1466
4 (1.21)	0.0107	0.0153	0.0201	0.0245	0.0290	0.0335	0.0379	0.0422	0.0464	0.0545	0.0657	0.0761	0.0853	0.0936	0.1008
5 (1.52)	0.0071	0.0102	0.0134	0.0163	0.0194	0.0224	0.0254	0.0283	0.0312	0.0369	0.0449	0.0525	0.0595	0.0661	0.0720
6 (1.82)	0.0050	0.0072	0.0095	0.0116	0.0138	0.0159	0.0181	0.0202	0.0223	0.0264	0.0323	0.0381	0.0435	0.0486	0.0534
7 (2.13)	0.0038	0.0054	0.0071	0.0087	0.0103	0.0119	0.0135	0.0151	0.0167	0.0198	0.0243	0.0288	0.0329	0.0370	0.0409
8 (2.43)	0.0029	0.0042	0.0055	0.0067	0.0079	0.0092	0.0104	0.0117	0.0129	0.0154	0.0189	0.0224	0.0258	0.0290	0.0322
9 (2.74)	0.0023	0.0033	0.0043	0.0053	0.0063	0.0073	0.0083	0.0093	0.0103	0.0122	0.0151	0.0179	0.0207	0.0233	0.0259
10 (3.04)	0.0019	0.0027	0.0035	0.0043	0.0051	0.0060	0.0068	0.0076	0.0084	0.0100	0.0123	0.0147	0.0169	0.0191	0.0213
12 (3.65)	0.0013	0.0019	0.0025	0.0030	0.0036	0.0042	0.0047	0.0053	0.0059	0.0070	0.0086	0.0103	0.0119	0.0135	0.0151
14 (4.26)	0.0010	0.0014	0.0018	0.0022	0.0027	0.0031	0.0035	0.0039	0.0043	0.0052	0.0064	0.0076	0.0088	0.0100	0.0112
16 (4.87)	0.0007	0.0011	0.0014	0.0017	0.0020	0.0024	0.0027	0.0030	0.0033	0.0040	0.0049	0.0059	0.0068	0.0077	0.0087
20 (6.09)	0.0005	0.0007	0.0009	0.0011	0.0013	0.0015	0.0017	0.0019	0.0021	0.0025	0.0032	0.0038	0.0044	0.0050	0.0056
24 (7.31)	0.0003	0.0005	0.0006	0.0008	0.0009	0.0011	0.0012	0.0013	0.0015	0.0018	0.0022	0.0026	0.0030	0.0035	0.0039
28 (8.53)	0.0002	0.0003	0.0005	0.0006	0.0007	0.0008	0.0009	0.0010	0.0011	0.0013	0.0016	0.0019	0.0022	0.0026	0.0029
32 (9.75)	0.0002	0.0003	0.0004	0.0004	0.0005	0.0006	0.0007	0.0008	0.0008	0.0010	0.0012	0.0015	0.0017	0.0020	0.0022

A 746

TABLE 7 Diameter-Thickness Ratios for Laying Condition Type 1

NOTE — $E' = 150$ $K_b = 0.235$ $K_x = 0.108$

Bending Stress Design	Trench Load P_v, psi		D/t^C or D/t_1	Bending Stress Design	Trench Load P_v, psi		D/t^C or D/t_1
	Deflection Design				Deflection Design		
	3 %[A] max	5 %[B] max			3 %[A] max	5 %[B] max	
5.17	3.89	6.48	150	8.86	7.12	11.87	100
5.21	3.91	6.52	149	8.99	7.26	12.11	99
5.26	3.94	6.57	148	9.13	7.41	12.35	98
5.30	3.97	6.62	147	9.27	7.57	12.61	97
5.35	4.00	6.67	146	9.41	7.73	12.88	96
5.40	4.03	6.72	145	9.56	7.89	13.15	95
5.45	4.06	6.77	144	9.71	8.07	13.45	94
5.49	4.09	6.82	143	9.87	8.25	13.75	93
5.54	4.13	6.88	142	10.03	8.44	14.07	92
5.59	4.16	6.94	141	10.20	8.64	14.40	91
5.65	4.20	6.99	140	10.37	8.85	14.74	90
5.70	4.23	7.05	139	10.55	9.06	15.11	89
5.75	4.27	7.12	138	10.74	9.29	15.48	88
5.80	4.31	7.18	137	10.93	9.53	15.88	87
5.86	4.35	7.25	136	11.13	9.78	16.30	86
5.91	4.39	7.31	135	11.34	10.04	16.73	85
5.97	4.43	7.38	134	11.55	10.31	17.19	84
6.03	4.47	7.46	133	11.78	10.60	17.67	83
6.09	4.52	7.53	132	12.01	10.90	18.17	82
6.15	4.56	7.61	131	12.25	11.22	18.70	81
6.21	4.61	7.69	130	12.50	11.56	19.26	80
6.27	4.66	7.77	129	12.76	11.91	19.85	79
6.33	4.71	7.85	128	13.03	12.28	20.46	78
6.40	4.76	7.94	127	13.31	12.67	21.11	77
6.46	4.82	8.03	126	13.60	13.08	21.79	76
6.53	4.87	8.12	125	13.91	13.51	22.52	75
6.60	4.93	8.22	124	14.23	13.97	23.28	74
6.67	4.99	8.32	123	14.56	14.45	24.08	73
6.74	5.05	8.42	122	14.91	14.96	24.93	72
6.82	5.11	8.52	121	15.27	15.50	25.83	71
6.89	5.18	8.63	120	15.65	16.07	26.78	70
6.97	5.25	8.74	119	16.05	16.68	27.79	69
7.05	5.32	8.86	118	16.46	17.32	28.86	68
7.13	5.39	8.98	117	16.89	18.00	30.00	67
7.21	5.46	9.11	116	17.35	18.73	31.21	66
7.29	5.54	9.24	115	17.83	19.50	32.49	65
7.38	5.62	9.37	114	18.33	20.32	33.86	64
7.47	5.71	9.51	113	18.85	21.19	35.32	63
7.56	5.79	9.65	112	19.40	22.12	36.87	62
7.65	5.88	9.80	111	19.98	23.12	38.53	61
7.75	5.97	9.96	110	20.59	24.18	40.30	60
7.85	6.07	10.12	109	21.23	25.32	42.20	59
7.95	6.17	10.28	108	21.91	26.54	44.23	58
8.05	6.27	10.46	107	22.63	27.85	46.42	57
8.16	6.38	10.63	106	23.38	29.26	48.76	56
8.27	6.49	10.82	105	24.18	30.77	51.28	55
8.38	6.61	11.01	104	25.02	32.39	53.99	54
8.49	6.73	11.22	103	25.92	34.15	56.92	53
8.61	6.86	11.43	102	26.86	36.05	60.08	52
8.74	6.99	11.64	101	27.87	38.10	63.50	51

2
4
3

11

TABLE 7 *Continued*

Trench Load P_v, psi				Trench Load P_r, psi			
Bending Stress Design	Deflection Design		D/t^C or D/t_1	Bending Stress Design	Deflection Design		D/t^C or D/t_1
	3 %[A] max	5 %[B] max			3 %[A] max	5 %[B] max	
28.94	40.32	67.20	50	46.84	83.54	139.23	39
30.07	42.73	71.22	49	49.30	90.28	150.47	38
31.28	45.35	75.58	48	51.96	97.80	163.00	37
32.57	48.20	80.34	47	54.86	106.20	177.00	36
33.95	51.31	85.52	46				
				58.02	115.62	192.70	35
35.42	54.72	91.19	45	61.46	126.21	210.36	34
37.00	58.44	97.40	44	65.23	138.18	230.29	33
38.69	62.53	104.22	43	69.36	151.73	252.88	32
40.50	67.03	111.71	42	73.92	167.15	278.58	31
42.46	71.99	119.98	41	78.94	184.77	307.96	30
44.56	77.47	129.11	40				

2
4
4

[A] Maximum 3 % deflection is recommended for rigid or semirigid linings such as cement mortar and most epoxies.
[B] Maximum 5 % deflection is recommended for flexible linings such as bituminous and plastic.
[C] The D/t for the tabulated P_r nearest to the calculated P_r is selected. When the calculated P_r is halfway between two tabulated values, the smaller D/t should be used.

TABLE 8 Diameter-Thickness Ratios for Laying Condition Type 2

NOTE—$E' = 300$ $K_b = 0.210$ $K_x = 0.105$

Bending Stress Design	Trench Load P_v, psi		D/t^C or D/t_1	Bending Stress Design	Trench Load, P_r, psi		D/t^C or D/t_1
	Deflection Design				Deflection Design		
	3 %^{A} max	5 %^{B} max			3 %^{A} max	5 %^{B} max	
7.42	6.61	11.02	150	12.01	9.94	16.57	100
7.48	6.64	11.06	149	12.16	10.09	16.81	99
7.54	6.67	11.11	148	12.31	10.24	17.06	98
7.61	6.70	11.16	147	12.46	10.40	17.33	97
7.67	6.73	11.21	146	12.62	10.56	17.60	96
7.74	6.76	11.27	145	12.79	10.73	17.89	95
7.80	6.79	11.32	144	12.96	10.91	18.19	94
7.87	6.83	11.38	143	13.13	11.10	18.50	93
7.94	6.86	11.43	142	13.31	11.29	18.82	92
8.01	6.89	11.49	141	13.49	11.50	19.17	91
8.08	6.93	11.55	140	13.68	11.71	19.52	90
8.15	6.97	11.61	139	13.88	11.94	19.89	89
8.22	7.01	11.68	138	14.08	12.17	20.28	88
8.29	7.05	11.74	137	14.30	12.42	20.69	87
8.37	7.09	11.81	136	14.51	12.67	21.12	86
8.44	7.13	11.88	135	14.74	12.94	21.57	85
8.52	7.17	11.95	134	14.97	13.22	22.04	84
8.59	7.22	12.03	133	15.21	13.52	22.53	83
8.67	7.26	12.10	132	15.46	13.83	23.05	82
8.75	7.31	12.18	131	15.72	14.16	23.60	81
8.83	7.36	12.26	130	15.99	14.50	24.17	80
8.91	7.41	12.35	129	16.28	14.86	24.77	79
8.99	7.46	12.43	128	16.57	15.24	25.40	78
9.07	7.51	12.52	127	16.87	15.64	26.07	77
9.16	7.57	12.62	126	17.19	16.06	26.77	76
9.25	7.63	12.71	125	17.52	16.51	27.52	75
9.33	7.69	12.81	124	17.86	16.98	28.30	74
9.42	7.75	12.91	123	18.22	17.48	29.13	73
9.51	7.81	13.02	122	18.59	18.00	30.00	72
9.60	7.87	13.12	121	18.98	18.56	30.93	71
9.70	7.94	13.24	120	19.39	19.14	31.91	70
9.79	8.01	13.35	119	19.82	19.77	32.95	69
9.89	8.08	13.47	118	20.27	20.43	34.05	68
9.99	8.16	13.60	117	20.73	21.13	35.22	67
10.09	8.23	13.72	116	21.23	21.87	36.46	66
10.19	8.31	13.86	115	21.74	22.67	37.78	65
10.29	8.40	13.99	114	22.28	23.51	39.18	64
10.40	8.48	14.14	113	22.85	24.41	40.68	63
10.51	8.57	14.29	112	23.45	25.37	42.28	62
10.62	8.66	14.44	111	24.07	26.39	43.99	61
10.73	8.76	14.60	110	24.74	27.49	45.81	60
10.84	8.86	14.76	109	25.43	28.66	47.76	59
10.96	8.96	14.93	108	26.17	29.91	49.86	58
11.08	9.07	15.11	107	26.95	31.26	52.10	57
11.21	9.18	15.30	106	27.77	32.71	54.51	56
11.33	9.29	15.49	105	28.64	34.26	57.10	55
11.46	9.41	15.69	104	29.56	35.93	59.89	54
11.59	9.54	15.89	103	30.53	37.74	62.90	53
11.73	9.67	16.11	102	31.57	39.69	66.15	52
11.87	9.80	16.33	101	32.67	41.80	69.67	51

2
4
5

TABLE 8 *Continued*

Trench Load P_v, psi				Trench Load, P_r, psi			
Bending Stress Design	Deflection Design		D/t^C or D/t_1	Bending Stress Design	Deflection Design		D/t^C or D/t_1
	3 %[A] max	5 %[B] max			3 %[A] max	5 %[B] max	
33.84	44.09	73.48	50	51.06	82.29	137.16	40
35.08	46.56	77.61	49	53.57	88.54	147.57	39
36.41	49.26	82.10	48	56.30	95.48	159.13	38
37.83	52.19	86.99	47	59.25	103.21	172.02	37
39.34	55.40	92.33	46	62.46	111.85	186.42	36
40.96	58.89	98.16	45	65.96	121.54	202.56	35
42.70	62.73	104.54	44	69.79	132.44	220.73	34
44.57	66.93	111.55	43	73.98	144.74	241.23	33
46.57	71.56	119.26	42	78.57	158.68	264.46	32
48.73	76.66	127.76	41	83.64	174.54	290.90	31
				89.23	192.67	321.11	30

2
4
6

[A] Maximum 3 % deflection is recommended for rigid or semirigid linings such as cement mortar and most epoxies.
[B] Maximum 5 % deflection is recommended for flexible linings such as bituminous and plastic.
[C] The D/t for the tabulated P_r nearest to the calculated P_r is selected. When the calculated P_v is halfway between two tabulated values, the smaller D/t should be used.

TABLE 9 Diameter-Thickness Ratios for Laying Condition Type 3

NOTE—$E' = 400$ $K_b = 0.189$ $K_x = 0.103$

Trench Load P_v, psi				Trench Load P_r, psi			
Bending Stress Design	Deflection Design		D/t^C or D/t_1	Bending Stress Design	Deflection Design		D/t^C or D/t_1
	3 %[A] max	5 %[B] max			3 %[A] max	5 %[B] max	
10.00	8.52	14.19	150	15.71	12.21	20.35	98
10.08	8.54	14.24	149	15.88	12.37	20.62	97
10.16	8.57	14.29	148	16.06	12.54	20.90	96
10.24	8.60	14.34	147				
10.33	8.64	14.39	146	16.23	12.72	21.20	95
				16.42	12.90	21.50	94
10.41	8.67	14.45	145	16.61	13.09	21.82	93
10.49	8.70	14.50	144	16.80	13.29	22.15	92
10.58	8.73	14.56	143	17.00	13.50	22.50	91
10.66	8.77	14.62	142				
10.75	8.81	14.68	141	17.21	13.72	22.86	90
				17.42	13.95	23.24	89
10.83	8.84	14.74	140	17.64	14.18	23.64	88
10.92	8.88	14.80	139	17.86	14.43	24.06	87
11.01	8.92	14.87	138	18.10	14.70	24.49	86
11.10	8.96	14.93	137				
11.19	9.00	15.00	136	18.34	14.97	24.95	85
				18.59	15.26	25.43	84
11.28	9.04	15.07	135	18.85	15.56	25.93	83
11.37	9.09	15.15	134	19.12	15.88	26.46	82
11.46	9.13	15.22	133	19.40	16.21	27.01	81
11.56	9.18	15.30	132				
11.65	9.23	15.38	131	19.68	16.56	27.60	80
				19.99	16.93	28.21	79
11.75	9.28	15.46	130	20.30	17.31	28.86	78
11.84	9.33	15.55	129	20.62	17.72	29.54	77
11.94	9.38	15.64	128	20.96	18.15	30.26	76
12.04	9.44	15.73	127				
12.14	9.49	15.82	126	21.31	18.61	31.01	75
				21.68	19.09	31.81	74
12.25	9.55	15.92	125	22.07	19.59	32.65	73
12.35	9.61	16.02	124	22.47	20.13	33.55	72
12.45	9.67	16.12	123	22.88	20.69	34.49	71
12.56	9.74	16.23	122				
12.67	9.80	16.34	121	23.32	21.29	35.49	70
				23.78	21.93	36.55	69
12.78	9.87	16.45	120	24.26	22.60	37.67	68
12.89	9.94	16.57	119	24.76	23.32	38.86	67
13.00	10.02	16.69	118	25.29	24.08	40.13	66
13.11	10.09	16.82	117				
13.23	10.17	16.95	116	25.85	24.88	41.47	65
				26.43	25.74	42.91	64
13.34	10.25	17.09	115	27.04	26.66	44.43	63
13.46	10.34	17.23	114	27.68	27.64	46.06	62
13.58	10.42	17.37	113	28.36	28.68	47.80	61
13.71	10.51	17.52	112				
13.83	10.61	17.68	111	29.08	29.80	49.66	60
				29.83	30.99	51.65	59
13.96	10.71	17.84	110	30.63	32.27	53.78	58
14.09	10.81	18.01	109	31.47	33.64	56.07	57
14.22	10.91	18.18	108	32.36	35.12	58.53	56
14.36	11.02	18.37	107				
14.50	11.13	18.55	106	33.31	36.70	61.17	55
				34.30	38.41	64.02	54
14.64	11.25	18.75	105	35.37	40.25	67.08	53
14.78	11.37	18.95	104	36.49	42.24	70.40	52
14.93	11.50	19.16	103	37.69	44.39	73.98	51
15.08	11.63	19.38	102				
15.23	11.77	19.61	101	38.97	46.72	77.86	50
				40.33	49.25	82.08	49
15.39	11.91	19.85	100	41.78	5i.99	86.65	48
15.55	12.06	20.10	99				

TABLE 9 *Continued*

Trench Load P_v, psi				Trench Load P_v, psi			
Bending Stress Design	Deflection Design		D/t^C or D/t_1	Bending Stress Design	Deflection Design		D/t^C or D/t_1
	3 %[A] max	5 %[B] max			3 %[A] max	5 %[B] max	
43.33	54.98	91.64	47	63.61	99.11	165.18	38
44.98	58.25	97.08	46	66.86	106.99	178.32	37
				70.40	115.80	193.00	36
46.76	61.81	103.02	45				
48.66	65.72	109.53	44	74.27	125.67	209.46	35
50.71	70.01	116.68	43	78.49	136.78	227.97	34
52.91	74.72	124.54	42	83.11	149.32	248.87	33
55.28	79.92	133.20	41	88.19	163.54	272.56	32
				93.79	179.71	299.51	31
57.84	85.67	142.78	40	99.97	198.18	330.31	30
60.61	92.04	153.39	39				

[A] Maximum 3 % deflection is recommended for rigid or semirigid linings such as cement mortar and most epoxies.

[B] Maximum 5 % deflection is recommended for flexible linings such as bituminous and plastic.

[C] The D/t for the tabulated P_v nearest to the calculated P_v is selected. When the calculated P_v is halfway between two tabulated values, the smaller D/t should be used.

TABLE 10 Diameter-Thickness Ratios for Laying Condition Type 4

NOTE — $E' = 500$ $K_b = 0.157$ $K_x = 0.096$

Bending Stress Design	Trench Load P_v, psi Deflection Design 3 %[A] max	Trench Load P_v, psi Deflection Design 5 %[B] max	D/t^C or D/t_1	Bending Stress Design	Trench Load P_v, psi Deflection Design 3 %[A] max	Trench Load P_v, psi Deflection Design 5 %[B] max	D/t^C or D/t_1
16.34	11.04	18.40	150	22.83	15.01	25.02	98
16.45	11.07	18.46	149	23.01	15.18	25.30	97
16.55	11.11	18.51	148	23.20	15.36	25.61	96
16.65	11.14	18.56	147				
16.76	11.17	18.62	146	23.38	15.55	25.92	95
				23.58	15.75	26.25	94
16.86	11.21	18.68	145	23.78	15.95	26.59	93
16.96	11.24	18.74	144	23.99	16.17	26.94	92
17.07	11.28	18.80	143	24.20	16.39	27.32	91
17.18	11.31	18.86	142				
17.28	11.35	18.92	141	24.42	16.62	27.71	90
				24.64	16.87	28.11	89
17.39	11.39	18.99	140	24.88	17.12	28.54	88
17.50	11.43	19.06	139	25.12	17.39	28.99	87
17.60	11.48	19.13	138	25.37	17.67	29.45	86
17.71	11.52	19.20	137				
17.82	11.56	19.27	136	25.63	17.97	29.95	85
				25.90	18.28	30.46	84
17.93	11.61	19.35	135	26.18	18.60	31.00	83
18.04	11.66	19.43	134	26.47	18.94	31.57	82
18.15	11.71	19.51	133	26.77	19.30	32.16	81
18.26	11.76	19.59	132				
18.37	11.81	19.68	131	27.09	19.67	32.79	80
				27.42	20.07	33.45	79
18.49	11.86	19.77	130	27.76	20.48	34.14	78
18.60	11.92	19.86	129	28.11	20.92	34.87	77
18.72	11.97	19.95	128	28.49	21.38	35.64	76
18.83	12.03	20.05	127				
18.95	12.09	20.15	126	28.87	21.87	36.45	75
				29.28	22.38	37.31	74
19.06	12.15	20.26	125	29.70	22.93	38.21	73
19.18	12.22	20.36	124	30.15	23.50	39.17	72
19.30	12.28	20.47	123	30.62	24.11	40.18	71
19.42	12.35	20.59	122				
19.54	12.42	20.71	121	31.11	24.75	41.25	70
				31.62	25.43	42.39	69
19.66	12.50	20.83	120	32.16	26.16	43.59	68
19.78	12.57	20.96	119	32.72	26.92	44.87	67
19.91	12.65	21.09	118	33.32	27.74	46.23	66
20.04	12.73	21.22	117				
20.16	12.82	21.36	116	33.95	28.60	47.67	65
				34.61	29.53	49.21	64
20.29	12.91	21.51	115	35.30	30.51	50.85	63
20.42	13.00	21.66	114	36.04	31.56	52.60	62
20.55	13.09	21.82	113	36.81	32.68	54.47	61
20.69	13.19	21.98	112				
20.82	13.29	22.15	111	37.63	33.88	56.46	60
				38.50	35.16	58.60	59
20.96	13.39	22.32	110	39.42	36.53	60.88	58
21.10	13.50	22.50	109	40.39	38.00	63.34	57
21.24	13.61	22.69	108	41.42	39.58	65.97	56
21.39	13.73	22.88	107				
21.54	13.85	23.08	106	42.51	41.28	68.81	55
				43.67	43.12	71.86	54
21.69	13.98	23.29	105	44.91	45.09	75.15	53
21.84	14.11	23.51	104	46.22	47.22	78.71	52
22.00	14.24	23.74	103	47.62	49.53	82.55	51
22.16	14.38	23.97	102				
22.32	14.53	24.22	101	49.11	52.03	86.72	50
				50.70	54.74	91.24	49
22.49	14.68	24.47	100	52.41	57.69	96.15	48
22.66	14.84	24.74	99				

TABLE 10 *Continued*

Trench Load P_v, psi				Trench Load P_v, psi			
Bending Stress Design	Deflection Design		D/t^C or D/t_1	Bending Stress Design	Deflection Design		D/t^C or D/t_1
	3 %[A] max	5 %[B] max			3 %[A] max	5 %[B] max	
54.23	60.90	101.50	47	78.24	108.24	180.40	38
56.18	64.40	107.33	46	82.11	116.70	194.50	37
				86.33	126.15	210.25	36
58.27	68.23	113.71	45				
60.52	72.42	120.70	44	90.93	136.74	227.91	35
62.93	77.02	128.36	43	95.97	148.66	247.77	34
65.54	82.08	136.80	42	101.49	162.12	270.20	33
68.35	87.66	146.09	41	107.56	177.37	295.61	32
				114.25	194.72	324.53	31
71.39	93.82	156.37	40	121.65	214.54	357.57	30
74.67	100.65	167.75	39				

[A] Maximum 3 % deflection is recommended for rigid or semirigid linings such as cement mortar and most epoxies.
[B] Maximum 5 % deflection is recommended for flexible linings such as bituminous and plastic.
[C] The D/t for the tabulated P_v nearest to the calculated P_v is selected. When the calculated P_v is halfway between two tabulated values, the smaller D/t should be used.

TABLE 11 Diameter-Thickness Ratios for Laying Condition Type 5

NOTE $-E' = 700$ $K_b = 0.128$ $K_x = 0.085$

Bending Stress Design	Deflection Design		D/t^C or D/t_1	Bending Stress Design	Deflection Design		D/t^C or D/t_1
	3 %^{A} max	5 %^{B} max			3 %^{A} max	5 %^{B} max	
30.21	16.78	27.96	150	36.54	20.89	34.82	100
30.34	16.81	28.02	149	36.69	21.07	35.12	99
30.48	16.85	28.08	148	36.85	21.26	35.43	98
30.61	16.89	28.14	147	37.01	21.45	35.76	97
30.74	16.92	28.20	146	37.17	21.66	36.10	96
30.87	16.96	28.27	145	37.34	21.87	36.45	95
30.99	17.00	28.34	144	37.52	22.09	36.82	94
31.12	17.04	28.40	143	37.70	22.32	37.20	93
31.25	17.09	28.48	142	37.89	22.56	37.61	92
31.38	17.13	28.55	141	38.08	22.82	38.03	91
31.50	17.17	28.62	140	38.28	23.08	38.47	90
31.63	17.22	28.70	139	38.49	23.36	38.93	89
31.76	17.27	28.78	138	38.71	23.65	39.41	88
31.88	17.32	28.86	137	38.93	23.95	39.91	87
32.01	17.37	28.94	136	39.17	24.27	40.44	86
32.13	17.42	29.03	135	39.41	24.60	41.00	85
32.25	17.47	29.12	134	39.67	24.95	41.58	84
32.38	17.53	29.21	133	39.94	25.31	42.19	83
32.50	17.58	29.30	132	40.22	25.70	42.83	82
32.62	17.64	29.40	131	40.51	26.10	43.50	81
32.75	17.70	29.50	130	40.82	26.52	44.21	80
32.87	17.76	29.61	129	41.14	26.97	44.95	79
32.99	17.83	29.71	128	41.48	27.44	45.73	78
33.11	17.89	29.82	127	41.84	27.93	46.56	77
33.23	17.96	29.94	126	42.21	28.46	47.43	76
33.35	18.03	30.05	125	42.60	29.01	48.34	75
33.47	18.11	30.18	124	43.02	29.59	49.31	74
33.59	18.18	30.30	123	43.45	30.20	50.33	73
33.71	18.26	30.43	122	43.92	30.85	51.41	72
33.83	18.34	30.56	121	44.40	31.53	52.56	71
33.95	18.42	30.70	120	44 91	32.26	53.77	70
34.07	18.51	30.85	119	45.46	33.03	55.05	69
34.19	18.60	30.99	118	46.03	33.85	56.41	68
34.31	18.69	31.15	117	46.64	34.71	57.85	67
34.43	18.78	31.31	116	47.28	35.63	59.39	66
34.55	18.88	31.47	115	.47.96	36.61	61.02	65
34.68	18.98	31.64	114	48.68	37.65	62.76	64
34.80	19.09	31.82	113	49.44	38.77	64.61	63
34.92	19.20	32.00	112	50.25	39.95	66.58	62
35.05	19.31	32.19	111	51.11	41.21	68.69	61
35.17	19.43	32.39	110	52.02	42.57	70.94	60
35.30	19.55	32.59	109	52.99	44.01	73.36	59
35.43	19.68	32.80	108	54.02	45.56	75.94	58
35.56	19.81	33.02	107	55.12	47.23	78.71	57
35.69	19.95	33.25	106	56.28	49.01	81.69	56
35.83	20.09	33.48	105	57.53	50.93	84.89	55
35.96	20.24	33.73	104	58.86	53.00	88.34	54
36.10	20.39	33.99	103	60.28	55.23	92.05	53
36.25	20.55	34.25	102	61.79	57.64	96.07	52
36.39	20.72	34.53	101	63.41	60.25	100.41	51

2
5
1

19

TABLE 11 *Continued*

Trench Load P_v, psi				Trench Load, P_v, psi			
Bending Stress Design	Deflection Design		D/t^C or D/t_1	Bending Stress Design	Deflection Design		D/t^C or D/t_1
	3 %^{A} max	5 %^{B} max			3 %^{A} max	5 %^{B} max	
65.14	63.07	105.12	50	91.47	110.27	183.78	40
67.00	66.13	110.22	49	95.40	117.98	196.64	39
68.99	69.46	115.77	48	99.67	126.56	210.93	38
71.12	73.09	121.81	47	104.32	136.11	226.84	37
73.41	77.04	128.40	46	109.40	146.78	244.63	36
75.88	81.36	135.61	45	114.94	158.75	264.58	35
78.54	86.10	143.49	44	121.02	172.21	287.01	34
81.40	91.29	152.15	43	127.69	187.41	312.34	33
84.50	97.01	161.68	42	135.03	204.63	341.04	32
87.85	103.31	172.18	41	143.14	224.22	373.70	31
				152.11	246.61	411.02	30

^{A} Maximum 3 % deflection is recommended for rigid or semirigid linings such as cement mortar and most epoxies.
^{B} Maximum 5 % deflection is recommended for flexible linings such as bituminous and plastic.
^{C} The D/t for the tabulated P_v nearest to the calculated P_v is selected. When the calculated P_v is halfway between two tabulated values, the smaller D/t should be used.

2 5 2

TABLE 12 Pipe Selection Table (Cement-Lined Pipe)

Pipe Size, in.	Thickness Class	Nominal Thickness, in.	Laying Condition				
			Type 1	Type 2	Type 3	Type 4	Type 5
			Maximum Depth of Cover, ft (m)[A]				
4	51	0.26	76 (23.1)	86 (26.0)	96 (29.2)	B	B
	52	0.29	B	B	B	B	B
6	50	0.25	32 (9.7)	38 (11.6)	44 (13.4)	56 (17.0)	75 (22.9)
	51	0.28	49 (14.9)	57 (17.4)	64 (19.5)	80 (24.3)	B
	52	0.31	67 (20.4)	77 (23.5)	86 (26.0)	B	B
8	50	0.27	25 (7.6)	30 (9.1)	36 (11.0)	46 (14.0)	64 (19.5)
	51	0.30	36 (10.9)	42 (12.9)	49 (14.9)	61 (18.6)	81 (24.7)
	52	0.33	47 (14.3)	54 (16.5)	62 (18.9)	77 (23.5)	99 (30.1)
10	50	0.29	19 (5.8)	24 (7.3)	29 (8.9)	38 (11.6)	55 (16.8)
	51	0.32	27 (8.2)	32 (9.8)	38 (11.6)	49 (15.0)	66 (20.1)
	52	0.35	35 (10.6)	41 (12.5)	47 (14.3)	59 (18.0)	79 (24.0)
12	50	0.31	17 (5.1)	22 (6.7)	27 (8.2)	36 (11.0)	52 (15.9)
	51	0.34	23 (7.0)	28 (8.5)	33 (10.0)	43 (13.1)	60 (18.2)
	52	0.37	30 (9.1)	35 (10.7)	41 (12.5)	53 (16.1)	71 (21.6)
14	50	0.33	15 (4.6)	19 (5.8)	24 (7.3)	33 (10.0)	49 (14.9)
	51	0.36	19 (5.8)	23 (7.0)	28 (8.5)	38 (11.6)	55 (16.8)
	52	0.39	24 (7.3)	29 (8.9)	34 (10.3)	44 (13.4)	62 (18.9)
16	50	0.34	13 (4.0)	17 (5.1)	21 (6.4)	30 (9.1)	47 (14.3)
	51	0.37	16 (4.9)	21 (6.4)	25 (7.6)	34 (10.4)	51 (15.5)
	52	0.40	20 (6.1)	25 (7.6)	30 (9.1)	40 (12.1)	57 (17.3)
18	50	0.35	11 (3.3)	15 (4.6)	20 (6.1)	29 (8.9)	42 (12.8)
	51	0.38	14 (4.2)	19 (5.8)	23 (7.0)	32 (9.7)	49 (14.9)
	52	0.41	18 (5.5)	22 (6.7)	27 (8.2)	36 (11.0)	53 (16.2)
20	50	0.36	10 (3.0)	14 (4.3)	18 (5.5)	27 (8.2)	38 (11.6)
	51	0.39	13 (4.0)	17 (5.1)	21 (6.4)	30 (9.1)	44 (13.4)
	52	0.42	16 (4.9)	20 (6.1)	25 (7.6)	34 (10.4)	50 (15.2)
24	50	0.38	8 (2.4)	12 (3.7)	17 (5.1)	23 (7.0)	31 (9.4)
	51	0.41	10 (3.0)	15 (4.6)	19 (5.8)	27 (8.2)	36 (11.0)
	52	0.44	13 (4.0)	17 (5.1)	21 (6.4)	30 (9.1)	41 (12.5)
30	50	0.39	C	10 (3.5)	14 (3.7)	18 (5.5)	25 (7.6)
	51	0.43		12 (3.7)	16 (4.9)	21 (6.4)	29 (8.9)
	52	0.47		14 (4.3)	19 (5.8)	24 (7.3)	33 (10.0)
36	50	0.43	C	10 (3.5)	13 (4.0)	17 (5.1)	25 (7.6)
	51	0.48		12 (3.7)	16 (4.9)	20 (6.0)	28 (8.5)
	52	0.53		15 (4.6)	19 (5.8)	24 (7.3)	32 (9.8)
42	50	0.47	C	9 (2.7)	13 (4.0)	16 (4.9)	24 (7.3)
	51	0.53		12 (3.7)	15 (4.6)	19 (5.8)	27 (8.2)
	52	0.59		14 (4.3)	18 (5.5)	22 (6.7)	30 (9.1)
48	50	0.51	C	9 (2.7)	12 (3.7)	15 (4.6)	23 (7.0)
	51	0.58		12 (3.7)	14 (4.3)	18 (5.5)	26 (7.9)
	52	0.65		14 (4.3)	18 (5.5)	21 (6.4)	30 (9.1)
54	50	0.57	C	9 (2.7)	12 (3.7)	15 (4.6)	23 (7.0)
	51	0.65		12 (3.7)	14 (4.3)	18 (5.5)	25 (7.6)
	52	0.73		14 (4.3)	17 (5.1)	21 (6.4)	29 (8.9)

[A] These pipes are adequate for depths of cover from 2.5 ft (0.76 m) up to the maximum shown including an allowance for single H-20 truck with 1.5 impact factor.
[B] Calculated maximum depth of cover exceeds 100 ft (30.5 m).
[C] Laying Condition Type 1 is limited to 24 in. and smaller pipe.

2
5
3

TABLE 13 Pipe Selection Table (Pipe with Flexible Lining)

Pipe Size, in.	Thickness Class	Nominal Thickness, in.	Laying Condition				
			Type 1	Type 2	Type 3	Type 4	Type 5
			Maximum Depth of Cover, ft (m) [A]				
4	51	0.26	76 (23.1)	86 (26.0)	96 (29.2)	[B]	[B]
	52	0.29	[B]	[B]	[B]	[B]	[B]
6	50	0.25	32 (9.7)	38 (11.6)	44 (13.4)	56 (17.0)	75 (22.9)
	51	0.28	49 (14.9)	57 (17.4)	64 (19.5)	80 (24.3)	[C]
	52	0.31	67 (20.4)	77 (23.5)	86 (26.0)	[C]	[C]
8	50	0.27	25 (7.6)	30 (9.1)	36 (11.0)	46 (14.0)	64 (19.5)
	51	0.30	36 (10.9)	42 (12.9)	49 (14.9)	61 (18.6)	81 (24.7)
	52	0.33	47 (14.3)	54 (16.5)	62 (18.9)	77 (23.5)	99 (30.1)
10	50	0.29	19 (5.8)	24 (7.3)	29 (8.9)	38 (11.6)	55 (16.8)
	51	0.32	27 (8.2)	32 (9.8)	38 (11.6)	49 (15.0)	66 (20.1)
	52	0.35	35 (10.6)	41 (12.5)	47 (14.3)	59 (18.0)	79 (24.0)
12	50	0.31	17 (5.1)	22 (6.7)	27 (8.2)	36 (11.0)	52 (15.9)
	51	0.34	23 (7.0)	28 (8.5)	33 (10.0)	43 (13.1)	60 (18.2)
	52	0.37	30 (9.1)	35 (10.7)	41 (12.5)	53 (16.1)	71 (21.6)
14	50	0.33	15 (4.6)	19 (5.8)	24 (7.3)	33 (10.0)	49 (14.9)
	51	0.36	19 (5.8)	23 (7.0)	28 (8.5)	38 (11.6)	55 (16.8)
	52	0.39	24 (7.3)	29 (8.9)	34 (10.3)	44 (13.4)	62 (18.9)
16	50	0.34	13 (4.0)	17 (5.1)	21 (6.4)	30 (9.1)	47 (14.3)
	51	0.37	16 (4.9)	21 (6.4)	25 (7.6)	34 (10.4)	51 (15.5)
	52	0.40	20 (6.1)	25 (7.6)	30 (9.1)	40 (12.1)	57 (17.3)
18	50	0.35	11 (3.3)	15 (4.6)	20 (6.1)	29 (8.9)	45 (13.7)
	51	0.38	14 (4.2)	19 (5.8)	23 (7.0)	32 (9.7)	49 (14.9)
	52	0.41	18 (5.5)	22 (6.7)	27 (8.2)	36 (11.0)	53 (16.2)
20	50	0.36	10 (3.0)	14 (4.3)	18 (5.5)	27 (8.2)	44 (13.4)
	51	0.39	13 (4.0)	17 (5.1)	21 (6.4)	30 (9.1)	47 (14.3)
	52	0.42	16 (4.9)	20 (6.1)	25 (7.6)	34 (10.4)	50 (15.2)
24	50	0.38	8 (2.4)	12 (3.7)	17 (5.1)	25 (7.6)	42 (12.8)
	51	0.41	10 (3.0)	15 (4.6)	19 (5.8)	28 (8.5)	45 (13.7)
	52	0.44	13 (4.0)	17 (5.1)	21 (6.4)	30 (9.1)	47 (14.3)
30	50	0.39	[C]	10 (3.5)	14 (3.7)	22 (6.7)	39 (11.9)
	51	0.43		12 (3.7)	16 (4.9)	25 (7.6)	42 (12.8)
	52	0.47		14 (4.3)	19 (5.8)	27 (8.2)	44 (13.4)
36	50	0.43	[C]	10 (3.5)	14 (3.7)	22 (6.7)	39 (11.9)
	51	0.48		12 (3.7)	16 (4.9)	25 (7.6)	42 (12.8)
	52	0.53		15 (4.6)	19 (5.8)	27 (8.2)	44 (13.4)
42	50	0.47	[C]	9 (2.7)	13 (4.0)	22 (6.7)	39 (11.9)
	51	0.53		12 (3.7)	16 (4.9)	25 (7.6)	42 (12.8)
	52	0.59		14 (4.3)	18 (5.5)	27 (8.2)	44 (13.4)
48	50	0.51	[C]	9 (2.7)	13 (4.0)	21 (6.4)	39 (11.9)
	51	0.58		12 (3.7)	16 (4.9)	24 (7.3)	41 (12.8)
	52	0.65		14 (4.3)	18 (5.5)	27 (8.2)	44 (13.4)
54	50	0.57	[C]	9 (2.7)	13 (4.0)	21 (6.4)	38 (11.6)
	51	0.65		12 (3.7)	16 (4.9)	24 (7.3)	41 (12.8)
	52	0.73		14 (4.3)	18 (5.5)	27 (8.2)	44 (13.4)

[A] These pipes are adequate for depths of cover from 2.5 ft (0.76 m) up to the maximum shown including an allowance for single H-20 truck with 1.5 impact factor.

[B] Calculated maximum depth of cover exceeds 100 ft (30.5 m).

[C] Laying Condition Type 1 is limited to 24 in. and smaller pipe.

TABLE 14 Standard Dimensions and Weights of Push-on-Joint Ductile Iron Pipe

Size, in.	Thickness Class	Thickness, in. (mm)	Outside Diameter,[A] in. (mm)	Weight of Barrel per ft, lb (kg)	Weight of Bell, lb (kg)	18-ft Laying Length		20-ft Laying Length	
						Weight per Length,[B] lb (kg)	Average Weight per ft,[C] lb (kg)	Weight per Length,[B] lb (kg)	Average Weight per ft,[C] lb (kg)
4	51	0.26 (6.6)	4.80 (121.9)	11.3 (5.12)	11 (4.98)	215 (97.52)	11.9 (5.39)	235 (106.59)	11.8 (5.35)
	52	0.29 (7.4)	4.80 (121.9)	12.6 (5.71)	11 (4.98)	240 (108.86)	13.2 (5.98)	265 (120.20)	13.2 (5.98)
6	50	0.25 (6.4)	6.90 (175.2)	16.0 (7.25)	18 (8.16)	305 (138.34)	17.0 (7.71)	340 (154.22)	16.9 (7.66)
	51	0.28 (7.1)	6.90 (175.2)	17.8 (8.07)	18 (8.16)	340 (154.22)	18.8 (8.52)	375 (170.10)	18.7 (8.48)
	52	0.31 (7.9)	6.90 (175.2)	19.6 (8.89)	18 (8.16)	370 (168.73)	20.6 (9.34)	410 (185.97)	20.5 (9.29)
8	50	0.27 (6.9)	9.05 (229.8)	22.8 (10.34)	26 (11.79)	435 (197.31)	24.2 (10.97)	480 (217.72)	24.1 (10.93)
	51	0.30 (7.6)	9.05 (229.8)	25.2 (11.43)	26 (11.79)	480 (217.78)	26.6 (12.06)	530 (240.40)	26.5 (12.03)
	52	0.33 (8.4)	9.05 (229.8)	27.7 (12.56)	26 (11.79)	525 (238.14)	29.1 (13.19)	580 (263.08)	29.0 (13.15)
10	50	0.29 (7.3)	11.10 (281.9)	30.1 (13.65)	34 (15.42)	575 (260.82)	32.0 (14.51)	635 (288.03)	31.8 (14.42)
	51	0.32 (8.1)	11.10 (281.9)	33.2 (15.05)	34 (15.42)	630 (285.76)	35.1 (15.92)	700 (317.52)	34.9 (15.83)
	52	0.35 (8.9)	11.10 (281.9)	36.2 (16.42)	34 (15.42)	685 (310.71)	38.1 (17.28)	760 (344.73)	37.9 (17.19)
12	50	0.31 (7.9)	13.20 (335.2)	38.4 (17.41)	43 (19.50)	735 (333.39)	40.8 (18.50)	810 (367.41)	40.6 (18.41)
	51	0.34 (8.6)	13.20 (335.2)	42.0 (19.05)	43 (19.50)	800 (362.88)	44.4 (20.13)	885 (401.43)	44.2 (20.04)
	52	0.37 (9.4)	13.20 (335.2)	45.6 (20.68)	43 (19.50)	865 (392.36)	48.0 (21.77)	955 (433.18)	47.8 (21.68)
14	50	0.33 (8.4)	15.30 (388.6)	47.5 (21.54)	63 (28.57)	920 (417.31)	51.0 (23.13)	1015 (460.40)	50.6 (22.95)
	51	0.36 (9.1)	15.30 (388.6)	51.7 (23.45)	63 (28.57)	995 (451.33)	55.2 (25.03)	1095 (496.69)	54.8 (24.85)
	52	0.39 (9.9)	15.30 (388.6)	55.9 (25.35)	63 (28.57)	1070 (485.35)	59.4 (26.94)	1180 (535.24)	59.0 (26.76)
16	50	0.34 (8.6)	17.40 (441.9)	55.8 (25.31)	76 (34.47)	1080 (489.88)	60.0 (27.21)	1190 (539.78)	59.6 (27.03)
	51	0.37 (9.4)	17.40 (441.9)	60.6 (27.48)	76 (34.47)	1165 (528.44)	64.8 (29.39)	1290 (585.14)	64.4 (29.21)
	52	0.40 (10.1)	17.40 (441.9)	65.4 (29.66)	76 (34.47)	1255 (569.26)	69.6 (31.57)	1385 (628.23)	69.2 (31.38)
18	50	0.35 (8.9)	19.50 (495.3)	64.4 (29.21)	87 (39.46)	1245 (564.73)	69.2 (31.38)	1375 (623.70)	68.8 (31.20)
	51	0.38 (9.7)	19.50 (495.3)	69.8 (31.66)	87 (39.46)	1345 (610.09)	74.6 (33.83)	1485 (673.59)	74.2 (33.65)
	52	0.41 (10.4)	19.50 (495.3)	75.2 (34.11)	87 (39.46)	1440 (653.18)	80.0 (36.28)	1590 (721.22)	79.6 (34.88)
20	50	0.36 (9.1)	21.60 (548.6)	73.5 (33.33)	97 (43.99)	1420 (644.11)	78.9 (35.78)	1565 (709.88)	78.4 (35.56)
	51	0.39 (9.9)	21.60 (548.6)	79.5 (36.06)	97 (43.99)	1530 (694.00)	84.9 (38.51)	1685 (764.31)	84.4 (38.28)
	52	0.42 (10.7)	21.60 (548.6)	85.5 (38.78)	97 (43.99)	1635 (741.63)	90.9 (41.23)	1805 (818.74)	90.4 (41.00)
24	50	0.38 (9.7)	25.80 (655.3)	92.9 (42.13)	120 (54.43)	1790 (811.94)	99.6 (45.17)	1980 (898.12)	98.9 (44.86)
	51	0.41 (10.4)	25.80 (655.3)	100.1 (45.40)	120 (54.43)	1920 (870.91)	106.8 (48.44)	2120 (961.63)	106.1 (48.12)
	52	0.44 (11.1)	25.80 (655.3)	107.3 (48.67)	120 (54.43)	2050 (929.88)	114.0 (51.71)	2265 (1027.40)	113.3 (51.39)
30	50	0.39 (9.9)	32.00 (812.8)	118.5 (53.75)	D	2350 (1065.96)	130.5 (59.19)	2535 (1149.87)	126.6 (57.42)
	51	0.43 (10.9)	32.00 (812.8)	130.5 (59.19)		2565 (1163.48)	142.5 (64.63)	2775 (1258.74)	138.6 (62.86)
	52	0.47 (11.9)	32.00 (812.8)	142.5 (64.63)		2780 (1261.00)	154.5 (70.08)	3015 (1367.60)	150.6 (68.31)

TABLE 14 *Continued*

Size, in.	Thickness Class	Thickness, in. (mm)	Outside Diameter,[A] in. (mm)	Weight of Barrel per ft, lb (kg)	Weight of Bell, lb (kg)	18-ft Laying Length		20-ft Laying Length	
						Weight per Length,[B] lb (kg)	Average Weight per ft,[C] lb (kg)	Weight per Length,[B] lb (kg)	Average Weight per ft,[C] lb (kg)
36	50	0.43 (10.9)	38.30 (972.8)	156.5 (70.98)	[E]	3110 (1410.69)	172.7 (78.33)	3345 (1517.29)	167.3 (75.88)
	51	0.48 (12.2)	38.30 (972.8)	174.5 (79.15)		3435 (1558.11)	190.7 (86.50)	3705 (1680.58)	185.3 (84.05)
	52	0.53 (13.5)	38.30 (972.8)	192.4 (87.27)		3755 (1703.26)	208.6 (94.62)	4065 (1843.88)	203.2 (92.17)
42	50	0.47 (11.9)	44.50 (1130.3)	198.9 (90.22)	261 (118.38)			4240 (1923.26)	212.0 (96.16)
	51	0.53 (13.5)	44.50 (1130.3)	224.0 (101.60)	261 (118.38)			4740 (2150.06)	237.0 (107.50)
	52	0.59 (15.0)	44.50 (1130.3)	249.1 (112.99)	261 (118.38)			5245 (2379.13)	262.2 (118.93)
48	50	0.51 (13.0)	50.80 (1290.3)	246.6 (111.85)	316 (143.33)			5250 (2381.40)	262.4 (119.02)
	51	0.58 (14.7)	50.80 (1290.3)	280.0 (127.00)	316 (143.33)			5915 (2683.04)	295.8 (134.17)
	52	0.65 (16.5)	50.80 (1290.3)	313.4 (142.15)	316 (143.33)			6585 (2986.95)	329.2 (149.32)
54	50	0.57 (14.5)	57.10 (1450.3)	309.8 (140.52)	370 (167.83)			6565 (2977.88)	328.3 (148.91)
	51	0.65 (16.5)	57.10 (1450.3)	352.7 (159.98)	370 (167.83)			7425 (3367.98)	371.2 (168.37)
	52	0.73 (18.5)	57.10 (1450.3)	395.6 (179.44)	370 (167.83)			8280 (3755.80)	414.1 (187.83)

[A] Tolerances of outside diameter of spigot end: 4 to 12 in., ±0.06 in. (±1.5 mm); 14 to 24 in., +0.05 in. (+1.3 mm), −0.08 in. (−2.0 mm); 30 to 54 in., +0.08 in. (+2.0 mm), −0.06 in. (−1.5 mm).

[B] Including bell; calculated weight of pipe rounded off to nearest 5 lb (2.3 kg).

[C] Including bell; average weight per foot, based on calculated weight of pipe before rounding.

[D] Weight of 30-in. bell is 216 lb (97.97 kg) for 18-ft pipe and 163 lb (73.93 kg) for 20-ft pipe.

[E] Weight of 36-in. bell is 292 lb (132.45 kg) for 18-ft pipe and 216 lb (97.97 kg) for 20-ft pipe.

24

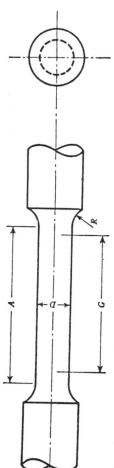

NOTE 1—The reduced section (A) may have a gradual taper from the ends toward the center with the ends not more than 0.005 in. (0.13 mm) larger in diameter than the center on the standard specimen and not more than 0.003 in. (0.08 mm) larger in diameter than the center on the small size specimens.

NOTE 2—If desired, on the small size specimens the length of the reduced section may be increased to accommodate an extensometer. However, reference marks for the measurement of elongation should nevertheless be spaced at the indicated gage length (G).

NOTE 3—The gage length and fillets shall be as shown, but the ends may be of any form to fit the holders of the testing machine in such a way that the load shall be axial. If the ends are to be held in grips it is desirable, if possible to make the length of the grip section great enough to allow the specimen to extend into the grips a distance equal to two thirds or more of the length of the grips.

Dimension	Standard Specimen 0.50-in. (12.7-mm) Round	Small-Size Specimens Proportional to Standard			
		0.350-in. (8.89-mm) Round	0.250-in. (6.35-mm) Round	0.175-in. (4.45-mm) Round	0.125-in. (3.18-mm) Round
		Dimensions, in. (mm)			
G	2.000 ± 0.005 (50.80 ± 0.13)	1.400 ± 0.005 (35.56 ± 0.13)	1.000 ± 0.005 (25.40 ± 0.13)	0.700 ± 0.005 (17.78 ± 0.13)	0.500 ± 0.005 (12.70 ± 0.13)
D	0.500 ± 0.010 (12.70 ± 0.25)	0.350 ± 0.007 (8.89 ± 0.18)	0.250 ± 0.005 (6.35 ± 0.13)	0.175 ± 0.005 (4.44 ± 0.13)	0.125 ± 0.005 (3.18 ± 0.13)
R, min	3/8 (9.5)	1/4 (6.4)	3/16 (4.8)	3/32 (2.4)	3/32 (2.4)
A, min	2 1/4 (57.2)	1 3/4 (44.4)	1 1/4 (31.8)	3/4 (19)	5/8 (15.9)
Tᵃ	0.71 and greater (18.0)	0.50 to 0.70 (12.2 to 17.8)	0.35 to 0.49 (8.9 to 12.4)	0.25 to 0.34 (6.4 to 8.6)	0.18 to 0.24 (4.6 to 6.1)

ᵃ Thickness of the section from the wall of the pipe from which the tension specimen is to be machined.

FIG. 1 Tension Test Specimen.

25

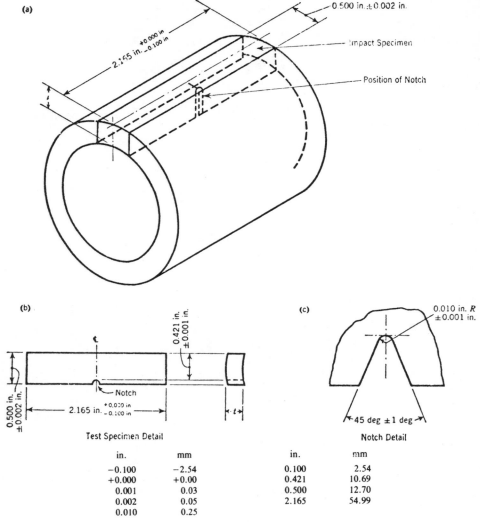

(a)

0.500 in. ±0.002 in.

2.165 in. $^{+0.000\ in.}_{-0.100\ in.}$

Impact Specimen

Position of Notch

2
5
8

(b)

0.421 in. ±0.001 in.

Notch

0.500 in. ±0.002 in.

2.165 in. $^{+0.000\ in.}_{-0.100\ in.}$

t

Test Specimen Detail

(c)

0.010 in. R ±0.001 in.

45 deg ±1 deg

Notch Detail

in.	mm	in.	mm
−0.100	−2.54	0.100	2.54
+0.000	+0.00	0.421	10.69
0.001	0.03	0.500	12.70
0.002	0.05	2.165	54.99
0.010	0.25		

NOTE—t = pipe-wall thickness.

FIG. 2 Impact Test Specimen.

SECTION V
GRAY AND DUCTILE IRON FITTINGS

Introduction

Fittings are produced in accordance with ANSI/AWWA Standard C110, ANSI Standard B16.1 as well as manufacturers' standards. Due to their irregular shapes, fittings are statically cast. They are available in either gray or ductile iron and are equipped with mechanical, push-on, flanged joints, or plain ends.

Preparatory to pouring iron, molds forming the outside contours of the fittings are assembled with cores that form the openings through the fittings. Iron is poured into the mold assembly and flows into the void surrounding the core. After cooling, the fittings are removed from the mold, cleaned, inspected, gauged for dimensional accuracy, weighed, lined and coated as required.

In general, gray and ductile iron fittings of the following configurations are furnished in accordance with the ANSI/AWWA C110 Standard: bends; tees; crosses; base bends; reducers; sleeves; caps; plugs; offsets; and tapped tees. Included in the fittings manufactured in accordance with ANSI B16.1 are the following: long-radius fittings, reducing elbows, reducing on-the-run tees; side outlet fittings; eccentric reducers and laterals; and true wyes. Manufacturers' standards govern other fittings and in some cases fittings are produced with ANSI B16.1 overall dimensions and ANSI/AWWA C110 thicknesses. All flanged fittings manufactured using ANSI/AWWA C110 have a minimum safety factor of 3.0 times the rated working pressure. Pressure ratings shown in ANSI/AWWA C110 should not be confused with 250-lb. flange ratings covered by ANSI B16.1 which are rated for steam working pressure.

The minimum grade of cast iron used in fittings is 25,000 psi iron strength with higher grades where necessary to secure higher pressure ratings. Ductile iron used in fittings must have an ultimate tensile strength of 70,000 psi, a yield strength of 50,000 psi and a minimum elongation of 5%.

for

GRAY-IRON AND DUCTILE-IRON FITTINGS,

3 in. THROUGH 48 in.,

FOR WATER AND OTHER LIQUIDS

2
6
1

Revised edition approved by American National Standards Institute, Inc., Apr. 7, 1977.

Administrative Secretariat

AMERICAN WATER WORKS ASSOCIATION

Co-Secretariats

AMERICAN GAS ASSOCIATION
NEW ENGLAND WATER WORKS ASSOCIATION

NOTICE

AMERICAN WATER WORKS ASSOCIATION
6666 West Quincy Avenue, Denver, Colorado 80235

262

Committee Personnel

Subcommittee 3, Fittings, which reviewed this standard, had the following personnel at that time:

HAROLD KENNEDY JR., *Chairman*
WALTER AMORY, *Vice-Chairman*

User Members	*Producer Members*
WALTER AMORY	ABRAHAM FENSTER
GEORGE F. KEENAN	W. D. GOODE
LEONARD ORLANDO JR.	THOMAS D. HOLMES
ARNOLD M. TINKEY	HAROLD KENNEDY JR.
ROBERT D. WILROY	T. M. KILEY
	EDWARD C. SEARS

Standards Committee A21, Cast-Iron Pipe and Fittings, which reviewed and approved this standard, had the following personnel at the time of approval:

LLOYD W. WELLER, *Chairman*
EDWARD C. SEARS, *Vice-Chairman*
PAUL A. SCHULTE, *Secretary*

Organization Represented	*Name of Representative*
American Gas Association	LEONARD ORLANDO JR.
American Society of Civil Engineers	KENNETH W. HENDERSON
American Society of Mechanical Engineers	JAMES S. VANICK
American Society for Testing and Materials	BEN C. HELTON
American Water Works Association	RAYMOND J. KOCOL
	ARNOLD M. TINKEY
	LLOYD W. WELLER
Cast Iron Pipe Research Association	THOMAS D. HOLMES
	HAROLD KENNEDY JR.
	EDWARD C. SEARS
	W. HARRY SMITH
Individual Producer	ALFRED F. CASE
Manufacturers' Standardization Society of the Valve and Fittings Industry	ABRAHAM FENSTER
New England Water Works Association	WALTER AMORY
Naval Facilities Engineering Command	STANLEY C. BAKER
Underwriters' Laboratories, Inc.	JOHN E. PERRY
Canadian Standards Association	W. F. SEMENCHUK*

---

* Liaison representative without vote.

iii

Table of Contents

Foreword

This foreword is provided for information only and is not a part of ANSI/AWWA C110.

American National Standards Committee A21 on Cast-Iron Pipe and Fittings was organized in 1926 under the sponsorship of the American Gas Association, the American Society for Testing and Materials, the American Water Works Association, and the New England Water Works Association. Since 1972, the Co-Secretariats have been A.G.A., AWWA, and NEWWA, with AWWA serving as Administrative Secretariat. The present scope of Committee A21 activity is

Standardization of specifications for cast-iron and ductile-iron pressure pipe for gas, water, and other liquids, and fittings for use with such pipe. These specifications to include design, dimensions, materials, coatings, linings, joints, accessories, and methods of inspection and tests.

The work of Committee A21 is conducted by subcommittees. The directive of Subcommittee 3—Fittings is that

The scope of the subcommittee activity shall include the periodic review of all current A21 standards for fittings and the preparation of revisions and new standards, when needed, for fittings to be used with cast-iron and ductile-iron pressure pipe included in A21 standards.

I. History of Standard

The evolution of AWWA and ANSI standards for fittings is presented in this foreword to provide information relative to systems having aged cast-iron pipe and fittings still in service.

The earliest record of an AWWA standard for cast-iron pipe is contained in the proceedings for 1890.

In 1902, NEWWA adopted a more detailed standard entitled "Standard Specification for Cast Iron Pipe and Special Castings."

The next AWWA standard for pipe and fittings, 7C.1–1908, was approved May 12, 1908. A second edition, C100–52T, was approved by AWWA Dec. 31, 1952, and by NEWWA Jan. 23, 1953. The third edition, C100–54T, was approved by AWWA Oct. 25, 1954, and finally issued as C100–55, having been advanced from tentative to standard without change Jun. 17, 1955. Standard C100–55 covered fittings in the size range 4–60 in. The fittings were all bell and spigot (caulked joint) of the so-called long-radius design. The outside diameter (OD) for spigots varied with wall thicknesses, which were designated classes A, B, C, and D. Fittings 4–12 in. were made to class D patterns, having only one OD and pressure rating. Fittings 14–24 in. were furnished in class B and D, and fittings 30–60 in. were furnished in classes A, B, C, and D. All fittings made to AWWA 7C.1–1908 and C100–55 had the class identification cast on the fitting.

ASA A21.10–1952 (AWWA C110–52) was approved by ASA Sep. 30, 1952. The standard covered 3–12-in. fittings of the so-called "short-body" design and were the subject of extensive research and tests by Committee A21. The rated pressure given by the standard was 250 psi plus water hammer. The standard provided a 2.5 safety factor plus water hammer based on burst tests. Hydraulic losses were determined and compared with those found with AWWA long-radius fittings. The minimum grade of cast iron in the standard was 25 000 psi tensile strength.

ASA A21.10–1964 (AWWA C110–64) was approved by ASA Jan. 9, 1964. The revision covered 2–48-in. fittings. The design of the 14–48-in. fittings in the revision was based on an exhaustive series of burst tests. The minimum grade of cast iron (25 000 psi tensile strength) was retained and higher grades up to 35 000 psi tensile strength were used to secure higher pressure ratings without changing radically the thicknesses. Ductile iron, grade 80-60-03, was also added in the 14–48-in. sizes with a rated working pressure of 250 psi having the same wall thicknesses as 150-psi rated gray-iron fittings. The minimum safety factor based on burst tests of representative fittings of the weakest type was 3× the rated working pressure. Tables for flanged fittings and mechanical-joint fittings were added for the first time.

ANSI A21.10–1971 (AWWA C110–71) was approved by ANSI Jul. 14, 1971. Ductile-iron fittings were added in sizes 3–12 in. and were rated for 350 psi working pressure. The grade of ductile iron was changed to 70-50-05 to provide greater toughness. The safety factor against bursting was 3× the rated working pressure. If required by the purchaser on special order, fittings were required to withstand a hydrostatic proof test not to exceed 1.5× the rated working pressure without leaks or permanent distortions.

ANSI A21.10a–1972 (AWWA C110a–72) was approved Dec. 17, 1972 as a supplement to ANSI

A21.10–71. The pressure rating for 14–24-in. ductile-iron fittings was increased to 350 psi.

II. Latest Revision of A21.10

At the meeting of Standards Committee A21 in 1974, Subcommittee 3 was assigned the task of reviewing and updating A21.10–1971 and A21.10a–1972. Accordingly, Subcommittee 3 undertook a study to determine the necessary and desirable revisions. An extensive review of ANSI A21.10–1971 and A21.10a–1972 was conducted in an effort to update and comply with other current A21 standards, particularly ANSI A21.15.

Subcommittee 3 completed its study and submitted the proposed revisions to ANSI Standards Committee A21 in April 1976.

The usage of the bell and spigot (caulked joints) has steadily declined, until presently it has become a rarity. Subcommittee 3 concluded that bell-and-spigot fittings should be deleted. Bell-and-spigot fittings are still available from some foundries on special order.

NOTE: Care should be used when connecting mechanical-joint fittings to aged existing cast-iron pipe. The outside diameter of aged pipe should be measured prior to cutting since some pipe were manufactured to a larger diameter than is presently specified in A21 standards. Mechanical-joint sleeves or bell-and-spigot (caulked joint) sleeves are available to provide transition from existing cast-iron pipe; however, they must be specified on the purchase order. The following standards contain reference dimensions useful in classifying existing cast-iron pipe:

AWWA 7C.1–1908 (AWWA C100–55) "Cast Iron Pressure Fittings," Table 1.

ASA A21.2–1953 (AWWA C102–53) "American Standard for Cast Iron Pit Cast Pipe," Tables 2.1 and 2.2.

ANSI A21.6–1975 (AWWA C106–1975) "American Standard for Cast Iron Pipe Centrifugally Cast in Metal Molds," Tables 6.4, 6.5, and 6.6.

ANSI A21.8–1975 (AWWA C108–75) "American National Standard for Cast Iron Pipe Centrifugally Cast in Sand-Lined Molds," Tables 8.4, 8.5, and 8.6.

ANSI A21.51–1976 (AWWA C151–76) "American National Standard for Ductile Iron Pipe Centrifugally Cast in Metal Molds or Sand-Lined Molds for Water or Other Liquids," Tables 51.4 and 51.5.

Center to bottom of socket dimensions (dimension A in Table 10.3 and dimension J in Table 10.4) for A21.10 mechanical-joint fittings are the same as the center to bottom of socket dimensions for bell-and-spigot (caulked joint) fittings specified in previous editions of A21.10.

ANSI A21.10a–1972 published as a supplement to ANSI A21.10–1971 is incorporated into this revision.

Cast-iron pipe and fittings in the 2- and $2\frac{1}{4}$-in. sizes are no longer manufactured in the US. These sizes are deleted in this revision of A21.10.

This revision includes 3–48-in. mechanical-joint and flanged fittings only. At least one manufacturer offers 54-in. fittings in flanged and push-on joints; however, 54-in. fittings are not included as a part of this standard.

Flanged fittings are listed without change; however, bolt-length specifications have been revised to conform to ANSI A21.15–1975. Refer to Appendix A for information on the use of flanged fittings.

III. Major Revisions

1. The scope was revised to incorporate ANSI A21.10a–1972.

2. Bell-and-spigot fittings (caulked joints) were discontinued as a part of ANSI A21.10. Bolt lengths for flanged fittings were revised to comply with ANSI A21.15–1975.

3. 2- and $2\frac{1}{4}$-in. fittings have been deleted.

4. The revisions in ANSI A21.10a–1972 have been incorporated into the tables and all tables concerning bell-and-spigot (caulked joints) fittings were deleted.

5. Three appendices were added for information: Appendix A covers bolts, gaskets, and the installation of flanged fittings; Appendix B is a listing of special fittings that are available but are not a part of the standard. These include reducing bends, Y branches, blind flanges, reducing tees, bull head tees, flared fittings, side outlet tees, and side outlet elbows and wall pipe. Appendix C states the position of the Committee with regard to metrication.

IV. Options

This standard includes certain options which, if desired, must be specified in the invitation for bids and on the purchase order. Also, a number of items must be specified to describe completely the fittings required. The following summarizes the details and available options and lists the sections of the standard where they are listed:

1. Size, joint type, pressure rating (Sec. 10–1 and tables.)

2. Joint specifications (Sec. 10–3.1.)

3. Type of iron (Sec. 10–3.2.)
4. End combinations (Sec. 10–3.3.)
5. Certification by manufacturer (Sec. 10–4.3.)
6. Inspection by purchaser (Sec. 10–5.)
7. Cement lining* (Sec. 10–8.2.)
8. Special coatings and linings (Sec. 10–8.4.)
9. Acceptance tests (Sec. 10–10.1.)
10. Special tests (Sec. 10–12.)
11. Special flange bolt-hole alignment (Sec. 10–14.3.)

V. Special Service Requirements

The following special service requirements should be noted:

1. The fittings for which this standard is intended are those normally

_____

* Experience has indicated the bituminous inside coating is not complete protection against loss in pipe capacity caused by tuberculation. Cement linings are recommended for most waters.

used for water and sanitary sewer systems. Fittings for other services may require special consideration by the purchaser.

2. Although this standard does not specify orientation of bolt holes in the flanges of the mechanical joint, it is at times convenient or necessary to have the bolt holes oriented. The normal but not universal practice is to have the bolt holes straddle the vertical centerline of the fittings, valves, and hydrants. (The vertical centerline of a fitting is determined when the fitting is in the position to change the direction of fluid flowing in a horizontal plane. With standard base bends and standard base tees, the vertical centerline is determined when the fitting is in a position to change the fluid flowing in a vertical plane.) If orientation is known to be necessary, it should be stated on the purchase order.

Revision of
A21.10–1971
(AWWA C110–71)

for

Gray-Iron and Ductile-Iron Fittings,

3 in. Through 48 in.,

for Water and Other Liquids

Sec. 10–1—Scope

This standard covers 3–48-in. gray-iron and/or ductile-iron fittings to be used with gray-iron or ductile-iron pipe for water and other liquids. Specifications for fittings with mechanical joints and flange joints are listed in the tables. This standard may also be used for fittings with push-on joints or such other joints as may be agreed upon at the time of purchase. For the 3–24-in. size range, ductile-iron mechanical-joint fittings are rated for 350 psi working pressure; ductile-iron flange-joint fittings are rated for 250 psi working pressure; and gray-iron fittings having all types of joints covered by this standard are rated for 150 or 250 psi working pressures, as shown in the tables. For the 30–48-in. size range, fittings of all types of joints covered by this standard are shown in the tables as gray-iron and/or ductile-iron for rated working pressures of 150 or 250 psi, as shown in the tables.

Sec. 10–2—Definitions

Under this standard the following definitions shall apply:

10–2.1 *Purchaser.* The party entering into a contract or agreement to purchase fittings according to this standard.

10–2.2 *Manufacturer.* The party that produces the fittings.

10–2.3 *Inspector.* The representative of the purchaser, authorized to inspect on behalf of the purchaser to determine whether or not the fittings meet this standard.

10–2.4 *Mechanical joint.* A bolted joint of the stuffing box type as detailed in Table 10.1 and as described in ANSI A21.11 (AWWA C111) of latest revision.

10–2.5 *Push-on joint.* The single rubber gasket joint as described in ANSI A21.11 (AWWA C111) of latest revision.

10–2.6 *Flange joint.* The flanged and bolted joint as detailed in Table 10.14.

1

10–2.7 *Gray iron.* The cast ferrous material in which a major part of the carbon content occurs as free carbon or graphite in the form of flakes interspersed throughout the metal.

10–2.8 *Ductile iron.* The cast ferrous material in which the free graphite present is in a spheroidal form.

Sec. 10–3—General Requirements

10–3.1 Fittings with mechanical joints and flange joints shall conform to the dimensions and weights shown in the tables in this standard, unless otherwise agreed upon at the time of purchase. The mechanical joint shall also conform in all respects to ANSI A21.11 (AWWA C111) of latest revision. Unless otherwise specified, the mechanical-joint gland shall be gray iron in accordance with ANSI A21.11 of latest revision, and bolts and gaskets shall conform to the requirements of the same standard.

10–3.2 Fittings shall be cast from gray iron or ductile iron, as shown in the tables. When both are shown in the tables, either may be used at the manufacturer's option unless otherwise specified on the purchase order. All fittings shall be capable of withstanding, without bursting, hydrostatic tests of $3\times$ the rated water working pressure.

10–3.3 Standard fittings shall be furnished with end combinations shown in the tables. When fittings of other designs or dimensions are purchased under this standard, it is the obligation of the purchaser to supply with each order specific details for each size, pressure rating, or type of fitting. Plain ends of mechanical-joint fittings may be furnished with bevels for assembly with push-on joint bells.

NOTE: All bell fittings, without plain ends, are preferred.

Sec. 10–4—Inspection and Certification by Manufacturer

10–4.1 The manufacturer shall establish the necessary quality control and inspection practice to ensure compliance with this standard. All fittings shall be clean and sound without defects that could impair their service.

10–4.2 Repairing of defects by welding or other methods shall not be allowed if such repairs could adversely affect the serviceability of the fitting or its capability to meet strength requirements of this standard.

10–4.3 The manufacturer shall, if required on the purchase order, furnish a sworn statement that the inspection and all the specified tests have been made and the results thereof comply with the requirements of this standard.

Sec. 10–5—Inspection by Purchaser

10–5.1 If the purchaser desires to inspect fittings at the manufacturer's plant, he shall so specify on the purchase order, stating the conditions (such as time and the extent of the inspection) under which the inspection shall be made.

10–5.2 The inspector shall have free access to those areas of the manufacturer's plant that are necessary to ensure compliance with this standard. The manufacturer shall make available for the inspector's use such gages as are necessary for inspection. The manufacturer shall provide the inspector with assistance as necessary for the handling of fittings.

Sec. 10–6—Delivery and Acceptance

All fittings and accessories shall comply with this standard. Fittings or accessories that do not comply with this standard shall be replaced by the manufacturer at the agreed point of delivery. The manufacturer shall not be liable for shortages or damaged fittings or accessories after acceptance at the agreed point of delivery except as recorded on the delivery receipt or similar document by the carrier's agent.

Sec. 10–7—Tolerances or Permitted Variations

10–7.1 *Dimensions.* Fittings shall be gaged with suitable gages at sufficiently frequent intervals to ensure that the dimensions comply with the requirements of this standard. The smallest inside diameter of the sockets and the outside diameter of the plain ends shall be tested with circular gages. Other socket dimensions shall be gaged as is appropriate.

10–7.2 *Thickness.* Minus tolerances for metal thicknesses, except those shown in Tables 10.1 and 10.2, shall not be more than the following:

Fitting Size *in.*	Minus Tolerance *in.*
3–6	0.10
8–20	0.12
24–48	0.15

An additional tolerance shall be permitted over areas not exceeding 8 in. in any direction as follows: for 3–12-in. fittings, 0.02 in.; for 14–48-in. fittings, 0.03 in.

10–7.3 *Weight.* The weight of any fitting shall not be less than the nominal tabulated weight by more than 10 per cent for fittings 12 in. or smaller in diameter or by more than 8 per cent for fittings larger than 12 in. in diameter. The nominal tabu-

lated weight is the weight of the fitting before the application of any lining or coating other than the standard coatings.

Sec. 10–8—Coatings and Linings

10–8.1 *Outside coating.* The outside coating for general use under all normal conditions shall be a bituminous coating approximately 1 mil thick. The coating shall be applied to the outside of all fittings, unless otherwise specified. The finished coating shall be continuous, smooth, neither brittle when cold nor sticky when exposed to the sun, and strongly adherent to the fitting.

10–8.2 *Cement–mortar linings.* Cement linings shall be in accordance with ANSI A21.4 (AWWA C104) of latest revision. If desired, cement linings shall be specified in the invitation for bids and on the purchase order.

10–8.3 *Inside coating.* Unless otherwise specified, the inside coating for fittings that are not cement-lined shall be a bituminous material as thick as practicable (at least 1 mil) and conform to all appropriate requirements for sealcoat in ANSI A21.4 (AWWA C104) of latest revision.

10–8.4 *Special coatings and linings.* For special conditions, other types of coatings and linings may be available. Such special coatings and linings shall be specified in the invitation for bids and on the purchase order.

Sec. 10–9—Markings on Fittings

Fittings shall have distinctly cast upon them the manufacturer's identification, pressure rating, nominal diameters of openings, and the number of degrees or fraction of the circle

271

on all bends. Ductile-iron fittings shall have the letters "DI" or "Ductile" cast on them. Cast letters and figures shall be on the outside and shall have dimensions no smaller than the following:

Size in.	Height of Letters in.	Relief in.
Less than 8	As large as practical	As large as practical
8–10	$\frac{3}{4}$	$\frac{3}{32}$
12–48	$1\frac{1}{4}$	$\frac{3}{32}$

Sec. 10–10—Acceptance Tests

10–10.1 *Physical test—gray-iron fittings.* The standard acceptance test for the physical characteristics of gray-iron fittings shall be one of the following:

1. Transverse test conducted in accordance with ASTM A438-62 (1974).
2. Tensile test conducted in accordance with ASTM A48–74.

10–10.1.1 *Choice of test.* Unless specified by the purchaser, either the tensile test or the transverse test, at the option of the manufacturer, shall be used as the acceptance test. The acceptance values for tensile and transverse tests shall be as follows:

Iron Strength psi (1000's)	Fitting Size in.	Bar Diam. in.	Span* in.	Min. Breaking Load lb.	Min. Tensile Strength† psi (1000's)
25	3–14	1.20	18	2000	25
30	14–24	1.20	18	2200	30
30	20–48	2.00	24	7600	30
35	16–24	1.20	18	2400	35
35	20–36	2.00	24	8300	35

* ASTM A438–62 (1974).
† ASTM A48–74.

For 20-in. and 24-in. fittings with a body thickness greater than 1 in., a 2-in. test bar shall be used.

10–10.2 *Physical test—ductile-iron fittings.* The standard acceptance test for the physical characteristics of ductile-iron fittings shall be a tensile test from coupons cast from the same iron as the fittings. The coupons shall be cast and the tests made in accordance with ASTM A536–72 except the grade shall be 70–50–05. Either the keel block or Y block shall be used as the test coupons at the option of the manufacturer. The acceptance shall be as follows: minimum tensile strength, 70 000 psi; minimum yield strength, 50 000 psi; minimum elongation, 5 per cent.

10–10.3 *Sampling.* At least one sample shall be taken during each period of approximately 3 hr while the melting unit is operated continuously.

Sec. 10–11—Chemical Limitations for Gray-Iron Fittings

Analyses of the iron in gray-iron fittings shall be made at sufficiently frequent intervals to ensure compliance with the following limits: phosphorus, 0.90 per cent maximum; sulfur, 0.15 per cent maximum.

Control of the other chemical constituents shall be maintained to meet the physical property requirements of this standard. Samples for chemical analyses shall be representative and shall be obtained from either acceptance test specimens or specimens cast for this purpose.

Sec. 10–12—Additional Tests Required by the Purchaser

If tests other than those provided in this standard are required by the purchaser, such tests shall be specified in the invitation for bids and on the purchase order. Although it is not customary to make hydrostatic proof tests of fittings at the foundry, such tests may be made on special order at additional cost. If proof tests at the foundry are required by the purchaser for an order of fittings, the fittings shall withstand, without leaks or permanent distortion, hydrostatic test pressures not to exceed 1.5× the rated water working pressures.

Sec. 10–13—Defective Specimens and Retests

When any physical test specimen shows defective machining or lack of continuity of metal, it shall be discarded and replaced by another specimen cast in the same sampling period as the specimen that failed.

Sec. 10–14—Special Requirements for Flanged Fittings

10–14.1 *Flanges.* Flanges shall conform to the dimensions shown in Table 10.14, which are adequate for water service of 250 psi working pressure.

NOTE: The bolt circle and bolt holes of these flanges match those of the class 125 flanges shown in ANSI B16.1 and can be joined with class 125 B16.1 flanges. Flanges in A21.10 cannot be joined with class 250 B16.1 flanges.

10–14.2 *Facing.* Flanges shall be plain faced without projection or raised face and shall be furnished smooth or with shallow serrations. Flanges may be back faced or spot faced for compliance with the flange thickness tolerance specified in this standard. Bearing surfaces for bolting shall be parallel to the flange face within 3 deg.

10–14.3 *Bolt holes.* Bolt holes shall be in accordance with the dimensions shown in Table 10.14. They shall be equally spaced and shall straddle the centerline of the fitting.

10–14.3.1 Misalignment of corresponding bolt holes of two opposing flanges shall not exceed 0.12 in.

10–14.3.2 If bolt-hole alignment other than provided for in this standard is required by the purchaser, it shall be specified in the invitation for bids and on the purchase order.

10–14.4 *Laying-length dimensions.* Face-to-face dimensions shall conform to a tolerance of ±0.06 in. for sizes 3–10 in. and ±0.12 in. for sizes 12–48 in. Center-to-face tolerances shall be one half those of face-to-face tolerances. The largest opening shall govern the tolerance for all openings.

Index to Tables

TABLE 10.1

Standard Mechanical-Joint Dimensions—in.*

Size *in.*	Plain End *A*	*B*	*C*	*D*	*F*	*φ* *deg*	*X*	*J*	*K*₁	*K*₂	*L*	*M*	*N*	*O*	*P*	*S*	*Y*	Bolts			Weight—*lb*	
																		No.	Size	Lgth.	Bell	Gland Bolts Gasket
3	3.96 ±.06	2.50	4.84 ±.04	4.94 +.06 −.04	4.06 +.07 −.03	28	3/4 +.06 −.0	6.19 ±.06	7.69 −.12	7.69 −.12	.94 −.06	.62 −.06	.75	.31	.63	.52 −.10	.12	4	5/8	3	11	7
4	4.80 ±.06	2.50	5.92 ±.04	6.02 +.06 −.04	4.90 +.07 −.03	28	7/8 +.06 −.0	7.50 ±.06	9.12 −.12	9.12 −.12	1.00 −.06	.75 −.06	.75	.31	.75	.65 −.10	.12	4	3/4	3 1/2	16	10
6	6.90 ±.06	2.50	8.02 ±.04	8.12 +.06 −.04	7.00 +.07 −.03	28	7/8 +.06 −.0	9.50 ±.06	11.12 −.12	11.12 −.12	1.06 −.06	.88 −.06	.75	.31	.75	.70 −.10	.12	6	3/4	3 1/2	23	16
8	9.05 ±.06	2.50	10.17 ±.04	10.27 +.06 −.04	9.15 +.07 −.03	28	7/8 +.06 −.0	11.75 ±.06	13.37 −.12	13.37 −.12	1.12 −.08	1.00 −.08	.75	.31	.75	.75 −.12	.12	6	3/4	4	31	25
10	11.10 ±.06	2.50	12.22 +.06 −.04	12.34 +.06 −.04	11.20 +.07 −.03	28	7/8 +.06 −.0	14.00 ±.06	15.69 −.12	15.62 −.12	1.19 −.08	1.00 −.08	.75	.31	.75	.80 −.12	.12	8	3/4	4	41	30
12	13.20 ±.06	2.50	14.32 +.06 −.04	14.44 +.06 −.04	13.30 +.07 −.03	28	7/8 +.06 −.0	16.25 ±.06	17.94 −.12	17.88 −.12	1.25 −.08	1.00 −.08	.75	.31	.75	.85 −.12	.12	8	3/4	4	51	40
14	15.30 +.05 −.08	3.50	16.40 +.07 −.05	16.54 +.07 −.05	15.44 +.06 −.07	28	7/8 +.06 −.0	18.75 ±.06	20.31 −.12	20.25 −.12	1.31 −.12	1.25 −.12	.75	.31	.75	.89 −.12	.12	10	3/4	4 1/2	79	45
16	17.40 +.05 −.08	3.50	18.50 +.07 −.05	18.64 +.07 −.05	17.54 +.06 −.07	28	7/8 +.06 −.0	21.00 ±.06	22.56 −.12	22.50 −.12	1.38 −.12	1.31 −.12	.75	.31	.75	.97 −.12	.12	12	3/4	4 1/2	97	55
18	19.50 +.05 −.08	3.50	20.60 +.07 −.05	20.74 +.07 −.05	19.64 +.06 −.07	28	7/8 +.06 −.0	23.25 ±.06	24.83 −.15	24.75 −.15	1.44 −.12	1.38 −.12	.75	.31	.75	1.05 −.15	.12	12	3/4	4 1/2	117	65
20	21.60 +.05 −.08	3.50	22.70 +.07 −.05	22.84 +.07 −.05	21.74 +.06 −.07	28	7/8 +.06 −.0	25.50 ±.06	27.08 −.15	27.00 −.15	1.50 −.12	1.44 −.12	.75	.31	.75	1.12 −.15	.12	14	3/4	4 1/2	140	85
24	25.80 +.05 −.08	3.50	26.90 +.07 −.05	27.04 +.07 −.05	25.94 +.06 −.07	28	7/8 +.06 −.0	30.00 ±.06	31.58 −.15	31.50 −.15	1.62 −.12	1.56 −.12	.75	.31	.75	1.22 −.15	.12	16	3/4	5	185	105
30	32.00 +.08 −.06	4.00	33.29 +.08 −.06	33.46 +.08 −.06	32.17 +.08 −.06	20	1 1/8 +.06 −.0	36.88 ±.06	39.12 −.18	39.12 −.18	1.81 −.12	2.00 −.12	.75	.38	1.00	1.50 −.15	.12	20	1	6	315	220
36	38.30 +.08 −.06	4.00	39.59 +.08 −.06	39.76 +.08 −.06	38.47 +.08 −.06	20	1 1/8 +.06 −.0	43.75 ±.06	46.00 −.18	46.00 −.18	2.00 −.12	2.00 −.12	.75	.38	1.00	1.80 −.15	.12	24	1	6	445	285
42	44.50 +.08 −.06	4.00	45.79 +.08 −.06	45.96 +.08 −.06	44.67 +.08 −.06	20	1 3/8 +.06 −.0	50.62 ±.06	53.12 −.18	53.12 −.18	2.00 −.12	2.00 −.12	.75	.38	1.00	1.95 −.15	.12	28	1 1/4	6	570	400
48	50.80 +.08 −.06	4.00	52.09 +.08 −.06	52.26 +.08 −.06	50.97 +.08 −.06	20	1 3/8 +.06 −.0	57.50 ±.06	60.00 −.18	60.00 −.18	2.00 −.12	2.00 −.12	.75	.38	1.00	2.20 −.15	.12	32	1 1/4	6	725	475

* See Fig. 10.1.

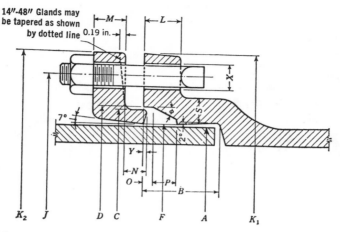

Fig. 10.1. 3–48 in. Standard Mechanical-Joint Dimensions (See Table 10.1)

1. *Diameter of cored holes may be tapered an additional 0.06 in.*

2. *Dimension A in Table 10.1 is the outside diameter of the plain end of the fitting.*

3. *In the event of ovalness of the plain end OD, the mean diameter measured by a circumferential tape shall not be less than the minimum diameter shown in the table. The minor axis shall not be less than the above minimum diameter plus an additional minus tolerance of 0.04 in. for sizes 8–12 in., 0.07 in. for sizes 14–24 in., and 0.10 in. for sizes 30–48 in.*

4. *K_1 and K_2 are the dimensions across the bolt holes. For sizes 3–48 in., the gland may be polygon shaped.*

TABLE 10.2

Plain-end Dimensions and Tolerances for Mechanical-Joint Fittings

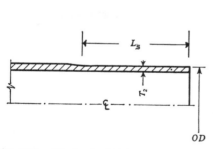

Fig. 10.2. Mechanical-Joint Plain-End Dimensions and Tolerances (See Table 10.2)

1. *All sizes of fittings with plain ends have 8 in. of added laying length as compared with the laying length of standard all-bell fittings.*

2. *In Fig. 10.2, dimension L is minimum length of the plain end which must be gaged to ensure that the outside diameter is within the dimensions and tolerances specified in Table 10.2.*

Size in.	OD—in.	T_2—in.	L_s—in.
3	3.96±0.06	0.48−0.10	5.5
4	4.80±0.06	0.47−0.10	5.5
6	6.90±0.06	0.50−0.10	5.5
8	9.05±0.06	0.54−0.12	5.5
10	11.10±0.06	0.60−0.12	5.5
12	13.20±0.06	0.68−0.12	5.5
14	15.30+0.05 −0.08	0.66−0.12	8.0
16	17.40+0.05 −0.08	0.70−0.12	8.0
18	19.50+0.05 −0.08	0.75−0.12	8.0
20	21.60+0.05 −0.08	0.80−0.12	8.0
24	25.80+0.05 −0.08	0.89−0.15	8.0
30	32.00+0.08 −0.06	1.03−0.15	8.0
36	38.30+0.08 −0.06	1.15−0.15	8.0
42	44.50+0.08 −0.06	1.28−0.15	8.0
48	50.80+0.08 −0.06	1.42−0.15	8.0

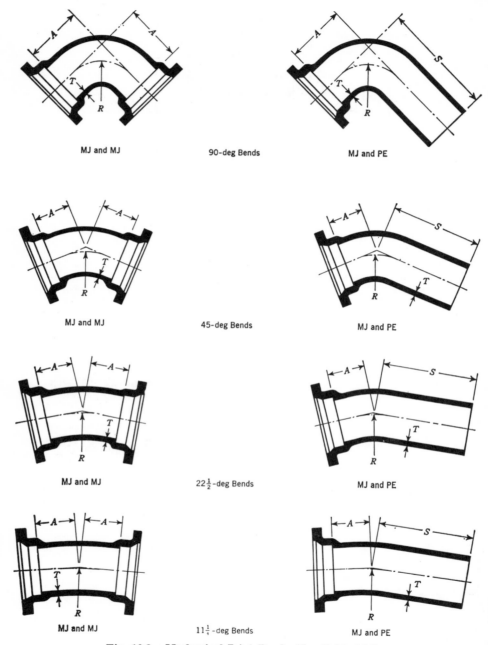

MJ and MJ   90-deg Bends   MJ and PE

MJ and MJ   45-deg Bends   MJ and PE

MJ and MJ $22\frac{1}{2}$-deg Bends   MJ and PE

MJ and MJ $11\frac{1}{4}$-deg Bends   MJ and PE

Fig. 10.3.   Mechanical-Joint Bends (See Table 10.3)

TABLE 10.3

*Mechanical-Joint Bends**

Size in.	Pressure Rating psi	Iron Strength psi (1000's)	Dimensions—in.				Weight—lb†	
			T	A	S	R	MJ & MJ	MJ & PE
90-deg Bends								
3	250	25	0.48	5.5	13.5	4.0	35	35
3	350	DI‡	0.48	5.5	13.5	4.0	35	35
4	250	25	0.52	6.5	14.5	4.5	55	50
4	350	DI	0.52	6.5	14.5	4.5	55	50
6	250	25	0.55	8.0	16.0	6.0	85	80
6	350	DI	0.55	8.0	16.0	6.0	85	80
8	250	25	0.60	9.0	17.0	7.0	125	120
8	350	DI	0.60	9.0	17.0	7.0	125	120
10	250	25	0.68	11.0	19.0	9.0	190	190
10	350	DI	0.68	11.0	19.0	9.0	190	190
12	250	25	0.75	12.0	20.0	10.0	255	255
12	350	DI	0.75	12.0	20.0	10.0	255	255
14	150	25	0.66	14.0	22.0	11.5	340	325
14	250	25	0.82	14.0	22.0	11.5	380	365
14	350	DI	0.66	14.0	22.0	11.5	340	325
16	150	30	0.70	15.0	23.0	12.5	430	410
16	250	30	0.89	15.0	23.0	12.5	490	470
16	350	DI	0.70	15.0	23.0	12.5	430	410
18	150	30	0.75	16.5	24.5	14.0	545	520
18	250	30	0.96	16.5	24.5	14.0	625	600
18	350	DI	0.75	16.5	24.5	14.0	545	520
20	150	30	0.80	18.0	26.0	15.5	680	650
20	250	30	1.03	18.0	26.0	15.5	790	755
20	350	DI	0.80	18.0	26.0	15.5	680	650
24	150	30	0.89	22.0	30.0	18.5	1,025	985
24	250	30	1.16	22.0	30.0	18.5	1,215	1,175
24	350	DI	0.89	22.0	30.0	18.5	1,025	985
30	150	30	1.03	25.0	33.0	21.5	1,690	1,585
30	250	30	1.37	25.0	33.0	21.5	2,030	1,920
30	250	DI	1.03	25.0	33.0	21.5	1,690	1,585
36	150	30	1.15	28.0	36.0	24.5	2,475	2,310
36	250	30	1.58	28.0	36.0	24.5	3,045	2,880
36	250	DI	1.15	28.0	36.0	24.5	2,475	2,310
42	150	30	1.28	31.0	39.0	27.5	3,410	3,200
42	250	30	1.78	31.0	39.0	27.5	4,255	4,050
42	250	DI	1.28	31.0	39.0	27.5	3,410	3,200
48	150	30	1.42	34.0	42.0	30.5	4,595	4,330
48	250	30	1.96	34.0	42.0	30.5	5,745	5,475
48	250	DI	1.42	34.0	42.0	30.5	4,595	4,330

* Dimension details of mechanical-joint bells are shown in Table 10.1; dimension details of plain ends are shown in Table 10.2.
† Weight does not include accessory weights. See Table 10.1 for accessory weights.
‡ Ductile Iron.

TABLE 10.3

Mechanical-Joint Bends (*contd.*)

Size in.	Pressure Rating psi	Iron Strength psi (1000's)	Dimensions—in.				Weight—lb†	
			T	A	S	R	MJ & MJ	MJ & PE
45-deg Bends								
3	250	25	0.48	3.0	11.0	3.62	30	30
3	350	DI‡	0.48	3.0	11.0	3.62	30	30
4	250	25	0.52	4.0	12.0	4.81	50	45
4	350	DI	0.52	4.0	12.0	4.81	50	45
6	250	25	0.55	5.0	13.0	7.25	75	70
6	350	DI	0.55	5.0	13.0	7.25	75	70
8	250	25	0.60	5.5	13.5	8.44	110	105
8	350	DI	0.60	5.5	13.5	8.44	110	105
10	250	25	0.68	6.5	14.5	10.88	155	155
10	350	DI	0.68	6.5	14.5	10.88	155	155
12	250	25	0.75	7.5	15.5	13.25	215	215
12	350	DI	0.75	7.5	15.5	13.25	215	215
14	150	25	0.66	7.5	15.5	12.06	270	255
14	250	25	0.82	7.5	15.5	12.06	300	280
14	350	DI	0.66	7.5	15.5	12.06	270	255
16	150	30	0.70	8.0	16.0	13.25	340	320
16	250	30	0.89	8.0	16.0	13.25	380	360
16	350	DI	0.70	8.0	16.0	13.25	340	320
18	150	30	0.75	8.5	16.5	14.50	420	395
18	250	30	0.96	8.5	16.5	14.50	470	445
18	350	DI	0.75	8.5	16.5	14.50	420	395
20	150	30	0.80	9.5	17.5	16.88	530	500
20	250	30	1.03	9.5	17.5	16.88	595	565
20	350	DI	0.80	9.5	17.5	16.88	530	500
24	150	30	0.89	11.0	19.0	18.12	755	715
24	250	30	1.16	11.0	19.0	18.12	865	825
24	350	DI	0.89	11.0	19.0	18.12	755	715
30	150	30	1.03	15.0	23.0	27.75	1,380	1,275
30	250	30	1.37	15.0	23.0	27.75	1,620	1,510
30	250	DI	1.03	15.0	23.0	27.75	1,380	1,275
36	150	30	1.15	18.0	26.0	35.00	2,095	1,930
36	250	30	1.58	18.0	26.0	35.00	2,525	2,360
36	250	DI	1.15	18.0	26.0	35.00	2,095	1,930
42	150	30	1.28	21.0	29.0	42.25	2,955	2,745
42	250	30	1.78	21.0	29.0	42.25	3,635	3,425
42	250	DI	1.28	21.0	29.0	42.25	2,955	2,745
48	150	30	1.42	24.0	32.0	49.50	4,080	3,815
48	250	30	1.96	24.0	32.0	49.50	5,040	4,770
48	250	DI	1.42	24.0	32.0	49.50	4,080	3,815

* Dimension details of mechanical-joint bells are shown in Table 10.1; dimension details of plain ends are shown in Table 10.2.
† Weight does not include accessory weights. See Table 10.1 for accessory weights.
‡ Ductile Iron.

TABLE 10.3

Mechanical-Joint Bends (contd.)*

Size in.	Pressure Rating psi	Iron Strength psi (1000's)	Dimensions—in.				Weight—lb†	
			T	A	S	R	MJ & MJ	MJ & PE
22½-deg Bends								
3	250	25	0.48	3.0	11.0	7.56	30	30
3	350	DI‡	0.48	3.0	11.0	7.56	30	30
4	250	25	0.52	4.0	12.0	10.06	50	45
4	350	DI	0.52	4.0	12.0	10.06	50	45
6	250	25	0.55	5.0	13.0	15.06	75	70
6	350	DI	0.55	5.0	13.0	15.06	75	70
8	250	25	0.60	5.5	13.5	17.62	110	105
8	350	DI	0.60	5.5	13.5	17.62	110	105
10	250	25	0.68	6.5	14.5	22.62	160	160
10	350	DI	0.68	6.5	14.5	22.62	160	160
12	250	25	0.75	7.5	15.5	27.62	220	220
12	350	DI	0.75	7.5	15.5	27.62	220	220
14	150	25	0.66	7.5	15.5	25.12	275	260
14	250	25	0.82	7.5	15.5	25.12	300	285
14	350	DI	0.66	7.5	15.5	25.12	275	260
16	150	30	0.70	8.0	16.0	27.62	345	325
16	250	30	0.89	8.0	16.0	27.62	385	365
16	350	DI	0.70	8.0	16.0	27.62	345	325
18	150	30	0.75	8.5	16.5	30.19	430	405
18	250	30	0.96	8.5	16.5	30.19	480	455
18	350	DI	0.75	8.5	16.5	30.19	430	405
20	150	30	0.80	9.5	17.5	35.19	535	505
20	250	30	1.03	9.5	17.5	35.19	605	575
20	350	DI	0.80	9.5	17.5	35.19	535	505
24	150	30	0.89	11.0	19.0	37.69	765	725
24	250	30	1.16	11.0	19.0	37.69	880	840
24	350	DI	0.89	11.0	19.0	37.69	765	725
30	150	30	1.03	15.0	23.0	57.81	1,400	1,295
30	250	30	1.37	15.0	23.0	57.81	1,650	1,540
30	250	DI	1.03	15.0	23.0	57.81	1,400	1,295
36	150	30	1.15	18.0	26.0	72.88	2,135	1,970
36	250	30	1.58	18.0	26.0	72.88	2,580	2,410
36	250	DI	1.15	18.0	26.0	72.88	2,135	1,970
42	150	30	1.28	21.0	29.0	88.00	3,020	2,810
42	250	30	1.78	21.0	29.0	88.00	3,720	3,510
42	250	DI	1.28	21.0	29.0	88.00	3,020	2,810
48	150	30	1.42	24.0	32.0	103.06	4,170	3,905
48	250	30	1.96	24.0	32.0	103.06	5,160	4,895
48	250	DI	1.42	24.0	32.0	103.06	4,170	3,905

* Dimension details of mechanical-joint bells are shown in Table 10.1; dimension details of plain ends are shown in Table 10.2.
† Weight does not include accessory weights. See Table 10.1 for accessory weights.
‡ Ductile Iron.

TABLE 10.3

Mechanical-Joint Bends* (contd.)

Size *in.*	Pressure Rating *psi*	Iron Strength *psi (1000's)*	Dimensions—*in.*				Weight—*lb*†	
			T	A	S	R	MJ & MJ	MJ & PE
				11¼-deg Bends				
3	250	25	0.48	3.0	11.0	15.25	30	30
3	350	DI‡	0.48	3.0	11.0	15.25	30	30
4	250	25	0.52	4.0	12.0	20.31	50	45
4	350	DI	0.52	4.0	12.0	20.31	50	45
6	250	25	0.55	5.0	13.0	30.50	75	70
6	350	DI	0.55	5.0	13.0	30.50	75	70
8	250	25	0.60	5.5	13.5	35.50	110	105
8	350	DI	0.60	5.5	13.5	35.50	110	105
10	250	25	0.68	6.5	14.5	45.69	160	160
10	350	DI	0.68	6.5	14.5	45.69	160	160
12	250	25	0.75	7.5	15.5	55.81	220	220
12	350	DI	0.75	7.5	15.5	55.81	220	220
14	150	25	0.66	7.5	15.5	50.75	275	260
14	250	25	0.82	7.5	15.5	50.75	305	285
14	350	DI	0.66	7.5	15.5	50.75	275	260
16	150	30	0.70	8.0	16.0	55.81	345	325
16	250	30	0.89	8.0	16.0	55.81	385	365
16	350	DI	0.70	8.0	16.0	55.81	345	325
18	150	30	0.75	8.5	16.5	60.94	430	405
18	250	30	0.96	8.5	16.5	60.94	480	455
18	350	DI	0.75	8.5	16.5	60.94	430	405
20	150	30	0.80	9.5	17.5	71.06	540	510
20	250	30	1.03	9.5	17.5	71.06	610	575
20	350	DI	0.80	9.5	17.5	71.06	540	510
24	150	30	0.89	11.0	19.0	76.12	770	730
24	250	30	1.16	11.0	19.0	76.12	885	845
24	350	DI	0.89	11.0	19.0	76.12	770	730
30	150	30	1.03	15.0	23.0	116.75	1,410	1,305
30	250	30	1.37	15.0	23.0	116.75	1,655	1,550
30	250	DI	1.03	15.0	23.0	116.75	1,410	1,305
36	150	30	1.15	18.0	26.0	147.25	2,145	1,980
36	250	30	1.58	18.0	26.0	147.25	2,595	2,425
36	250	DI	1.15	18.0	26.0	147.25	2,145	1,980
42	150	30	1.28	21.0	29.0	177.69	3,035	2,825
42	250	30	1.78	21.0	29.0	177.69	3,740	3,535
42	250	DI	1.28	21.0	29.0	177.69	3,035	2,825
48	150	30	1.42	24.0	32.0	208.12	4,190	3,925
48	250	30	1.96	24.0	32.0	208.12	5,195	4,925
48	250	DI	1.42	24.0	32.0	208.12	4,190	3,925

* Dimension details of mechanical-joint bells are shown in Table 10.1; dimension details of plain ends are shown in Table 10.2.
† Weight does not include accessory weights. See Table 10.1 for accessory weights.
‡ Ductile Iron.

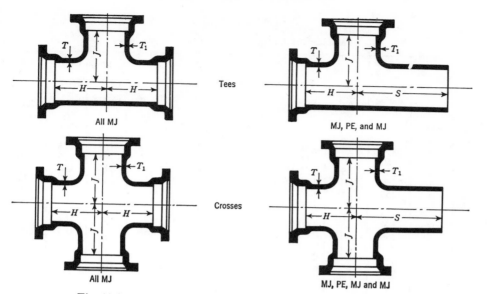

Fig. 10.4. **Mechanical-Joint Tees and Crosses (See Table 10.4)**

TABLE 10.4

*Mechanical-Joint Tees and Crosses**

Size in.		Pressure Rating psi	Iron Strength psi (1000's)	Dimensions—in.					Weight—lb†			
									Tee		Cross	
Run	Branch			T	T_1	H	J	S	All MJ	MJ, PE & MJ	All MJ	MJ, PE, MJ & MJ
3	3	250	25	0.48	0.48	5.5	5.5	13.5	55	55	70	70
3	3	350	DI‡	0.48	0.48	5.5	5.5	13.5	55	55	70	70
4	3	250	25	0.52	0.48	6.5	6.5	14.5	75	70	90	85
4	3	350	DI	0.52	0.48	6.5	6.5	14.5	75	70	90	85
4	4	250	25	0.52	0.52	6.5	6.5	14.5	80	75	105	100
4	4	350	DI	0.52	0.52	6.5	6.5	14.5	80	75	105	100

* Dimension details of mechanical-joint bells are shown in Table 10.1; dimension details of plain ends are shown in Table 10.2.
† Weight does not include accessory weights. See Table 10.1 for accessory weight.
‡ Ductile Iron.

TABLE 10.4

Mechanical-Joint Tees and Crosses (contd.)

Size in.		Pressure Rating psi	Iron Strength psi (1000's)	Dimensions—in.					Weight—lb†			
									Tee		Cross	
Run	Branch			T	T₁	H	J	S	All MJ	MJ, PE & MJ	All MJ	MJ, PE, MJ & MJ
6	3	250	25	0.55	0.48	8.0	8.0	16.0	110	105	125	120
6	3	350	DI‡	0.55	0.48	8.0	8.0	16.0	110	105	125	120
6	4	250	25	0.55	0.52	8.0	8.0	16.0	115	110	140	135
6	4	350	DI	0.55	0.52	8.0	8.0	16.0	115	110	140	135
6	6	250	25	0.55	0.55	8.0	8.0	16.0	125	120	160	155
6	6	350	DI	0.55	0.55	8.0	8.0	16.0	125	120	160	155
8	4	250	25	0.60	0.52	9.0	9.0	17.0	165	160	185	180
8	4	350	DI	0.60	0.52	9.0	9.0	17.0	165	160	185	180
8	6	250	25	0.60	0.55	9.0	9.0	17.0	175	170	205	200
8	6	350	DI	0.60	0.55	9.0	9.0	17.0	175	170	205	200
8	8	250	25	0.60	0.60	9.0	9.0	17.0	185	180	235	230
8	8	350	DI	0.60	0.60	9.0	9.0	17.0	185	180	235	230
10	4	250	25	0.68	0.52	11.0	11.0	19.0	235	235	260	260
10	4	350	DI	0.68	0.52	11.0	11.0	19.0	235	235	260	260
10	6	250	25	0.68	0.55	11.0	11.0	19.0	250	250	285	285
10	6	350	DI	0.68	0.55	11.0	11.0	19.0	250	250	285	285
10	8	250	25	0.68	0.60	11.0	11.0	19.0	260	260	310	310
10	8	350	DI	0.68	0.60	11.0	11.0	19.0	260	260	310	310
10	10	250	25	0.80	0.80	11.0	11.0	19.0	310	310	380	380
10	10	350	DI	0.80	0.80	11.0	11.0	19.0	310	310	380	380
12	4	250	25	0.75	0.52	12.0	12.0	20.0	315	315	340	340
12	4	350	DI	0.75	0.52	12.0	12.0	20.0	315	315	340	340
12	6	250	25	0.75	0.55	12.0	12.0	20.0	325	325	360	360
12	6	350	DI	0.75	0.55	12.0	12.0	20.0	325	325	360	360
12	8	250	25	0.75	0.60	12.0	12.0	20.0	340	340	385	385
12	8	350	DI	0.75	0.60	12.0	12.0	20.0	340	340	385	385
12	10	250	25	0.87	0.80	12.0	12.0	20.0	390	390	460	460
12	10	350	DI	0.87	0.80	12.0	12.0	20.0	390	390	460	460
12	12	250	25	0.87	0.87	12.0	12.0	20.0	410	410	495	495
12	12	350	DI	0.87	0.87	12.0	12.0	20.0	410	410	495	495
14	6	150	25	0.66	0.55	14.0	14.0	22.0	435	420	475	460
14	6	250	25	0.82	0.55	14.0	14.0	22.0	485	470	525	505
14	6	350	DI	0.66	0.55	14.0	14.0	22.0	435	420	475	460
14	8	150	25	0.66	0.60	14.0	14.0	22.0	450	435	500	485
14	8	250	25	0.82	0.60	14.0	14.0	22.0	500	480	550	535
14	8	350	DI	0.66	0.60	14.0	14.0	22.0	450	435	500	485
14	10	150	25	0.66	0.68	14.0	14.0	22.0	465	450	540	525
14	10	250	25	0.82	0.68	14.0	14.0	22.0	515	500	585	570
14	10	350	DI	0.66	0.68	14.0	14.0	22.0	465	450	540	525
14	12	150	25	0.82	0.75	14.0	14.0	22.0	540	525	630	615
14	12	250	30	0.82	0.75	14.0	14.0	22.0	540	525	630	615
14	12	350	DI	0.66	0.75	14.0	14.0	22.0	495	475	585	570

* Dimension details of mechanical-joint bells are shown in Table 10.1; dimension details of plain ends are shown in Table 10.2.
† Weight does not include accessory weights. See Table 10.1 for accessory weights.
‡ Ductile Iron.

TABLE 10.4

Mechanical-Joint Tees and Crosses* (contd.)

Size in. Run	Branch	Pressure Rating psi	Iron Strength psi (1000's)	T	T₁	H	J	S	Tee All MJ	Tee MJ, PE & MJ	Cross All MJ	Cross MJ, PE MJ & MJ
14	14	150	25	0.82	0.82	14.0	14.0	22.0	585	570	710	695
14	14	250	30	0.82	0.82	14.0	14.0	22.0	585	570	710	695
14	14	350	DI‡	0.66	0.66	14.0	14.0	22.0	520	500	635	620
16	6	150	30	0.70	0.55	15.0	15.0	23.0	540	520	575	555
16	6	250	30	0.89	0.55	15.0	15.0	23.0	615	590	650	630
16	6	350	DI	0.70	0.55	15.0	15.0	23.0	540	520	575	555
16	8	150	30	0.70	0.60	15.0	15.0	23.0	550	530	605	585
16	8	250	30	0.89	0.60	15.0	15.0	23.0	625	605	675	655
16	8	350	DI	0.70	0.60	15.0	15.0	23.0	550	530	605	585
16	10	150	30	0.70	0.68	15.0	15.0	23.0	570	550	645	625
16	10	250	30	0.89	0.68	15.0	15.0	23.0	645	620	710	690
16	10	350	DI	0.70	0.68	15.0	15.0	23.0	570	550	645	625
16	12	150	30	0.70	0.75	15.0	15.0	23.0	590	570	685	665
16	12	250	30	0.89	0.75	15.0	15.0	23.0	660	640	745	725
16	12	350	DI	0.70	0.75	15.0	15.0	23.0	590	570	685	665
16	14	150	30	0.89	0.82	15.0	15.0	23.0	710	690	830	810
16	14	250	35	0.89	0.82	15.0	15.0	23.0	710	690	830	810
16	14	350	DI	0.70	0.66	15.0	15.0	23.0	620	600	735	715
16	16	150	30	0.89	0.89	15.0	15.0	23.0	740	720	895	875
16	16	250	35	0.89	0.89	15.0	15.0	23.0	740	720	895	875
16	16	350	DI	0.70	0.70	15.0	15.0	23.0	650	625	790	770
18	6	150	30	0.75	0.55	13.0	15.5	21.0	590	565	625	600
18	6	250	30	0.96	0.55	13.0	15.5	21.0	670	645	705	680
18	6	350	DI	0.75	0.55	13.0	15.5	21.0	590	565	625	600
18	8	150	30	0.75	0.60	13.0	15.5	21.0	605	580	655	630
18	8	250	30	0.96	0.60	13.0	15.5	21.0	685	655	730	705
18	8	350	DI	0.75	0.60	13.0	15.5	21.0	605	580	655	630
18	10	150	30	0.75	0.68	13.0	15.5	21.0	620	595	685	660
18	10	250	30	0.96	0.68	13.0	15.5	21.0	700	670	760	735
18	10	350	DI	0.75	0.68	13.0	15.5	21.0	620	595	685	660
18	12	150	30	0.75	0.75	13.0	15.5	21.0	640	615	725	700
18	12	250	30	0.96	0.75	13.0	15.5	21.0	715	690	790	765
18	12	350	DI	0.75	0.75	13.0	15.5	21.0	640	615	725	700
18	14	150	30	0.75	0.66	16.5	16.5	24.5	755	730	870	845
18	14	250	30	0.96	0.82	16.5	16.5	24.5	865	840	990	965
18	14	350	DI	0.75	0.66	16.5	16.5	24.5	755	730	870	845
18	16	150	30	0.96	0.89	16.5	16.5	24.5	905	880	1,060	1,035
18	16	250	35	0.96	0.89	16.5	16.5	24.5	905	880	1,060	1,035
18	16	350	DI	0.75	0.70	16.5	16.5	24.5	785	760	930	905
18	18	150	30	0.96	0.96	16.5	16.5	24.5	945	920	1,130	1,105
18	18	250	35	0.96	0.96	16.5	16.5	24.5	945	920	1,130	1,105
18	18	350	DI	0.75	0.75	16.5	16.5	24.5	820	795	995	965
20	6	150	30	0.80	0.55	14.0	17.0	22.0	725	695	760	730
20	6	250	30	1.03	0.55	14.0	17.0	22.0	830	800	865	835
20	6	350	DI	0.80	0.55	14.0	17.0	22.0	725	695	760	730
20	8	150	30	0.80	0.60	14.0	17.0	22.0	735	705	790	760
20	8	250	30	1.03	0.60	14.0	17.0	22.0	845	810	890	860
20	8	350	DI	0.80	0.60	14.0	17.0	22.0	735	705	790	760

* Dimension details of mechanical-joint bells are shown in Table 10.1; dimension details of plain ends are shown in Table 10.2.

† Weight does not include accessory weights. See Table 10.1 for accessory weights.

‡ Ductile Iron.

TABLE 10.4

Mechanical-Joint Tees and Crosses* (contd.)

Size in.		Pressure Rating psi	Iron Strength psi (1000's)	Dimensions—in.					Weight—lb†			
									Tee		Cross	
Run	Branch			T	T₁	H	J	S	All MJ	MJ, PE & MJ	All MJ	MJ, PE, MJ & MJ
20	10	150	30	0.80	0.68	14.0	17.0	22.0	755	725	820	790
20	10	250	30	1.03	0.68	14.0	17.0	22.0	860	825	920	890
20	10	350	DI‡	0.80	0.68	14.0	17.0	22.0	755	725	820	790
20	12	150	30	0.80	0.75	14.0	17.0	22.0	775	745	860	830
20	12	250	30	1.03	0.75	14.0	17.0	22.0	875	840	955	920
20	12	350	DI	0.80	0.75	14.0	17.0	22.0	775	745	860	830
20	14	150	30	0.80	0.66	14.0	17.0	22.0	795	765	905	875
20	14	250	35	1.03	0.82	14.0	17.0	22.0	910	875	1,025	990
20	14	350	DI	0.80	0.66	14.0	17.0	22.0	795	765	905	875
20	16	150	30	0.80	0.70	18.0	18.0	26.0	945	915	1,085	1,055
20	16	250	35	1.03	0.89	18.0	18.0	26.0	1,095	1,060	1,245	1,215
20	16	350	DI	0.80	0.70	18.0	18.0	26.0	945	915	1,085	1,055
20	18	150	35	1.03	0.96	18.0	18.0	26.0	1,140	1,110	1,330	1,300
20	18	350	DI	0.80	0.75	18.0	18.0	26.0	985	950	1,155	1,120
20	20	150	35	1.03	1.03	18.0	18.0	26.0	1,185	1,155	1 415	1,385
20	20	350	DI	0.80	0.80	18.0	18.0	26.0	1,020	990	1 230	1,200
24	6	150	30	0.89	0.55	15.0	19.0	23.0	985	945	1,025	985
24	6	250	30	1.16	0.55	15.0	19.0	23.0	1,145	1,105	1,180	1,140
24	6	350	DI	0.89	0.55	15.0	19.0	23.0	985	945	1,025	985
24	8	150	30	0.89	0.60	15.0	19.0	23.0	1,000	960	1,045	1,005
24	8	250	30	1.16	0.60	15.0	19.0	23.0	1,160	1,115	1,200	1,160
24	8	350	DI	0.89	0.60	15.0	19.0	23.0	1,000	960	1,045	1,005
24	10	150	30	0.89	0.68	15.0	19.0	23.0	1,020	980	1,085	1,045
24	10	250	30	1.16	0.68	15.0	19.0	23.0	1,170	1,130	1,230	1,190
24	10	350	DI	0.89	0.68	15.0	19.0	23.0	1,020	980	1,085	1,045
24	12	150	30	0.89	0.75	15.0	19.0	23.0	1,030	990	1,110	1,070
24	12	250	30	1.16	0.75	15.0	19.0	23.0	1,185	1,145	1,260	1,220
24	12	350	DI	0.89	0.75	15.0	19.0	23.0	1,030	990	1,110	1,070
24	14	150	30	0.89	0.66	15.0	19.0	23.0	1,055	1,015	1,155	1,115
24	14	250	30	1.16	0.82	15.0	19.0	23.0	1,220	1,180	1,325	1,285
24	14	350	DI	0.89	0.66	15.0	19.0	23.0	1,055	1,015	1,155	1,115
24	16	150	30	0.89	0.70	15.0	19.0	23.0	1,075	1,035	1,200	1,160
24	16	250	35	1.16	0.89	15.0	19.0	23.0	1,245	1,200	1,375	1,335
24	16	350	DI	0.89	0.70	15.0	19.0	23.0	1,075	1,035	1,200	1,160
24	18	150	30	0.89	0.75	22.0	22.0	30.0	1,400	1,360	1,590	1,550
24	18	250	35	1.16	0.96	22.0	22.0	30.0	1,660	1,615	1,865	1,820
24	18	350	DI	0.89	0.75	22.0	22.0	30.0	1,400	1,360	1,590	1,550
24	20	150	35	1.16	1.03	22.0	22.0	30.0	1,720	1,680	1,965	1,925
24	20	350	DI	0.89	0.80	22.0	22.0	30.0	1,450	1,410	1,675	1,630
24	24	150	35	1.16	1.16	22.0	22.0	30.0	1,815	1,775	2,155	2,115
24	24	350	DI	0.89	0.89	22.0	22.0	30.0	1,535	1,490	1,835	1,795
30	6	150	30	1.03	0.55	18.0	23.0	26.0	1,730	1,615	1,770	1,655
30	6	250	30	1.37	0.55	18.0	23.0	26.0	2,050	1,935	2,085	1,970
30	6	250	DI	1.03	0.55	18.0	23.0	26.0	1,730	1,615	1,770	1,655
30	8	150	30	1.03	0.60	18.0	23.0	26.0	1,745	1,630	1,795	1,680
30	8	250	30	1.37	0.60	18.0	23.0	26.0	2,060	1,945	2,110	1,990
30	8	250	DI	1.03	0.60	18.0	23.0	26.0	1,745	1,630	1,795	1,680

* Dimension details of mechanical-joint bells are shown in Table 10.1; dimension details of plain ends are shown in Table 10.2.
† Weight does not include accessory weights. See Table 10.1 for accessory weights.
‡ Ductile Iron.

TABLE 10.4

Mechanical-Joint Tees and Crosses (contd.)

Size in.		Pressure Rating psi	Iron Strength psi (1000's)	Dimensions—in.					Weight—lb†			
									Tee		Cross	
Run	Branch			T	T₁	H	J	S	All MJ	MJ, PE & MJ	All MJ	MJ, PE, MJ & MJ
30	10	150	30	1.03	0.68	18.0	23.0	26.0	1,760	1,645	1,830	1,715
30	10	250	30	1.37	0.68	18.0	23.0	26.0	2,075	1,960	2,135	2,020
30	10	250	DI‡	1.03	0.68	18.0	23.0	26.0	1,760	1,645	1,830	1,715
30	12	150	30	1.03	0.75	18.0	23.0	26.0	1,780	1,665	1,865	1,750
30	12	250	30	1.37	0.75	18.0	23.0	26.0	2,090	1,970	2,165	2,045
30	12	250	DI	1.03	0.75	18.0	23.0	26.0	1,780	1,665	1,865	1,750
30	14	150	30	1.03	0.66	18.0	23.0	26.0	1,800	1,685	1,905	1,790
30	14	250	30	1.37	0.82	18.0	23.0	26.0	2,120	2,005	2,230	2,115
30	14	250	DI	1.03	0.66	18.0	23.0	26.0	1,800	1,685	1,905	1,790
30	16	150	30	1.03	0.70	18.0	23.0	26.0	1,820	1,705	1,950	1,835
30	16	250	30	1.37	0.89	18.0	23.0	26.0	2,145	2,030	2,280	2,165
30	16	250	DI	1.03	0.70	18.0	23.0	26.0	1,820	1,705	1,950	1,835
30	18	150	30	1.03	0.75	18.0	23.0	26.0	1,845	1,730	2,000	1,885
30	18	250	35	1.37	0.96	18.0	23.0	26.0	2,170	2,055	2,330	2,215
30	18	250	DI	1.03	0.75	18.0	23.0	26.0	1,845	1,730	2,000	1,885
30	20	150	30	1.03	0.80	18.0	23.0	26.0	1,875	1,760	2,060	1,945
30	20	250	35	1.37	1.03	18.0	23.0	26.0	2,205	2,090	2,395	2,280
30	20	250	DI	1.03	0.80	18.0	23.0	26.0	1,875	1,760	2,060	1,945
30	24	150	35	1.37	1.16	25.0	25.0	33.0	2,880	2,765	3,180	3,065
30	24	250	DI	1.03	0.89	25.0	25.0	33.0	2,400	2,280	2,675	2,560
30	30	150	35	1.37	1.37	25.0	25.0	33.0	3,105	2,990	3,640	3,520
30	30	250	DI	1.03	1.03	25.0	25.0	33.0	2,595	2,480	3,075	2,955
36	8	150	30	1.15	0.60	20.0	26.0	28.0	2,520	2,345	2,565	2,390
36	8	250	30	1.58	0.60	20.0	26.0	28.0	3,050	2,870	3,095	2,915
36	8	250	DI	1.15	0.60	20.0	26.0	28.0	2,520	2,345	2,565	2,390
36	10	150	30	1.15	0.68	20.0	26.0	28.0	2,535	2,360	2,600	2,425
36	10	250	30	1.58	0.68	20.0	26.0	28.0	3,065	2,885	3,120	2,940
36	10	250	DI	1.15	0.68	20.0	26.0	28.0	2,535	2,360	2,600	2,425
36	12	150	30	1.15	0.75	20.0	26.0	28.0	2,550	2,375	2,630	2,455
36	12	250	30	1.58	0.75	20.0	26.0	28.0	3,075	2,895	3,140	2,960
36	12	250	DI	1.15	0.75	20.0	26.0	28.0	2,550	2,375	2,630	2,455
36	14	150	30	1.15	0.66	20.0	26.0	28.0	2,570	2,395	2,665	2,490
36	14	250	30	1.58	0.82	20.0	26.0	28.0	3,105	2,925	3,205	3,025
36	14	250	DI	1.15	0.66	20.0	26.0	28.0	2,570	2,395	2,665	2,490
36	16	150	30	1.15	0.70	20.0	26.0	28.0	2,585	2,410	2,705	2,530
36	16	250	30	1.58	0.89	20.0	26.0	28.0	3,125	2,945	3,245	3,065
36	16	250	DI	1.15	0.70	20.0	26.0	28.0	2,585	2,410	2,705	2,530
36	18	150	30	1.15	0.75	20.0	26.0	28.0	2,610	2,435	2,750	2,575
36	18	250	30	1.58	0.96	20.0	26.0	28.0	3,150	2,970	3,290	3,110
36	18	250	DI	1.15	0.75	20.0	26.0	28.0	2,610	2,435	2,750	2,575
36	20	150	30	1.15	0.80	20.0	26.0	28.0	2,635	2,460	2,805	2,630
36	20	250	35	1.58	1.03	20.0	26.0	28.0	3,175	2,995	3,345	3,165
36	20	250	DI	1.15	0.80	20.0	26.0	28.0	2,635	2,460	2,805	2,630
36	24	150	30	1.15	0.89	20.0	26.0	28.0	2,690	2,515	2,910	2,735
36	24	250	35	1.58	1.16	20.0	26.0	28.0	3,230	3,050	3,450	3,270
36	24	250	DI	1.15	0.89	20.0	26.0	28.0	2,690	2,515	2,910	2,735
36	30	150	35	1.58	1.37	28.0	28.0	36.0	4,345	4,170	4,790	4,615
36	30	250	DI	1.15	1.03	28.0	28.0	36.0	3,545	3,365	3,965	3,790
36	36	150	35	1.58	1.58	28.0	28.0	36.0	4,590	4,410	5,280	5,105
36	36	250	DI	1.15	1.15	28.0	28.0	36.0	3,745	3,565	4,370	4,190

* Dimension details of mechanical-joint bells are shown in Table 10.1; dimension details of plain ends are shown in Table 10.2.
† Weight does not include accessory weights. See Table 10.1 for accessory weights.
‡ Ductile Iron.

TABLE 10.4

Mechanical-Joint Tees and Crosses (contd.)

Size in.		Pressure Rating psi	Iron Strength psi (1000's)	Dimensions—in.					Weight—lb†			
									Tee		Cross	
Run	Branch			T	T₁	H	J	S	All MJ	MJ, PE & MJ	All MJ	MJ, PE MJ & MJ
42	12	150	30	1.28	0.75	23.0	30.0	31.0	3,555	3,335	3,640	3,420
42	12	250	30	1.78	0.75	23.0	30.0	31.0	4,385	4,160	4,450	4,225
42	12	250	DI‡	1.28	0.75	23.0	30.0	31.0	3,555	3,335	3,640	3,420
42	14	150	30	1.28	0.66	23.0	30.0	31.0	3,575	3,355	3,675	3,455
42	14	250	30	1.78	0.82	23.0	30.0	31.0	4,415	4,190	4,515	4,290
42	14	250	DI	1.28	0.66	23.0	30.0	31.0	3,575	3,355	3,675	3,455
42	16	150	30	1.28	0.70	23.0	30.0	31.0	3,595	3,375	3,715	3,495
42	16	250	30	1.78	0.89	23.0	30.0	31.0	4,435	4,210	4,550	4,325
42	16	250	DI	1.28	0.70	23.0	30.0	31.0	3,595	3,375	3,715	3,495
42	18	150	30	1.28	0.75	23.0	30.0	31.0	3,615	3,395	3,755	3,535
42	18	250	30	1.78	0.96	23.0	30.0	31.0	4,455	4,230	4,595	4,370
42	18	250	DI	1.28	0.75	23.0	30.0	31.0	3,615	3,395	3,755	3,535
42	20	150	30	1.28	0.80	23.0	30.0	31.0	3,640	3,420	3,810	3,590
42	20	250	30	1.78	1.03	23.0	30.0	31.0	4,480	4,255	4,645	4,420
42	20	250	DI	1.28	0.80	23.0	30.0	31.0	3,640	3,420	3,810	3,590
42	24	150	30	1.78	1.16	23.0	30.0	31.0	4,530	4,305	4,745	4,520
42	24	250	DI	1.28	0.89	23.0	30.0	31.0	3,690	3,470	3,910	3,690
42	30	150	30	1.78	1.37	31.0	31.0	39.0	5,800	5,575	6,210	5,985
42	30	250	DI	1.28	1.03	31.0	31.0	39.0	4,650	4,425	5,040	4,815
42	36	150	DI	1.28	1.15	31.0	31.0	39.0	4,880	4,655	5,425	5,200
42	36	250	DI	1.78	1.58	31.0	31.0	39.0	6,075	5,850	6,655	6,430
42	42	150	DI	1.28	1.28	31.0	31.0	39.0	5,085	4,860	5,840	5,615
42	42	250	DI	1.78	1.78	31.0	31.0	39.0	6,320	6,095	7,145	6,920
48	12	150	30	1.42	0.75	26.0	34.0	34.0	4,870	4,580	4,955	4,665
48	12	250	30	1.96	0.75	26.0	34.0	34.0	6,025	5,735	6,095	5,805
48	12	250	DI	1.42	0.75	26.0	34.0	34.0	4,870	4,580	4,955	4,665
48	14	150	30	1.42	0.66	26.0	34.0	34.0	4,885	4,595	4,985	4,695
48	14	250	30	1.96	0.82	26.0	34.0	34.0	6,055	5,770	6,155	5,865
48	14	250	DI	1.42	0.66	26.0	34.0	34.0	4,885	4,595	4,985	4,695
48	16	150	30	1.42	0.70	26.0	34.0	34.0	4,905	4,615	5,025	4,735
48	16	250	30	1.96	0.89	26.0	34.0	34.0	6,075	5,785	6,195	5,905
48	16	250	DI	1.42	0.70	26.0	34.0	34.0	4,905	4,615	5,025	4,735
48	18	150	30	1.42	0.75	26.0	34.0	34.0	4,925	4,635	5,065	4,775
48	18	250	30	1.96	0.96	26.0	34.0	34.0	6,095	5,805	6,235	5,945
48	18	250	DI	1.42	0.75	26.0	34.0	34.0	4,925	4,635	5,065	4,775
48	20	150	30	1.42	0.80	26.0	34.0	34.0	4,950	4,660	5,115	4,825
48	20	250	30	1.96	1.03	26.0	34.0	34.0	6,120	5,830	6,285	5,995
48	20	250	DI	1.42	0.80	26.0	34.0	34.0	4,950	4,660	5,115	4,825
48	24	150	30	1.42	0.89	26.0	34.0	34.0	4,995	4,705	5,210	4,920
48	24	250	30	1.96	1.16	26.0	34.0	34.0	6,165	5,880	6,375	6,085
48	24	250	DI	1.42	0.89	26.0	34.0	34.0	4,995	4,705	5,210	4,920
48	30	150	30	1.96	1.37	26.0	34.0	34.0	6,315	6,025	6,670	6,385
48	30	250	DI	1.42	1.03	26.0	34.0	34.0	5,140	4,855	5,495	5,210
48	36	150	30	1.96	1.58	34.0	34.0	42.0	7,835	7,545	8,360	8,075
48	36	250	DI	1.42	1.15	34.0	34.0	42.0	6,280	5,995	6,790	6,500
48	42	150	DI	1.42	1.28	34.0	34.0	42.0	6,510	6,225	7,150	6,860
48	42	250	DI	1.96	1.78	34.0	34.0	42.0	8,130	7,845	8,815	8,530
48	48	150	DI	1.42	1.42	34.0	34.0	42.0	6,765	6,475	7,655	7,370
48	48	250	DI	1.96	1.96	34.0	34.0	42.0	8,420	8,135	9,380	9,095

* Dimension details of mechanical-joint bells are shown in Table 10.1; dimension details of plain ends are shown in Table 10.2.
† Weight does not include accessory weights. See Table 10.1 for accessory weights.
‡ Ductile Iron.

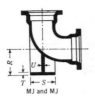
MJ and MJ

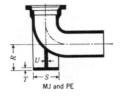
MJ and PE

Fig. 10.5. Mechanical-Joint Base Bends (See Table 10.5)

For other dimensions of base bends, see Table 10.3

TABLE 10.5

Mechanical-Joint Base Bends

| Size in. | Pressure Rating psi | Iron Strength psi (1000's) | Dimensions—in. | | | | Weight—lb† | | |
			R*	S Diam.	T	U	MJ & MJ	MJ & PE	Base Only
3	250	25	4.88	5.00	0.56	0.50	45	45	10
3	350	DI‡	4.88	5.00	0.56	0.50	45	45	10
4	250	25	5.50	6.00	0.62	0.50	65	60	10
4	350	DI	5.50	6.00	0.62	0.50	65	60	10
6	250	25	7.00	7.00	0.69	0.62	105	100	20
6	350	DI	7.00	7.00	0.69	0.62	105	100	20
8	250	25	8.38	9.00	0.94	0.88	165	160	40
8	350	DI	8.38	9.00	0.94	0.88	165	160	40
10	250	25	9.75	9.00	0.94	0.88	235	235	45
10	350	DI	9.75	9.00	0.94	0.88	235	235	45
12	250	25	11.25	11.00	1.00	1.00	320	320	65
12	350	DI	11.25	11.00	1.00	1.00	320	320	65
14	150	25	12.50	11.00	1.00	1.00	410	395	70
14	250	25	12.50	11.00	1.00	1.00	450	435	70
14	350	DI	12.50	11.00	1.00	1.00	410	395	70
16	150	30	13.75	11.00	1.00	1.00	505	485	75
16	250	30	13.75	11.00	1.00	1.00	565	545	75
16	350	DI	13.75	11.00	1.00	1.00	505	485	75
18	150	30	15.00	13.50	1.12	1.12	660	635	115
18	250	30	15.00	13.50	1.12	1.12	740	715	115
18	350	DI	15.00	13.50	1.12	1.12	660	635	115
20	150	30	16.00	13.50	1.12	1.12	800	770	120
20	250	30	16.00	13.50	1.12	1.12	910	875	120
20	350	DI	16.00	13.50	1.12	1.12	800	770	120
24	150	30	18.50	13.50	1.12	1.12	1,155	1,115	130
24	250	30	18.50	13.50	1.12	1.12	1,345	1,305	130
24	350	DI	18.50	13.50	1.12	1.12	1,155	1,115	130
30	150	30	23.00	16.00	1.19	1.15	1,880	1,775	190
30	250	30	23.00	16.00	1.19	1.15	2,220	2,110	190
30	250	DI	23.00	16.00	1.19	1.15	1,880	1,775	190
36	150	30	26.00	19.00	1.25	1.15	2,725	2,560	250
36	250	30	26.00	19.00	1.25	1.15	3,295	3.130	250
36	250	DI	26.00	19.00	1.25	1.15	2,725	2,560	250
42	150	30	30.00	23.50	1.44	1.28	3,820	3,610	410
42	250	30	30.00	23.50	1.44	1.28	4,665	4,460	410
42	250	DI	30.00	23.50	1.44	1.28	3,820	3,610	410
48	150	30	34.00	25.00	1.56	1.42	5,110	4,845	515
48	250	30	34.00	25.00	1.56	1.42	6,260	5,990	515
48	250	DI	34.00	25.00	1.56	1.42	5,110	4,845	515

* Dimension R is a finished dimension; unfinished bases will be ⅛ in. longer; for base drilling see Table 10.15.
† Weight does not include accessory weights. See Table 10.1 for accessory weights.
‡ Ductile Iron.

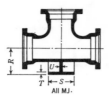 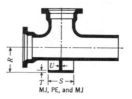

Fig. 10.6. Mechanical-Joint Base Tees (See Table 10.6)

For other dimensions of base tees, see Table 10.4.

TABLE 10.6

Mechanical-Joint Base Tees

Size in.	Pressure Rating psi	Iron Strength psi (1000's)	Dimensions—in.				Weight—lb†		
			R*	S Diam.	T	U	All MJ	MJ, PE & MJ	Base Only
3	250	25	4.88	5.00	0.56	0.50	60	60	5
3	350	DI‡	4.88	5.00	0.56	0.50	60	60	5
4	250	25	5.50	6.00	0.62	0.50	90	85	10
4	350	DI	5.50	6.00	0.62	0.50	90	85	10
6	250	25	7.00	7.00	0.69	0.62	140	135	15
6	350	DI	7.00	7.00	0.69	0.62	140	135	15
8	250	25	8.38	9.00	0.94	0.88	215	210	30
8	350	DI	8.38	9.00	0.94	0.88	215	210	30
10	250	25	9.75	9.00	0.94	0.88	340	340	30
10	350	DI	9.75	9.00	0.94	0.88	340	340	30
12	250	25	11.25	11.00	1.00	1.00	455	455	45
12	350	DI	11.25	11.00	1.00	1.00	455	455	45
14	150	25	12.50	11.00	1.00	1.00	635	620	50
14	250	30	12.50	11.00	1.00	1.00	635	620	50
14	350	DI	12.50	11.00	1.00	1.00	570	550	50
16	150	30	13.75	11.00	1.00	1.00	790	770	50
16	250	35	13.75	11.00	1.00	1.00	790	770	50
16	350	DI	13.75	11.00	1.00	1.00	700	675	50
18	150	30	15.00	13.50	1.12	1.12	1,020	995	75
18	250	35	15.00	13.50	1.12	1.12	1,020	995	75
18	350	DI	15.00	13.50	1.12	1.12	895	870	75
20	150	35	16.00	13.50	1.12	1.12	1,260	1,230	75
20	350	DI	16.00	13.50	1.12	1.12	1,095	1,065	75
24	150	35	18.50	13.50	1.12	1.12	1,895	1,855	80
24	350	DI	18.50	13.50	1.12	1.12	1,615	1,570	80
30	150	35	23.00	16.00	1.19	1.15	3,225	3,110	120
30	250	DI	23.00	16.00	1.19	1.15	2,715	2,600	120
36	150	35	26.00	19.00	1.25	1.15	4,750	4,570	160
36	250	DI	26.00	19.00	1.25	1.15	3,905	3,725	160
42	150	DI	30.00	23.50	1.44	1.28	5,355	5,130	270
42	250	DI	30.00	23.50	1.44	1.28	6,590	6,365	270
48	150	DI	34.00	25.00	1.56	1.42	7,100	6,810	335
48	250	DI	34.00	25.00	1.56	1.42	8,755	8,470	335

* Dimension R is a finished dimension; unfinished bases will be ⅛ in. longer; for base drilling see Table 10.15.
† Weight does not include accessory weights. See Table 10.1 for accessory weights.
‡ Ductile Iron.

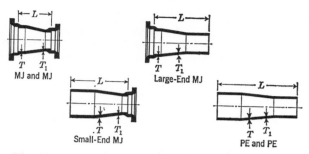

Fig. 10.7. Mechanical-Joint Reducers (See Table 10.7)

TABLE 10.7

*Mechanical-Joint Reducers**

Size in.		Pressure Rating psi	Iron Strength psi (1000's)	Thickness in.		MJ & MJ		Small-End MJ		Large-End MJ		PE & PE	
Large End	Small End			T Large End	T₁ Small End	L in.	Weight† lb	L in.	Weight† lb	L in.	Weight† lb	L in.	Weight† lb
4	3	250	25	0.52	0.48	7	40	15	35	15	40	23	35
4	3	350	DI‡	0.52	0.48	7	40	15	35	15	40	23	35
6	3	250	25	0.55	0.48	9	55	17	50	17	55	25	50
6	3	350	DI	0.55	0.48	9	55	17	50	17	55	25	50
6	4	250	25	0.55	0.52	9	60	17	60	17	60	25	55
6	4	350	DI	0.55	0.52	9	60	17	60	17	60	25	55
8	4	250	25	0.60	0.52	11	80	19	80	19	80	27	75
8	4	350	DI	0.60	0.52	11	80	19	80	19	80	27	75
8	6	250	25	0.60	0.55	11	95	19	90	19	90	27	85
8	6	350	DI	0.60	0.55	11	95	19	90	19	90	27	85
10	4	250	25	0.68	0.52	12	105	20	100	20	100	28	100
10	4	350	DI	0.68	0.52	12	105	20	100	20	100	28	100
10	6	250	25	0.68	0.55	12	115	20	115	20	115	28	115
10	6	350	DI	0.68	0.55	12	115	20	115	20	115	28	115
10	8	250	25	0.68	0.60	12	135	20	130	20	130	28	130
10	8	350	DI	0.68	0.60	12	135	20	130	20	130	28	130
12	4	250	25	0.75	0.52	14	135	22	130	22	130	30	130
12	4	350	DI	0.75	0.52	14	135	22	130	22	130	30	130
12	6	250	25	0.75	0.55	14	150	22	150	22	145	30	145
12	6	350	DI	0.75	0.55	14	150	22	150	22	145	30	145

* For dimension details of mechanical-joint bells, see Table 10.1; for dimension details of plain ends, see Table 10.2. Eccentric reducers with the same dimensions and weights given for concentric reducers are available from most manufacturers if specified on the purchase order.
† Weight does not include accessory weights. See Table 10.1 for accessory weights.
‡ Ductile Iron.

TABLE 10.7

Mechanical-Joint Reducers* (contd.)

Size in.		Pressure Rating psi	Iron Strength psi (1000's)	Thickness in.		MJ & MJ		Small-End MJ		Large-End MJ		PE & PE	
Large End	Small End			T Large End	T₁ Small End	L in.	Weight† lb	L in.	Weight† lb	L in.	Weight† lb	L in.	Weight† lb
12	8	250	25	0.75	0.60	14	165	22	165	22	165	30	165
12	8	350	DI‡	0.75	0.60	14	165	22	165	22	165	30	165
12	10	250	25	0.75	0.68	14	190	22	190	22	185	30	185
12	10	350	DI	0.75	0.68	14	190	22	190	22	185	30	185
14	6	150	25	0.66	0.55	16	190	24	175	24	185	32	170
14	6	250	25	0.82	0.55	16	200	24	185	24	200	32	185
14	6	350	DI	0.66	0.55	16	190	24	175	24	185	32	170
14	8	150	25	0.66	0.60	16	210	24	190	24	205	32	190
14	8	250	25	0.82	0.60	16	220	24	205	24	220	32	205
14	8	350	DI	0.66	0.60	16	210	24	190	24	205	32	190
14	10	150	25	0.66	0.68	16	230	24	215	24	230	32	215
14	10	250	25	0.82	0.68	16	245	24	230	24	245	32	230
14	10	350	DI	0.66	0.68	16	230	24	215	24	230	32	215
14	12	150	25	0.66	0.75	16	255	24	240	24	255	32	240
14	12	250	25	0.82	0.75	16	270	24	255	24	275	32	260
14	12	350	DI	0.66	0.75	16	255	24	240	24	255	32	240
16	6	150	30	0.70	0.55	18	230	26	210	26	230	34	210
16	6	250	30	0.89	0.55	18	250	26	230	26	250	34	230
16	6	350	DI	0.70	0.55	18	230	26	210	26	230	34	210
16	8	150	30	0.70	0.60	18	250	26	230	26	250	34	230
16	8	250	30	0.89	0.60	18	270	26	250	26	270	34	250
16	8	350	DI	0.70	0.60	18	250	26	230	26	250	34	230
16	10	150	30	0.70	0.68	18	280	26	255	26	275	34	255
16	10	250	30	0.89	0.68	18	300	26	280	26	300	34	280
16	10	350	DI	0.70	0.68	18	280	26	255	26	275	34	255
16	12	150	30	0.70	0.75	18	305	26	285	26	305	34	285
16	12	250	30	0.89	0.75	18	325	26	305	26	330	34	310
16	12	350	DI	0.70	0.75	18	305	26	285	26	305	34	285
16	14	150	30	0.70	0.66	18	335	26	310	26	315	34	295
16	14	250	30	0.89	0.82	18	370	26	350	26	355	34	335
16	14	350	DI	0.70	0.66	18	335	26	310	26	315	34	295
18	8	150	30	0.75	0.60	19	295	27	270	27	295	35	270
18	8	250	30	0.96	0.60	19	320	27	295	27	320	35	295
18	8	350	DI	0.75	0.60	19	295	27	270	27	295	35	270
18	10	150	30	0.75	0.68	19	325	27	300	27	320	35	295
18	10	250	30	0.96	0.68	19	350	27	325	27	350	35	325
18	10	350	DI	0.75	0.68	19	325	27	300	27	320	35	295
18	12	150	30	0.75	0.75	19	350	27	325	27	350	35	325
18	12	250	30	0.96	0.75	19	380	27	355	27	385	35	360
18	12	350	DI	0.75	0.75	19	350	27	325	27	350	35	325
18	14	150	30	0.75	0.66	19	380	27	355	27	365	35	340
18	14	250	30	0.96	0.82	19	425	27	400	27	410	35	385
18	14	350	DI	0.75	0.66	19	380	27	355	27	365	35	340
18	16	150	30	0.75	0.70	19	415	27	390	27	395	35	370
18	16	250	30	0.96	0.89	19	465	27	440	27	445	35	420
18	16	350	DI	0.75	0.70	19	415	27	390	27	395	35	370
20	10	150	30	0.80	0.68	20	375	28	345	28	375	36	345
20	10	250	30	1.03	0.68	20	410	28	380	28	410	36	380
20	10	350	DI	0.80	0.68	20	375	28	345	38	375	36	345

* For dimension details of mechanical-joint bells, see Table 10.1; for dimension details of plain ends, see Table 10.2. Eccentric reducers with the same dimensions and weights given for concentric reducers are available from most manufacturers if specified on the purchase order
† Weight does not include accessory weights. See Table 10.1 for accessory weights.
‡ Ductile Iron.

TABLE 10.7

Mechanical-Joint Reducers* (contd.)

Size in. Large End	Small End	Pressure Rating psi	Iron Strength psi (1000's)	Thickness in. T Large End	T₁ Small End	MJ & MJ L in.	Weight† lb	Small-End MJ L in.	Weight† lb	Large-End MJ L in.	Weight† lb	PE & PE L in.	Weight† lb
20	12	150	30	0.80	0.75	20	405	28	375	28	405	36	375
20	12	250	30	1.03	0.75	20	440	28	410	28	445	36	415
20	12	350	DI‡	0.80	0.75	20	405	28	375	28	405	36	375
20	14	150	30	0.80	0.66	20	430	28	400	28	415	36	385
20	14	250	30	1.03	0.82	20	485	28	455	28	470	36	440
20	14	350	DI	0.80	0.66	20	430	28	400	28	415	36	385
20	16	150	30	0.80	0.70	20	470	28	435	28	445	36	415
20	16	250	30	1.03	0.89	20	530	28	500	28	510	36	475
20	16	350	DI	0.80	0.70	20	470	28	435	28	445	36	415
20	18	150	30	0.80	0.75	20	510	28	475	28	485	36	455
20	18	250	30	1.03	0.96	20	575	28	545	28	550	36	520
20	18	350	DI	0.80	0.75	20	510	28	475	28	485	36	455
24	12	150	30	0.89	0.75	24	550	32	510	32	550	40	510
24	12	250	30	1.16	0.75	24	610	32	570	32	615	40	575
24	12	350	DI	0.89	0.75	24	550	32	510	32	550	40	510
24	14	150	30	0.89	0.66	24	575	32	535	32	560	40	520
24	14	250	30	1.16	0.82	24	660	32	620	32	645	40	605
24	14	350	DI	0.89	0.66	24	575	32	535	32	560	40	520
24	16	150	30	0.89	0.70	24	615	32	575	32	595	40	555
24	16	250	30	1.16	0.89	24	705	32	665	32	685	40	645
24	16	350	DI	0.89	0.70	24	615	32	575	32	595	40	555
24	18	150	30	0.89	0.75	24	660	32	620	32	635	40	595
24	18	250	30	1.16	0.96	24	760	32	720	32	735	40	695
24	18	350	DI	0.89	0.75	24	660	32	620	32	635	40	595
24	20	150	30	0.89	0.80	24	705	32	665	32	675	40	635
24	20	250	30	1.16	1.03	24	815	32	775	32	785	40	745
24	20	350	DI	0.89	0.80	24	705	32	665	32	675	40	635
30	18	150	30	1.03	0.75	30	990	38	885	38	965	46	860
30	18	250	30	1.37	0.96	30	1,160	38	1,050	38	1,130	46	1,025
30	18	250	DI	1.03	0.75	30	990	38	885	38	965	46	860
30	20	150	30	1.03	0.80	30	1,050	38	945	38	1,020	46	915
30	20	250	30	1.37	1.03	30	1,225	38	1,120	38	1,195	46	1,090
30	20	250	DI	1.03	0.80	30	1,050	38	945	38	1,020	46	915
30	24	150	30	1.03	0.89	30	1,165	38	1,060	38	1,125	46	1,020
30	24	250	30	1.37	1.16	30	1,360	38	1,255	38	1,320	46	1,215
30	24	250	DI	1.03	0.89	30	1,165	38	1,060	38	1,125	46	1,020
36	20	150	30	1.15	0.80	36	1,450	44	1,285	44	1,420	52	1,255
36	20	250	30	1.58	1.03	36	1,730	44	1,560	44	1,695	52	1,530
36	20	250	DI	1.15	0.80	36	1,450	44	1,285	44	1,420	52	1,255
36	24	150	30	1.15	0.89	36	1,580	44	1,410	44	1,535	52	1,370
36	24	250	30	1.58	1.16	36	1,885	44	1,720	44	1,845	52	1,680
36	24	250	DI	1.15	0.89	36	1,580	44	1,410	44	1,535	52	1,370
36	30	150	30	1.15	1.03	36	1,855	44	1,690	44	1,750	52	1,585
36	30	250	30	1.58	1.37	36	2,225	44	2,060	44	2,120	52	1,950
36	30	250	DI	1.15	1.03	36	1,855	44	1,690	44	1,750	52	1,585
42	20	150	30	1.28	0.80	42	1,915	50	1,705	50	1,880	58	1,670
42	20	250	30	1.78	1.03	42	2,320	50	2,110	50	2,285	58	2,080
42	20	250	DI	1.28	0.80	42	1,915	50	1,705	50	1,880	58	1,670

* For dimension details of mechanical-joint bells, see Table 10.1; for dimension details of plain ends, see Table 10.2. Eccentric reducers with the same dimensions and weights given for concentric reducers are available from most manufacturers if specified on the purchase order.

† Weight does not include accessory weights. See Table 10.1 for accessory weights.

‡ Ductile Iron.

TABLE 10.7

Mechanical-Joint Reducers* (contd.)

Size in. Large End	Size in. Small End	Pressure Rating psi	Iron Strength psi (1000's)	Thickness in. T Large End	Thickness in. T₁ Small End	MJ & MJ L in.	MJ & MJ Weight† lb	Small-End MJ L in.	Small-End MJ Weight† lb	Large-End MJ L in.	Large-End MJ Weight† lb	PE & PE L in.	PE & PE Weight† lb
42	24	150	30	1.28	0.89	42	2,060	50	1,855	50	2,020	58	1,810
42	24	250	30	1.78	1.16	42	2,495	50	2,285	50	2,455	58	2,245
42	24	250	DI‡	1.28	0.89	42	2,060	50	1,855	50	2,020	58	1,810
42	30	150	30	1.28	1.03	42	2,370	50	2,165	50	2,265	58	2,055
42	30	250	30	1.78	1.37	42	2,885	50	2,675	50	2,780	58	2,570
42	30	250	DI	1.28	1.03	42	2,370	50	2,165	50	2,265	58	2,055
42	36	150	30	1.28	1.15	42	2,695	50	2,485	50	2,530	58	2,320
42	36	250	30	1.78	1.58	42	3,310	50	3,100	50	3,145	58	2,935
42	36	250	DI	1.28	1.15	42	2,695	50	2,485	50	2,530	58	2,320
48	30	150	30	1.42	1.03	48	3,005	56	2,740	56	2,900	64	2,635
48	30	250	30	1.96	1.37	48	3,680	56	3,410	56	3,570	64	3,305
48	30	250	DI	1.42	1.03	48	3,005	56	2,740	56	2,900	64	2,635
48	36	150	30	1.42	1.15	48	3,370	56	3,100	56	3,205	64	2,940
48	36	250	30	1.96	1.58	48	4,160	56	3,890	56	3,990	64	3,725
48	36	250	DI	1.42	1.15	48	3,370	56	3,100	56	3,205	64	2,940
48	42	150	30	1.42	1.28	48	3,750	56	3,480	56	3,540	64	3,275
48	42	250	30	1.96	1.78	48	4,655	56	4,390	56	4,445	64	4,180
48	42	250	DI	1.42	1.28	48	3,750	56	3,480	56	3,540	64	3,275

* For dimension details of mechanical-joint bells, see Table 10.1; for dimension details of plain ends, see Table 10.2. Eccentric reducers with the same dimensions and weights given for concentric reducers are available from most manufacturers if specified on the purchase order.
† Weight does not include accessory weights. See Table 10.1 for accessory weights.
‡ Ductile Iron.

TABLE 10.8

Mechanical-Joint Tapped Tees*

Size in.	Pressure Rating psi	Iron Strength psi (1000's)	T in.	L in.	Max. Tap in Boss in.	Weight† lb
3	250	25	0.48	8	2½	35
3	350	DI‡	0.48	8	2½	35
4	250	25	0.52	8	2½	45
4	350	DI	0.52	8	2½	45
6	250	25	0.55	8	2½	70
6	350	DI	0.55	8	2½	70
8	250	25	0.60	8	2½	95
8	350	DI	0.60	8	2½	95
10	250	25	0.68	8	2½	130
10	350	DI	0.68	8	2½	130
12	250	25	0.75	8	2½	165
12	350	DI	0.75	8	2½	165

* Two bosses can be used to make a tapped cross. For dimension details of mechanical-joint bells, see Table 10.1.
† Weight does not include accessory weights. See Table 10.1 for accessory weights.
‡ Ductile Iron.

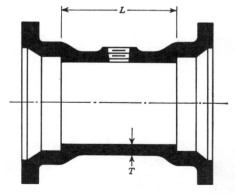

Fig. 10.8. Mechanical-Joint Tapped Tees (See Table 10.8)

TABLE 10.9

*Mechanical-Joint Offsets**

Size in.	PR† psi	Iron Strength psi (1000's)	D in.	T in.	MJ & MJ		MJ & PE	
					L in.	Wgt.‡ lb	L in.	Wgt.‡ lb
3	250	25	6	0.48	19	50	27	50
3	350	DI§	6	0.48	19	50	27	50
3	250	25	12	0.48	22	60	30	60
3	350	DI	12	0.48	22	60	30	60
3	250	25	18	0.48	30	75	38	75
3	350	DI	18	0.48	30	75	38	75
4	250	25	6	0.52	19	75	27	70
4	350	DI	6	0.52	19	75	27	70
4	250	25	12	0.52	22	85	30	80
4	350	DI	12	0.52	22	85	30	80
4	250	25	18	0.52	30	105	38	100
4	350	DI	18	0.52	30	105	38	100
6	250	25	6	0.55	20	110	28	105
6	350	DI	6	0.55	20	110	28	105
6	250	25	12	0.55	26	135	34	130
6	350	DI	12	0.55	26	135	34	130
6	250	25	18	0.55	33	165	41	160
6	350	DI	18	0.55	33	165	41	160
8	250	25	6	0.60	21	160	29	155
8	350	DI	6	0.60	21	160	29	155
8	250	25	12	0.60	28	200	36	195
8	350	DI	12	0.60	28	200	36	195
8	250	25	18	0.60	35	245	43	240
8	350	DI	18	0.60	35	245	43	240
10	250	25	6	0.68	22	220	30	220
10	350	DI	6	0.68	22	220	30	220
10	250	25	12	0.68	30	280	38	280
10	350	DI	12	0.68	30	280	38	280
10	250	25	18	0.68	38	340	46	340
10	350	DI	18	0.68	38	340	46	340
12	250	25	6	0.75	26	320	34	320
12	350	DI	6	0.75	26	320	34	320
12	250	25	12	0.75	37	420	45	420
12	350	DI	12	0.75	37	420	45	420
12	250	25	18	0.75	48	520	56	520
12	350	DI	18	0.75	48	520	56	520
14	150	25	6	0.66	27	380	35	365
14	250	25	6	0.82	27	435	35	420
14	350	DI	6	0.66	27	380	35	365
14	150	25	12	0.66	38	480	46	465
14	250	25	12	0.82	38	560	46	545
14	350	DI	12	0.66	38	480	46	465
14	150	25	18	0.66	49	585	57	570
14	250	25	18	0.82	49	680	57	665
14	350	DI	18	0.66	49	585	57	570
16	150	30	6	0.70	27	460	35	440
16	250	30	6	0.89	27	535	35	515
16	350	DI	6	0.70	27	460	35	440
16	150	30	12	0.70	40	600	48	580
16	250	30	12	0.89	40	715	48	690
16	350	DI	12	0.70	40	600	48	580
16	150	30	18	0.70	50	710	58	690
16	250	30	18	0.89	50	850	58	830
16	350	DI	18	0.70	50	710	58	690

* For dimension details of mechanical-joint bells, see Table 10.1; for dimension details of plain ends, see Table 10.2.
† Pressure rating.
‡ Weight does not include accessory weights. See Table 10.1 for accessory weights.
§ Ductile iron.

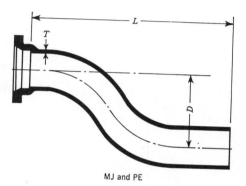

MJ and MJ

MJ and PE

Fig. 10.9. Mechanical-Joint Offsets (See Table 10.9)

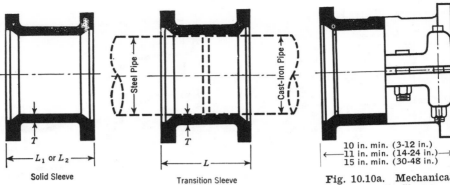

Fig. 10.10. Mechanical-Joint Sleeves
(See Table 10.10)

Solid Sleeve

Transition Sleeve

Fig. 10.10a. Mechanical Joint Split Sleeves

10 in. min. (3-12 in.)
11 in. min. (14-24 in.)
15 in. min. (30-48 in.)

Split sleeves are furnished with a pressure rating of 150 psi and can be furnished with boss and tap for service connections. Consult manufacturers for details.

TABLE 10.10

Mechanical-Joint Sleeves

Size in.	Pressure Rating psi	Iron Strength psi (1000's)	T in.	Solid Sleeves				Transition Sleeves*		
				L_1		L_2		Fits OD Steel Pipe	L	
				Length in.	Wgt.† lb	Length in.	Wgt.† lb		Length in.	Wgt.† lb
3	250	25	0.48	7.5	25	12	30	3.50	7.5	25
3	350	DI‡	0.48	7.5	25	12	30	3.50	7.5	25
4	250	25	0.52	7.5	35	12	45	4.50	7.5	35
4	350	DI	0.52	7.5	35	12	45	4.50	7.5	35
6	250	25	0.55	7.5	45	12	65	6.62	7.5	45
6	350	DI	0.55	7.5	45	12	65	6.62	7.5	45
8	250	25	0.60	7.5	65	12	85	8.62	7.5	65
8	350	DI	0.60	7.5	65	12	85	8.62	7.5	65
10	250	25	0.68	7.5	85	12	115	10.75	7.5	85
10	350	DI	0.68	7.5	85	12	115	10.75	7.5	85
12	250	25	0.75	7.5	110	12	145	12.75	7.5	110
12	350	DI	0.75	7.5	110	12	145	12.75	7.5	110
14	250	30	0.82	9.5	165	15	225			
14	350	DI	0.82	9.5	165	15	225			
16	250	30	0.89	9.5	200	15	275			
16	350	DI	0.89	9.5	200	15	275			
18	250	30	0.96	9.5	240	15	330			
18	350	DI	0.96	9.5	240	15	330			
20	250	30	1.03	9.5	275	15	380			
20	350	DI	1.03	9.5	275	15	380			
24	250	30	1.16	9.5	360	15	505			
24	350	DI	1.16	9.5	360	15	505			
30	250	30	1.37	15	745	24	1,085			
30	250	DI	1.37	15	745	24	1,085			
36	250	30	1.58	15	1,030	24	1,495			
36	250	DI	1.58	15	1,030	24	1,495			
42	250	30	1.78	15	1,330	24	1,940			
42	250	DI	1.78	15	1,330	24	1,940			
48	250	30	1.96	15	1,645	24	2,405			
48	250	DI	1.96	15	1,645	24	2,405			

* Transition sleeves are furnished with one end designed to fit standard steel pipe and the other to fit plain-end cast-iron pipe.
† Weight does not include accessory weights. See Table 10.1 for accessory weights.
‡ Ductile Iron.

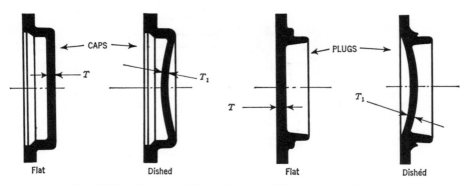

Fig. 10.11. Mechanical-Joint Caps and Plugs (See Table 10.11)

TABLE 10.11

Mechanical-Joint Caps and Plugs

Size in.	Pressure Rating psi	Iron Strength psi (1000's)	Caps				Plugs			
			Dimensions—in.		Weight—lb		Dimensions—in.		Weight—lb	
			T	T_1	Flat	Dished*	T	T_1	Flat	Dished*
3	250	25	0.50	0.48	12	12	0.50	0.48	10	10
3	350	DI†	0.50	0.48	12	12	0.50	0.48	10	10
4	250	25	0.60	0.52	15	15	0.60	0.52	15	15
4	350	DI	0.60	0.52	15	15	0.60	0.52	15	15
6	250	25	0.65	0.55	25	25	0.65	0.55	25	25
6	350	DI	0.65	0.55	25	25	0.65	0.55	25	25
8	250	25	0.70	0.60	45	45	0.70	0.60	45	45
8	350	DI	0.70	0.60	45	45	0.70	0.60	45	45
10	250	25	0.75	0.68	60	60	0.75	0.68	65	70
10	350	DI	0.75	0.68	60	60	0.75	0.68	65	70
12	250	25	0.75	0.75	80	80	0.75	0.75	85	90
12	350	DI	0.75	0.75	80	80	0.75	0.75	85	90
14	250	30	1.00	0.82	130	115	1.00	0.82	120	120
14	250	DI	0.82	0.66	120	110	0.82	0.66	115	115
16	250	30	1.11	0.89	175	155	1.11	0.89	155	150
16	250	DI	0.89	0.70	155	150	0.89	0.70	145	145
18	250	30	1.25	0.96	225	215	1.25	0.96	200	190
18	250	DI	0.96	0.75	195	185	0.96	0.75	185	180
20	250	30	1.40	1.03	285	250	1.40	1.03	255	215
20	250	DI	1.03	0.80	240	200	1.03	0.80	225	200
24	250	30	1.50	1.16	400	370	1.50	1.16	390	350
24	250	DI	1.16	0.89	345	300	1.16	0.89	335	290
30	150	30		1.37		680		1.37		660
30	250	DI		1.03		590		1.03		575
36	150	30		1.58		1,005		1.58		975
36	250	DI		1.15		850		1.15		815
42	150	30		1.78		1,535		1.78		1,355
42	250	DI		1.28		1,180		1.28		1,110
48	150	30		1.96		1,950		1.96		1,810
48	250	DI		1.42		1,595		1.42		1,455

* All plugs and caps 30 in. and larger are "dished."
† Ductile Iron.

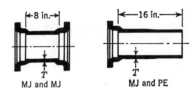

Fig. 10.12. Mechanical-Joint Connecting Pieces (See Table 10.12)

TABLE 10.12

Mechanical-Joint Connecting Pieces*

Size in.	Pressure Rating psi	Iron Strength psi (1000's)	T—in.	Weight—lb† MJ & MJ	Weight—lb† MJ & PE‡
3	250	25	0.48	35	35
3	350	DI§	0.48	35	35
4	250	25	0.52	45	45
4	350	DI	0.52	45	45
6	250	25	0.55	70	65
6	350	DI	0.55	70	65
8	250	25	0.60	95	95
8	350	DI	0.60	95	95
10	250	25	0.68	130	125
10	350	DI	0.68	130	125
12	250	25	0.75	165	165
12	350	DI	0.75	165	165
14	150	25	0.66	220	205
14	250	25	0.82	235	220
14	350	DI	0.66	220	205
16	150	30	0.70	270	250
16	250	30	0.89	290	270
16	350	DI	0.70	270	250
18	150	30	0.75	325	300
18	250	30	0.96	350	325
18	350	DI	0.75	325	300
20	150	30	0.80	390	360
20	250	30	1.03	420	385
20	350	DI	0.80	390	360
24	150	30	0.89	515	475
24	250	30	1.16	555	515
24	350	DI	0.89	515	475
30	150	30	1.03	840	730
30	250	30	1.37	905	795
30	250	DI	1.03	840	730
36	150	30	1.15	1,170	1,005
36	250	30	1.58	1,270	1,105
36	250	DI	1.15	1,170	1,005
42	150	30	1.28	1,500	1,295
42	250	30	1.78	1,635	1,430
42	250	DI	1.28	1,500	1,295
48	150	30	1.42	1,910	1,640
48	250	30	1.96	2,075	1,810
48	250	DI	1.42	1,910	1,640

* For dimensional details of mechanical joints, see Table 10.1; plain ends, Table 10.2.
† Weight does not include accessory weights. See Table 10.1 for accessory weights.
‡ May be furnished from centrifugally cast pipe.
§ Ductile iron.

Fig. 10.13. Connecting Pieces, One-End Flanged (See Table 10.13)

TABLE 10.13

Connecting Pieces, One-End Flanged*

Size in.	Pressure Rating psi	Iron Strength psi (1000's)	T—in.	Weight—lb†	
				MJ & Flg.‡	Flg. & PE‡
3	250	25	0.48	30	30
3	250	DI§	0.48	30	30
4	250	25	0.52	40	40
4	250	DI	0.52	40	40
6	250	25	0.55	60	55
6	250	DI	0.55	60	55
8	250	25	0.60	85	85
8	250	DI	0.60	85	85
10	250	25	0.68	115	115
10	250	DI	0.68	115	115
12	250	25	0.75	155	155
12	250	DI	0.75	155	155
14	150	25	0.66	195	180
14	250	25	0.82	210	195
14	250	DI	0.66	195	180
16	150	30	0.70	240	220
16	250	30	0.89	260	240
16	250	DI	0.70	240	220
18	150	30	0.75	280	255
18	250	30	0.96	305	280
18	250	DI	0.75	280	255
20	150	30	0.80	340	305
20	250	30	1.03	365	335
20	250	DI	0.80	340	305
24	150	30	0.89	455	415
24	250	30	1.16	495	455
24	250	DI	0.89	455	415
30	150	30	1.03	760	600
30	250	30	1.37	840	665
30	250	DI	1.03	760	600
36	150	30	1.15	1,070	830
36	250	30	1.58	1,195	930
36	250	DI	1.15	1,070	830
42	150	30	1.28	1,505	1,115
42	250	30	1.78	1,685	1,250
42	250	DI	1.28	1,505	1,115
48	150	30	1.42	1,885	1,390
48	250	30	1.96	2,140	1,560
48	250	DI	1.42	1,885	1,390

* For dimensional details of mechanical joints, see Table 10.1; plain ends, Table 10.3; flanges, Table 10.14.
† Weight does not include accessory weights. See Table 10.1 for accessory weights.
‡ May be furnished from centrifugally cast pipe.
§ Ductile Iron.

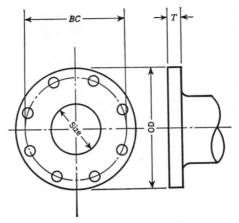

Fig. 10.14. Flange Details (See Table 10.14)

TABLE 10.14

Flange Details

Size in.	Dimensions—in.					No. of Bolts
	OD	BC	T	Bolt Hole Diam.	Bolt Diam. & Length—in.	
3	7.50	6.00	$0.75 \pm .12$	$\frac{3}{4}$	$\frac{5}{8} \times 2\frac{1}{2}$	4
4	9.00	7.50	$0.94 \pm .12$	$\frac{3}{4}$	$\frac{5}{8} \times 3$	8
6	11.00	9.50	$1.00 \pm .12$	$\frac{7}{8}$	$\frac{3}{4} \times 3\frac{1}{2}$	8
8	13.50	11.75	$1.12 \pm .12$	$\frac{7}{8}$	$\frac{3}{4} \times 3\frac{1}{2}$	8
10	16.00	14.25	$1.19 \pm .12$	1	$\frac{7}{8} \times 4$	12
12	19.00	17.00	$1.25 \pm .12$	1	$\frac{7}{8} \times 4$	12
14	21.00	18.75	$1.38 \pm .19$	$1\frac{1}{8}$	$1 \times 4\frac{1}{2}$	12
16	23.50	21.25	$1.44 \pm .19$	$1\frac{1}{8}$	$1 \times 4\frac{1}{2}$	16
18	25.00	22.75	$1.56 \pm .19$	$1\frac{1}{4}$	$1\frac{1}{8} \times 5$	16
20	27.50	25.00	$1.69 \pm .19$	$1\frac{1}{4}$	$1\frac{1}{8} \times 5$	20
24	32.00	29.50	$1.88 \pm .19$	$1\frac{3}{8}$	$1\frac{1}{4} \times 5\frac{1}{2}$	20
30	38.75	36.00	$2.12 \pm .25$	$1\frac{3}{8}$	$1\frac{1}{4} \times 6\frac{1}{2}$	28
36	46.00	42.75	$2.38 \pm .25$	$1\frac{5}{8}$	$1\frac{1}{2} \times 7$	32
42	53.00	49.50	$2.62 \pm .25$	$1\frac{5}{8}$	$1\frac{1}{2} \times 7\frac{1}{2}$	36
48	59.50	56.00	$2.75 \pm .25$	$1\frac{5}{8}$	$1\frac{1}{2} \times 8$	44

NOTE: For other requirements see Sec. 10–14.

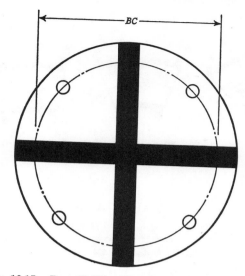

Fig. 10.15. Base Drilling Details (See Table 10.15)

TABLE 10.15

*Base Drilling Details**

Nom. Diam. in.	Dimensions—in.			Approx. Wgt. of Base—lb.	
	BC	Bolt Hole Diam.	No. of Bolts	Bends	Tees
3	3.88	$\frac{5}{8}$	4	10	5
4	4.75	$\frac{3}{4}$	4	10	10
6	5.50	$\frac{3}{4}$	4	20	10
8	7.50	$\frac{3}{4}$	4	40	15
10	7.50	$\frac{3}{4}$	4	45	30
12	9.50	$\frac{7}{8}$	4	65	30
14	9.50	$\frac{7}{8}$	4	70	45
16	9.50	$\frac{7}{8}$	4	75	50
18	11.75	$\frac{7}{8}$	4	115	50
20	11.75	$\frac{7}{8}$	4	120	75
24	11.75	$\frac{7}{8}$	4	130	75
30	14.25	1	4	190	80
36	17.00	1	4	250	120
42	21.25	$1\frac{1}{8}$	4	410	160
48	22.75	$1\frac{1}{4}$	4	515	270
					335

* Bases are not faced or drilled unless so specified in the purchase order.

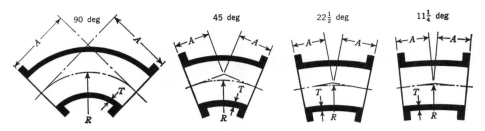

Fig. 10.16. Flanged Bends (See Table 10.16)

3
0
0

TABLE 10.16

*Flanged Bends**

Size in.	PR† psi	Iron Strength psi (1000's)	T	90 deg A	90 deg R	45 deg A	45 deg R	22½ deg A	22½ deg R	11¼ deg A	11¼ deg R	90 deg	45 deg	22½ deg	11¼ deg
							Dimensions—in.						Weight—lb		
3	250	25	0.48	5.5	4.0	3.0	3.62	3.0	7.56	3.0	15.25	25	20	20	20
4	250	25	0.52	6.5	4.5	4.0	4.81	4.0	10.06	4.0	20.31	45	40	40	40
6	250	25	0.55	8.0	6.0	5.0	7.25	5.0	15.06	5.0	30.50	65	55	55	55
8	250	25	0.60	9.0	7.0	5.5	8.44	5.5	17.62	5.5	35.50	105	90	90	90
10	250	25	0.68	11.0	9.0	6.5	10.88	6.5	22.62	6.5	45.69	165	130	135	135
12	250	25	0.75	12.0	10.0	7.5	13.25	7.5	27.62	7.5	55.81	235	195	205	205
14	150	25	0.66	14.0	11.5	7.5	12.06	7.5	25.12	7.5	50.75	290	220	225	225
14	250	25	0.82	14.0	11.5	7.5	12.06	7.5	25.12	7.5	50.75	330	245	250	255
14	250	DI‡	0.66	14.0	11.5	7.5	12.06	7.5	25.12	7.5	50.75	290	220	225	225
16	150	30	0.70	15.0	12.5	8.0	13.25	8.0	27.62	8.0	55.81	370	280	285	285
16	250	30	0.89	15.0	12.5	8.0	13.25	8.0	27.62	8.0	55.81	430	315	325	325
16	250	DI	0.70	15.0	12.5	8.0	13.25	8.0	27.62	8.0	55.81	370	280	285	285
18	150	30	0.75	16.5	14.0	8.5	14.50	8.5	30.19	8.5	60.94	450	325	335	335
18	250	30	0.96	16.5	14.0	8.5	14.50	8.5	30.19	8.5	60.94	530	375	385	385
18	250	DI	0.75	16.5	14.0	8.5	14.50	8.5	30.19	8.5	60.94	450	325	335	335
20	150	30	0.80	18.0	15.5	9.5	16.88	9.5	35.19	9.5	71.06	580	430	435	435
20	250	30	1.03	18.0	15.5	9.5	16.88	9.5	35.19	9.5	71.06	685	485	505	505
20	250	DI	0.80	18.0	15.5	9.5	16.88	9.5	35.19	9.5	71.06	580	430	435	435
24	150	30	0.89	22.0	18.5	11.0	18.12	11.0	37.69	11.0	76.12	900	630	640	645
24	250	30	1.16	22.0	18.5	11.0	18.12	11.0	37.69	11.0	76.12	1,085	730	755	760
24	250	DI	0.89	22.0	18.5	11.0	18.12	11.0	37.69	11.0	76.12	900	630	640	645
30	150	30	1.03	25.0	21.5	15.0	27.75	15.0	57.81	15.0	116.75	1,430	1,120	1,135	1,150
30	250	30	1.37	25.0	21.5	15.0	27.75	15.0	57.81	15.0	116.75	1,755	1,335	1,385	1,395
30	250	DI	1.03	25.0	21.5	15.0	27.75	15.0	57.81	15.0	116.75	1,430	1,120	1,135	1,150
36	150	30	1.15	28.0	24.5	18.0	35.00	18.0	72.88	18.0	147.25	2,135	1,755	1,790	1,805
36	250	30	1.58	28.0	24.5	18.0	35.00	18.0	72.88	18.0	147.25	2,690	2,155	2,235	2,250
36	250	DI	1.15	28.0	24.5	18.0	35.00	18.0	72.88	18.0	147.25	2,135	1,755	1,790	1,805
42	150	30	1.28	31.0	27.5	21.0	42.25	21.0	88.00	21.0	177.69	3,055	2,600	2,665	2,680
42	250	30	1.78	31.0	27.5	21.0	42.25	21.0	88.00	21.0	177.69	3,880	3,240	3,365	3,390
42	250	DI	1.28	31.0	27.5	21.0	42.25	21.0	88.00	21.0	177.69	3,055	2,600	2,665	2,680
48	150	30	1.42	34.0	30.5	24.0	49.50	24.0	103.06	24.0	208.12	4,095	3,580	3,665	3,695
48	250	30	1.96	34.0	30.5	24.0	49.50	24.0	103.06	24.0	208.12	5,210	4,485	4,660	4,690
48	250	DI	1.42	34.0	30.5	24.0	49.50	24.0	103.06	24.0	208.12	4,095	3,580	3,665	3,695

* Dimension details of flanges are shown in Table 10.14.
† Pressure rating.
‡ Ductile iron.

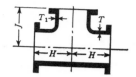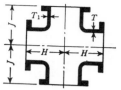

Fig. 10.17. Flanged Tees and Crosses (See Table 10.17)

TABLE 10.17

*Flanged Tees and Crosses**

Size—*in.*		Pressure Rating *psi*	Iron Strength *psi* (*1000's*)	Dimensions—*in.*				Weight—*lb*	
Run	Branch			T	T₁	H	J	Tee	Cross
3	3	250	25	0.48	0.48	5.5	5.5	40	50
4	3	250	25	0.52	0.48	6.5	6.5	60	70
4	4	250	25	0.52	0.52	6.5	6.5	65	80
6	3	250	25	0.55	0.48	8.0	8.0	85	95
6	4	250	25	0.55	0.52	8.0	8.0	90	110
6	6	250	25	0.55	0.55	8.0	8.0	95	120
8	4	250	25	0.60	0.52	9.0	9.0	140	155
8	6	250	25	0.60	0.55	9.0	9.0	145	165
8	8	250	25	0.60	0.60	9.0	9.0	155	195
10	4	250	25	0.68	0.52	11.0	11.0	205	220
10	6	250	25	0.68	0.55	11.0	11.0	215	240
10	8	250	25	0.68	0.60	11.0	11.0	225	265
10	10	250	25	0.80	0.80	11.0	11.0	270	330
12	4	250	25	0.75	0.52	12.0	12.0	290	310
12	6	250	25	0.75	0.55	12.0	12.0	295	320
12	8	250	25	0.75	0.60	12.0	12.0	310	345
12	10	250	25	0.87	0.80	12.0	12.0	360	415
12	12	250	25	0.87	0.87	12.0	12.0	385	460
14	6	150	25	0.66	0.55	14.0	14.0	375	400
14	6	250	25	0.82	0.55	14.0	14.0	420	450
14	6	250	DI†	0.66	0.55	14.0	14.0	375	400
14	8	150	25	0.66	0.60	14.0	14.0	390	425
14	8	250	25	0.82	0.60	14.0	14.0	435	475
14	8	250	DI	0.66	0.60	14.0	14.0	390	425
14	10	150	25	0.66	0.68	14.0	14.0	400	460
14	10	250	25	0.82	0.68	14.0	14.0	450	505
14	10	250	DI	0.66	0.68	14.0	14.0	400	460
14	12	150	25	0.82	0.75	14.0	14.0	470	555
14	12	250	30	0.82	0.75	14.0	14.0	470	555
14	12	250	DI	0.66	0.75	14.0	14.0	425	505
14	14	150	25	0.82	0.82	14.0	14.0	500	595
14	14	250	30	0.82	0.82	14.0	14.0	500	595
14	14	250	DI	0.66	0.66	14.0	14.0	435	530
16	6	150	30	0.70	0.55	15.0	15.0	465	490
16	6	250	30	0.89	0.55	15.0	15.0	540	565
16	6	250	DI	0.70	0.55	15.0	15.0	465	490
16	8	150	30	0.70	0.60	15.0	15.0	475	520
16	8	250	30	0.89	0.60	15.0	15.0	555	590

* Dimension details of flanges are in Table 10.14.
† Ductile Iron.

TABLE 10.17

Flanged Tees and Crosses* (contd.)

Size—in.		Pressure Rating psi	Iron Strength psi (1000's)	Dimensions—in.				Weight—lb	
Run	Branch			T	T₁	H	J	Tee	Cross
16	8	250	DI†	0.70	0.60	15.0	15.0	475	520
16	10	150	30	0.70	0.68	15.0	15.0	495	555
16	10	250	30	0.89	0.68	15.0	15.0	565	620
16	10	250	DI	0.70	0.68	15.0	15.0	495	555
16	12	150	30	0.70	0.75	15.0	15.0	520	605
16	12	250	30	0.89	0.75	15.0	15.0	590	665
16	12	250	DI	0.70	0.75	15.0	15.0	520	605
16	14	150	30	0.89	0.82	15.0	15.0	610	700
16	14	250	35	0.89	0.82	15.0	15.0	610	700
16	14	250	DI	0.70	0.66	15.0	15.0	530	620
16	16	150	30	0.89	0.89	15.0	15.0	635	755
16	16	250	35	0.89	0.89	15.0	15.0	635	755
16	16	250	DI	0.70	0.70	15.0	15.0	550	665
18	6	150	30	0.75	0.55	13.0	15.5	480	505
18	6	250	30	0.96	0.55	13.0	15.5	560	585
18	6	250	DI	0.75	0.55	13.0	15.5	480	505
18	8	150	30	0.75	0.60	13.0	15.5	495	535
18	8	250	30	0.96	0.60	13.0	15.5	570	605
18	8	250	DI	0.75	0.60	13.0	15.5	495	535
18	10	150	30	0.75	0.68	13.0	15.5	510	560
18	10	250	30	0.96	0.68	13.0	15.5	585	630
18	10	250	DI	0.75	0.68	13.0	15.5	510	560
18	12	150	30	0.75	0.75	13.0	15.5	535	610
18	12	250	30	0.96	0.75	13.0	15.5	605	670
18	12	250	DI	0.75	0.75	13.0	15.5	535	610
18	14	150	30	0.75	0.66	16.5	16.5	630	720
18	14	250	30	0.96	0.82	16.5	16.5	740	830
18	14	250	DI	0.75	0.66	16.5	16.5	630	720
18	16	150	30	0.96	0.89	16.5	16.5	760	880
18	16	250	35	0.96	0.89	16.5	16.5	760	880
18	16	250	DI	0.75	0.70	16.5	16.5	650	765
18	18	150	30	0.96	0.96	16.5	16.5	785	915
18	18	250	35	0.96	0.96	16.5	16.5	785	915
18	18	250	DI	0.75	0.75	16.5	16.5	665	795
20	6	150	30	0.80	0.55	14.0	17.0	610	635
20	6	250	30	1.03	0.55	14.0	17.0	710	735
20	6	250	DI	0.80	0.55	14.0	17.0	610	635
20	8	150	30	0.80	0.60	14.0	17.0	620	665
20	8	250	30	1.03	0.60	14.0	17.0	720	755
20	8	250	DI	0.80	0.60	14.0	17.0	620	665
20	10	150	30	0.80	0.68	14.0	17.0	635	685
20	10	250	30	1.03	0.68	14.0	17.0	735	780
20	10	250	DI	0.80	0.68	14.0	17.0	635	685
20	12	150	30	0.80	0.75	14.0	17.0	660	735
20	12	250	30	1.03	0.75	14.0	17.0	755	820
20	12	250	DI	0.80	0.75	14.0	17.0	660	735
20	14	150	30	0.80	0.66	14.0	17.0	665	745
20	14	250	35	1.03	0.82	14.0	17.0	770	850
20	14	250	DI	0.80	0.66	14.0	17.0	665	745
20	16	150	30	0.80	0.70	18.0	18.0	810	915
20	16	250	35	1.03	0.89	18.0	18.0	950	1,065
20	16	250	DI	0.80	0.70	18.0	18.0	810	915
20	18	150	35	1.03	0.96	18.0	18.0	965	1,100
20	18	250	DI	0.80	0.75	18.0	18.0	820	945
20	20	150	35	1.03	1.03	18.0	18.0	1,005	1,175
20	20	250	DI	0.80	0.80	18.0	18.0	855	1,015

* Dimension details of flanges are in Table 10.14.
† Ductile Iron.

TABLE 10.17

Flanged Tees and Crosses (contd.)

Size—in.		Pressure Rating psi	Iron Strength psi (1000's)	Dimensions—in.				Weight—lb	
Run	Branch			T	T₁	H	J	Tee	Cross
24	6	150	30	0.89	0.55	15.0	19.0	845	875
24	6	250	30	1.16	0.55	15.0	19.0	1,000	1,025
24	6	250	DI†	0.89	0.55	15.0	19.0	845	875
24	8	150	30	0.89	0.60	15.0	19.0	860	895
24	8	250	30	1.16	0.60	15.0	19.0	1,010	1,045
24	8	250	DI	0.89	0.60	15.0	19.0	860	895
24	10	150	30	0.89	0.68	15.0	19.0	880	930
24	10	250	30	1.16	0.68	15.0	19.0	1,020	1,065
24	10	250	DI	0.89	0.68	15.0	19.0	880	930
24	12	150	30	0.89	0.75	15.0	19.0	890	960
24	12	250	30	1.16	0.75	15.0	19.0	1,040	1,100
24	12	250	DI	0.89	0.75	15.0	19.0	890	960
24	14	150	30	0.89	0.66	15.0	19.0	900	975
24	14	250	30	1.16	0.82	15.0	19.0	1,050	1,125
24	14	250	DI	0.89	0.66	15.0	19.0	900	975
24	16	150	30	0.89	0.70	15.0	19.0	915	1,010
24	16	250	35	1.16	0.89	15.0	19.0	1,070	1,160
24	16	250	DI	0.89	0.70	15.0	19.0	915	1,010
24	18	150	30	0.89	0.75	22.0	22.0	1,220	1,365
24	18	250	35	1.16	0.96	22.0	22.0	1,470	1,620
24	18	250	DI	0.89	0.75	22.0	22.0	1,220	1,365
24	20	150	35	1.16	1.03	22.0	22.0	1,510	1,695
24	20	250	DI	0.89	0.80	22.0	22.0	1,255	1,430
24	24	150	35	1.16	1.16	22.0	22.0	1,585	1,850
24	24	250	DI	0.89	0.89	22.0	22.0	1,330	1,570
30	12	150	30	1.03	0.75	18.0	23.0	1,490	1,565
30	12	250	30	1.37	0.75	18.0	23.0	1,780	1,840
30	12	250	DI	1.03	0.75	18.0	23.0	1,490	1,565
30	14	150	30	1.03	0.66	18.0	23.0	1,490	1,570
30	14	250	30	1.37	0.82	18.0	23.0	1,790	1,865
30	14	250	DI	1.03	0.66	18.0	23.0	1,490	1,570
30	16	150	30	1.03	0.70	18.0	23.0	1,505	1,605
30	16	250	30	1.37	0.89	18.0	23.0	1,810	1,900
30	16	250	DI	1.03	0.70	18.0	23.0	1,505	1,605
30	18	150	30	1.03	0.75	18.0	23.0	1,515	1,615
30	18	250	35	1.37	0.96	18.0	23.0	1,815	1,910
30	18	250	DI	1.03	0.75	18.0	23.0	1,515	1,615
30	20	150	30	1.03	0.80	18.0	23.0	1,540	1,670
30	20	250	35	1.37	1.03	18.0	23.0	1,840	1,960
30	20	250	DI	1.03	0.80	18.0	23.0	1,540	1,670
30	24	150	35	1.37	1.16	25.0	25.0	2,475	2,695
30	24	250	DI	1.03	0.89	25.0	25.0	2,025	2,245
30	30	150	35	1.37	1.37	25.0	25.0	2,615	2,980
30	30	250	DI	1.03	1.03	25.0	25.0	2,150	2,500
36	12	150	30	1.15	0.75	20.0	26.0	2,170	2,240
36	12	250	30	1.58	0.75	20.0	26.0	2,670	2,725
36	12	250	DI	1.15	0.75	20.0	26.0	2,170	2,240
36	14	150	30	1.15	0.66	20.0	26.0	2,175	2,240
36	14	250	30	1.58	0.82	20.0	26.0	2,680	2,740
36	14	250	DI	1.15	0.66	20.0	26.0	2,175	2,240
36	16	150	30	1.15	0.70	20.0	26.0	2,185	2,270
36	16	250	30	1.58	0.89	20.0	26.0	2,690	2,765
36	16	250	DI	1.15	0.70	20.0	26.0	2,185	2,270
36	18	150	30	1.15	0.75	20.0	26.0	2,190	2,280
36	18	250	30	1.58	0.96	20.0	26.0	2,695	2,770
36	18	250	DI	1.15	0.75	20.0	26.0	2,190	2,280
36	20	150	30	1.15	0.80	20.0	26.0	2,210	2,325
36	20	250	35	1.58	1.03	20.0	26.0	2,715	2,810

* Dimension details of flanges are in Table 10.14.
† Ductile Iron.

TABLE 10.17

Flanged Tees and Crosses* (contd.)

Size—*in.*		Pressure Rating *psi*	Iron Strength *psi* (1000's)	Dimensions—*in.*				Weight—*lb*	
Run	Branch			T	T₁	H	J	Tee	Cross
36	20	250	DI†	1.15	0.80	20.0	26.0	2,210	2,325
36	24	150	30	1.15	0.89	20.0	26.0	2,255	2,405
36	24	250	35	1.58	1.16	20.0	26.0	2,750	2,880
36	24	250	DI	1.15	0.89	20.0	26.0	2,255	2,405
36	30	150	35	1.58	1.37	28.0	28.0	3,745	4,025
36	30	250	DI	1.15	1.03	28.0	28.0	3,000	3,300
36	36	150	35	1.58	1.58	28.0	28.0	3,930	4,405
36	36	250	DI	1.15	1.15	28.0	28.0	3,160	3,620
42	12	150	30	1.28	0.75	23.0	30.0	3,165	3,240
42	12	250	30	1.78	0.75	23.0	30.0	3,950	4,005
42	12	250	DI	1.28	0.75	23.0	30.0	3,165	3,240
42	14	150	30	1.28	0.66	23.0	30.0	3,170	3,240
42	14	250	30	1.78	0.82	23.0	30.0	3,960	4,020
42	14	250	DI	1.28	0.66	23.0	30.0	3,170	3,240
42	16	150	30	1.28	0.70	23.0	30.0	3,180	3,270
42	16	250	30	1.78	0.89	23.0	30.0	3,970	4,045
42	16	250	DI	1.28	0.70	23.0	30.0	3,180	3,270
42	18	150	30	1.28	0.75	23.0	30.0	3,185	3,275
42	18	250	30	1.78	0.96	23.0	30.0	3,970	4,045
42	18	250	DI	1.28	0.75	23.0	30.0	3,185	3,275
42	20	150	30	1.28	0.80	23.0	30.0	3,205	3,320
42	20	250	30	1.78	1.03	23.0	30.0	3,990	4,080
42	20	250	DI	1.28	0.80	23.0	30.0	3,205	3,320
42	24	150	30	1.78	1.16	23.0	30.0	4,020	4,135
42	24	250	DI	1.28	0.89	23.0	30.0	3,245	3,395
42	30	150	30	1.78	1.37	31.0	31.0	5,225	5,445
42	30	250	DI	1.28	1.03	31.0	31.0	4,125	4,375
42	36	150	DI	1.28	1.15	31.0	31.0	4,265	4,655
42	36	250	DI	1.78	1.58	31.0	31.0	5,360	5,720
42	42	150	DI	1.28	1.28	31.0	31.0	4,470	5,065
42	42	250	DI	1.78	1.78	31.0	31.0	5,580	6,155
48	12	150	30	1.42	0.75	26.0	34.0	4,315	4,390
48	12	250	30	1.96	0.75	26.0	34.0	5,425	5,480
48	12	250	DI	1.42	0.75	26.0	34.0	4,315	4,390
48	14	150	30	1.42	0.66	26.0	34.0	4,315	4,385
48	14	250	30	1.96	0.82	26.0	34.0	5,435	5,495
48	14	250	DI	1.42	0.66	26.0	34.0	4,315	4,385
48	16	150	30	1.42	0.70	26.0	34.0	4,330	4,415
48	16	250	30	1.96	0.89	26.0	34.0	5,445	5,515
48	16	250	DI	1.42	0.70	26.0	34.0	4,330	4,415
48	18	150	30	1.42	0.75	26.0	34.0	4,330	4,420
48	18	250	30	1.96	0.96	26.0	34.0	5,445	5,515
48	18	250	DI	1.42	0.75	26.0	34.0	4,330	4,420
48	20	150	30	1.42	0.80	26.0	34.0	4,350	4,460
48	20	250	30	1.96	1.03	26.0	34.0	5,460	5,545
48	20	250	DI	1.42	0.80	26.0	34.0	4,350	4,460
48	24	150	30	1.42	0.89	26.0	34.0	4,385	4,535
48	24	250	30	1.96	1.16	26.0	34.0	5,485	5,595
48	24	250	DI	1.42	0.89	26.0	34.0	4,385	4,535
48	30	150	30	1.96	1.37	26.0	34.0	5,540	5,705
48	30	250	DI	1.42	1.03	26.0	34.0	4,455	4,670
48	36	150	30	1.96	1.58	34.0	34.0	7,035	7,310
48	36	250	DI	1.42	1.15	34.0	34.0	5,555	5,880
48	42	150	DI	1.42	1.28	34.0	34.0	5,720	6,215
48	42	250	DI	1.96	1.78	34.0	34.0	7,195	7,630
48	48	150	DI	1.42	1.42	34.0	34.0	5,900	6,570
48	48	250	DI	1.96	1.96	34.0	34.0	7,385	8,005

* Dimension details of flanges are in Table 10.14.
† Ductile Iron.

Fig. 10.18. Flanged Base Bends (See Table 10.18)

For other dimensions of base bends, see Table 10.16.

TABLE 10.18

Flanged Base Bends

Size in.	Pressure Rating psi	Iron Strength psi (1000's)	Dimensions—in.				Weight—lb	
			R*	S Diam.	T	U	Base Fitting	Base Only
3	250	25	4.88	5.00	0.56	0.50	35	10
4	250	25	5.50	6.00	0.62	0.50	55	10
6	250	25	7.00	7.00	0.69	0.62	85	20
8	250	25	8.38	9.00	0.94	0.88	145	40
10	250	25	9.75	9.00	0.94	0.88	210	45
12	250	25	11.25	11.00	1.00	1.00	300	65
14	150	25	12.50	11.00	1.00	1.00	360	70
14	250	25	12.50	11.00	1.00	1.00	400	70
14	250	DI†	12.50	11.00	1.00	1.00	360	70
16	150	30	13.75	11.00	1.00	1.00	445	75
16	250	30	13.75	11.00	1.00	1.00	505	75
16	250	DI	13.75	11.00	1.00	1.00	445	75
18	150	30	15.00	13.50	1.12	1.12	565	115
18	250	30	15.00	13.50	1.12	1.12	645	115
18	250	DI	15.00	13.50	1.12	1.12	565	115
20	150	30	16.00	13.50	1.12	1.12	700	120
20	250	30	16.00	13.50	1.12	1.12	805	120
20	250	DI	16.00	13.50	1.12	1.12	700	120
24	150	30	18.50	13.50	1.12	1.12	1,030	130
24	250	30	18.50	13.50	1.12	1.12	1,215	130
24	250	DI	18.50	13.50	1.12	1.12	1,030	130
30	150	30	23.00	16.00	1.19	1.15	1,625	190
30	250	30	23.00	16.00	1.19	1.15	1,945	190
30	250	DI	23.00	16.00	1.19	1.15	1,625	190
36	150	30	26.00	19.00	1.25	1.15	2,385	250
36	250	30	26.00	19.00	1.25	1.15	2,940	250
36	250	DI	26.00	19.00	1.25	1.15	2,385	250
42	150	30	30.00	23.50	1.44	1.28	3,465	410
42	250	30	30.00	23.50	1.44	1.28	4,290	410
42	250	DI	30.00	23.50	1.44	1.28	3,465	410
48	150	30	34.00	25.00	1.56	1.42	4,610	515
48	250	30	34.00	25.00	1.56	1.42	5,725	515
48	250	DI	34.00	25.00	1.56	1.42	4,610	515

* Dimension R is a finished dimension; unfinished bases will be $\frac{1}{8}$ in. longer; details for base drilling are given in Table 10.15.
† Ductile Iron.

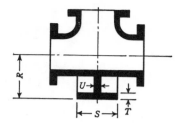

Fig. 10.19. Flanged Base Tees (See Table 10.19)

For other dimensions of base tees, see Table 10.17.

TABLE 10.19

Flanged Base Tees

Size in.	Pressure Rating psi	Iron Strength psi (1000's)	Dimensions—in.				Weight—lb	
			R*	S Diam.	T	U	Base Fitting	Base Only
3	250	25	4.88	5.00	0.56	0.50	45	5
4	250	25	5.50	6.00	0.62	0.50	75	10
6	250	25	7.00	7.00	0.69	0.62	110	15
8	250	25	8.38	9.00	0.94	0.88	185	30
10	250	25	9.75	9.00	0.94	0.88	300	30
12	250	25	11.25	11.00	1.00	1.00	430	45
14	150	25	12.50	11.00	1.00	1.00	550	50
14	250	30	12.50	11.00	1.00	1.00	550	50
14	250	DI†	12.50	11.00	1.00	1.00	485	50
16	150	30	13.75	11.00	1.00	1.00	685	50
16	250	35	13.75	11.00	1.00	1.00	685	50
16	250	DI	13.75	11.00	1.00	1.00	600	50
18	150	30	15.00	13.50	1.12	1.12	860	75
18	250	35	15.00	13.50	1.12	1.12	860	75
18	250	DI	15.00	13.50	1.12	1.12	740	75
20	150	35	16.00	13.50	1.12	1.12	1,080	75
20	250	DI	16.00	13.50	1.12	1.12	930	75
24	150	35	18.50	13.50	1.12	1.12	1,665	80
24	250	DI	18.50	13.50	1.12	1.12	1,410	80
30	150	35	23.00	16.00	1.19	1.15	2,735	120
30	250	DI	23.00	16.00	1.19	1.15	2,270	120
36	150	35	26.00	19.00	1.25	1.15	4,090	160
36	250	DI	26.00	19.00	1.25	1.15	3,320	160
42	150	DI	30.00	23.50	1.44	1.28	4,740	270
42	250	DI	30.00	23.50	1.44	1.28	5,850	270
48	150	DI	34.00	25.00	1.56	1.42	6,235	335
48	250	DI	34.00	25.00	1.56	1.42	7,720	335

* Dimension R is a finished dimension; unfinished bases will be ⅛ in. longer; details for base drilling are given in Table 10.15.
† Ductile Iron.

TABLE 10.20
*Flanged Reducers**

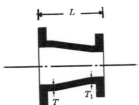

Fig. 10.20. Flanged Reducers
(See Table 10.20)

Size—in. Large	Small	Pressure Rating psi	Iron Strength psi (1000's)	T Large End	T₁ Small End	L	Wgt. lb
4	3	250	25	0.52	0.48	7	30
6	3	250	25	0.55	0.48	9	40
6	4	250	25	0.55	0.52	9	45
8	4	250	25	0.60	0.52	11	65
8	6	250	25	0.60	0.55	11	75
10	4	250	25	0.68	0.52	12	85
10	6	250	25	0.68	0.55	12	90
10	8	250	25	0.68	0.60	12	110
12	4	250	25	0.75	0.52	14	120
12	6	250	25	0.75	0.55	14	130
12	8	250	25	0.75	0.60	14	145
12	10	250	25	0.75	0.68	14	170
14	6	150	25	0.66	0.55	16	155
14	6	250	25	0.82	0.55	16	165
14	6	250	DI†	0.66	0.55	16	155
14	8	150	25	0.66	0.60	16	175
14	8	250	25	0.82	0.60	16	185
14	8	250	DI	0.66	0.60	16	175
14	10	150	25	0.66	0.68	16	190
14	10	250	25	0.82	0.68	16	205
14	10	250	DI	0.66	0.68	16	190
14	12	150	25	0.66	0.75	16	220
14	12	250	25	0.82	0.75	16	235
14	12	250	DI	0.66	0.75	16	220
16	6	150	30	0.70	0.55	18	190
16	6	250	30	0.89	0.55	18	210
16	6	250	DI	0.70	0.55	18	190
16	8	150	30	0.70	0.60	18	210
16	8	250	30	0.89	0.60	18	230
16	8	250	DI	0.70	0.60	18	210
16	10	150	30	0.70	0.68	18	235
16	10	250	30	0.89	0.68	18	255
16	10	250	DI	0.70	0.68	18	235
16	12	150	30	0.70	0.75	18	265
16	12	250	30	0.89	0.75	18	285
16	12	250	DI	0.70	0.75	18	265
16	14	150	30	0.70	0.66	18	280
16	14	250	30	0.89	0.82	18	315
16	14	250	DI	0.70	0.66	18	280
18	8	150	30	0.75	0.60	19	240
18	8	250	30	0.96	0.60	19	265
18	8	250	DI	0.75	0.60	19	240
18	10	150	30	0.75	0.68	19	265
18	10	250	30	0.96	0.68	19	290
18	10	250	DI	0.75	0.68	19	265
18	12	150	30	0.75	0.75	19	295
18	12	250	30	0.96	0.75	19	320
18	12	250	DI	0.75	0.75	19	295
18	14	150	30	0.75	0.66	19	310
18	14	250	30	0.96	0.82	19	350
18	14	250	DI	0.75	0.66	19	310
18	16	150	30	0.75	0.70	19	340
18	16	250	30	0.96	0.89	19	385
18	16	250	DI	0.75	0.70	19	340
20	10	150	30	0.80	0.68	20	310
20	10	250	30	1.03	0.68	20	340
20	10	250	DI	0.80	0.68	20	310
20	12	150	30	0.80	0.75	20	345
20	12	250	30	1.03	0.75	20	375
20	12	250	DI	0.80	0.75	20	345
20	14	150	30	0.80	0.66	20	355
20	14	250	30	1.03	0.82	20	405
20	14	250	DI	0.80	0.66	20	355
20	16	150	30	0.80	0.70	20	390
20	16	250	30	1.03	0.89	20	445
20	16	250	DI	0.80	0.70	20	390

Size Large	Small	Pressure Rating psi	Iron Strength psi (1000's)	T Large End	T₁ Small End	L	Wgt. lb
20	18	150	30	0.80	0.75	20	410
20	18	250	30	1.03	0.96	20	470
20	18	250	DI	0.80	0.75	20	410
24	12	150	30	0.89	0.75	24	480
24	12	250	30	1.16	0.75	24	535
24	12	250	DI	0.89	0.75	24	480
24	14	150	30	0.89	0.66	24	490
24	14	250	30	1.16	0.82	24	565
24	14	250	DI	0.89	0.66	24	490
24	16	150	30	0.89	0.70	24	525
24	16	250	30	1.16	0.89	24	610
24	16	250	DI	0.89	0.70	24	525
24	18	150	30	0.89	0.75	24	550
24	18	250	30	1.16	0.96	24	645
24	18	250	DI	0.89	0.75	24	550
24	20	150	30	0.89	0.80	24	590
24	20	250	30	1.16	1.03	24	695
24	20	250	DI	0.89	0.80	24	590
30	18	150	30	1.03	0.75	30	810
30	18	250	30	1.37	0.96	30	970
30	18	250	DI	1.03	0.75	30	810
30	20	150	30	1.03	0.80	30	870
30	20	250	30	1.37	1.03	30	1,035
30	20	250	DI	1.03	0.80	30	870
30	24	150	30	1.03	0.89	30	970
30	24	250	30	1.37	1.16	30	1,155
30	24	250	DI	1.03	0.89	30	970
36	20	150	30	1.15	0.80	36	1,230
36	20	250	30	1.58	1.03	36	1,495
36	20	250	DI	1.15	0.80	36	1,230
36	24	150	30	1.15	0.89	36	1,345
36	24	250	30	1.58	1.16	36	1,635
36	24	250	DI	1.15	0.89	36	1,345
36	30	150	30	1.15	1.03	36	1,555
36	30	250	30	1.58	1.37	36	1,905
36	30	250	DI	1.15	1.03	36	1,555
42	24	150	30	1.28	0.89	42	1,820
42	24	250	30	1.78	1.16	42	2,235
42	24	250	DI	1.28	0.89	42	1,820
42	30	150	30	1.28	1.03	42	2,060
42	30	250	30	1.78	1.37	42	2,555
42	30	250	DI	1.28	1.03	42	2,060
42	36	150	30	1.28	1.15	42	2,345
42	36	250	30	1.78	1.58	42	2,935
42	36	250	DI	1.28	1.15	42	2,345
48	30	150	30	1.42	1.03	48	2,625
48	30	250	30	1.96	1.37	48	3,270
48	30	250	DI	1.42	1.03	48	2,625
48	36	150	30	1.42	1.15	48	2,950
48	36	250	30	1.96	1.58	48	3,710
48	36	250	DI	1.42	1.15	48	2,950
48	42	150	30	1.42	1.28	48	3,320
48	42	250	30	1.96	1.78	48	4,190
48	42	250	DI	1.42	1.28	48	3,320

* Dimension details of flanges are given in Table 10.14. Eccentric reducers with the same dimensions and weights given for concentric reducers are available from most manufacturers if specified on the purchase order
† Ductile Iron.

Appendix A

Flanged Fittings—Bolts, Gaskets, and Installation

*This appendix is for information only and is not a part
of ANSI/AWWA C110.*

The bolts and gaskets to be used with the flanged fittings are to be selected by the purchaser with due consideration for the particular pressure service and installation requirements.

Bolts and nuts. Size, length, and number of bolts are shown in Table 10.14. Bolts conform to ANSI B18.2.1 and nuts conform to ANSI B18.2.2. Bolts smaller than ¾ in. have either standard square or heavy hex heads and heavy hex nuts. Bolts ¾ in. and larger have either square or hex heads and either hex or heavy hex nuts. Bolts and nuts are threaded in accordance with American National Standard B1.1 for "Screw Threads—Coarse Thread Series," class 2A external and class 2B internal. Bolts and nuts are low-carbon steel and conform to the chemical and mechanical requirements of ASTM A307, grade B. The carbon steel bolts should be used where gray-iron flanges are installed with flat ring gaskets that extend only to the bolts. Higher strength bolts may properly be used where gray-iron flanges are installed with full-face gaskets. Higher strength bolts may be used where ductile flanges are installed with either ring- or full-face gaskets.

Gaskets. Gaskets are rubber, either ring or full face, and are ⅛ in. thick, unless otherwise specified by the purchaser, conforming to the dimensions shown in Table A1, "Flange Gasket Details."

Installation. The design, assembly, and installation of the flanged piping system are the responsibility of the purchaser. The following suggestions are for general guidance:

a. The underground use of the flanged joint is generally not desirable because of the rigidity of the joint.

b. Flange faces should bear uniformly on the gasket, and bolts should be tightened uniformly.

c. Users of flanged fittings should be careful to prevent bending or torsional strains from being applied to cast flanges or flanged fittings. Piping systems must be designed so that piping connected to flanges or flanged fittings is properly anchored, supported, or restrained to prevent breakage of fittings and flanges.

TABLE A1

Flange Gasket Details

Nom. Size in.	Ring		Full Face				No. of Holes
	Nom. ID	OD	Nom. ID	OD	BC	Bolt Hole Diam.	
3	3	$5\frac{3}{8}$	3	$7\frac{1}{2}$	6	$\frac{3}{4}$	4
4	4	$6\frac{7}{8}$	4	9	$7\frac{1}{2}$	$\frac{3}{4}$	8
6	6	$8\frac{3}{4}$	6	11	$9\frac{1}{2}$	$\frac{7}{8}$	8
8	8	11	8	$13\frac{1}{2}$	$11\frac{3}{4}$	$\frac{7}{8}$	8
10	10	$13\frac{3}{8}$	10	16	$14\frac{1}{4}$	1	12
12	12	$16\frac{1}{8}$	12	19	17	1	12
14	14	$17\frac{3}{4}$	14	21	$18\frac{3}{4}$	$1\frac{1}{8}$	12
16	16	$20\frac{1}{4}$	16	$23\frac{1}{2}$	$21\frac{1}{4}$	$1\frac{1}{8}$	16
18	18	$21\frac{5}{8}$	18	25	$22\frac{3}{4}$	$1\frac{1}{4}$	16
20	20	$23\frac{7}{8}$	20	$27\frac{1}{2}$	25	$1\frac{1}{4}$	20
24	24	$28\frac{1}{4}$	24	32	$29\frac{1}{2}$	$1\frac{3}{8}$	20
30	30	$34\frac{3}{4}$	30	$38\frac{3}{4}$	36	$1\frac{3}{8}$	28
36	36	$41\frac{1}{4}$	36	46	$42\frac{3}{4}$	$1\frac{5}{8}$	32
42	42	48	42	53	$49\frac{1}{2}$	$1\frac{5}{8}$	36
48	48	$54\frac{1}{2}$	48	$59\frac{1}{2}$	56	$1\frac{5}{8}$	44

Appendix B

Special Fittings

This appendix is for information only and is not a part of ANSI/AWWA C110.

These special fittings are not a part of the standard but are available. For dimensions and pressure ratings, consult your supplier.

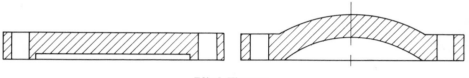

Blind Flanges

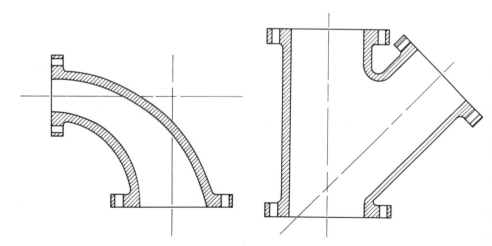

Flanged 90-deg Reducing Bends   Flange Wye Branches

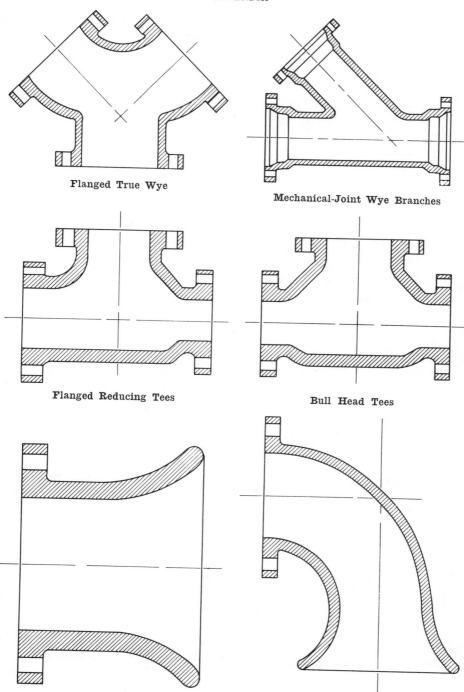

Flanged True Wye

Mechanical-Joint Wye Branches

Flanged Reducing Tees

Bull Head Tees

Flange and Flares

Flange and Flare 90-deg Bends

312

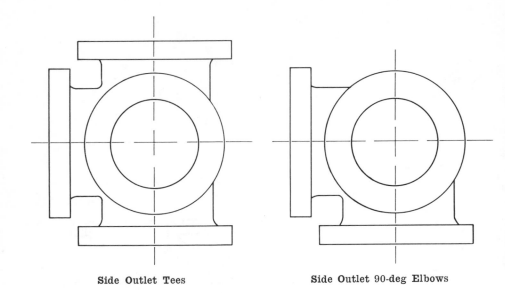

Side Outlet Tees   Side Outlet 90-deg Elbows

Wall Pipe

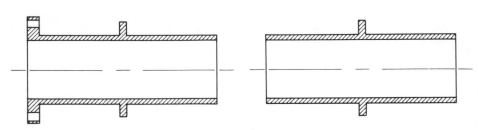

Flange and Plain End   Plain End and Plain End

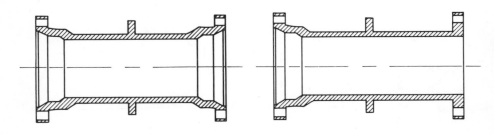

MJ and MJ   MJ and Flange

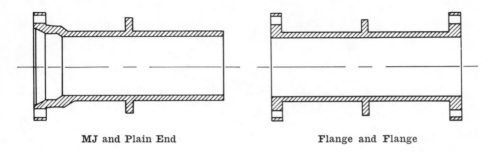

| MJ and Plain End | Flange and Flange |

NOTE: Wall pipe can be furnished with tapped holes in flanges or mechanical-joint bells if required. Wall sleeves (other than those shown in Fig. 10.10) are also available on special request.

Appendix C

Metrication

This appendix is for information only and is not a part of ANSI/AWWA C110.

It is recommended that metric units not be shown opposite US customary units in this standard because of the extensive space that would be required. Of the 54 pages in this standard, 40 include tables and, therefore, it might be expected that metrication would result in a total printed standard of 100 or more pages. This would increase the price of the standard drastically and it might well serve to delay its publication. It is hoped that in future revisions some of the material can be consolidated.

This recommendation is based on the interests of those who use the standard from the standpoints of economy, efficiency, and practicality.

SECTION VI
Joints for Gray & Ductile Iron Pipe

Introduction

Mechanical and push-on joints covered by ANSI Standard A21.11 (AWWA C111) comply with Federal Specification WW-P-421c.

The American National Standard for Flanged Pipe with Threaded Flanges, ANSI A21.15, (AWWA C115) was adopted in 1975.

Special types of pipe and fittings shown in this Section are available from individual manufacturers for special application. The manufacturer should be consulted for information on design and capabilities of these special products.

PUSH ON
JOINT
ASSEMBLY

ANSI A21.11–1972
(AWWA C111–72)
Revision of
ANSI A21.11–1964
(AWWA C111–64)

AMERICAN NATIONAL STANDARD

for

RUBBER–GASKET JOINTS FOR CAST–IRON AND DUCTILE–IRON PRESSURE PIPE AND FITTINGS

PUBLISHED BY AMERICAN WATER WORKS ASSOCIATION, INC.

SPONSORS

AMERICAN GAS ASSOCIATION
AMERICAN SOCIETY FOR TESTING AND MATERIALS
AMERICAN WATER WORKS ASSOCIATION
NEW ENGLAND WATER WORKS ASSOCIATION

Revised Edition Approved by the American National Standards Institute, Inc., Mar. 2, 1972

NOTICE

This Standard has been especially printed by the American Water Works Association for incorporation into this volume. It is current as of December 1, 1975. It should be noted, however, that all AWWA Standards are updated at least once in every five years. Therefore, before applying this Standard it is suggested that you confirm its currency with the American Water Works Association.

AMERICAN WATER WORKS ASSOCIATION
6666 West Quincy Avenue, Denver, Colorado 80235

American National Standard

Table of Contents

3
1
9

NOTICE

The corrections noted in the Errata Notice to the first printing of ANSI A21.11–1972 have been incorporated in this printing.

Foreword

This foreword is provided for information only and is not a part of ANSI A21.11–1972 (AWWA C111–72).

American National Standards Committee A21 on Cast-Iron Pipe and Fittings was organized in 1926 under the sponsorship of the American Gas Association, the American Society for Testing and Materials, the American Water Works Association, and the New England Water Works Association. The present scope of Committee A21 is

> Standardization of specifications for cast-iron and ductile-iron pressure pipe for gas, water, and other liquids, and fittings for use with such pipe; these specifications to include design, dimensions, materials, coatings, linings, joints, accessories, and methods of inspection and test.

The work of Committee A21 is carried out by subcommittees. The scope of Subcommittee 2—Joints for Pipe and Fittings is:

> To include an examination of all present A21 standards for joints for pipe and fittings to determine what is needed to bring them up to date. The examinations shall include all related matters concerning joints for cast-iron pipe and fittings.

Subcommittee 2 reviewed the 1964 edition of A21.11 and submitted a proposed revision to American National Standards Committee in 1971.

Major Revisions

Scope. The name of the standard has been changed to "Rubber-Gasket Joints for Cast-Iron and Ductile-Iron Pressure Pipe and Fittings." The standard now also includes 54-in.-diameter push-on joints.

Mechanical joint glands. Physical acceptance test standards for ductile-iron mechanical joint glands are now included in the standard. However, because of their excellent service record and because neither additional strength nor ductility is required for the service intended, cast-iron glands will be furnished with ductile-iron pipe and fittings unless ductile-iron glands are specified.

Gaskets. Requirements for maximum compression set and resistance to ozone cracking of gasket rubber for mechanical and push-on joints have been added to the standard. ASTM methods of testing for these requirements are also listed in the standard.

Performance requirements of the push-on joint. Because the push-on joints covered by this standard are not standardized, but are furnished in accordance with the manufacturer's standard design dimensions and tolerances, performance requirements are considered appropriate for inclusion in this standard. These performance requirements include the adoption of minimum working-pressure ratings for each joint size. The long and excellent service record of the mechanical joint with standardized dimensions has proved its performance ability; therefore, performance criteria are not necessary.

Mechanical joint dimensions. The manufacturer has been provided with an option to furnish ductile-iron pipe, 30 in. and larger in diameter, with bell thicknesses compatible with the wall thickness of the pipe. The bolt length for 14-in. mechanical joints has been changed from $\frac{3}{4}$ in. by 4 in. to $\frac{3}{4}$ in. by $4\frac{1}{2}$ in. to assure full nut engagement.

Note: Push-on joints for cast-iron pipe are so designed that a negative pressure cannot pull the gasket into the pipe. Testing has been performed to confirm this design parameter for joint-sealing capability under the condition of negative pressure within the pipe.

Options

This standard includes certain options that, if desired, must be specified in the invitation for bids and on the purchase order. These options are found in the following sections of the standard:

1. Inspection, subsection 11–4.2
2. Certification and Test Records, subsection 11–5.1
3. Special Requirements for the Mechanical Joint, subsections 11–6.1 and 11–6.3

Note: Subsection 11–6.3 provides for tapped holes in the bells of mechanical joints for stud bolts. This option is intended for use where headed bolts or slotted holes will not suffice, as, for example, where the bell is to be embedded in a concrete wall.

Committee Personnel

Subcommittee 2, which developed this standard, had the following personnel at that time:

FRANK J. CAMEROTA, *Chairman*   MILES R. SUCHOMEL, *Vice-Chairman*

Consumer or General Interest Members

STANLEY C. BAKER   CHARLES E. SMITH
KENNETH J. CARL   MAURICE C. STOUT
JOHN W. CARROLL

Producer Members

BRUCE I. DEDMAN   RICHARD C. WETZEL
EDWARD C. SEARS   CHARLES W. WRIGHT
ALBERT H. SMITH, JR.

Standards Committee A21 (Cast-Iron Pipe and Fittings), which reviewed and approved this standard, had the following personnel at the time of approval:

WALTER AMORY, *Chairman*
CARL A. HENRIKSON, *Vice-Chairman*
JAMES B. RAMSEY, *Secretary*
CHARLES R. VELZY, *Treasurer*

Organization Represented	*Name of Representative*
American Gas Association	LEONARD ORLANDO, JR.
American Society of Civil Engineers	KENNETH W. HENDERSON
American Society of Mechanical Engineers	CHARLES R. VELZY
American Society for Testing and Materials	(Vacant)
American Water Works Association	VANCE C. LISCHER
Cast Iron Pipe Research Association	CARL A. HENRIKSON
	EDWARD C. SEARS
	W. HARRY SMITH
Individual Producers	FRANK J. CAMEROTA
	WILLIAM T. MAHER
Manufacturers' Standardization Society of the Valve and Fittings Industry	ABRAHAM FENSTER
Naval Facilities Engineering Command	STANLEY C. BAKER
New England Water Works Association	WALTER AMORY
Underwriters' Laboratories, Inc.	MILES R. SUCHOMEL
Canadian Standards Association	W. F. SEMENCHUK*

* Liaison representative without vote.

American National Standard for

RUBBER–GASKET JOINTS FOR CAST–IRON AND DUCTILE–IRON PRESSURE PIPE AND FITTINGS

11–1—Scope

This standard covers rubber-gasket joints of the following types for cast-iron and ductile-iron pressure pipe and fittings.

11–1.1 *Mechanical joint.* The mechanical joint is designed for pipe and fittings in sizes 2–48 in. for conveying gas, water, or other liquids.

11–1.2. *Push-on joint.* The push-on joint is designed for pipe and fittings in sizes 2–54 in. for conveying water or other liquids.

11–2—Definitions

11–2.1 *Joints and accessories.* For the purpose of this standard the word "joint" includes accessories.

11–2.2. *Mechanical joint.* The mechanical joint is a bolted joint of the stuffing-box type, as shown in Fig. 11.1. Each joint shall consist of: (1) a bell, cast integrally with the pipe or fitting and provided with an exterior flange having bolt holes or slots, and a socket with annular recesses for the sealing gasket and the plain end of the pipe or fitting; (2) a pipe or fitting plain end; (3) a sealing gasket; (4) a follower gland with bolt holes; and (5) tee-head bolts and hexagonal nuts.

11–2.3. *Push-on joint.* The push-on joint is a single rubber-gasket joint. It is assembled by the positioning of a continuous, molded, rubber ring gasket in an annular recess in the pipe or fitting socket and the forcing of the plain end of the entering pipe or fitting into the socket. The plain end compresses the gasket radially to form a positive seal. The gasket and the annular recess are so designed and shaped that the gasket is locked in place against displacement.

11–2.4. *Purchaser.* The purchaser is the party entering into a contract or agreement to purchase joints according to this standard.

11–2.5. *Manufacturer.* The manufacturer is the party entering into a contract or agreement to furnish joints according to this standard.

11–2.6. *Inspector.* The inspector is the representative of the purchaser, authorized to inspect in behalf of the purchaser to determine whether the requirements of this standard have been met.

11–3—General Requirements

11–3.1. Joints made in conformance with the provisions of this standard are intended for use on cast-iron and ductile-iron pipe and fittings manufactured in accordance with the following standards, where applicable:

ANSI A21.6–1970 (AWWA C106–70)—Cast-Iron Pipe Centrifugally

1

Cast in Metal Molds, for Water or Other Liquids

ANSI A21.7–1970—Cast-Iron Pipe Centrifugally Cast in Metal Molds, for Gas

ANSI A21.8–1970 (AWWA C108–70)—Cast-Iron Pipe Centrifugally Cast in Sand-Lined Molds, for Water or Other Liquids

ANSI A21.9–1970—Cast-Iron Pipe Centrifugally Cast in Sand-lined Molds, for Gas

Federal Specification WW–P–421c—Pipe, Cast Gray and Ductile Iron, Pressure (for Water and Other Liquids)

Federal Specification WW–P–360b—Pipe, Cast Iron, Pressure (for Gas, Water, or other Liquids)

ANSI A21.10–1971 (AWWA C110–71)—Gray-Iron and Ductile-Iron Fittings, 2–48 in., for Water and Other Liquids

ANSI A21.12–1971 (AWWA C112–71)—2-in. and 2¼-in. Cast-Iron Pipe, Centrifugally Cast, for Water or Other Liquids

ANSI A21.14–1968—Gray-Iron and Ductile-Iron Fittings, 3–24 in., for Gas

ANSI A21.51–1971 (AWWA C151–71)—Ductile-Iron Pipe, Centrifugally Cast in Metal Molds or Sand-Lined Molds, for Water or Other Liquids

ANSI A21.52–1971—Ductile-Iron Pipe, Centrifugally Cast, in Metal Molds or Sand-Lined Molds for Gas

11–3.2. The joints shall have the same pressure rating as the pipe or fitting of which they are a part.

11–3.3. A recommended method of joint assembly shall be furnished by the manufacturer on request of the purchaser.

11–3.4. Unless otherwise specified, gaskets, glands, bolts, and nuts shall be furnished with mechanical joints,

and gaskets and lubricant shall be furnished with push-on joints, all in sufficient quantity for assembly of each joint.

11–4—Inspection

11–4.1. Inspection shall be made in accordance with the provisions of the standard under which the pipe or fittings are purchased.

11–4.2. If the purchaser wishes to inspect the manufacture of glands, bolts, or gaskets that may be made by subcontractors, special arrangements therefor must be made at the time the order is placed.

11–5—Certification and Test Records

11–5.1. The manufacturer shall, if required on the purchase order, furnish a sworn statement that the inspection and all of the tests specified have been made and that the results thereof comply with the requirements of this standard.

11–5.2. A record of the specified tests of glands, bolts, and gaskets shall be retained for 1 yr and shall be available to the purchaser at the foundry.

11–6—Special Requirements for the Mechanical Joint

11–6.1. *Glands.* Unless otherwise specified, cast-iron glands shall be furnished with ductile-iron pipe and fittings. Glands shall have a bituminous coating unless otherwise specified and shall have cast or stamped upon them the manufacturer's identification, the nominal size, and the letters "DI" or word "ductile" if made of ductile iron.

11–6.1.1. *Cast-iron glands.* The acceptability of the cast iron used in the glands shall be determined by

2

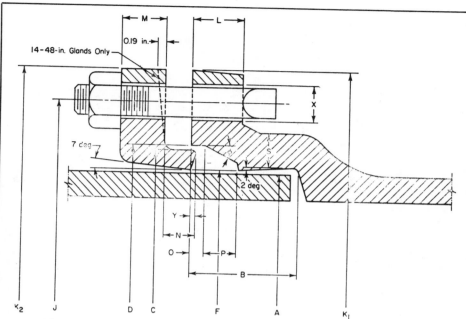

Fig. 11.1. Mechanical-Joint Dimensions for Sizes 2–48 in. (See Table 11.1)

Notes

1. The nominal thickness of the pipe bell "S" shall not be less than the nominal wall thickness of the pipe of which it is a part. The "S" dimensions shown in Table 11.1 for centrifugal pipe are for reference and were used to calculate bell weights.

2. The diameter of cored holes may be tapered an additional 0.06 in.

3. In the event of an ovalness to the outside diameter of the plain end, the mean diameter measured by a circumferential tape shall not be less than the minimum diameter shown in the table. The minor axis shall not be less than the foregoing minimum diameter plus an additional minus tolerance of 0.04 in. for sizes 8–12 in., 0.07 in. for sizes 14–24 in., and 0.10 in. for sizes 30–48 in.

4. K_1 and K_2 are the dimensions across the bolt holes. For sizes 2 and $2\frac{1}{4}$ in., both flange and gland may be oval in shape. For sizes 3–48 in., the gland may be polygon in shape.

5. Gland thickness "M" for sizes 14–48 in. may be tapered as shown at the option of the manufacturer.

6. The "L" dimension shown in Table 11.1 for 2-in. and $2\frac{1}{4}$-in. sizes applies to mechanical-joint pipe. The "L" dimension for 2-in. and $2\frac{1}{4}$-in. fittings is 0.75 in. (−0.05 in.).

7. At the manufacturer's option, ductile-iron pipe in sizes 30-in. and larger may be furnished with the bell-flange thicknesses shown in Table 11.1 (dimension L), or with reduced bell-flange thicknesses that are compatible with the wall thickness of the pipe. When reduced bell-flange thicknesses are furnished, the manufacturer shall provide details of the joint and accessories if requested by the purchaser.

TABLE 11.1

Mechanical-Joint Dimensions—in.

Size	A Plain End	B	C	D	F	φ	X	J	K₁ Centrifugal Pipe	K₁ Pit Cast Pipe and Fittings	K₂	L*	M	N	O	P	S Centrifugal Pipe	S Pit Cast Pipe and Fittings	Y	Bolts No.	Bolts Size	Bolts Length
2	±0.05 2.50	2.50	±0.05 3.39	±0.05 3.50	±0.05 2.61	28°	+0.06−0.0 3/4	±0.05 4.75	−0.05 6.00	−0.10 6.25	−0.10 6.25	−0.05 0.56	−0.05 0.62	0.50	0.31	0.63	0.37	−0.07 0.44	0.08	2	5/8	2 1/2
2 1/4	±0.05 2.75	2.50	±0.05 3.64	±0.05 3.75	±0.05 2.86	28°	+0.06−0.0 3/4	±0.05 5.00	−0.05 6.25	−0.10 6.50	−0.10 6.50	−0.05 0.56	−0.05 0.62	0.50	0.31	0.63	0.37	−0.07 0.44	0.08	2	5/8	2 1/2
3	±0.06 3.96	2.50	±0.04 4.84	+0.06−0.04 4.94	+0.07−0.03 4.06	28°	+0.06−0.0 3/4	±0.06 6.19	−0.06 7.62	−0.12 7.69	−0.12 7.69	−0.06 0.94	−0.06 0.62	0.75	0.31	0.63	0.47	−0.10 0.52	0.12	4	5/8	3
4	±0.06 4.80	2.50	±0.04 5.92	+0.06−0.04 6.02	+0.07−0.03 4.90	28°	+0.06−0.0 7/8	±0.06 7.50	−0.06 9.06	−0.12 9.12	−0.12 9.12	−0.06 1.00	−0.06 0.75	0.7	0.31	0.75	0.55	−0.10 0.65	0.12	4	3/4	3 1/2
6	±0.06 6.90	2.50	±0.04 8.02	+0.06−0.04 8.12	+0.07−0.03 7.00	28°	+0.06−0.0 7/8	±0.06 9.50	−0.06 11.06	−0.12 11.12	−0.12 11.12	−0.06 1.06	−0.06 0.88	0.75	0.31	0.75	0.60	−0.10 0.70	0.12	6	3/4	3 1/2
8	±0.06 9.05	2.50	±0.04 10.17	+0.06−0.04 10.27	+0.07−0.03 9.15	28°	+0.06−0.0 7/8	±0.06 11.75	−0.06 13.31	−0.12 13.37	−0.12 13.37	−0.08 1.12	−0.08 1.00	0.75	0.31	0.75	0.66	−0.12 0.75	0.12	6	3/4	4
10	±0.06 11.10	2.50	+0.06−0.04 12.22	+0.06−0.04 12.34	+0.07−0.03 11.20	28°	+0.06−0.0 7/8	±0.06 14.00	−0.06 15.62	−0.12 15.69	−0.12 15.62	−0.08 1.19	−0.08 1.00	0.75	0.31	0.75	0.72	−0.12 0.80	0.12	8	3/4	4
12	±0.06 13.20	2.50	+0.06−0.04 14.32	+0.06−0.04 14.44	+0.07−0.03 13.30	28°	+0.06−0.0 7/8	±0.06 16.25	−0.06 17.88	−0.12 17.94	−0.12 17.88	−0.08 1.25	−0.08 1.00	0.75	0.31	0.75	0.79	−0.12 0.85	0.12	8	3/4	4
14	+0.05−0.08 15.30	3.50	+0.07−0.05 16.40	+0.07−0.05 16.54	+0.06−0.07 15.44	28°	+0.06−0.0 7/8	±0.06 18.75	−0.08 20.25	−0.12 20.31	−0.12 20.25	−0.12 1.31	−0.12 1.25	0.75	0.31	0.75	0.85	−0.12 0.89	0.12	10	3/4	4 1/2
16	+0.05−0.08 17.40	3.50	+0.07−0.05 18.50	+0.07−0.05 18.64	+0.06 0.07 17.54	28°	+0.06−0.0 7/8	±0.06 21.00	−0.08 22.50	−0.12 22.56	−0.12 22.50	−0.12 1.38	−0.12 1.31	0.75	0.31	0.75	0.91	−0.12 0.97	0.12	12	3/4	4 1/2
18	+0.05−0.08 19.50	3.50	+0.07−0.05 20.60	+0.07−0.05 20.74	+0.06−0.07 19.64	28°	+0.06−0.0 7/8	±0.06 23.25	−0.08 24.75	−0.15 24.83	−0.15 24.75	−0.12 1.44	−0.12 1.38	0.75	0.31	0.75	0.97	−0.15 1.05	0.12	12	3/4	4 1/2
20	+0.05−0.08 21.60	3.50	+0.07−0.05 22.70	+0.07−0.05 22.84	+0.06−0.07 21.74	28°	+0.06−0.0 7/8	±0.06 25.50	−0.08 27.00	−0.15 27.08	−0.15 27.00	−0.12 1.50	−0.12 1.44	0.75	0.31	0.75	1.03	−0.15 1.12	0.12	14	3/4	4 1/2
24	+0.05−0.08 25.80	3.50	+0.07−0.05 26.90	+0.07−0.05 27.04	+0.06−0.07 25.94	28°	+0.06−0.0 7/8	±0.06 30.00	−0.08 31.50	−0.15 31.58	−0.15 31.50	−0.12 1.62	−0.12 1.56	0.75	0.31	0.75	1.08	−0.15 1.22	0.12	16	3/4	5
30	+0.08−0.06 32.00	4.00	+0.08−0.06 33.29	+0.08−0.06 33.46	+0.08−0.06 32.17	20°	+0.06−0.0 1 1/8	±0.06 36.88	−0.12 39.12	−0.18 39.12	−0.18 39.12	−0.12 1.81	−0.12 2.00	0.75	0.38	1.00	1.20	−0.15 1.50	0.12	20	1	6
36	+0.08−0.06 38.30	4.00	+0.08−0.06 39.59	+0.08−0.06 39.76	+0.08−0.06 38.47	20°	+0.06−0.0 1 1/8	±0.06 43.75	−0.12 46.00	−0.18 46.00	−0.18 46.00	−0.12 2.00	−0.12 2.00	0.75	0.38	1.00	1.35	−0.15 1.80	0.12	24	1	6
42	+0.08−0.06 44.50	4.00	+0.08−0.06 45.79	+0.08−0.06 45.96	+0.08−0.06 44.67	20°	+0.06−0.0 1 3/8	±0.06 50.62	−0.12 53.12	−0.18 53.12	−0.18 53.12	−0.12 2.00	−0.12 2.00	0.75	0.38	1.00	1.48	−0.15 1.95	0.12	28	1 1/4	6
48	+0.08−0.06 50.80	4.00	+0.08−0.06 52.09	+0.08−0.06 52.26	+0.08−0.06 50.97	20°	+0.06−0.0 1 3/8	±0.06 57.50	−0.12 60.00	−0.18 60.00	−0.18 60.00	−0.12 2.00	−0.12 2.00	0.75	0.38	1.00	1.61	−0.15 2.20	0.12	32	1 1/4	6

tests made on bars cast from the same iron as the glands. The test bars shall be ASTM standard bars, cast and tested in accordance with ASTM A48–64 for tensile strength or A438–68 for transverse breaking load. At the option of the manufacturer, either the tensile test or the transverse test may be used as the acceptance test. The required properties are given in the following table:

Class of Iron	Bar Diam. in.	Span in.	Min. Breaking Load—lb (1,000's)	Min. Tensile psi (1,000's)
25	1.2	18	2	25
25	2	24	6.8	25

11–6.1.2. *Ductile-iron glands.* The standard acceptance test for the physical characteristics of ductile-iron glands shall be a tensile test from coupons cast from the same iron as the glands. The coupons shall be cast and the tests made in accordance with ASTM A536–67. The ductile iron from which the glands are cast shall have a minimum elongation of 3 per cent.

11–6.2. *Dimensions and tolerances.* The dimensions of the bell, socket, plain end, and gland lip and the diameter and location of the bolt holes shall be gauged at sufficiently frequent intervals to assure compliance with the dimensions shown in Fig. 11.1 and Table 11.1.

11–6.3. *Bolt holes or slots.* When necessary for the insertion of bolts, the bell flange shall have slots of the same width as the diameter of the bolt holes. When specified, the bell flange shall be furnished with holes tapped for stud bolts.

11–6.4. *Gaskets.* Gasket dimensions shall conform to the dimensions and tolerances shown in Fig. 11.2 and Table 11.2. The size, mold number, gasket manufacturer's mark, the letters "MJ" (or "SM"), and the year of manufacture shall be molded in the rubber as shown in Fig. 11.2.

Rubber gaskets shall be vulcanized natural or vulcanized synthetic rubber, free of porous areas, foreign materials, and visible defects. No reclaimed rubber shall be used.

Quality-control procedures shall be utilized to assure that gaskets meet the requirements of this standard. The manufacturer shall retain monthly reports of representative quality-control tests results for gaskets manufactured that month.

The required properties of the gasket rubber and the required methods of test are given in the following table:

Property	ASTM Test Method	Required Value
Hardness Durometer "A"	D2240-68	75 ± 5
Min. ultimate tensile—*psi*	D412-68	1,500
Min. ultimate elongation—*per cent**	D412-68	150
Min. aging—*per cent†*	D572-67‡	60
Max. compression set—*per cent*	D395-67 method B	20
Resistance to ozone cracking	D-1149-64§	No cracking

* Of original length.
† Of original values for tensile and ultimate elongation.
‡ Oxygen pressure method, after 96 hr at 70C ± 1 deg at 300 psi ± 10.
§ After a minimum of 25 hr exposure in 50-pphm ozone concentration at 104F on a loop-mounted gasket with approximately 20 per cent elongation at outer surface.

11–6.5. *Bolts and nuts.* Dimensions of tee-head bolts and hexagonal nuts shall comply with the dimensions and tolerances shown in Fig. 11.3 and Table 11.3. At the manufacturer's option, they shall be made of either

3
2
7

5

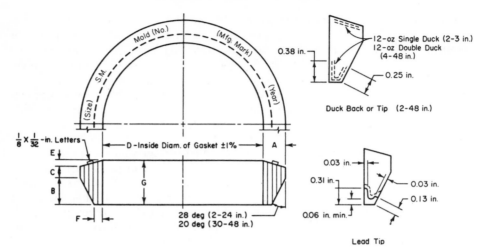

Fig. 11.2. Mechanical-Joint Gasket, 2–48 in. (See Table 11.2 and Notes)

Notes

1. Tipped or backed gaskets may be made in the same mold as plain rubber gaskets, but the inside diameter of such reinforced portions shall not exceed the "pipe OD."

2. The duck for tips and backs shall be frictioned before molding.

TABLE 11.2

2–48-in. Mechanical-Joint Gasket Dimensions—in.

Pipe Size	Pipe OD	Dimensions of Plain Rubber Gaskets						
		A ±0.01 in.	B	C	D +1 per cent −1 per cent	E	F ±0.01 in.	G ±0.02 in.
2	2.50	0.48	0.62	0.31	2.48	0.12	0.15	1.05
2¼	2.75	0.48	0.62	0.31	2.72	0.12	0.15	1.05
3	3.96	0.48	0.62	0.31	3.86	0.12	0.15	1.05
4	4.80	0.62	0.75	0.31	4.68	0.16	0.22	1.22
6	6.90	0.62	0.75	0.31	6.73	0.16	0.22	1.22
8	9.05	0.62	0.75	0.31	8.85	0.16	0.22	1.22
10	11.10	0.62	0.75	0.31	10.87	0.16	0.22	1.22
12	13.20	0.62	0.75	0.31	12.95	0.16	0.22	1.22
14	15.30	0.62	0.75	0.31	14.99	0.16	0.22	1.22
16	17.40	0.62	0.75	0.31	17.07	0.16	0.22	1.22
18	19.50	0.62	0.75	0.31	19.13	0.16	0.22	1.22
20	21.60	0.62	0.75	0.31	21.20	0.16	0.22	1.22
24	25.80	0.62	0.75	0.31	25.34	0.16	0.22	1.22
30	32.00	0.73	1.00	0.38	31.47	0.16	0.37	1.54
36	38.30	0.73	1.00	0.38	37.67	0.16	0.37	1.54
42	44.50	0.73	1.00	0.38	43.78	0.16	0.37	1.54
48	50.80	0.73	1.00	0.38	49.98	0.16	0.37	1.54

high-strength cast iron containing a minimum of 0.50 per cent copper, or high-strength, low-alloy steel. The steel shall have the following characteristics:

Characteristic	Value
Min. yield strength—*psi*	45,000
Min. elongation in 2 in.—*per cent*	20
Max. content—*per cent*	
Carbon	0.20
Manganese	1.25
Sulfur	0.05
Min. content—*per cent*	
Nickel	0.25
Copper	0.20
Combined (Ni, Cu, Cr)	1.25

11–6.5.1. *Threads.* The design of internal and external threads shall conform to ANSI B1.1–1960—Unified Screw Threads, and to B1.2–1966—Screw Thread Gages and Gaging. Thread form shall conform to the standards and to the dimensions of the coarse-thread series (UNC) Unified Coarse; external threads shall be made in compliance with Class 2A limits, and internal threads shall be made in compliance with Class 2B limits. Bolts shall be threaded concentric to the longitudinal axis of the shank. Nuts shall be tapped concentric to the vertical axis and at right angles to the load surfaces within a tolerance of 2 deg to insure axial loading.

11–6.5.2. *Proof test.* Statistical quality-control procedures shall be utilized to assure that bolts and nuts meet the specified test loads without permanent stretch. Samples of assembled bolts and nuts shall be proof-tested in tension to the load values designated. For testing, the nuts shall be assembled flush with the end of the bolts. The load shall be applied without impact between the nut and the bolt head in a suitable machine that will insure axial loading. The

specified test loads shall not break the nut or bolt or permanently stretch the bolt. Permanent stretch is defined as 0.002 in./in. of bolt length. Assembled bolts and nuts shall be tested at the following load values, which have been determined on the basis of a 45,000-psi stress at the root of the thread:

Bolt Diameter *in.*	Load *lb*
$\frac{5}{8}$	9,000
$\frac{3}{4}$	13,500
1	24,500
$1\frac{1}{4}$	40,000

11–6.5.3. *Workmanship.* Bolt shanks shall be straight within $\frac{1}{16}$ in./6 in. of length. The two load-bearing surfaces of the bolt heads shall be in a common plane that shall be at right angles to the bolt shank.

Bolts and nuts shall be sound, clean, and coated with a rust-resistant lubricant; their surfaces shall be free of objectionable protrusions that would interfere with their fit in the made-up mechanical joint.

11–6.5.4. *Packing.* The nuts shall be assembled on the bolts for packing. They shall be packed in suitable containers that shall be plainly marked with the manufacturer's name, and the size, quantity, and weight of the contents.

11–7—Special Requirements for the Push-on Joint

11–7.1. *Drawings.* The manufacturer shall furnish drawings of the joint and gasket, if requested by the purchaser.

11–7.2. *Dimensions and tolerances.* The dimensions of the bell, socket, and plain end shall be in accordance with the manufacturer's standard design dimensions and tolerances. Such di-

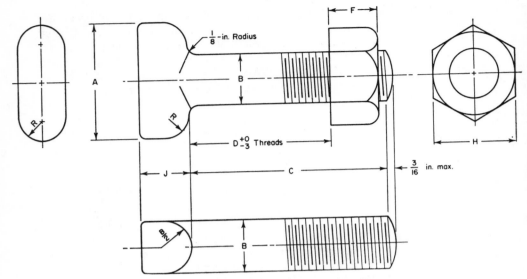

Fig. 11.3. Mechanical-Joint Bolts and Nuts

Notes

1. Dimension B is unthreaded shank.
2. Dimension D is measured to face of nut run up finger tight.
3. Draft, when required to be 6 deg maximum, may be deducted from bolt head dimensions, and radius $B/2$ may be changed to suit draft.
4. Gates, if required, may protrude a maximum of $\frac{1}{8}$ in. above the top of the bolt head.

TABLE 11.3

Mechanical-Joint Bolt and Nut Dimensions—in.

Nom. Size	A ± 0.05	B ± 0.03	$C +.25$ -0.06	D	$E*$	F	H	$J +0.15$ -0.03	R Max.
$\frac{5}{8} \times 2\frac{1}{2}$	1.50	0.625	2.5	1.25	11	0.625 ± 0.04	$1.062 +0.00$ -0.04	0.625	0.312
$\frac{5}{8} \times 3$	1.50	0.625	3.0	1.75	11	0.625 ± 0.04	$1.062 +0.00$ -0.04	0.625	0.312
$\frac{3}{4} \times 3\frac{1}{2}$	1.75	0.750	3.5	1.75	10	0.750 ± 0.06	$1.250 +0.00$ -0.06	0.750	0.375
$\frac{3}{4} \times 4$	1.75	0.750	4.0	2.25	10	0.750 ± 0.06	$1.250 +0.00$ -0.06	0.750	0.375
$\frac{3}{4} \times 4\frac{1}{2}$	1.75	0.750	4.5	2.75	10	0.750 ± 0.06	$1.250 +0.00$ -0.06	0.750	0.375
$\frac{3}{4} \times 5$	1.75	0.750	5.0	3.25	10	0.750 ± 0.06	$1.250 +0.00$ -0.06	0.750	0.375
1×6	2.25	1.000	6.0	3.75	8	1.000 ± 0.08	$1.625 +0.00$ -0.08	1.000	0.500
$1\frac{1}{4} \times 6$	2.50	1.250	6.0	3.75	7	1.250 ± 0.08	$2.000 +0.00$ -0.08	1.250	0.625

* Number of threads per inch [Coarse-Thread Series (ANSIB1.1—Unified Standard for Screw Threads) Class 2A, External Fit UNC2A and Class 2B, UNC2B(ANSIB1.2—Standard for Gages and Gaging)].

mensions shall be gaged at sufficiently frequent intervals to assure dimensional control.

11–7.3. *Gaskets.* Gasket dimensions shall be in accordance with the manufacturer's standard design dimensions and tolerances. The gasket shall be of such size and shape as to provide an adequate compressive force against the spigot and socket after assembly to effect a positive seal under all combinations of joint and gasket tolerances. The trade name or trademark, size, mold number, gasket manufacturer's mark, and year of manufacture shall be molded on the gaskets. Markings shall not be located on the sealing surfaces.

Gaskets shall be vulcanized natural or vulcanized synthetic rubber. No reclaimed rubber shall be used. When two hardnesses of rubber are included in a gasket, the soft and hard portions shall be integrally molded and joined in a strong vulcanized bond. Gaskets shall be free of porous areas, foreign material, and visible defects.

The required properties of the gasket rubber and the required method of test are given in the following table:

ard. The manufacturer shall retain monthly reports of representative quality-control test results for gaskets manufactured in that month.

11–7.4. *Lubricant.* The lubricant shall be suitable for lubricating the parts of the joint for assembly. The lubricant shall be nontoxic, shall not support the growth of bacteria, and shall have no deteriorating effects on the gasket material. It shall not impart taste or odor to water in a pipe that has been flushed in accordance with AWWA C601–68—Standard for Disinfecting Water Mains. The lubricant containers shall be labeled with the trade name or trademark and the pipe manufacturer's name.

11–7.5. *Marking.* Pipe and fittings having push-on joints shall be marked with the proprietary name or trademark of the joint.

11–8—Performance Requirements of the Push-on Joint

The manufacturer shall have qualified the design of his joint by having performed the tests given in this section and shall have records to show the results of these tests.

Property	ASTM Test Method	Main Body of Gasket	Harder Portion (if Used)
Nominal hardness Durometer "A"	D2240-68	50–65	80–85
Tolerance on nominal hardness		±5	±5
Min. ultimate tensile—*psi*	D412-68	2,000	1,200
Min. ultimate elongation—*per cent**	D412-68	300	125
Min. aging—*per cent*†	D572-67‡	60	—
Max. compression set—*per cent*	D395-67 method B	20	—
Resistance to ozone cracking	D-1149-64§	No cracking	—

* Of original length.
† Of original values of tensile and ultimate elongation.
‡ Oxygen pressure method: after 96 hr at 70 C ± 1° at 300 psi ± 10.
§ After a minimum of 25 hr exposure in 50-pphm ozone concentration at 104 F on a loop-mounted gasket with approximately 20 per cent elongation at outer surface.

Quality-control procedures shall be utilized to assure that the gaskets will meet the requirements of this stand-

11–8.1. *Working-pressure ratings.* The working-pressure rating of the push-on joint is established by subjecting representative sizes to hydro-static pressures of twice the rated working pressure of the pipe or fitting with which the joint is to be used but in no event less than twice the minimum working-pressure rating shown in the table at the end of this paragraph. The hydrostatic pressure shall be applied to joint samples having maximum clearances between the plain end and socket as allowed by the specified tolerances. In addition to those samples tested in the undeflected condition, some samples shall be tested while deflected to the angle recommended as a maximum by the manufacturer. At least one sample shall be tested while the plain end is offset laterally within the socket to the maximum extent permitted by the joint design. The joint shall not leak during any of these tests. The minimum working pressure ratings shall be as follows:

Joint Size—*in.*	Minimum Working-Pressure Rating—*psi*
16 and smaller	350
18	300
20–24	250
30–54	200

11–8.2. *Assembly.* Assembly of representative sizes shall be possible with the joints having minimum clearances between the plain end and socket as allowed by the specified tolerances. These joints shall be capable of being assembled and deflected to the maximum attainable angle—but not in excess of the maximum recommended by the manufacturer—without damage to the gaskets, pipe, or fittings, and without displacement of the gasket from its intended position.

11–8.3. *Gaskets.* The gaskets used for qualifying the joint shall be of the same form and material as the gaskets intended for use in service.

Appendix A

Notes on Installation of Mechanical Joints

These notes are not part of the standard, but are given for information.

The successful operation of the mechanical joint specified requires that the spigot be centrally located in the bell and that adequate anchorage shall be provided where abrupt changes in direction and dead ends occur.

The rubber gasket seals most effectively (particularly when sealing gas) if the surfaces with which it comes in contact are cleaned thoroughly (for example, with a wire brush) just prior to assembly to remove all loose rust or foreign material. Lubrication and additional cleaning are provided by brushing both the gasket and the spigot (as with soapy water for example) just prior to slipping the gasket onto the spigot and assembling the joint.

For water and gas service the normal range of bolt torques to be applied and the lengths of wrenches, that should satisfactorily produce the ranges of torques are given in the table.

When tightening bolts, it is essential that the gland be brought up toward the pipe flange evenly, maintaining approximately the same distance between the gland and the face of the flange at all points around the socket. This may be done by partially tightening the bottom bolt first, then the top bolt, next the bolts at either side, and finally the remaining bolts. This cycle should be repeated until all bolts are within the range of torques shown in the table. If effective sealing is not attained at the maximum torque indicated, the joint should be disassembled, thoroughly cleaned, and reassembled. Overstressing of bolts to compensate for poor installation practice is to be avoided.

Pipe Size— in.	Bolt Size— in.	Range of Torque— ft-lb	Length of Wrench— in.*
2–3	$\frac{5}{8}$	45–60	8
4–24	$\frac{3}{4}$	75–90	10
30–36	1	85–100	12
42–48	$1\frac{1}{4}$	105–120	14

* The torque loads may be applied with torque-measuring or torque-indicating wrenches, which may also be used to check the application of approximate torque loads applied by a man trained to give an average pull on a definite length of regular socket wrench.

ANSI A21.15–1975
(AWWA C115–75)

First Edition

AMERICAN NATIONAL STANDARD

for

FLANGED CAST-IRON AND DUCTILE-IRON PIPE WITH THREADED FLANGES

ADMINISTRATIVE SECRETARIAT

AMERICAN WATER WORKS ASSOCIATION

CO-SECRETARIATS

AMERICAN GAS ASSOCIATION
NEW ENGLAND WATER WORKS ASSOCIATION

Approved by American National Standards Institute, Inc. May 28, 1975.

NOTICE

This Standard has been especially printed by the American Water Works Association for incorporation into this volume. It is current as of December 1, 1975. It should be noted, however, that all AWWA Standards are updated at least once in every five years. Therefore, before applying this Standard it is suggested that you confirm its currency with the American Water Works Association.

PUBLISHED BY

AMERICAN WATER WORKS ASSOCIATION
6666 West Quincy Avenue, Denver, Colorado 80235

3
3
6

American National Standard

An American National Standard implies a consensus of those substantially concerned with its scope and provisions. An American National Standard is intended as a guide to aid the manufacturer, the consumer, and the general public. The existence of an American National Standard does not in any respect preclude anyone, whether he has approved the standard or not, from manufacturing, marketing, purchasing, or using products, processes, or procedures not conforming to the standard. American National Standards are subject to periodic review, and users are cautioned to obtain the latest editions. Producers of goods made in conformity with an American National Standard are encouraged to state on their own responsibility in advertising, promotion material, or on tags or labels that the goods are produced in conformity with particular American National Standards.

CAUTION NOTICE. This American National Standard may be revised or withdrawn at any time. The procedures of the American National Standards Institute require that action be taken to reaffirm, revise, or withdraw this standard no later than five (5) years from the date of publication. Purchasers of American National Standards may receive current information on all standards by calling or writing the American National Standards Institute, 1430 Broadway, New York, N. Y. 10018, (212) 868-1220.

Table of Contents

Committee Personnel

Subcommittee 1, Pipe, which developed this standard, had the following personnel at that time:

EDWARD C. SEARS, *Chairman*
WALTER AMORY, *Vice-Chairman*

User Members

ROBERT S. BRYANT
FRANK E. DOLSON
GEORGE F. KEENAN
LEONARD ORLANDO JR.
JOHN E. PERRY

Producer Members

W. D. GOODE
CARL A. HENRIKSON
THOMAS D. HOLMES
SIDNEY P. TEAGUE

Standards Committee A21, Cast-Iron Pipe and Fittings, which reviewed and approved this standard, had the following personnel at the time of approval:

LLOYD W. WELLER, *Chairman*
CARL A. HENRIKSON, *Vice-Chairman*
JAMES B. RAMSEY, *Secretary*

Organization Represented	*Name of Representative*
American Gas Association	LEONARD ORLANDO JR.
American Society of Civil Engineers	KENNETH W. HENDERSON
American Society of Mechanical Engineers	JAMES S. VANICK
American Society for Testing and Materials	ALBERT H. SMITH JR.
American Water Works Association	ARNOLD M. TINKEY
	LLOYD W. WELLER
Cast Iron Pipe Research Association	CARL A. HENRIKSON
	EDWARD C. SEARS
	W. HARRY SMITH
	ALFRED F. CASE
Individual Producer	
	ABRAHAM FENSTER
Manufacturers' Standardization Society of the Valve and Fittings Industry	
New England Water Works Association	WALTER AMORY
Naval Facilities Engineering Command	STANLEY C. BAKER
Underwriters' Laboratories, Inc.	JOHN E. PERRY
Canadian Standards Association	W. F. SEMENCHUK*

---

* Liaison representative without vote.

Foreword

This foreword is provided for information only and is not a part of ANSI A21.15 (AWWA C115).

I—History of Standard

American National Standard Committee A21 on Cast-Iron Pipe and Fittings was organized in 1926 under the sponsorship of the American Gas Association, the American Society for Testing and Materials, the American Water Works Association, and the New England Water Works Association. The present scope of Committee A21 activity is

Standardization of specifications for cast-iron and ductile-iron pressure pipe for gas, water, and other liquids, and fittings for use with such pipe. These specifications to include design, dimensions, materials, coatings, linings, joints, accessories, and methods of inspection and test.

The work of Committee A21 is conducted by subcommittees. The directive to Subcommittee 1—Pipe is that

The scope of the subcommittee activity shall include the periodic review of all current A21 standards for pipe, the preparation of revisions and new standards where needed, as well as other matters pertaining to pipe standards.

Flanged fittings are covered in ANSI-A21.10 (AWWA C110). The flanged pipe used with these fittings has been purchased for many years in accordance with users' and manufacturers' standards. A need for an ANSI standard for flanged pipe has been indicated. Consequently, Subcommittee 1 submitted a proposed standard for flanged pipe to Committee A21 in 1974.

II—Options

This standard includes certain options that, if desired, must be specified on the purchase order. Also, a number of items must be specified to describe completely the pipe required. The following summarizes these details and available options and lists the sections of the standard where they can be found:

1. Size, whether cast-iron or ductile-iron, thickness or class, and finished length
2. Certification by manufacturer—Sec. 15-3.2
3. Inspection by purchaser—Sec. 15-4.1
4. Flanges for ductile-iron pipe—Sec. 15-7.3
5. Bolt-hole alignment—Sec. 15-8.4
6. Cement lining—Sec. 15-9.2. Experience has indicated that bituminous inside coating is not complete protection against loss in pipe capacity because of tuberculation. Cement linings are recommended for most waters.
7. Special coatings and linings—Sec. 15-9.5.

v

First Edition

American National Standard for

Flanged Cast-Iron and Ductile-Iron Pipe With Threaded Flanges

Sec. 15-1—Scope

This standard pertains to 3–48 in. gray-iron and 3–54 in. ductile-iron flanged pipe with threaded flanges for water or other liquids. The flanged pipe are rated for a maximum working pressure of either 150 or 250 psi as specified in the tables. All flanges are rated for a maximum working pressure of 250 psi.

Sec. 15-2—Definitions

Under this standard the following definitions shall apply :

15-2.1. *Purchaser*. The party entering into a contract or agreement to purchase flanged pipe according to this standard.

15-2.2. *Manufacturer*. The party that produces the flanged pipe.

15-2.3. *Inspector*. The representative of the purchaser, authorized to inspect in behalf of the purchaser to determine whether or not the flanged pipe meet this standard.

15-2.4. *Gray iron*. Gray iron is a cast ferrous material in which a major part of the carbon content occurs as graphite in the form of flakes interspersed throughout the metal.

15-2.5. *Ductile iron*. Ductile iron is a cast ferrous material in which the graphite present is in a spheroidal form rather than in flake form.

Sec. 15-3—Inspection and Certification by Manufacturer

15-3.1. The manufacturer shall establish the necessary quality control and inspection practice to ensure compliance with this standard.

15-3.2. The manufacturer shall, if required on the purchase order, furnish a sworn statement that the flanged pipe comply with the requirements of this standard.

15-3.3. All flanged pipe shall be clean and sound without defects that will impair their service. Repairing of defects by welding or other methods shall not be allowed if such repairs will adversely affect the serviceability of the flanged pipe or its capability to meet strength requirements of this standard.

Sec. 15-4—Inspection by Purchaser

15-4.1. If the purchaser desires to inspect flanged pipe at the manufacturer's plant, the purchaser shall so specify on the purchase order, stating the conditions (such as time and the extent of inspection) under which the inspection shall be made.

15-4.2. The inspector shall have free access to those parts of the manufacturer's plant that are necessary to ensure compliance with this standard. The manufacturer shall make available for the Inspector's use such gages as are necessary for inspection. The manufacturer shall provide the inspector with assistance as necessary for the handling of flanged pipe.

Sec. 15-5—Delivery and Acceptance

All flanged pipe shall comply with this standard. Flanged pipe not complying with this standard shall be replaced by the manufacturer at the agreed point of delivery. The manufacturer shall not be liable for shortages or damaged pipe after acceptance at the agreed point of delivery except

as recorded on the delivery receipt or similar document by the carrier's agent.

Sec. 15-6—Pipe Barrel

15-6.1. Gray-iron pipe barrels shall conform to the requirements of ANSI A21.6 (AWWA C106) or ANSI A21.8 (AWWA C108) of latest revision, and ductile-iron pipe barrels shall conform to the requirements of ANSI A21.51 (AWWA C151) of latest revision.

15-6.2. The nominal thicknesses of gray-iron and ductile-iron flanged pipe shall not be less than those shown in Tables 15.1 and 15.2.

3
4
2

TABLE 15.1

Gray-Iron Flanged Pipe With Threaded Flanges

Nominal Pipe Size *in.*	Thickness Class*	Maximum Working Pressure *psi*	Nominal Thickness *in.*	OD *in.*	Weight—*lb*	
					One Flange Only	Pipe Barrel per ft
3	24	250	0.38	3.96	7	13.3
4	23	250	0.38	4.80	13	16.5
6	22	250	0.38	6.90	17	24.3
8	22	250	0.41	9.05	27	34.7
10	22	250	0.44	11.10	38	46.0
12	22	150	0.48	13.20	58	59.8
12	23	250	0.52	13.20	58	64.6
14	22	150	0.51	15.30	72	73.9
14	24	250	0.59	15.30	72	85.1
16	22	150	0.54	17.40	90	89.2
16	24	250	0.63	17.40	90	103.6
18	22	150	0.58	19.50	90	107.6
18	24	250	0.68	19.50	90	125.4
20	22	150	0.62	21.60	115	127.5
20	24	250	0.72	21.60	115	147.4
24	23	150	0.73	25.80	160	179.4
24	24	250	0.79	25.80	160	193.7
30	23	150	0.85	32.00	240	259.5
30	25	250	0.99	32.00	240	300.9
36	23	150	0.94	38.30	350	344.2
36	25	250	1.10	38.30	350	401.1
42	23	150	1.05	44.50	500	447.2
42	25	250	1.22	44.50	500	517.6
48	23	150	1.14	50.80	625	554.9
48	25	250	1.33	50.80	625	644.9

* ANSI-A21.6 (AWWA C106) and A21.8 (AWWA C108).
NOTE: The nominal thicknesses of gray-iron flanged pipe shall not be less than those shown in this table.

THREADED FLANGED PIPE

15-6.3. The minus thickness tolerances of pipe shall not exceed the following:

Pipe Size in.	Gray-Iron Pipe (A21.6, A21.8) in.	Ductile-Iron Pipe (A21.51) in.
3–8	0.05	0.05
10–12	0.06	0.06
14–24	0.08	0.07
30–42	0.10	0.07
48	0.10	0.08
54	—	0.09

15-6.4. Threads on the pipe barrel shall be taper pipe threads (NPT) in accordance with ANSI B2.1 adapted to the gray-iron and ductile-iron pipe outside diameters shown in Tables 15.1 and 15.2.

Sec. 15-7—Flanges

15-7.1. Flanges shall conform to the dimensions shown in Table 15.3.

15-7.2. All flanges shall have a taper pipe thread (NPT) in accordance with ANSI B2.1 adapted to the gray-iron and ductile-iron pipe outside diameters shown in Tables 15.1 and 15.2.

15-7.3. Unless otherwise specifically directed by the purchaser, flanges for ductile-iron pipe shall be ductile iron. Flanges for gray-iron pipe shall be gray iron. Flanges shall conform to the respective chemical and physical properties specified for gray-iron and ductile-iron fittings in ANSI A21.10 (AWWA C110).

Sec. 15-8—Fabrication

15-8.1. *Bolt holes.* Bolt holes shall be in accordance with the dimensions shown in Table 15.3. The bolt holes shall be equally spaced.

15-8.2. *Assembly.* Both flange and pipe threads shall be clean prior to application of thread compound. The thread compound shall give adequate lubrication and sealing properties to provide satisfactory pressure-tight

TABLE 15.2
Ductile-Iron Flanged Pipe With Threaded Flanges

Nominal Pipe Size in.	Thickness Class*	Maximum Working Pressure psi	Nominal Thickness in.	OD in.	Weight—lb	
					One Flange Only	Pipe Barrel per ft
3	3	250	0.31	3.96	7	10.9
4	3	250	0.32	4.80	13	13.8
6	3	250	0.34	6.90	17	21.4
8	3	250	0.36	9.05	27	30.1
10	3	250	0.38	11.10	38	39.2
12	3	250	0.40	13.20	58	49.2
14	3	250	0.42	15.30	72	60.1
16	3	250	0.43	17.40	90	70.1
18	3	250	0.44	19.50	90	80.6
20	3	250	0.45	21.60	115	91.5
24	3	250	0.47	25.80	160	114.4
30	3	250	0.51	32.00	240	154.4
36	3	250	0.58	38.30	350	210.3
42	3	250	0.65	44.50	500	274.0
48	3	250	0.72	50.80	625	346.6
54	3	250	0.81	57.10	760	438.3

* ANSI A21.51 (AWWA C151).
NOTE: The nominal thicknesses of ductile-iron flanged pipe shall not be less than those shown in this table.

3

joints. Threaded flanges shall be individually fitted and machine-tightened on the threaded pipe by the manufacturer.

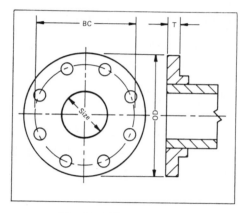

Fig. 15.1 Flange Details
See Table 15.3

NOTE: Flanges are not interchangeable in the field.

15-8.3. *Facing.* The flanges and pipe ends shall be faced after fabrication. Flanges shall be plain-faced without projection or raised-face and shall be furnished smooth or with shallow serrations. Flanges may be back-faced or spot-faced for compliance with the flange thickness tolerance specified in this standard. Bearing surfaces for bolting shall be parallel to the flange face within 3 deg.

15-8.4. *Flange alignment.* When pipe is furnished with two flanges, the bolt holes shall be aligned unless otherwise specified. Misalignment of corresponding bolt holes of the two flanges shall not exceed 0.12 in. The machined flange faces shall be perpendic-

TABLE 15.3

Threaded Flange Detail

Nominal Pipe Size in.	OD in.	BC in.	T in.	Bolt Hole Diameter in.	Bolt Diameter & Length in.	Number of Bolts
3	7.50	6.00	0.75 ± 0.12	$\frac{3}{4}$	$\frac{5}{8} \times 2\frac{1}{2}$	4
4	9.00	7.50	0.94 ± 0.12	$\frac{3}{4}$	$\frac{5}{8} \times 3$	8
6	11.00	9.50	1.00 ± 0.12	$\frac{7}{8}$	$\frac{3}{4} \times 3\frac{1}{2}$	8
8	13.50	11.75	1.12 ± 0.12	$\frac{7}{8}$	$\frac{3}{4} \times 3\frac{1}{2}$	8
10	16.00	14.25	1.19 ± 0.12	1	$\frac{7}{8} \times 4$	12
12	19.00	17.00	1.25 ± 0.12	1	$\frac{7}{8} \times 4$	12
14	21.00	18.75	1.38 ± 0.19	$1\frac{1}{8}$	$1 \times 4\frac{1}{2}$	12
16	23.50	21.25	1.44 ± 0.19	$1\frac{1}{8}$	$1 \times 4\frac{1}{2}$	16
18	25.00	22.75	1.56 ± 0.19	$1\frac{1}{4}$	$1\frac{1}{8} \times 5$	16
20	27.50	25.00	1.69 ± 0.19	$1\frac{1}{4}$	$1\frac{1}{8} \times 5$	20
24	32.00	29.50	1.88 ± 0.19	$1\frac{3}{8}$	$1\frac{1}{4} \times 5\frac{1}{2}$	20
30	38.75	36.00	2.12 ± 0.25	$1\frac{3}{8}$	$1\frac{1}{4} \times 6\frac{1}{2}$	28
36	46.00	42.75	2.38 ± 0.25	$1\frac{5}{8}$	$1\frac{1}{2} \times 7$	32
42	53.00	49.50	2.62 ± 0.25	$1\frac{5}{8}$	$1\frac{1}{2} \times 7\frac{1}{2}$	36
48	59.50	56.00	2.75 ± 0.25	$1\frac{5}{8}$	$1\frac{1}{2} \times 8$	44
54	66.25	62.75	3.00 ± 0.25	2	$1\frac{3}{4} \times 8\frac{1}{2}$	44

Facing: Flanges are plain-faced without projection or raised-face and are furnished smooth or with shallow serrations.

Back facing: Flanges may be back-faced or spot-faced for compliance with the flange thickness tolerances.

Flanges: The flanges are adequate for water service of 250-psi working pressure. The bolt circle and bolt holes of these flanges match those of Class 125 flanges shown in ANSI B16.1 and can be joined with Class 125 B16.1 flanges or Class 150 ANSI B16.5 flanges. The flanges do not match the Class 250 flanges shown in ANSI B16.1 and cannot be joined with Class 250 B16.1 flanged fittings and valves.

ular to the pipe center line and shall be parallel such that any two face-to-face dimensions 180 deg apart at the flange OD shall not differ by more than 0.06 in.

15-8.5. *Finished pipe length.* Flanged pipe shall be furnished to the lengths specified on the order. When pipe is furnished with two flanges, the face-to-face dimensions shall conform to a tolerance of ±0.12 in. The overall length of flange and plain-end pipe shall conform to a tolerance of ±0.25 in.

15-8.6. *Finished pipe weight.* The weight of any single pipe shall not be less than the calculated weight by more than 10 per cent.

Sec. 15-9—Coatings and Linings

15-9.1. *Outside coating.* Unless otherwise specified, the outside coating shall be a bituminous coating approximately 1 mil thick. The coating shall be applied to the outside of all pipe, unless otherwise specified. The finished coating shall be continuous, smooth, neither brittle when cold nor sticky when exposed to the sun, and shall be strongly adherent to the pipe.

15-9.2. *Cement–mortar linings.* If desired, cement linings shall be specified in the invitation for bids and on the purchase order. Cement linings shall be in accordance with ANSI

A21.4 (AWWA C104) of latest revision.

15-9.3. *Inside coating.* Unless otherwise specified, the inside coating for pipe that is not cement-lined shall be a bituminous material as thick as practicable (at least 1 mil) and conforming to all appropriate requirements for seal coat in ANSI A21.4 (AWWA C104) of latest revision.

15-9.4. *Flange coatings.* A rust-preventive coating shall be applied to the machined faces of the flanges. The rust-preventive coating shall be soluble in commercial solvent for ready removal before pipe installation. Unless otherwise specified, the back of the flanges and the bolt holes shall be coated with standard outside coating (see Sec. 15-9.1.).

15-9.5. *Special coatings and linings.* For special conditions, other types of coatings and linings may be available. Such special coatings and linings shall be specified in the invitation for bids and on the purchase order.

Sec. 15-10—Marking

The length and the weight shall be shown on each pipe. The manufacturer's mark and the letters *DI,* if ductile iron, shall be cast or stamped on the flanges.

Appendix

This appendix is provided for information only and is not a part of ANSI A21.15 (AWWA C115).

A1—Bolts, Gaskets, and Installation

The bolts and gaskets to be used with the flanged pipe are to be selected by the purchaser with due consideration for the particular pressure service and installation requirements.

Bolts and nuts. Size, length, and number of bolts are shown in Table 15.3. Bolts conform to ANSI B18.2.1 and nuts conform to ANSI B18.2.2. Bolts smaller than ¾ in. have either standard square or heavy hex heads and heavy hex nuts. Bolts ¾ in. and larger have either square or hex heads and either hex or heavy hex nuts. Bolts and nuts are threaded in accordance with American National Standard B1.1 for Screw Threads—Coarse Thread Series, Class 2A, external, and Class 2B, internal. Bolts and nuts of low-carbon steel conforming to the chemical and mechanical requirements of ASTM A307, Grade B are suitable for use with the flanges covered by this standard when used with the rubber gaskets covered in this appendix. Higher strength bolts should not be used when a cast-iron flange is used with a flat ring gasket.

Gaskets. Gaskets are rubber, either ring or full face, and are ⅛ in. thick, unless otherwise specified by the owner or consulting engineer, conforming to the dimensions shown in Table A1 of this appendix.

Installation. The design, assembly, and installation of the flanged piping system are the responsibility of the purchaser. The following suggestions are for general guidance:

(a) The underground use of the flanged joint is generally not desirable because of the rigidity of the joint.

(b) Flanged joints should be fitted so that the contact faces bear uniformly on the gasket and then are made up with relatively uniform bolt stress.

TABLE A1
Flange Gasket Details

Nominal Pipe Size in.	Ring		Full Face				
	Nominal ID	OD	Nominal ID	OD	BC	Bolt Hole Diameter	Number of Holes
3	3	5⅜	3	7½	6	¾	4
4	4	6⅞	4	9	7½	¾	8
6	6	8¾	6	11	9½	⅞	8
8	8	11	8	13½	11¾	⅞	8
10	10	13⅜	10	16	14¼	1	12
12	12	16⅛	12	19	17	1	12
14	14	17¾	14	21	18¾	1⅛	12
16	16	20¼	16	23½	21¼	1⅛	16
18	18	21⅝	18	25	22¾	1¼	16
20	20	23⅞	20	27½	25	1¼	20
24	24	28¼	24	32	29½	1⅜	20
30	30	34¾	30	38¾	36	1⅜	28
36	36	41¼	36	46	42¾	1⅝	32
42	42	48	42	53	49½	1⅝	36
48	48	54½	48	59½	56	1⅝	44
54	54	61	54	66¼	62¾	2	44

END OF STANDARD

6

THRUST RESTRAINT for
Underground Piping Systems

Fundamental design principles of fluid mechanics recognize the presence of unbalanced thrust forces in pressure piping systems. These forces, resulting from static and dynamic fluid action on the pipe, require physical restraint for system stabilization.

Locations where unbalanced thrust forces commonly occur are:

Bends	Wyes
Reducers	Offsets
Tees	Dead-ends
Valves	Hydrants

In addition, installations on steep slopes, in swamps, marshes, muck or peat bogs frequently require special restraining techniques for efficient anchorage.

Adequate restraint is generally achieved for ductile or cast iron piping systems by employing one or more of the following methods:

Restrained Joints
Thrust Blocks
Tie Rods

Combined Systems and Structural Connections

Soil characteristics are of prime importance in the design of thrust restraining systems. Accepted principles of soil mechanics have been applied in the derivation and formulation of the design procedures discussed herein.

Thrust Blocking

Concrete thrust blocks are the most common method of restraint now in use, providing stable soil conditions prevail and space requirements permit placement. Successful blocking is dependent upon factors such as location, availability and placement of concrete, and possible disturbance through future excavation. Concrete blocks are readily utilized in combination with tie rods, structural anchoring, thrust collars and restrained joints.

Thrust blocks are generally cate-

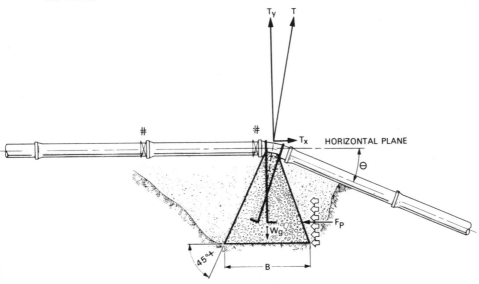

Figure 1—GRAVITY THRUST BLOCK
Restrained joints may be used when $T_x > F_p$

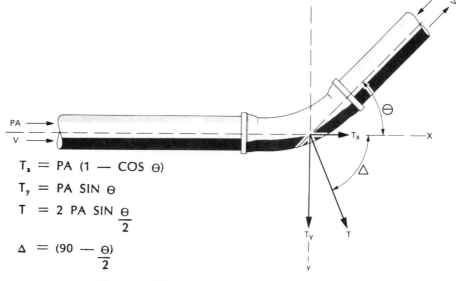

$$T_x = PA (1 - COS \theta)$$

$$T_y = PA\ SIN\ \theta$$

$$T = 2\ PA\ SIN\ \frac{\theta}{2}$$

$$\Delta = \frac{(90 - \theta)}{2}$$

Figure 2—THRUST FORCES ACTING ON A BEND

gorized into two groups: gravity and bearing blocks.

Gravity Blocks (Figure 1): Important factors considered in design are:

—Horizontal and vertical thrust components
—Allowable bearing value of soil
—Combined weight of pipe, water and soil prism
—Density of block material
—Block dimensions and volume
—A thrust force analysis is conducted similar to Figure 2.

Physical characteristics of the block are determined from the following formulas:

$$(1) \quad V_G = \frac{PA\ Sin\ \theta}{W_m}$$

(neglecting W_y)

$$V_G = \frac{T_y - W_y}{W_m}$$

(including W_y)

where $W_y = 1/2\ W_e L_x$

Earth cover (W_e) is neglected, when determining (W_e), if unstable conditions are anticipated. The horizontal thrust component (T_x) is counteracted by soil pressure on the vertical face of the block (F_p) or by joint restraint.

Allowable soil bearing pressure determines the minimum size of the block base.

Bearing Blocks (Figure 3):

Significant design criteria for bearing blocks include the following factors:

—Passive soil pressure
—Placement of bearing surface against undisturbed soil
—Block height (h) should be equal to or less than one half the total depth to the block base (H_T) except (h) should not be less than (D). Thus $h \leq 1/2\ H_T$ or $h \geq D$, whichever is greater.
—Block width (b) usually varies

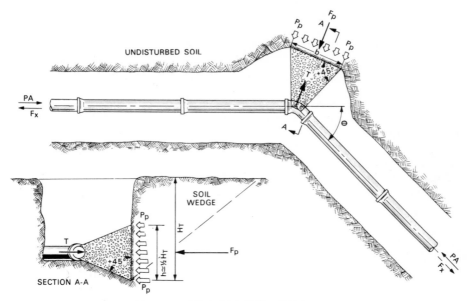

Figure 3—BEARING THRUST BLOCK

from one to two times the height (h).

—Concrete should not be poured on joints, limiting flexibility.

The required block bearing area, based on passive soil pressure, is expressed as follows:

$$(2) \quad A_b = hb = \frac{T}{P_p}$$

For the case where $h = 1/2\ H_T$,

(3)

$$b = \frac{2PA\ \text{Sin}\ \dfrac{\theta}{2}}{3/8\ w\ H_T^2\ N_\phi + C_s\ H_T\ \sqrt{N_\phi}}$$

(H_T) is estimated, permitting calculation of (b). Dimensions are selected by trial and error.

Pipelines under shallow cover are frequently deepened at the bends, increasing the depth of cover, to achieve more efficient block design. Colinear positioning of (T) and (F_p) is required to eliminate overturning moment on the block.

Tie Rods

Restraint with tie rods is versatile and relatively easy. Locations where tie rods are readily used include:

—Anchorage to structure, thrust collars, "deadman" anchors, and superstructures

—Joint restraint by utilizing clamps, pipe flange

—holes, or lugs cast on fittings.

—Restraint for field-cut, make-up sections

Tie rods on exposed piping systems must counteract total resultant thrust forces. However, on buried systems employing soil friction and lateral soil resistance, the effective thrust force at a joint (T_j) is proportional to its distance from the bend (L_j) and the restrained length (L).

$$(4) \quad T_j = F_s(L-L_j)$$

$$F_s = A_p C + W \tan \delta$$

where $C = f_c C_s$ and
$\delta = f_\phi \phi$

(see Table 1)

The required number of rods (N) is

(5) $\quad N = \dfrac{S_f T_j}{F}\quad$ where $F = SA_r$

and $S_f = 1.5$

Coating or wrapping is recommended for buried tie rods to prevent corrosion attack from corrosive soils.

Combined Systems and Structural Connections

Several restraining techniques are frequently required for thrust stabilization. Typical combinations include concrete blocks and tie rods, restrained joints and tie rods, or restrained joints, tie rods and thrust anchors.

Low head, in-plant piping is conveniently restrained and supported by attachments to nearby structures. Typical anchoring devices include the use of wall brackets, U-bolts and clamps, base elbows and tees, wall sleeves, structural steel frames, concrete supports and anchor bolts or straps.

Selection becomes a matter of preference depending upon convenience.

Summary

The proper restraint of unbalanced thrust forces is an important consideration in pressure piping design. Functional methods employed for cast iron piping systems are restrained joints, thrust blocks, tie rods or any combination thereof.

DEFINITION OF TERMS

A = Pipe cross-sectional area (in²) ($36\pi D^2$) with "D" in (ft)
A_b = Minimum bearing area of block base (ft²)
A_r = Cross-sectional area of rod (in²)
b = Width of thrust block (ft)
B = Gravity block base dimension (ft)
C_s = Soil cohesion (psf)

D = Conduit outside diameter (ft)
F = Force developed per rod (lbs)
f_c = Ratio of pipe cohesion/soil cohesion
F_p = Resisting force developed by P_p (lbs)
F_s = Conduit frictional resistance neglecting bell resistance (plf)
f_ϕ = Ratio of pipe friction angle/soil friction angle
h = Height of thrust block (ft)
H_T = Depth to bottom of block (ft)
L = Restrained pipe length each side (ft)
L_p = Nominal pipe length adj. to fitting (ft)
L_T = Length of tee (ft)
L_j = Distance from bend to joint (ft)
L_x = $L_T + 2L_p$ (ft)
N = Number of rods
N_ϕ = $\text{Tan}^2 (45° + \phi/2)$
P = Max. sustained pressure (psi) (test pressure or sustained surge pressure)
P_p = Passive soil pressure (psf)
S = Tensile stress of rod material (psi)
S_f = Safety factor (usually 1.25)
T = Resultant thrust force (lbs)
T_j = Thrust force at joint (lbs)
T_x = x thrust force component (lbs)
T_y = y thrust force component (lbs)
V = Fluid velocity (fps)
V_G = Volume of gravity block (ft³)
w = Soil unit weight (pcf)
W_e = Prism earthload (plf)
W_c = $W_e + W_p + W_w$ (plf)
W_g = Weight of gravity block (lbs)
W_m = Density of block material (pcf)
W_y = Effective weight of soil, pipe and water (lbs)
Θ = Bend deflection angle (degrees)
ϕ = Soil internal friction angle (degrees)
Δ = Angle between T and x-axis (degrees)

Table 1

Soil Friction and Cohesion Factors

Soil Description	Friction Angle ϕ(Degrees)	Cohesion C_s (psf)	f_ϕ	f_c
Well graded sand:				
Dry	44.5	0	0.76	0
Saturated	39	0	0.80	0
Silt (passing 200 sieve)				
Dry	40	0	0.95	0
Saturated	32	0	0.75	0
Cohesive granular soil				
Wet to moist	13–22	385–920	0.65	0.35
Clay				
Wet to moist	11.5–16.5	460–1,175	0.50	0.50
At maximum compaction			0.50	0.80

SECTION VII
BASIC PROCEDURES FOR THE INSTALLATION
OF
DUCTILE AND CAST IRON PIPE

Basic Procedures for the Installation of Ductile and Cast Iron Pipe

For the purposes of this section, the word "pipe" includes ductile and cast iron pipe: where a procedure refers to one of these materials only, it is so specified. It should be noted that experience with ductile iron pipe since the early 1950's has demonstrated its superior resistance to impact, beam loads and ring crushing loads. **Therefore, handling and installation procedures, in general, are much less critical for this pipe material. This fact can result in considerable savings in the installed cost of a pipeline project.**

Introduction

Proper installation procedures will enhance the long and useful life of both ductile and cast iron pipe. The information presented in this paper may be useful as general guidelines for the installation of these materias. More specific data is available in The Guides for the Installation of Ductile Iron Pipe and Gray Cast Iron Water Mains which are available from CIPRA.

Receiving, Handling and Storage

It is important that all pipe be carefully inspected for damage that may have occurred in transit. Unloading may be accomplished using slings, hooks, pipe tongs or skids. Under no circumstances should pipe be dropped on old automobile tires or other cushions and when handled on skidways, it should not be skidded or rolled against pipe already on the ground. Care should be exercised to avoid injury to the coating or lining; if damage occurs, repairs must be made.

Manufacturers who employ special methods of packaging pipe for shipment will gladly send instructions for unloading.

Proper storage procedures are important and warrant special consideration:

—Suggested maximum allowable stacking heights are available and should be observed; each pipe size should be stacked separately.

—Lubricant for rubber joints should be kept in a sanitary condition as an aid in disinfection of the main.

—Rubber gaskets should be used on a first-in, first-out basis and should be stored in a cool, dark location to avoid deterioration.

The Trench

Trench location is of prime importance in urban areas where water mains are installed to a line and grade established by the engineer to avoid damage to other subsurface utilities. In these areas no excavation should be attempted before obtaining clearance from other utilities.

If a gas service pipe is broken, the gas utility should be notified immediately and an experienced gas service man should supervise repairs.

House sewers must be returned to good condition after installation of the water main. Any installation in the proximity of underground telephone or power conduits should proceed only after notification of the appropriate utility.

Trees, shrubs, lawns, fences, etc. must be protected during construction; where removal is necessary, permission must be obtained and replacement made. Pavement, sidewalks and curbs must be replaced according to local standards.

The required earth cover over pipe varies depending on pipe size and geographical location. Generally a minimum cover of 4 to 5 feet is required for large diameter mains; for small diameter mains, cover varies from 2½ to 4 feet in the southern states to as much as 7 or 8 feet in the northern states because of the depth of frost penetration.

One of the most important requirements of a good trench is the preparation of the trench bottom; it should be true and even in order to insure soil support for the full length of the pipe barrel. In this phase of the excavation, the following should be observed:

—A trench for cast iron pipe passing over a sewer or previous excavation requires compaction to provide support equal to native soil.

—If a sling is used to lower the pipe, an indentation not exceeding 18 inches in length should be made at the middle of the trench bed to facilitate the removal of the sling.

—Soft subgrade requires the addition of crushed stone; in extreme cases, piling may be necessary for proper support.

—Large bell holes are not required for push-on or mechanical joint pipe; however, a small depression is necessary to permit the pipe barrel to lie flat and to allow space for joint assembly.

The following suggested trench widths are sufficient to permit proper pipe installation with room for joint assembly and backfill tamping around the pipe:

TABLE 1

SUGGESTED TRENCH WIDTH

Nominal Pipe Size (Inches)	Trench Width (Inches)	Nominal Pipe Size (Inches)	Trench Width (Inches)
4	28	20	44
6	30	24	48
8	32	30	54
10	34	36	60
12	36	42	66
14	38	48	72
16	40	54	78
18	42		

The trench should never be wider than the width used as design criteria.

When rock excavation is necessary, certain precautions must be taken. Any rock encountered must be removed so that it will not be closer than 6 inches to the bottom and sides of pipe in sizes up to 24 inches in diameter and no closer than 9 inches for pipe in sizes 30 inches or larger in diameter. Following rock excavation, a bed of approved material should be placed on the bottom of the trench to the above-mentioned distances, leveled and tamped. A straight-edge may be used for checking the trench bottom to detect high points of rock that may protrude through the cushion.

City, state, or Federal regulations usually govern requirements for blasting; barricade placement and other warning devices for public safety; shoring; storage of excavated material; and protection of underground and surface structures.

If it is determined that soils are corrosive or expansive (some dense clays expand when saturated and have been known to exert as much as 17,500 lbs. of pressure per square foot), special care should be exercised. Economical protection against corrosive soils is available through the use of 8-mil thick, loose polyethylene encasement as outlined in American National Standards Institute (ANSI) Standard A21.5. Because of its high strength, ductile iron pipe is recommended for expansive soil areas.

Installing The Pipe

Laying Conditions

Ductile and cast iron pipe, like any other pipe, are installed with respect to the trench bottom using specific laying or bedding conditions. The following conditions are recommended for cast iron pipe:

LAYING CONDITIONS

A	B	F
FLAT BOTTOM TRENCH BACKFILL NOT TAMPED	FLAT BOTTOM TRENCH BACKFILL TAMPED	PIPE BEDDED IN GRAVEL OR SAND. BACKFILL TAMPED

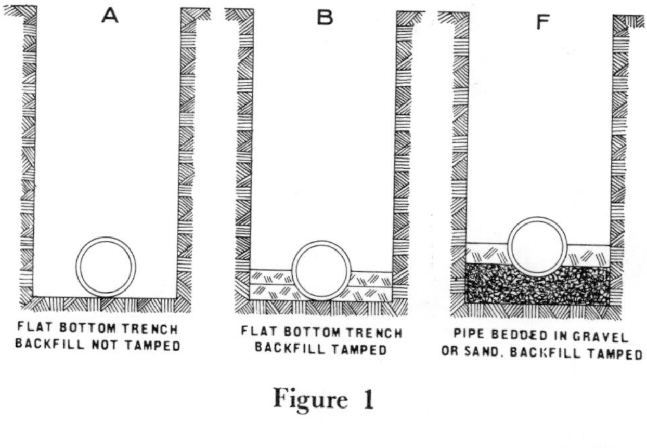

Figure 1

Each laying condition has its merits as well as its relationship to the ability of the pipe to carry loads placed upon it, with Laying Condition B being the most common.

For ductile iron pipe, laying condition types 1, 2, 3, 4 or 5 are recommended. For details see ANSI Standard A21.50 (Section III).

Pipe Placement

Proper equipment and procedures are necessary for safe and efficient pipe placement. Some of the more important considrations are:

—Bell and plain ends must be cleaned to prevent leaking joints and to assure proper seating of the gaskets.

—Before placement in the trench, pipe should be inspected for damage and the inside should be swabbed to remove loose dirt and foreign objects.

—Bells usually face the direction in which the work is progressing. When the main is being laid downhill, the pipes are frequently laid with the bells facing uphill for ease of installation.

—When laying push-on joint pipe, the correct gasket must be used for the type of joint being installed and the gasket must face the proper direction. Sand or grit must be removed from the gasket groove to assure watertightness.

Following are the steps in push-on joint assembly:

Figure 2

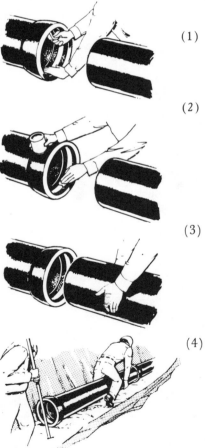

(1) Clean the groove and bell socket and insert the gasket, making sure that it faces the proper direction and is correctly seated.

(2) After cleaning any dirt or foreign material from the plain end, apply lubricant in accordance with the pipe manufacturer's recommendations. The lubricant is supplied in sterile cans and every effort should be made to keep it that way.

(3) Be sure that the plain end is beveled; square or sharp edges may damage the gasket and may cause a leak. Push the plain end into the bell of the pipe. Keep the joint straight while pushing. Make deflection after the joint is made.

(4) Small pipe can be pushed home with a long bar. Large pipe require additional power, such as a jack, lever puller or backhoe. The supplier will provide a jack or lever pullers on a rental basis. A timber header should be used between the pipe and jack or backhoe bucket to avoid damage to the pipe.

Joints and Fittings

Mechanical or push-on joint fittings can be used with push-on joint pipe. The plain end of the pipe usually is provided with 1 or 2 painted gauge lines which show whether it has been properly positioned in the bell socket after assembly. The pipe manufacturer's instructions as to the location of these lines should be followed.

Pit cast pipe was manufactured in 4 classifications (A, B, C and D), each having a different outside diameter than modern ductile or cast iron pipe. Before making extensions, existing pipe in a system should be measured to determine if transition fittings are required.

When laying mechanical joint pipe, the socket and plain ends should be clean. The assembly of the joint is simple and requires the use of an ordinary ratchet wrench. A torque wrench should be used for the first day or two of construction to accustom the workmen to the proper amount of pressure to apply to the wrench.

TABLE 2

SUGGESTED TORQUE

Bolt Size	Ft. Lbs.
⅝ inch	45-60
¾ inch	75-90
1 inch	85-100
1¼ inch	105-120

Flanged joints are seldom used in underground water mains except for valves and fittings for large meter settings, valve vaults and similar installations. Joint deflection is available in both push-on and mechanical joint pipe. Pipe should be assembled in a straight line both horizontally and vertically before defection is made. For mechanical joint pipe, the bolts should be partially tightened before the length of pipe is deflected.

Maximum allowable deflections are listed in the following table and should not be exceeded. For design purposes, deflection should be limited to 80% of the values shown in Table 3.

MAXIMUM DEFLECTION FULL LENGTH PIPE

Push-On Type Joint

Pipe Dia.	Deflec. Angle	Maximum Deflection Pipe Lengths of				Approximate Radius of Curve Produced by Succession of Joints, Pipe Lengths of			
		12'	16'	18'	20'	12'	16'	18'	20'
2"	5°	12"	17"	19"	21"	140'	185'	205'	230'
2 1/4"	5°	12"	17"	19"	21"	140'	185'	205'	230'
3"	5°	12"	17"	19"	21"	140'	185'	205'	230'
4"	5°	--	17"	19"	21"	---	185'	205'	230'
6"	5°	--	17"	19"	21"	---	185'	205'	230'
8"	5°	--	17"	19"	21"	---	185'	205'	230'
10"	5°	--	17"	19"	21"	---	185'	205'	230'
12"	5°	--	17"	19"	21"	---	185'	205'	230'
14"	3°	--	10"	11"	12"	---	300'	340'	380'
16"	3°	--	10"	11"	12"	---	300'	340'	380'
18"	3°	--	10"	11"	12"	---	300'	340'	380'
20"	3°	--	10"	11"	12"	---	300'	340'	380'
24"	3°	--	10"	11"	12"	---	300'	340'	380'
30"	3°	--	6"	11"	12"	---	450'	340'	380'
36"	3°	--	6"	11"	12"	---	450'	340'	380'
42"	2°	--	6"	7 1/2"	8"	---	450'	510'	570'
48"	2°	--	6"	7 1/2"	8"	---	450'	510'	570'
54"	1° -30'	--	5"	5 1/2"	6"	---	510'	680'	760'

TABLE 3

Frozen house services are often thawed by heat from high level electrical current. Current may be conducted across ductile or cast iron pipe joints by the use of special gaskets with metal tips or embedded metal conductors; wedges inserted at the joints; or conductive cables and metal strips.

It is important to use an adequate safety factor with regard to electrical current needs.

Cutting Pipe

Cast iron pipe may be cut with a hydraulic squeeze cutter (for pipe up to 20 inches in diameter); a rotary wheel hand cutter; an abrasive wheel cutter; a milling wheel; or a power driven hacksaw. Cast iron pipe should not be cut with an oxyacetylene torch.

Because of the nature of ductile iron, all cutters cannot be used, particularly the hydraulic squeeze cutter. Ductile iron pipe may be cut with an abrasive wheel cutter; a rotary wheel hand cutter (with carbide cutters); a guillotine pipe saw; a milling wheel; or an oxyacetylene torch. (If an oxyacetylene torch is used on ductile iron sewer or gas pipe, the presence of combustible gases could cause an explosion.)

Cut ends of ductile and cast iron pipe should be beveled and filed to prevent gasket damage in joint assembly.

Railroad and Highway Crossings

Regulations relative to the installation of water mains under highways and railroad crossings are specific and complete and often require a casing pipe. Mechanical or push-on joints should be used under railroads because of their ability to withstand vibration.

State, local and area regulations should be checked and the needed permits obtained well in advance of the actual work.

Valve and Hydrant Installation

To assure that the pipe will not be required to carry their weight, heavy valves and fittings should be supported by treated timbers, crushed stone, concrete pads or specially tamped trench bottoms.

When valves are placed in masonry vaults, special precautions must be taken to protect the pipeline. Cast iron valve boxes, when used, should rest above the valve body so that no weight is transferred to the valve itself.

New valves may be installed in existing mains by the use of cutting-in valves and sleeves or by use of a solid sleeve.

Blow-offs and drains should discharge above ground and should be installed so that there is no possibility of sewage or other contamination entering the main. Air release and/or vacuum vents should be provided at high points in the line and in areas of negative pressure.

Fire hydrants should be placed to provide maximum accessibility and minimum possibility of damage from vehicles or injury to pedestrians. In areas of the country where an undrained hydrant barrel would freeze, a drainage pit 2 feet in diameter and 3 feet deep should be excavated below the hydrant opening and filled with coarse sand to a depth of 6 inches above the hydrant opening but providing sufficient aggregate void space to more than equal the volume of the barrel. The drainage pit should neither be near, nor connected to, a sewer. Hydrants may be anchored by any one of several methods, such as thrust blocks, tie rods, or special restrained fittings (see Thrust Restraint), and special procedures are necessary for each method.

To prevent water hammer, valve and hydrants should be closed very slowly, especially for the last few turns near full closure.

Thrust Restraint

Thrust forces are created in a pipeline at changes in direction, tees, deadends or where changes in pipe size occur at reducers. Available restraint methods include concrete thrust blocks, restrained joints and tie rods. Forces to be restrained are given in Table 4.

Total Pounds					
Nom. Pipe Dia. In.	Dead End	90° Bend	45° Bend	22 1/2° Bend	11 1/4° Bend
4	1,810	2,559	1,385	706	355
6	3,739	5,288	2,862	1,459	733
8	6,433	9,097	4,923	2,510	1,261
10	9,677	13,685	7,406	3,776	1,897
12	13,685	19,353	10,474	5,340	2,683
14	18,385	26,001	14,072	7,174	3,604
16	23,779	33,628	18,199	9,278	4,661
18	29,865	42,235	22,858	11,653	5,855
20	36,644	51,822	28,046	14,298	7,183
24	52,279	73,934	40,013	20,398	10,249
30	80,425	113,738	61,554	31,380	15,766
36	115,209	162,931	88,177	44,952	22,585
42	155,528	219,950	119,036	60,684	30,489
48	202,683	286,637	155,127	79,083	39,733
54	256,072	362,140	195,989	99,914	50,199

NOTE:

To determine thrust at pressures other than 100 psi, multiply the thrust obtained in the table by the ratio of the pressure to 100.

For example, the thrust on a 12 inch, 90° bend at

125 psi is $19{,}353 \times \dfrac{125}{100} = 24{,}191$ pounds.

Table 4

To determine the size of a concrete thrust block and for further design information, please refer to the special section entitled Thrust Restraint for Underground Piping Systems on page 347.

The following precautions must be observed when constructing thrust blocks:
—Blocks must be poured against undisturbed soil.
—The pipe joint and bolts must be accessible.
—Concrete should be cured for at least 5 days and should have a compression strength of 2,000 lbs. at 28 days.
—Blocks must be positioned to counteract the direction of the resultant thrust force.

Restrained push-on and mechanical joints are available for all pipe sizes and present no installation problems. They are used for resisting thrust forces where there is a shortage of space or where the soil behind a fitting will not provide adequate support. This restraining method involves placement of these special joints at appropriate fittings and for a predetermined number of pipe lengths on each side. See Section VI for the design of thrust restraint systems utilizing restrained joints.

Tie rods may be used by themselves or in combination with other restraint devices. When tie rods are used with steel bands around the pipe barrel, only 1 rod should be attached to each band and the band should be cocked to prevent slippage along the pipe barrel. A band placed behind a bell may be used for 2 rods. For mechanical joint pipe, tie rods may be threaded through the bolt holes in the flange and secured by nuts. All rods and bands should be made of corrosion-resistant material or coated to prevent rust or deterioration.

Restraint may be necessary for more than 1 length of pipe on each side of any change in direction, or at any deadend or tee.

Backfilling

The purpose of backfilling is to fill the trench and to protect the pipe by providing support along and under it. It is one of the most important phases of water main construction and proper procedures are imperative.

Backfill material should be of good quality and should be free of cinders, frozen material, ashes, refuse, boulders, rocks or organic material. Soil containing stones up to 8 inches in their greatest dimension may be used from 1 foot above the top of the pipe to the ground surface.

Local authorities control requirements for backfilling under paved streets and the soil is usually compacted in 6-inch lifts, using air or gasoline-powered compactors. Many cities require that the entire trench be filled with compacted earth or select material, such as sand, gravel or limestone screenings. Compaction below and to the top of the pipe is intended to provide support for the water main and all tamping above the pipe is to support the new pavement.

Water jetting or trench flooding are sometimes used to obtain the necessary consolidation of the soil.

After backfilling operations have been completed, the work area must be restored to its original condition.

Testing

If possible, all new pipelines should be hydrostatically tested before backfilling is completed. Sufficient earth should be placed on the pipe between

FLUSHING AND DISINFECTING

TABLE 6

CIPRA RECOMMENDED ALLOWABLE LEAKAGE PER 1000-FT. OF PIPELINE*
(GALLONS PER HOUR)

Avg. Test Pressure PSI	NOMINAL PIPE DIAMETER - INCHES																
	2	3	4	6	8	10	12	14	16	18	20	24	30	36	42	48	54
450	0.32	0.48	0.64	0.95	1.27	1.59	1.91	2.23	2.55	2.87	3.18	3.82	4.78	5.73	6.69	7.64	8.60
400	0.30	0.45	0.60	0.90	1.20	1.50	1.80	2.10	2.40	2.70	3.00	3.60	4.50	5.41	6.31	7.21	8.11
350	0.28	0.42	0.56	0.84	1.12	1.40	1.69	1.97	2.25	2.53	2.81	3.37	4.21	5.06	5.90	6.74	7.58
300	0.26	0.39	0.52	0.78	1.04	1.30	1.56	1.82	2.08	2.34	2.60	3.12	3.90	4.68	5.46	6.24	7.02
275	0.25	0.37	0.50	0.75	1.00	1.24	1.49	1.74	1.99	2.24	2.49	2.99	3.73	4.48	5.23	5.98	6.72
250	0.24	0.36	0.47	0.71	0.95	1.19	1.42	1.66	1.90	2.14	2.37	2.85	3.56	4.27	4.99	5.70	6.41
225	0.23	0.34	0.45	0.68	0.90	1.13	1.35	1.58	1.80	2.03	2.25	2.70	3.38	4.05	4.73	5.41	6.03
200	0.21	0.32	0.43	0.64	0.85	1.06	1.28	1.48	1.70	1.91	2.12	2.55	3.19	3.82	4.46	5.09	5.73
175	0.20	0.30	0.40	0.59	0.80	0.99	1.19	1.39	1.59	1.79	1.98	2.38	2.98	3.58	4.17	4.77	5.36
150	0.19	0.28	0.37	0.55	0.74	0.92	1.10	1.29	1.47	1.66	1.84	2.21	2.76	3.31	3.86	4.41	4.97
125	0.17	0.25	0.34	0.50	0.67	0.84	1.01	1.18	1.34	1.51	1.68	2.01	2.52	3.02	3.53	4.03	4.53
100	0.15	0.23	0.30	0.45	0.60	0.75	0.90	1.05	1.20	1.35	1.50	1.80	2.25	2.70	3.15	3.60	4.05

*For Mechanical or push-on joint pipe with 18-ft. nominal lengths. To obtain the recommended allowable leakage for pipe with 20-ft. nominal lengths, multiply the leakage calculated from the above table by 0.9.

If the pipeline under test contains sections of various diameters, the allowable leakage will be the sum of the computed leakage for each size.

joints to prevent movement under test pressure. In city streets, heavy traffic demands may require backfilling after a few lengths of pipe have been laid. The pipeline should be filled slowly and care should be exercised to vent all high points and expel all air. All fittings and hydrants should be properly anchored and all valves completely closed before applying test pressure.

In performing the test, pressure is applied by means of an adequate pump connected to the pipe, bringing the main up to the test pressure, which is recommended as 1.5 times the working pressure at the point of testing. This test pressure should be held for 2 hours. Make-up water should be measured with a meter or by pumping water from a vessel of known volume.

FLUSHING AND DISINFECTING

All new water systems or extensions to existing systems should be thoroughly flushed and disinfected before being placed in service. Public health authorities require disinfection and bacteriological examination to assure freedom from contamination. (Refer to AWWA Standard C601 for Disinfecting Water Mains.)

The flushing velocity should be at least 2.5 fps for small mains. For mains larger than 18 inches in diameter, a lower rate may be used. Table 7 lists required openings to obtain the required velocity of 2.5 fps for flushing and is excerpted from AWWA Standard C601.

REQUIRED OPENINGS—ETC.

(40-psi Residual Pressure)

Pipe Size in.	Flow Required to Produce 2.5-fps Velocity gpm	Orifice Size in.	Hydrant Outlet Nozzles	
			Number	Size in.
4	100	15/16	1	2 1/2
6	220	1 3/8	1	2 1/2
8	390	1 7/8	1	2 1/2
10	610	2 5/16	1	2 1/2
12	880	2 13/16	1	2 1/2
14	1,200	3 1/4	2	2 1/2
16	1,565	3 5/8	2	2 1/2
18	1,980	4 3/16	2	2 1/2

*With 40 psi residual pressure, a 2 1/2 in. hydrant outlet nozzle will discharge approximately 1,000 gpm and a 4 1/2 in. hydrant nozzle will discharge approximately 2,500 gpm.

TABLE 7

Disinfection of mains can be accomplished by the addition of chlorine as a liquid, a hypochlorite solution or hypochlorite tablets. Liquid chlorine is injected into the main under pressure with a portable chlorinator to provide at least 50 ppm available chlorine. To insure that the required concentration is maintained, chlorine residuals should be checked. The chlorinated water solution should remain in the pipe for at least 24 hours, at the end of which period the chlorine concentration should be at least 25 ppm. Final flushing may then be accomplished.

The slug method of chlorination, which is used for large diameter water mains of long length, consists of moving a column of highly concentrated chlorine solution (at least 300 ppm) along the interior of the pipe with at least 3 hours contact with the pipe wall.

The tablet method is generally used for short extensions (no longer than 2,500 feet) of 12-inch and smaller diameter mains. The required number of tablets are placed in the crown of each pipe length and held in place by an approved mastic. The main is then filled with water at a velocity of less than 1 fps and the water is left in the main for 24 hours before flushing. Table 8, excerpted from AWWA Standard C601, indicates the number of tablets required for each size of pipe up to 12 inches in diameter.

NUMBER OF HYPOCHLORITE TABLETS OF 5-G REQUIRED FOR DOSE OF 50 Mg/1*

Length of Section Ft.	Diameter of Pipe in.					
	2	4	6	8	10	12
13 or less	1	1	2	2	3	5
18	1	1	2	3	5	6
20	1	1	2	3	5	7
30	1	2	3	5	7	10
40	1	2	4	6	9	14

*Based on 3¾ g available chlorine per tablet.

TABLE 8

Repairing Main Breaks

Many devices and materials are available to repair pipe breaks, including mechanical joint split sleeves, bolted repair clamps, bell repair clamps, solid sleeves, mechanical joint bell split sleeves, and repair clamps.

Following the repair of main breaks, proper disinfection procedures are necessary to provide protection from contamination and caution must be exercised to insure that a strong concentration of chlorine does not enter the customer's service lines.

American Water Works Association

ANSI/AWWA C600–77

Revision of
C600–64

AMERICAN NATIONAL
STANDARD

for

INSTALLATION OF GRAY AND DUCTILE CAST-IRON WATER MAINS AND APPURTENANCES

NOTICE

This Standard has been especially printed by the American Water Works Association for incorporation into this volume. It is current as of December 1, 1975. It should be noted, however, that all AWWA Standards are updated at least once in every five years. Therefore, before applying this Standard it is suggested that you confirm its currency with the American Water Works Association.

Approved by AWWA Board of Directors May 8, 1977.
Approved by American National Standards Institute Jun. 22, 1977.

AMERICAN WATER WORKS ASSOCIATION
6666 West Quincy Avenue, Denver, Colorado 80235

Committee Personnel

The AWWA Standards Committee on Installation of Cast-Iron Water Mains, which developed this standard, had the following personnel at the time of approval:

A. M. TINKEY, *Chairman*
W. HARRY SMITH, *Vice-Chairman*
ROBERT ZIMMERMAN, *Secretary*

Consumer Members

H. KENNETH ANDERSON, Water Bureau, Portland, OR	(AWWA)
E. M. BONADEO, Detroit Water and Sewage Department, Detroit, MI	(AWWA)
CHARLES A. FROMAN JR., Gary–Hobart Water Corp., Gary, IN	(AWWA)
ALBERT HELT, The Water Bureau, Metropolitan District, Hartford, CT	(NEWWA)
WILLIAM H. PRIESTER, Department of Water & Power, Los Angeles, CA	(AWWA)
JOHN E. READEY, Department of Water & Sewers, Chicago, IL	(AWWA)
MARK SADOLF, Utilities Department, Broward County, Ft. Lauderdale, FL	(AWWA)
WILLIAM R. THOMPSON,* Metropolitan District Commission, Boston, MA	(NEWWA)
A. M. TINKEY, St. Louis County Water Co., University City, MO	(AWWA)

General Interest Members

GORDON C. ANDERSON,* Insurance Services Office, New York, NY	(ISO)
WILLIAM A. BARKLEY, College of Engineering, Las Cruces, NM	(ASCE)
KENNETH J. CARL, Insurance Services Office, New York, NY	(ISO)
MARTIN P. DALY, Malcolm Pirnie, Inc., Paramus, NJ	(AWWA)
JOSEPH M. DENNIS, Department of Human Resources, Raleigh, NC	(CSSE)
EDWIN J. LASZEWSKI, City of Milwaukee, Milwaukee, WI	(APWA)
J. SAMUEL SLICER JR., Factory Mutual Research Corp., Norwood, MA	(FMR)
JOSEPH VELLANO, Latham, NY	(NUCA)
HENRY WILKENS JR., Water Division, Houston, TX	(AWWA)
ROBERT ZIMMERMAN, Greeley & Hansen, Chicago, IL	(AWWA)

Producer Members

T. C. JESTER, American–Darling Valve, Birmingham, AL	(MSS)
HAROLD KENNEDY JR., U.S. Pipe & Foundry Co., Birmingham, AL	(CIPRA)
EDWARD C. SEARS, American Cast Iron Pipe Co., Birmingham, AL	(CIPRA)
W. HARRY SMITH, Cast Iron Pipe Research Assn., Oak Brook, IL	(CIPRA)

* Alternate

Table of Contents

3
6
5

Foreword

This foreword is for information only and is not a part of AWWA C-600.

I. History of Standard

The first AWWA standard specifications, "Laying Cast Iron Pipe," (7D.1–1938) were adopted in Apr. 1938. They were intended as a guide in making extensions to existing distribution systems, and in preparing specifications for contracts for the construction of new systems or extensions. The standard was to be used as a guide for installing bell-and-spigot cast-iron pipe and did not cover the furnishing and delivery of material, any other type of pipe, or any other type of joint. The standard included a model addendum, which was to be used with project specifications, and was designed to be used as a part of the contract document. The standard was published in the Feb. 1938 edition of *Journal AWWA*.

The standard was revised in 1949, including a change of title to, "Standard Specifications for Installation of Cast-Iron Water Mains," (7D.1–T–1949 and C600–49T). The standard was expanded, adding numerous tables and installation guidelines. The model addendum was also expanded. The revised standard was published in the Dec. 1949 edition of *Journal AWWA*.

An additional section, Sec. 9b—Jointing of Mechanical-Joint Pipe—was added in May 1954. Section 9c—Joining of Push-on Joint Pipe—was added in 1964.

In 1975 the Standards Council formed the present C–600 Committee to revise C–600 to current practices and to add ductile iron as a pipe material.

In order to do this, the committee decided to completely change the character of the standard, removing the model addendum and making the standard more in compliance with the style of other AWWA standards.

II. Information Regarding the Use of This Standard

The AWWA standard, "Installation of Gray and Ductile Cast-Iron Water Mains and Appurtenances," can be used as a reference when making extensions to existing or constructing new distribution systems, using either ductile or gray cast-iron mains, with either mechanical- or push-on joints. It is not the intent for this standard to be used as a contract document but it may be used as a reference in the contract documents. It is based upon a consensus of the committee on the minimum practice consistent with sound, economical service under normal conditions, and its applicability under any circumstances must be reviewed by a responsible engineer. The standard is not intended to preclude the manufacture, marketing, purchase, or use of any product, process, or procedure.

III. Major Revisions

The standard has been rewritten completely and restructured to conform with the present style of AWWA standards. Ductile-iron pipe has been added to the standard, and normal installation practices have been updated completely. The addendum has been deleted.

Allowable leakage for both mechanical- and push-on joints has been reduced to one half of the value prescribed in the 1964 standard.

American Water Works Association

ANSI/AWWA C600–77

Revision of
C600–64

AMERICAN NATIONAL
STANDARD

for

Installation of Gray and Ductile Cast-Iron Water Mains and Appurtenances

Section 1—General

Sec. 1.1—Scope

This standard covers installation procedures for gray and ductile cast-iron pipe and appurtenances for water service.

1.1.1 *Conditions not covered.* Installations that require special attention, techniques, and materials are not covered. Each such installation requires special considerations based on many influencing factors and cannot be covered adequately in a single standard. This type of installation can best be accomplished by competent engineering design in consultation with representatives of the material manufacturing industry. Some of these typical installations are

1. Piping through rigid walls.
2. Subaqueous piping.
3. Piping on supports above or below ground.
4. Piping requiring insulation.
5. Plant- or pump-station piping.

Sec. 1.2—References

This standard references the following documents. They form a part of this standard to the extent specified herein. In any case of conflict, the requirements of this standard shall prevail.

AWWA C101, Thickness Design of Cast-Iron Pipe.

AWWA C104, Cement–Mortar Lining for Cast-Iron and Ductile-Iron Pipe and Fittings for Water.

AWWA C105, Polyethylene Encasement for Gray and Ductile Cast-Iron Piping for Water and Other Liquids.

AWWA C106, Cast-Iron Pipe Centrifugally Cast in Metal Molds, for Water or Other Liquids.

AWWA C108, Cast-Iron Pipe Centrifugally Cast in Sand-Lined Molds, for Water or Other Liquids.

AWWA C110, Gray-Iron and Ductile-Iron Fittings, 3 in. Through 48 in., for Water and Other Liquids.

AWWA C111, Rubber-Gasket Joints for Cast-Iron and Ductile-Iron Pressure Pipe and Fittings.

AWWA C115, Flanged Cast-Iron and Ductile-Iron Pipe With Threaded Flanges.

AWWA C150, Thickness Design of Ductile-Iron Pipe.

AWWA C151, Ductile-Iron Pipe, Centrifugally Cast in Metal Molds or Sand-Lined Molds, for Water or Other Liquids.

AWWA C500, Gate Valves—3-in. Through 48-in.—for Water and Other Liquids.

AWWA C502, Dry-Barrel Fire Hydrants.

AWWA C503, Wet-Barrel Fire Hydrants.

AWWA C504, Rubber Seated Butterfly Valves.

AWWA C601, Disinfecting Water Mains.

AWWA No. 20104, Handbook of Occupational Safety and Health Standards for Water Utilities.

AASHTO * T-99, Standard Method of Test for Moisture-Density Relationship for Soils.

Sec. 1.3—Definitions

Under this standard, the following definitions shall apply:

1.3.1 *Gray cast iron.* Cast ferrous material in which a major part of the carbon content occurs as free carbon in the form of flakes interspersed through the metal.

1.3.2 *Ductile cast iron.* Cast ferrous material in which a major part of the carbon content occurs as free carbon in nodules or spheroidal form.

1.3.3 *Owner.* The municipality or other organization that will own and operate the completed piping system. The owner may designate agents, such as an engineer, purchaser, or inspector for specific responsibilities with regard to piping construction projects.

1.3.4 *Contractor.* The party responsible for water main construction.

1.3.5 *Mechanical joint.* The gasketed and bolted joint as detailed in AWWA C111.

1.3.6 *Push-on joint.* The single rubber-gasket joint as described in AWWA C111.

Section 2—Inspection, Receiving, Handling, and Storage

Sec. 2.1—Inspection

At the discretion of the owner, all materials furnished by the contractor are subject to inspection and approval at the manufacturer's plant.

2.1.1 *Post delivery.* All pipe and appurtenances are subject to inspection at the point of delivery by the owner. Material found to be defective

---

* American Association of State Highway and Transportation Officials, 341 National Press Bldg., Washington, D.C. 20004.

due to manufacture or damage in shipment shall be rejected or recorded on the bill of lading and removed from the job site. The owner may perform tests as specified in the applicable AWWA standard to ensure conformance with the standard. In case of failure of the pipe or appurtenance to comply with such specifications, responsibility for replacement of the defective materials becomes that of the manufacturer.

2.1.2 *Workmanship.* All pipe and appurtenances shall be installed and joined in conformance with this standard and tested under pressure for defects and leaks in accordance with Sec. 4 of this standard.

Sec. 2.2—Handling and Storage

All pipe, fittings, valves, hydrants, and accessories shall be loaded and unloaded by lifting with hoists or skidding in order to avoid shock or damage. Under no circumstances shall such material be dropped. Pipe handled on skidways shall not be rolled or skidded against pipe on the ground.

2.2.1 *Padding.* Slings, hooks, or pipe tongs shall be padded and used in such a manner as to prevent damage to the exterior surface or internal lining of the pipe.

2.2.2 *Storage.* Materials, if stored, shall be kept safe from damage. The interior of all pipe, fittings, and other appurtenances shall be kept free from dirt or foreign matter at all times. Valves and hydrants shall be drained and stored in a manner that will protect them from damage by freezing.

2.2.2.1 Pipe shall not be stacked higher than the limits shown in Tables 1 and 2. The bottom tier shall be kept off the ground on timbers, rails, or concrete. Pipe in tiers shall be alternated: bell, plain end; bell, plain end. At least two rows of 4-in. × 4-in. timbers shall be placed between tiers and chocks affixed to each end in order to prevent movement.

2.2.2.2 Gaskets for mechanical- and push-on joints to be stored shall be placed in a cool location out of direct sunlight. Gaskets shall not come in contact with petroleum products. Gaskets shall be used on a first-in, first-out basis.

TABLE 1

Maximum Stacking Heights—Gray Cast-Iron Pipe

Pipe Size (in.)	Number of Tiers	
	18-ft length	20-ft length
3	18	18
4	16	16
6	13	13
8	11	11
10	9	8
12	8	7
14	7	7
16	6	6
18	6	5
20	5	4
24	4	3

TABLE 2

*Maximum Stacking Heights—Ductile Cast-Iron Pipe**

Pipe Size (in.)	Number of Tiers
3	18
4	16
6	13
8	11
10	10
12	9
14	8
16	7
18	6
20	6
24	5
30	4
36	4
42	3
48	3
54	3

* For 18- or 20-ft lengths.

2.2.2.3 Mechanical-joint bolts shall be handled and stored in such a manner that will ensure proper use with respect to types and sizes.

Section 3—Installation

Sec. 3.1—Alignment and Grade

The water mains shall be laid and maintained to lines and grades established by the plans and specifications with fittings, valves, and hydrants at the required locations unless otherwise approved by the owner. Valve-operating stems shall be oriented in a manner to allow proper operation. Hydrants shall be installed plumb.

3.1.1 *Prior investigation.* Prior to excavation, investigation shall be made to the extent necessary to determine the location of existing underground structures and conflicts. Care should be exercised by the contractor during excavation to avoid damage to existing structures.

3.1.2 *Unforeseen obstructions.* When obstructions that are not shown on the plans are encountered during the progress of work and interfere so that an alteration of the plans is required, the owner will alter the plans or order a deviation in line and grade or arrange for removal, relocation, or reconstruction of the obstructions.

3.1.3 *Clearance.* When crossing existing pipelines or other structures, alignment and grade shall be adjusted as necessary, with the approval of the owner, to provide clearance as required by federal, state, or local regulations or as deemed necessary by the owner to prevent future damage or contamination of either structure.

Sec. 3.2—Trench Construction

The trench shall be excavated to the required alignment, depth, and width and in conformance with all federal, state, and local regulations for the protection of the workmen.

3.2.1 *Trench preparation.* Trench preparation shall proceed in advance of pipe installation for only as far as stated in the specifications.

3.2.1.1 Discharge from any trench dewatering pumps shall be conducted to natural drainage channels, storm sewers, or an approved reservoir.

3.2.1.2 Excavated material shall be placed in a manner that will not obstruct the work nor endanger the workmen, obstruct sidewalks, driveways, or other structures and shall be done in compliance with federal, state, or local regulations.

3.2.2 *Pavement removal.* Removal of pavement and road surfaces shall be a part of the trench excavation and the amount removed shall depend upon the width of trench required for installation of the pipe and the dimensions of area required for the installation of valves, hydrants, specials, manholes, or other structures. The dimensions of pavement removed shall not exceed the dimensions of the opening required for installation of pipe, valves, hydrants, specials, manholes, and other structures by more than 6 in. in any direction unless otherwise required or approved by the owner. Methods, such as sawing, drilling, or chipping, shall be used to ensure the breakage of pavement along straight lines.

3.2.3 *Width.* The width of the trench at the top of the pipe shall be that of the single-pass capabilities of normally available excavating equipment and ample to permit the pipe to be laid and joined properly and allow the backfill to be placed as specified. Trench widths as shown in Table 3 may be used as a guide. Trenches shall be of such extra width, when required, to permit the placement of tim-

TABLE 3

Suggested Trench Widths at the Top of the Pipe*

Nominal Pipe Size in.	Trench Width in.
4	28
6	30
8	32
10	34
12	36
14	38
16	40
18	42
20	44
24	48
30	54
36	60
42	66
48	72
54	78

* The trench should never be wider than the width used as design criteria.

ber supports, sheeting, bracing, and appurtenances.

3.2.4 *Bell holes.* Holes for the bells shall be provided at each joint but shall be no larger than necessary for joint assembly and assurance that the pipe barrel will lie flat on the trench bottom. Other than noted previously, the trench bottom shall be true and even in order to provide support for the full length of the pipe barrel, except that a slight depression may be provided to allow withdrawal of pipe slings or other lifting tackle.

3.2.5 *Rock conditions.* When excavation of rock is encountered, all rock shall be removed to provide a clearance of at least 6 in. below and on each side of all pipe, valves, and fittings for pipe sizes 24 in. or smaller, and 9 in. for pipe sizes 30 in. and larger. When excavation is completed, a bed of sand, crushed stone, or earth that is free from stones, large clods, or frozen earth, shall be placed on the

bottom of the trench to the previously mentioned depths, leveled, and tamped.

3.2.5.1 These clearances and bedding procedures shall also be observed for pieces of concrete or masonry and other debris or subterranean structures, such as masonry walls, piers, or foundations that may be encountered during excavation.

3.2.5.2 This installation procedure shall be followed when gravel formations containing loose boulders greater than 8 in. in diameter are encountered.

3.2.5.3 In all cases, the specified clearances shall be maintained between the bottom of all pipe and appurtenances and any part, projection, or point of rock, boulder, or stones of sufficient size and placement which, in the opinion of the owner, could cause a fulcrum point.

3.2.6 *Previous excavations.* Should the trench pass over a sewer or other previous excavation, the trench bottom shall be sufficiently compacted to provide support equal to that of the native soil or conform to other regulatory requirements in a manner that will prevent damage to the existing installation.

3.2.7 *Blasting.* Blasting for excavation shall be permitted only after securing the approval of the owner who will establish the hours of blasting. The blasting procedure, including protection of persons and property, shall be in strict accordance with federal, state, and local regulations.

3.2.8 *Protection of property.* Trees, shrubs, fences, and all other property and surface structures shall be protected during construction unless their removal is shown in the plans and specifications or approved by the owner.

3.2.8.1 Any cutting of tree roots or branches shall be done only as approved by the owner.

3.2.8.2 Temporary support, adequate protection, and maintenance of all underground and surface structures, drains, sewers, and other obstructions encountered in the progress of the work shall be furnished by the contractor.

3.2.8.3 All properties that have been disturbed shall be restored as nearly as practical to their original condition.

3.2.9 *Unstable subgrade.* When the subgrade is found to be unstable or to include ashes, cinders, refuse, organic material, or other unsuitable material, such material shall be removed, to a minimum of at least 3 in., or to the depth ordered by the owner and replaced under the directions of the owner with clean, stable backfill material. When such materials are encountered, polyethylene encasement should be provided (see paragraph 3.4.5). The bedding shall be consolidated and leveled in order that the pipe may be installed in accordance with Sec. 3.2.4.

3.2.9.1 When the bottom of the trench or the subgrade is found to consist of material that is unstable to such a degree that, in the judgment of the owner it cannot be removed, a foundation for the pipe and/or appurtenance shall be constructed using piling, timber, concrete, or other materials at the direction of the owner.

3.2.10 *Safety.* Appropriate traffic control devices shall be provided in accordance with federal, state, or local regulations to regulate, warn, and guide traffic at the work site.

Sec. 3.3—Pipe Installation

Proper implements, tools, and facilities shall be provided and used for the safe and convenient performance of the work. All pipe, fittings, valves, and hydrants shall be lowered carefully into the trench by means of a derrick, ropes, or other suitable tools or equipment, in such a manner as to prevent damage to water-main materials and protective coatings and linings. Under no circumstances shall water-main materials be dropped or dumped into the trench. The trench should be dewatered prior to installation of the pipe.

3.3.1 *Examination of material.* All pipe fittings, valves, hydrants, and other appurtenances shall be examined carefully for damage and other defects immediately before installation. Defective materials shall be marked and held for inspection by the owner, who may prescribe corrective repairs or reject the materials.

3.3.2 *Pipe ends.* All lumps, blisters, and excess coating shall be removed from the socket and plain ends of each pipe, and the outside of the plain end and the inside of the bell shall be wiped clean and dry and be free from dirt, sand, grit, or any foreign material before the pipe is laid.

3.3.3 *Pipe cleanliness.* Foreign material shall be prevented from entering the pipe while it is being placed in the trench. During laying operations, no debris, tools, clothing, or other materials shall be placed in the pipe.

3.3.4 *Pipe placement.* As each length of pipe is placed in the trench, the joint shall be assembled and the pipe brought to correct line and grade. The pipe shall be secured in place with approved backfill material.

3.3.5 *Pipe plugs.* At times when pipe laying is not in progress, the open ends of pipe shall be closed by a watertight plug or other means approved by the owner. When practical, the plug shall remain in place until the trench is pumped completely dry. Care must be taken to prevent pipe flotation should the trench fill with water.

A

Flat bottom trench, untamped backfill.

B

Flat bottom trench, tamped backfill.

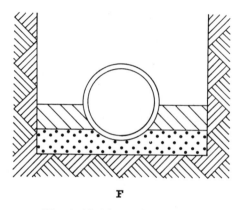

F

Pipe bedded in sand or gravel,
tamped backfill.

Fig. 1. Laying Conditions for Gray Cast-Iron Pipe

3.3.6 *Gray-iron laying conditions.* The specified laying conditions for gray cast-iron pipe shall be completed in accordance with AWWA C101 and as illustrated in Fig. 1.

3.3.7 *Ductile-iron laying conditions.* The specified laying conditions for ductile-iron pipe shall be completed in accordance with AWWA C150 and as illustrated in Fig. 2 (page 8).

Sec. 3.4—Joint Assembly

3.4.1 *Push-on joints.* Push-on joints shall be assembled as described and illustrated in Fig. 3 (page 9).

3.4.2 *Mechanical joints.* Mechanical joints shall be assembled as follows:

1. Wipe clean the socket and plain end. The plain end, socket, and gasket should be washed with a soap solution to improve gasket seating.

Type 1 *

Flat-bottom trench.† Loose backfill.

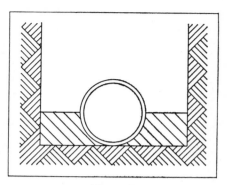

Type 2

Flat-bottom trench.† Backfill lightly consolidated to centerline of pipe.

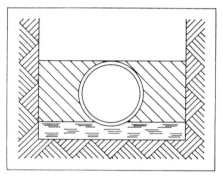

Type 3

*Pipe bedded in 4-in. minimum loose soil.‡
Backfill lightly consolidated to top of pipe.*

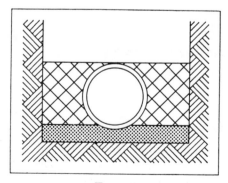

Type 4

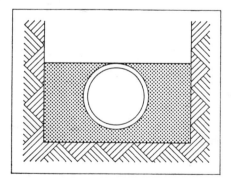

Type 5

Pipe bedded in sand, gravel, or crushed stone to depth of ⅛ pipe diameter, 4-in. minimum. Backfill compacted to top of pipe. (Approximately 80 per cent Standard Proctor, AASHTO T-99.)

Pipe bedded in compacted granular material to centerline of pipe. Compacted granular or select ‡ material to top of pipe. (Approximately 90 per cent Standard Proctor, AASHTO T-99.)

* For 30-in. and larger pipe, consideration should be given to the use of laying conditions other than Type 1.
† "Flat-bottom" is defined as undisturbed earth.
‡ "Loose soil" or "select material" is defined as native soil excavated from the trench, free of rocks, foreign materials, and frozen earth.

Fig. 2. Laying Conditions for Ductile Cast-Iron Pipe

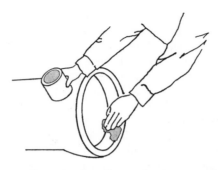

1. *Thoroughly clean the groove and bell socket and insert the gasket, making sure that it faces the proper direction and that it is correctly seated.*

2. *After cleaning dirt or foreign material from the plain end, apply lubricant in accordance with the pipe manufacturer's recommendations. The lubricant is supplied in sterile cans and every effort should be made to keep it sterile.*

3. *Be sure that the plain end is beveled; square or sharp edges may damage or dislodge the gasket and cause a leak. When pipe is cut in the field, bevel the plain end with a heavy file or grinder to remove all sharp edges. Push the plain end into the bell of the pipe. Keep the joint straight while pushing. Make deflection after the joint is assembled.*

4. *Small pipe can be pushed into the bell socket with a long bar. Large pipe require additional power, such as a jack, lever puller, or backhoe. The supplier may provide a jack or lever pullers on a rental basis. A timber header should be used between the pipe and jack or backhoe bucket to avoid damage to the pipe.*

Fig. 3. Push-on Joint Assembly

2. Place the gland on the plain end with the lip extension toward the plain end, followed by the gasket with the narrow edge of the gasket toward the plain end of the pipe.

3. Insert the pipe into the socket and press the gasket firmly and evenly into the gasket recess. Keep the joint straight during assembly. Make deflection after joint assembly but before tightening the bolts.

4. Push the gland toward the bell

and center it around the pipe with the gland lip against the gasket.

5. Align bolt holes and insert bolts, with bolt heads behind the bell flange, and tighten opposite nuts to keep the gland square with the socket.

6. Tighten the nuts in accordance with Table 4.

3.4.3 *Pipe deflection.* When it is necessary to deflect pipe from a straight line in either the vertical or horizontal plane, or where long radius curves are permitted, the amount of deflection shall not exceed that shown in Tables 5 or 6.

3
7
6

TABLE 4

Mechanical Joint—Bolt Torques

Bolt Diameter in.	Torque ft–lb
$\frac{5}{8}$	45–60
$\frac{3}{4}$	75–90
1	85–100
$1\frac{1}{4}$	105–120

3.4.4 *Pipe cutting.* Cutting pipe for the insertion of valves, fittings, or closure pieces shall be done in a neat, workmanlike manner without creating damage to the pipe or cement–mortar lining.

3.4.4.1 Gray cast-iron pipe may be cut using a hydraulic squeeze cutter, abrasive pipe saw, rotary wheel cutter, guillotine pipe saw, or milling wheel saw.

3.4.4.2 Ductile cast iron may be cut using an abrasive pipe saw, rotary wheel cutter, guillotine pipe saw, milling wheel saw, or oxyacetylene torch.

3.4.4.3 Cut ends and rough edges shall be ground smooth, and for push-on joint connections, the cut end shall be beveled.

3.4.5 *Polyethylene encasement.* When polyethylene encasement is specified for gray and ductile cast-iron pipe, it shall be installed in accordance with Sec. 5–4 of AWWA C105.

TABLE 5

Maximum Deflection Full Length Pipe—Push-on Type Joint

Pipe Diameter in.	Deflection Angle deg	Maximum Deflection—in.		Approx. Radius of Curve Produced by Succession of Joints—ft	
		(18-ft length)	(20-ft length)	(18-ft length)	(20-ft length)
3	5	19	21	205	230
4	5	19	21	205	230
6	5	19	21	205	230
8	5	19	21	205	230
10	5	19	21	205	230
12	5	19	21	205	230
14	3	11	12	340	380
16	3	11	12	340	380
18	3	11	12	340	380
20	3	11	12	340	380
24	3	11	12	340	380
30	3	11	12	340	380
36	3	11	12	340	380
42	2	$7\frac{1}{2}$	8	510	570
48	2	$7\frac{1}{2}$	8	510	570
54	$1\frac{1}{2}$	$5\frac{1}{2}$	6	680	760

TABLE 6

Maximum Deflection Full Length Pipe—Mechanical Joint Pipe

Size of Pipe in.	Deflection Angle deg–min	Maximum Deflection—in.		Approx. Radius of Curve Produced by Succession of Joints—ft	
		(18-ft length)	(20-ft length)	(18-ft length)	(20-ft length)
3	8–18	31	35	125	140
4	8–18	31	35	125	140
6	7–7	27	30	145	160
8	5–21	20	22	195	220
10	5–21	20	22	195	220
12	5–21	20	22	195	220
14	3–35	13½	15	285	320
16	3–35	13½	15	285	320
18	3–0	11	12	340	380
20	3–0	11	12	340	380
24	2–23	9	10	450	500
30	2–23	9	10	450	500
36	2–5	8	9	500	550
42	2–0	7½	8	510	570
48	2–0	7½	8	510	570

3
7
7

Sec. 3.5—Backfilling

Backfill shall be accomplished in accordance with the specified laying condition as described in Sec. 3.3.

3.5.1 *Backfill material.* All backfill material shall be free from cinders, ashes, refuse, vegetable or organic material, boulders, rocks or stones, frozen soil, or other material that, in the opinion of the owner, is unsuitable.

3.5.1.1 From 1 ft above the top of the pipe to the subgrade of the pavement, material containing stones up to 8 in. in their greatest dimension may be used, unless otherwise specified.

3.5.1.2 When the type of backfill material is not indicated on the drawings or is not specified, the excavated material may be used, provided that such material consists of loam, clay, sand, gravel, or other materials that, in the opinion of the owner, are suitable for backfilling.

3.5.1.3 If excavated material is indicated on the drawings or specified for backfill, and there is a deficiency due to a rejection of part thereof, the required amount of sand, gravel, or other approved material shall be provided.

3.5.1.4 All sand used for backfill shall be clean, graded from fine to coarse, not lumpy or frozen, and free from slag, cinders, ashes, rubbish, or other material that, in the opinion of the owner, is objectionable or deleterious. It should not contain a total of more than 10 per cent by weight of loam and clay, and all material must be capable of being passed through a ¾-in. sieve. Not more than 5 per cent shall remain on a No. 4 sieve.

3.5.1.5 Gravel used for backfill shall consist of clean gravel having durable particles graded from fine to coarse in a reasonably uniform combination with no boulders or stones larger than 2 in. in size. It shall be free from slag, cinders, ashes, refuse, or other deleterious or objectionable materials. It shall not contain excessive amounts of loam and clay and shall not be lumpy or frozen. No more than 15 per cent shall pass a No. 200 sieve.

3.5.1.6 Screenings used for backfill shall consist of the products obtained from crushing sound limestone or dolomite ledge rock and shall be free from shale, dust, excessive amounts of clay, and other undesirable materials. All materials shall pass a $\frac{1}{2}$-in. sieve, and no more than 25 per cent shall pass a No. 100 sieve.

3.5.2 *Compaction.* When special backfill compaction procedures are required, they shall be accomplished in accordance with project specifications or applicable federal, state, and local regulations.

3.5.3 *Partial backfilling during testing.* When specified by the owner, pressure and leakage testing may be accomplished before completion of backfilling and with pipe joints accessible for examination. In such cases, sufficient backfill material shall be placed over the pipe barrel between the joints to prevent movement.

Sec. 3.6—Valve-and-Fitting Installation

3.6.1 *Examination of material.* Prior to installation, valves shall be inspected for direction of opening, freedom of operation, tightness of pressure-containing bolting, cleanliness of valve ports and especially seating surfaces, handling damage, and cracks. Defective valves shall be corrected or held for inspection by the owner.

3.6.2 *Placement.* Valves, fittings, plugs, and caps shall be set and joined to the pipe in the manner specified in Sec. 3.3 for cleaning, laying, and joining pipe, except that 12-in. and larger valves should be provided with special support, such as treated timbers,

crushed stone, concrete pads, or sufficiently tamped trench bottom so that the pipe will not be required to support the weight of the valve.

3.6.3 *Valve location.* Valves in water mains shall, where practical, be located on the street property lines extended in unpaved areas unless shown otherwise on the plans.

3.6.3.1 Mains shall be drained through drainage branches or blowoffs. Drainage branches, blowoffs, air vents, and appurtenances shall be provided with valves and shall be located and installed as shown on the plans. Drainage branches or blowoffs shall not be directly connected to any storm or sanitary sewer, submerged in any stream, or be installed in any other manner that will permit back siphonage into the distribution system.

3.6.4 *Valve protection.* A valve box or a vault shall be provided for every valve.

3.6.4.1 A valve box shall be provided for every valve that has no gearing or operating mechanism or in which the gearing or operating mechanism is fully protected with a gear case. The valve box shall not transmit shock or stress to the valve and shall be centered over the operating nut of the valve, with the box cover flush with the surface of the finished area or such other level as may be directed by the owner.

3.6.4.2 A valve vault designed to prevent settling on the pipe shall be provided for every valve that has exposed gearing or operating mechanisms. The operating nut shall be readily accessible for operation through the opening in the valve vault which shall be set flush with the surface of the finished pavement or such other level as may be specified. Vaults shall be constructed to permit minor valve repairs and afford protection to the valve and pipe from impact where they pass through the vault walls.

3.6.4.3 In no case shall valves be used to bring misaligned pipe into alignment during installation. Pipe shall be supported in such a manner as to prevent stress on the valve.

3.6.5 *Plugs and caps.* All dead ends on new mains shall be closed with plugs or caps that are suitably restrained to prevent blowing off under test pressure. If a blowoff valve precedes the plug or cap, it too shall be restrained against blowing off. All dead ends shall be equipped with suitable blowoff facilities.

Sec. 3.7—Hydrant Installation

3.7.1 *Examination of material.* Prior to installation, inspect all hydrants for direction of opening, nozzle threading, operating-nut and cap-nut dimensions, tightness of pressure-containing bolting, cleanliness of inlet elbow, handling damage, and cracks. Defective hydrants shall be corrected or held for inspection by the owner.

3.7.2 *Placement.* All hydrants shall stand plumb and shall have their nozzles parallel with, or at right angles to, the curb, with the pumper nozzle facing the curb, except that hydrants having two-hose nozzles 90 deg apart shall be set with each nozzle facing the curb at an angle of 45 deg.

3.7.2.1 Hydrants shall be set to the established grade, with the centerline of the lowest nozzle at least 12 in. above the ground, or as directed by the owner.

3.7.2.2 Each hydrant shall be connected to the main with a 6-in. branch controlled by an independent 6-in. valve, unless otherwise specified by the owner.

3.7.2.3 When a dry-barrel hydrant is set in soil that is pervious, drainage shall be provided at the base of the hydrant by placing coarse gravel or crushed stone mixed with coarse sand, from the bottom of the trench to at least 6 in. above the waste opening in the hydrant and to a distance of 1 ft around the elbow. Where ground water rises above the drain port or when the hydrant is located within 8 ft of a sewer, the drain port shall be plugged and water pumped from the hydrant when freezing may occur.

3.7.2.4 When a dry-barrel hydrant with an open drain is set in clay or other impervious soil, a drainage pit 2 ft × 2 ft × 2 ft shall be excavated below each hydrant and filled with coarse gravel or crushed stone mixed with coarse sand, under and around the elbow of the hydrant and to a level of 6 in. above the drain port.

3.7.3 *Location.* Hydrants shall be located as shown on the plans or as directed by the owner.

3.7.4 *Protection.* In the case of hydrants that are intended to fail at the ground-line joint upon vehicle impact (traffic hydrants), specific care must be taken to provide adequate soil resistance to avoid transmitting shock moment to the lower barrel and inlet connection. In loose or poor load-bearing soil, this may be accomplished by pouring a concrete collar approximately 6 in. thick to a diameter of 2 ft at or near the ground line around the hydrant barrel.

Sec. 3.8—Thrust Restraint

3.8.1 *Hydrants.* The bowl of each hydrant shall be well braced against a sufficient area of unexcavated earth at the end of the trench with stone slabs or concrete backing, or it shall be tied to the pipe with suitable metal tie rods, clamps, or restrained joints as shown or directed by the owner.

3.8.1.1 Tie rods, clamps, or other components of dissimilar metal shall be protected against corrosion by hand application of a bituminous coating or by encasement of the entire assembly with 8-mil thick, loose polyethylene film in accordance with AWWA C105.

3.8.1.2 Thrust-restraint design pressure should be equal to the test pressure.

3.8.2 *Fittings.* All plugs, caps, tees, and bends, unless otherwise specified, shall be provided with reaction backing, or suitably restrained by attaching metal rods, clamps, or restrained joints as shown or specified by the owner.

3.8.3 *Restraint materials.* Vertical and horizontal reaction backing shall be made of concrete having a compressive strength of not less than 2000 psi after 28 days.

3.8.3.1 Backing shall be placed between solid ground and the fitting to be anchored; the area of bearing on the pipe and on the ground in each instance shall be that shown or directed by the owner. The backing shall, unless otherwise shown or directed, be so located as to contain the resultant thrust force and so that the pipe and fitting joints will be accessible for repair.

3.8.3.2 Restrained push-on joints, mechanical joints utilizing set-screw retainer glands or metal harness of tie rods, or clamps may be used instead of concrete backing if so indicated in the plans and specifications. Tie rods, clamps, or other components of dissimilar metal shall be protected against corrosion by hand application of a bituminous coating or by encasement of the entire assembly with 8-mil thick, loose polyethylene film in accordance with AWWA C105.

Section 4—Hydrostatic Testing

Sec. 4.1—Pressure Test

After the pipe has been laid, all newly laid pipe or any valved section thereof shall be subjected to a hydrostatic pressure of at least 1.5× the working pressure at the point of testing.

4.1.1 *Test pressure restrictions.* Test pressures shall

1. Not be less than 1.25× the working pressure at the highest point along the test section.

2. Not exceed pipe or thrust restraint design pressures.

3. Be of at least 2-hr duration.

4. Not vary by more than ±5 psi.

5. Not exceed twice the rated pressure of the valves or hydrants when

the pressure boundary of the test section includes closed gate valves or hydrants.

6. Not exceed the rated pressure of the valves if resilient-seated butterfly valves are used.

4.1.2 *Pressurization.* Each valved section of pipe shall be filled with water slowly and the specified test pressure, based on the elevation of the lowest point of the line or section under test and corrected to the elevation of the test gage, shall be applied by means of a pump connected to the pipe in a manner satisfactory to the owner.

4.1.3 *Air removal.* Before applying the specified test pressure, air shall be expelled completely from the pipe, valves, and hydrants. If permanent air vents are not located at all high points, the contractor shall install corporation cocks at such points so that the air can be expelled as the line is filled with water. After all the air has been expelled, the corporation cocks shall be closed and the test pressure applied. At the conclusion of the pressure test, the corporation cocks shall be removed and plugged, or left in place at the discretion of the owner.

4.1.4 *Examination.* All exposed pipe, fittings, valves, hydrants, and joints shall be examined carefully during the test. Any damage or defective pipe, fittings, valves, or hydrants that are discovered following the pressure test shall be repaired or replaced with sound material and the test shall be repeated until it is satisfactory to the owner.

Sec. 4.2—Leakage Test

A leakage test shall be conducted concurrently with the pressure test.

4.2.1 *Leakage defined.* Leakage shall be defined as the quantity of water that must be supplied into the newly laid pipe, or any valved section thereof, to maintain pressure within 5 psi of the specified test pressure after the air in the pipeline has been expelled and the pipe has been filled with water.

4.2.2 *Allowable leakage.* No pipe installation will be accepted if the leakage is greater than that determined by the following formula:

$$L = \frac{ND\sqrt{P}}{7400}$$

in which L is the allowable leakage, in gallons per hour; N is the number of joints in the length of pipeline tested; D is the nominal diameter of the pipe, in inches; and P is the average test pressure during the leakage test, in pounds per square inch gage.

4.2.2.1 Allowable leakage at various pressures is shown in Table 7 (page 16).

4.2.2.2 When testing against closed metal-seated valves, an additional leakage per closed valve of 0.0078 gal/hr/in. of nominal valve size shall be allowed.

4.2.2.3 When hydrants are in the test section, the test shall be made against the closed hydrant.

4.2.3 *Acceptance of installation.* Acceptance shall be determined on the basis of allowable leakage. If any test of pipe laid discloses leakage greater than that specified in Sec. 4.2.2, the contractor shall, at his own expense, locate and repair the defective material until the leakage is within the specified allowance.

4.2.3.1 All visible leaks are to be repaired regardless of the *amount* of leakage.

TABLE 7

Allowable Leakage per 1000 ft of Pipeline—gph*

Avg. Test Pressure psi	Nominal Pipe Diameter—in.																
	2	3	4	6	8	10	12	14	16	18	20	24	30	36	42	48	54
450	0.32	0.48	0.64	0.95	1.27	1.59	1.91	2.23	2.55	2.87	3.18	3.82	4.78	5.73	6.69	7.64	8.60
400	0.30	0.45	0.60	0.90	1.20	1.50	1.80	2.10	2.40	2.70	3.00	3.60	4.50	5.41	6.31	7.21	8.11
350	0.28	0.42	0.56	0.84	1.12	1.40	1.69	1.97	2.25	2.53	2.81	3.37	4.21	5.06	5.90	6.74	7.58
300	0.26	0.39	0.52	0.78	1.04	1.30	1.56	1.82	2.08	2.34	2.60	3.12	3.90	4.68	5.46	6.24	7.02
275	0.25	0.37	0.50	0.75	1.00	1.24	1.49	1.74	1.99	2.24	2.49	2.99	3.73	4.48	5.23	5.98	6.72
250	0.24	0.36	0.47	0.71	0.95	1.19	1.42	1.66	1.90	2.14	2.37	2.85	3.56	4.27	4.99	5.70	6.41
225	0.23	0.34	0.45	0.68	0.90	1.13	1.35	1.58	1.80	2.03	2.25	2.70	3.38	4.05	4.73	5.41	6.03
200	0.21	0.32	0.43	0.64	0.85	1.06	1.28	1.48	1.70	1.91	2.12	2.55	3.19	3.82	4.46	5.09	5.73
175	0.20	0.30	0.40	0.59	0.80	0.99	1.19	1.39	1.59	1.79	1.98	2.38	2.98	3.58	4.17	4.77	5.36
150	0.19	0.28	0.37	0.55	0.74	0.92	1.10	1.29	1.47	1.66	1.84	2.21	2.76	3.31	3.86	4.41	4.97
125	0.17	0.25	0.34	0.50	0.67	0.84	1.01	1.18	1.34	1.51	1.68	2.01	2.52	3.02	3.53	4.03	4.53
100	0.15	0.23	0.30	0.45	0.60	0.75	0.90	1.05	1.20	1.35	1.50	1.80	2.25	2.70	3.15	3.60	4.05

* For pipe with 18-ft nominal lengths. To obtain the recommended allowable leakage for pipe with 20-ft nominal lengths, multiply the leakage calculated from the table by 0.9. If the pipeline under test contains sections of various diameters, the allowable leakage will be the sum of the computed leakage for each size.

Section 5—Disinfection

Upon completion of a newly installed main or when repairs to an existing pipe are made, the main shall be disinfected according to instructions listed in AWWA C601 of latest revision.

Section 6—Highway and Railroad Crossings

Sec. 6.1—Casing Pipe

When casing pipe is specified for highways or railroad crossings, the project shall be completed in accordance with applicable federal, state, and local regulations. In the case of railroad crossings, the project should comply further with regulations established by the railroad company. General practice permits boring for casing diameters through 36 in. with maximum length of about 175 ft; jacking for diameters 30 in. through 60 in. with lengths of about 200 ft; and tunneling for pipes 48 in. and larger for longer lengths.

Sec. 6.2—Carrier Pipe

The casing pipe should be 6–8 in. larger than the outside diameter of the gray or ductile cast-iron pipe bells. Carrier pipe may be pushed or pulled through the completed casing pipe. Chocks or skids should be placed on the carrier pipe to ensure approximate centering within the casing pipe and to prevent damage during installation. Care must be exercised in order to avoid metal-to-metal contact. In order to avoid the transfer of earth and live loads to the carrier pipe, the space between the carrier and casing pipes should not be filled completely.

Section 7—Service Taps

Sec. 7.1—Tapping

Corporation stops may be installed either before or after pipe installation. Generally, they are located at ten or two o'clock on the circumference of the pipe and may be screwed directly into the tapped and threaded main without any additional appurtenances. When more than one tap in a gray cast-iron pipe is necessary to deliver the required flow, they should be staggered around the circumference at least 12 in. apart (not in a straight line). Ductile-iron pipe in all classes may be directly tapped with standard corporation stops; however, torque requirement for the installation may be effectively reduced by the application of two layers of 3-mil TFE tape to the male threads of the corporation stop.

3
8
3

SECTION VIII

ENVIRONMENTAL FACTORS RELATED TO THE DESIGN AND USE OF GRAY AND DUCTILE IRON PIPE

Flow of Liquids in Pipelines

Factors Affecting Flow Capacity in Water Mains

In some areas of the United States water mains have lost an appreciable part of their original carrying capacity after years of service, and for this reason consideration of the causes of this trouble and the remedies are of interest. Loss of carrying capacity results in increased costs either because of extra pumping expense or additional capital outlay if larger mains are required. There are many reasons for a reduction in the flow capacity of a pipeline. Increased head loss in a pipeline may be due to one or a combination of the following:

1. Sedimentation; mud, silt or sand.
2. Obstruction of the pipe due to debris: sticks, boards, stones, tools and other objects that may have gotten into the pipe during construction.
3. Partly closed valves.
4. Accumulation of air at summits.
5. Mineral deposits.
6. Slime growths on walls of pipe.
7. Tuberculation.

All of these difficulties can be remedied by proper design, operation and maintenance.

Sedimentation. Transmission mains that carry raw water from rivers or lakes are subject to heavy deposits of silt and sand, especially when the rivers are at flood stage, or the lakes are turbulent. Many of the older distribution systems were supplied with raw water for years before the construction of treatment plants. During low flow conditions, these waters deposit a layer of sediment along the bottom of the pipe. Sand may enter the raw water intake lines at most any time, and it may enter the distribution lines whenever the filters become defective or when the beds are inadequately maintained. If sedimentation has occurred, the remedy is to initiate and follow through with a main flushing program. If this type program is not effective, a pipe cleaning program may be necessary. It is important that the cause of the problem be determined and that remedial measures be undertaken to correct the situation; i.e., redesign of inlet works if required or initiate a preventive operating and maintenance program.

Obstructions in Pipe. Modern pipe laying specifications require that each length of pipe be cleaned out before installation in the line. They also require that the end of the pipeline be closed with a plug after each day's work. In spite of these provisions, it is a fact that at times undesirable objects are left or get placed in pipelines. Careful visual inspection of the pipe interior during installation and proper flushing technique upon completion of the pipeline should eliminate this difficulty.

Partly Closed Valves. In the ordinary operation of a water works system, it becomes necessary from time to time to close valves to carry on maintenance and extension work and in some systems, valves are throttled for pressure control purposes. Care should be taken to see that closed valves are opened after the construction work is completed and the location of throttle valves properly recorded so that in the event that future operation requires a full opening, these valves may be opened. The opening and closing of valves is an important part of distribution system maintenance. Records should be kept on each valve to ascertain its performance and to be certain that no valves are accidentally left closed or partly closed.

Accumulation of Air at Summits. In supply lines there is occasionally an opportunity for air to accumulate at a summit so that the water can occupy only a portion of the total area of the pipe. The remedy for this difficulty is to provide air release valves at summits of the pipeline. Air release valves are also necessary for filling the lines when they are first placed in service or after being shut down to make repairs. When testing aportion of a new installation, it is important to have air release valves in all summits of that section.

Mineral Deposits. In rare cases, waters are highly mineralized. These minerals are picked up from the rock formations through which the water moves in its underground passage. Some waters are super-saturated and the minerals only loosely held in solution. A small amount of air mixed with water in the pumping operation or a quick change in water temperature, may cause the mineral to be deposited. Natural lime waters usually form a hard deposit on the entire wall of the pipe and decrease the flow by reducing the diameter of the pipe. Mineral deposits in mains are difficult to remove, usually requiring special cleaning tools. Lime and alum deposits that result from softening and filtration processes are sometimes carried out into the mains. As a rule, these deposits are relatively soft and may be removed by ordinary pipe cleaning operations.

Slime Growths. Some water supplies are troubled with organic growths in the mains. Many of the growths may be due to the use of surface water containing microscopic organisms. While most of these organisms have little effect on the quality of the water, they may cling to the walls of the pipe, thereby reducing the rate of flow in the line. Organic growths may be removed by the use of chlorine, a combination of chlorine and ammonia, or copper sulfate. The nature of the treatment depends on each individual case and the application of chemicals should usually be started by the use of small dosages with gradual increase until the required effect results. A sudden change in chemical dosage could release many of the organisms into the distribution system, thereby causing taste, odor or turbidity complaints from the consumer.

Aggressive Water. Certain soft waters in the United States are aggressive to unprotected iron. Related problems virtually have been eliminated by factory installation of cement-mortar lining in almost all cast and ductile iron pipe used for water and wastewater. Older unlined pipelines may be cement-mortar lined in situ with good results.

Cement-Mortar Lined Gray and Ductile Cast Iron Pipe

Historical Development. The first cast iron water mains were not coated or lined, but were installed in the same condition in which they came from the casting molds. After many years' use, it became evident that the interior of the pipe was affected by certain types of water. The use of bituminous coating was proposed, and most of the cast iron pipe sold for water works service after about 1860 were provided with a hot dip bituminous lining and coating, usually of molten tar pitch. In those systems where the water was relatively hard and slightly alkaline, the bituminous linings were generally satisfactory. Where soft or acid waters were encountered, however, problems frequently arose, such as the water becoming red or rusty and a gradual reduction of the flow rate through the pipe. Corrosive water penetrated the pinholes in the tar coating and tuberculation ensued. The need of a better pipe lining to combat tuberculation led to experiments and research with cement mortar as a lining matrial.

In 1922, the first cement-lined cast iron pipe was installed in the water distribution system of Charleston, South Carolina. This pipe was lined by means of a projectile drawn through the pipe. After over 50 years of service, friction flow tests show that this original cement-lined cast iron pipe has retained a Hazen-Williams coefficient (C Factor) of 131.

This process, however, soon gave way to a centrifugal process. Since 1922 many improvements have ben made in the production of cement-lined pipe. Cement-mortar lined pipes are centrifugally lined at the foundry to assure that the best possible quality control is maintained, and that a uniform thickness of mortar is distributed throughout the entire length of the pipe. Cement linings prevent tuberculation by keeping the water from contacting the iron. The linings are smooth and offer very little frictional resistance to the flow of water. Almost all gray and ductile cast iron pipe installed in water systems today are cement lined.

Lining Process. The centrifugal process of applying cement-mortar linings is used in modern practice. By using this method, excellent quality control of the cement-mortar and the centrifugal lining operation can be maintained. Centrifugal lining enables the pipe manufacturer to produce cement-lined pipe of the highest quality—smooth, free of defects and meeting the rigid requirements of ANSI Standard A21.4, "Cement-Mortar Lining for Cast Iron and Ductile Iron Pipe and Fittings for Water."

The lined pipe are stored in a moist atmosphere during the curing period, or given a seal coating to prevent too rapid loss of moisture. The cement lining adheres to the wall of the pipe so that the pipe may be cut and tapped without damage to the lining.

Economics of Cement Lining. The advantages of cement-lined cast iron pipe go beyond the prevention of tuberculation and are clearly applicable to installations in territories where tuberculating waters do not exist. In order to fully understand the financial advantages of using cement linings, it is necessary to have some knowledge of certain hydraulic phenomena.

When water moves through pipe, friction is developed between the water and the inside of the pipe. The result is that, as the water travels through the pipe, some of the energy imparted to it by the pump is consumed by the friction, resulting in a loss of pressure. The amount of friction so developed is the criterion by which the size of pipe, and the amount of power required for pumping, are determined. When a given amount of water is to be transported, the total amount of friction developed depends on the diameter and length of the pipe and the condition of its interior.

The principal advantages of cement linings are higher flow coefficient when the pipe is new and maintained carrying capacity as the pipe grows older. The economy resulting from the prevention of tuberculation is obvious, but experience has shown that less friction results when cement linings are used even where non-tuberculating waters are transported.

For example, a test made on a new 36-inch bituminous lined cast iron supply line showed a coefficient of approximately 135. A test on a new 36-inch cement-lined cast iron line showed a coefficient of 145. Since new pipe was tested in both cases, the difference in values was due to the different conditions of the pipe interiors.

Financial Advantages. As a demonstration of the financial advantages accruing from the use of cement linings, consider a typical instance based on a 24″ pipe, 30,000 feet long carrying 8 mgd.

Tests made on numerous cement-lined pipe installations have established a "C" value of 140.

Tests made on bituminous lined pipe have established a "C" value of 130 when new, and a "C" value of approximately 100 after 30 years' service. Assuming no increase in demand and a pumping cost of $0.05 per million gallons to lift water one foot in elevation, the annual cost of pumping water against friction head only, if bituminous lined pipe were used (actual inside diameter 24.44

inches), would range from $8,322.00 per year when the pipe was new C=130 to $13,797.00 per year when the pipe was 30 years old C=100.

In the case of the 24-inch cement-lined pipe actual inside diameter=24.25", the pumping cost for the first year with a "C" value of 140, would be $7,533.60 and would remain at that figure throughout the 30-year period. The actual saving for this period, resulting from the use of cement-lined pipe, would be $105,780.00.

In the case of smaller diameter pipe used in distribution systems, sizes are usually determined by fire protection requirements. The additional volume of water available when cement linings are used may stop a fire in its early stages that would otherwise become a conflagration.

Where tuberculating waters are carried, the loss in capacity of smaller bituminous lined mains occurs at a faster rate than is the case with larger mains. This can mean that in a relatively short time the capacity of bituminous lined pipe is so reduced that cleaning or replacement becomes advisable. Cement lining is a minor part of the total cost of a pipeline project, and assures high carrying capacity for the life of the pipe.

Some old existing non-cement-mortar-lined pipelines have become tuberculated when exposed to very aggressive water. An effective correction of this condition is cleaning and cement-lining in place or cleaning followed by appropriate water treatment. No lasting value will result from cleaning a tuberculated pipeline unless it is followed by appropriate water treatment or installation of a lining. The advisability of pipeline rehabilitation will be dictated by an evaluation of the structural condition of the pipe and the adequacy of the pipe size following rehabilitation.

The nomograph (pg. 390) is based on the Hazen-Williams flow formula and shows relationships between flow coefficient, head loss, internal pipe diameter and discharge rate. If any three of these factors are known, the fourth may be determined by locating a point on the pivot line, which point lies on a common line with two of the known factors. Once the pivot point is established, the unknown factor will lie on a straight line between the pivot point and the third known factor. Arrows (ϕ) on the inside diameter line represent actual inside diameter of cement-mortar-lined ductile iron pipe Class 50.

Almost all gray and ductile iron pipe for water transmission and distribution are cement-mortar-lined in accordance with ANSI Standard A21.4 (AWWA C104) or Federal specifications.

The flow of water through this pipe is usually computed by the widely used Hazen-Williams formula:

$$Q = 0.006756 \times CD^{2.63} H^{.54}$$

Where:

Q = discharge in gallons per minute

C = Hazen-Williams flow coefficient

D = actual inside diameter

H = head loss in feet per 1,000 feet

NOMOGRAPH FOR PIPE SIZE, HEAD LOSS AND DISCHARGE

Based on Hazen-Williams Formula: $Q = 0.006756\,CD^{2.63}\,H^{.54}$
For Cement Mortar Lined Cast Iron Pipe $C = 140$.
(←) Shown are actual Inside Diameters of Cement
Mortar Lined Ductile Iron Pipe, Class 50

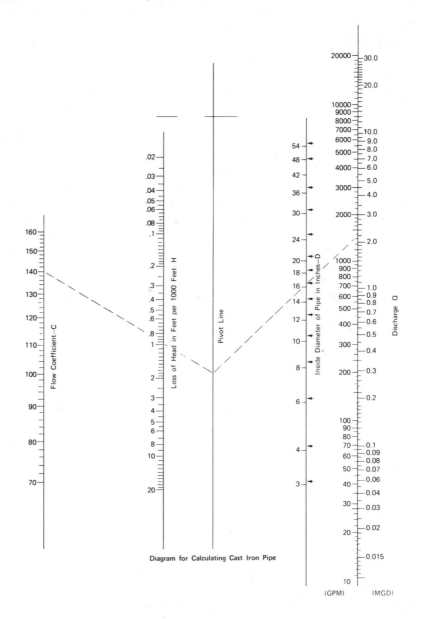

Diagram for Calculating Cast Iron Pipe

The flow coefficient "C" in the Hazen-Williams formula is in effect a measure of the condition of the pipe interior and is sometimes known as a friction coefficient. Tests employing this formula show that cement-mortar-lined gray or ductile iron pipe has a C factor of 140. Unlined pipe exposed to aggressive waters will suffer loss in C factor due to tuberculation.

Both gray and ductile iron pipe have flow advantages resulting from their greater than nominal internal diameters. For example, Class 52, 12-inch ductile iron pipe has an inside diameter of 12.46 inches. Cement-lined pipe of this same class has an inside diameter of 12.34 inches.

Example 1 — Maximum Delivery

To find the maximum delivery of an 8-inch, Class 50, cement-mortar-lined ductile iron pipe, 7,500 feet in length under 150 feet of head, the available head per 1,000 feet is $\frac{150}{7.5} = 20$ feet per 1,000 feet. By use of the nomograph the result is 1,200 gallons per minute or 1.73 million gallons per day.

Example 2 — Determination of Diameter

To find the diameter of pipe necessary to deliver 3,000,000 gallons per day through a pipeline 25,000 feet long under 150 feet of head, the available head per 1,000 feet is $\frac{150}{25} = 6$ feet per 1,000 feet. By use of the nomograph the result is 12-inch, Class 50 ductile iron pipe.

Example 3 — Friction Loss

To find the loss of head through a 10-inch pipeline 4,000 feet long, delivering 1,400,000 gallons per day, using a C factor of 140, the head loss is 4 feet per 1,000 feet, or 16 feet for the pipeline. If water is delivered at a point 100 feet above the pump, total head against the pump, is 100 feet (static) plus 16 feet (friction), or a total of 116 feet.

Example 4 — Delivery Determined from Pressure Reduction

Two accurate pressure gauges should be placed at a known distance apart and measurement of the difference in elevation recorded. If on a 12-inch pipe the pressure gauges are 500 feet apart and show a difference in pressure of 2 psi (4.6 feet of head) while one gauge is 1.8 feet above the other, the actual loss of head will be 4.6 plus or minus 1.8 = 6.4 or 2.8 feet per 500 feet or 12.8 or 5.6 feet per 1,000 feet, depending on elevation of the downstream gauge. Assuming that the downstream gauge is at the higher elevation, head loss due to friction is 5.6 feet per 1,000 feet. By use of the nomograph, the result is 1,900 gallons per minute, or 2.74 million gallons per day.

Water Hammer or Surge

Water hammer is a real force and has caused failure of many pipelines. It is the result of a sudden decrease in the pipeline fluid velocity. This rapid de-'celeration of the liquid mass sets up pressure waves which are transmitted through the pipeline system. It is estimated that in water transmission and distribution systems, a change in velocity of 1 fps in critical time can increase the pipeline pressure by approximately 50 psi. Water hammer can be caused by quick acting valves, check valves, rapid closure of fire hydrants, earthquakes, the sudden loss of power at pumping plants, and other situations. While there are surge suppression devices available, experience has shown that these devices, due to inactivity, rusting, silting, loss of power, overloading, etc., do not always properly arrest surge forces.

Therefore, water hammer must be considered in pipe thickness calculations. Procedures for the analysis of surge pressure in simple pipelines are presented in many reference works. The analysis of water hammer in a network of piping, such as that of a conventional water distribution system, is extremely complex. For this reason, standard water hammer allowances based on experience and good judgment are used in the Design Standards for Gray Cast Iron Pipe (ANSI A21.1) and for Ductile Iron Pipe (ANSI A21.50) as follows:

ALLOWANCES FOR SURGE PRESSURE

Pipe Size (in.)	Surge Pressure (psi) Gray Iron Pipe	Surge Pressure (psi) Ductile Iron Pipe
3-10	120	100
12-14	110	100
16-18	100	100
20	90	100
24	85	100
30	80	100
36	75	100
42-48	70	100

Each pipeline project is unique and the designer is cautioned to review possible causes of water hammer and to increase the allowance if circumstances dictate.

References

Hydraulics for Pipeliners by C. V. Lester. Pipeline News, September, 1955.

Water Hammer Nomograph. F. Caplan. Kaiser Engineers, Oakland, California.

Water Hammer Allowances in Pipe Design. Committee Report, JAWWA, Vol. 50, No. 3. March, 1958.

Standard Allowances for Water Hammer. Panel Discussion, JAWWA, Vol. 44, No. 11, November, 1952.

Practical Aspects of Water Hammer by S. Logan Kerr. JAWWA, Vol. 40, No. 6, May, 1948.

Water-Column Separation in Pump Discharge Lines by R. T. Richards. Transactions (ASME), August, 1954.

A Column Separation Accompanying Liquid Transients in Pipes by R. A. Faltzer. Transactions (ASME), December, 1967.

ANSI A21.4–1974
(AWWA C104–74)

Revision of
A21.4–1971
(AWWA C104–71)

AMERICAN NATIONAL STANDARD

for

CEMENT-MORTAR LINING FOR CAST-IRON AND DUCTILE-IRON PIPE AND FITTINGS FOR WATER

ADMINISTRATIVE SECRETARIAT

AMERICAN WATER WORKS ASSOCIATION

CO-SECRETARIATS
AMERICAN GAS ASSOCIATION
NEW ENGLAND WATER WORKS ASSOCIATION

Revised edition approved by American National Standards Institute, Inc., Mar. 7, 1974.

PUBLISHED BY

AMERICAN WATER WORKS ASSOCIATION

6666 West Quincy Avenue, Denver, Colo. 80235

Table of Contents

3
9
7

Committee Personnel

Subcommittee 4—Coatings and Linings—which reviewed and developed this revision, had the following personnel at the time of revision:

W. HARRY SMITH, *Chairman*
KENNETH W. HENDERSON, *Vice-Chairman*

User Members	*Producer Members*
STANLEY C. BAKER	FRANK J. CAMEROTA
KENNETH W. HENDERSON	W. D. GOODE
RAYMOND C. HOLMAN	WILLIAM U. MAHER
DAVID A. LINCOLN	JAMES H. SALE
R. E. MORRIS JR.	EDWARD C. SEARS
PETER E. PALLO	W. HARRY SMITH
EVERETT C. ROWLEY	ERNEST F. WAGNER

Standards Committee A21—Cast-Iron Pipe and Fittings—which reviewed and approved this standard, had the following personnel at the time of approval:

WALTER AMORY, *Chairman*
CARL A. HENRIKSON, *Vice-Chairman*
JAMES B. RAMSEY, *Secretary*

Organization Represented	*Name of Representative*
American Gas Association	LEONARD ORLANDO JR.
American Society of Civil Engineers	KENNETH W. HENDERSON
American Society of Mechanical Engineers	JAMES S. VANICK
American Society for Testing and Materials	ALBERT H. SMITH JR.
American Water Works Association	LLOYD W. WELLER
Cast Iron Pipe Research Association	CARL A. HENRIKSON
	EDWARD C. SEARS
	W. HARRY SMITH
Individual Producers	ALFRED F. CASE
	WILLIAM T. MAHER
Manufacturers' Standardization Society of the Valve and Fittings Industry	ABRAHAM FENSTER
New England Water Works Association	WALTER AMORY
Naval Facilities Engineering Command	STANLEY C. BAKER
Underwriters' Laboratories, Inc.	STANLEY E. AUCK
Canadian Standards Association	W. F. SEMENCHUK*

* Liaison representative without vote.

Foreword

This foreword is for information and is not a part of ANSI A21.4–1974 (AWWA C104–74).

I—History of Standard

The first recorded installation of cement-mortar linings in cast-iron pipe was in 1922 at Charleston, S.C., under the supervision of J. E. Gibson.

From 1922 to 1929, many installations were made under various manufacturers' specifications. In 1929, ASA (now ANSI) Sectional Committee A21 issued a tentative standard for cement-mortar linings. This was published as a tentative standard by AWWA in 1932. After many revisions and refinements, it was finally adopted by ASA in 1939 under the designation A21.4— Specifications for Cement-Mortar Lining for Cast-Iron Pipe and Fittings.

During the period 1940–52, much research was done on various types of cement, methods of manufacture, and methods of curing cement mortar to improve the quality of cement-mortar linings. As a result of this research, a revised edition of the 1939 standard was approved and issued in 1953.

The centrifugal process for lining was further developed during the 1940–52 period to provide the controls and techniques necessary for assurance of uniformity of thickness throughout the length of a pipe. Another major revision recognized the ability of cure-assist bituminous materials to provide controlled curing of the mortar. The use of this method was permitted as a substitute for the moist-curing process.

In 1958, Sectional Committee A21 was reorganized and subcommittees were established to study each group of standards in accordance with ASA's review and revision policy.

Subcommittee 4 (Coatings and Linings for Cast-Iron Pipe) was organized to examine the existing ASA A21.4–1953, "Standard for Cement-Mortar Lining for Cast-Iron Pipe and Fittings." This subcommittee completed its study of A21.4–53 and submitted a proposed revision to Sectional Committee A21 in 1963. The revised third edition was approved and issued in 1964.

The 1964 standard reduced the minimum permissible thickness of the lining. This reduction was based on more than 20 years of Cast Iron Pipe Research Association (CIPRA) studies of experimental test lines having cement-mortar linings varying from $\frac{1}{32}$ in. to $\frac{1}{4}$ in. in thickness, on field tests of linings of these thicknesses that had been in service for more than 30 years, and on the assurance of uniformity of thickness afforded by improvements in the centrifugal lining process.

Two thicknesses of lining were made available, and purchasers who required a lining thickness twice the standard thickness had the option of so specifying.

The cement linings were specified for use in water lines only. This qualification was made to avoid the use of cement-mortar linings in pipe carrying aggressive liquids, which would react with the lining to produce undesirable results.

The purchaser of cement-mortar-lined pipe or fittings for use with a water that is corrosive to calcium carbonate, such as a very soft water, is advised, before specifying the omission of the seal coat, to satisfy himself by appropriate test that such a lining will not impart objectionable hardness or alkalinity to the water. The procedure outlined in Sec. 4–14.4, modified by the substitution of the water with which the pipe is to be used for distilled water, is suggested as a convenient form of test.

The 1971 revision incorporated a standard test for toxicity of the seal coat material.

This standard does not include provisions for cement-mortar lining of pipelines in place.

II—Major Revision

The title of Sec. 4–13 has been changed from Finished Lining to Lining Quality, and the entire section has been rewritten to provide new requirements of acceptable lining.

III—Options

This standard includes certain options, which, if desired, *must* be specified. These are:

1. *Thickness of lining.* Two thicknesses of lining are available, and purchasers who require a lining thickness twice the standard thickness have the option of so specifying (Sec. 4–10).

2. *Seal coat.* As other seal coats than bituminous ones are available, this standard makes provision for their use (Sec. 4–14).

American National Standard for

Cement-Mortar Lining for Cast-Iron and Ductile-Iron Pipe and Fittings for Water

Sec. 4–1—Scope

This standard covers cement-mortar linings specified in the A21 series of ANSI Standards for Cast-Iron and Ductile-Iron Pipe and Fittings for Water and is intended for use as a supplement to those standards.

Sec. 4–2—Cement

The cement shall meet the requirements of "Standard Specifications for Portland Cement," ASTM Designation C150–73–a. The analysis and physical test records of each shipment shall be kept for reference for 1 year.

The type of cement selected shall be left to the option of the pipe and fittings manufacturer.

Sec. 4–3—Sand

4–3.1. *Type of sand.* The sand shall be well graded, from fine to coarse, and consist of inert granular material having hard, strong, durable, uncoated grains and meet the test requirements of Sec. 4–3.2.

4–3.2. *Testing of sand.* The sand shall be tested in accordance with the requirements of these sections:

4–3.2.1. *Sampling.* The sand to be tested shall be sampled according to Sections 14 and 15 of ASTM D75–71, "Standard Methods of Sampling Aggregates."

4–3.2.2. *Sieve tests.* The sand shall be tested with standard sieves, as defined in ASTM Designation E11–70, "Standard Specification for Wire-Cloth Sieves for Testing Purposes," and shall meet the requirements listed in Table 1. One sieve analysis shall be performed on each carload of sand delivered. For sand delivered by other means, one sieve analysis shall be made for each 50 tons.

4–3.2.3. *Colorimetric test.* The test for impurities shall be in accordance with ASTM C40–73, "Standard Method of Test for Organic Impurities in Sands for Concrete."

TABLE 1

Requirements for Sand Tested With Standard Sieves

Min. Thickness of Lining in.	Sieve Requirement*	
	100 Per Cent of Sand Shall Pass (Sieve No.)	75 Per Cent of Sand Shall Pass (Sieve No.)
$\frac{1}{16}$	12	20
$\frac{3}{32}$	12	16
$\frac{1}{8}$	12	†
$\frac{3}{16}$	8	†
$\frac{1}{4}$	6	†

* Not more than 10 per cent, by weight, of any sand shall pass through sieve No. 100.
† Not applicable.

Under this test, the sand shall not produce a color darker than required in the standard. The sand shall be acceptable, however, if it is shown by adequate test that the impurities causing the color are not harmful to the strength or other specified properties of the finished lining.

The colorimetric tests of sand from an established source of supply shall be made once each 6 months. For sand from a new source, these tests shall be made not less than once a month for a period of 6 months.

4–3.2.4. *Decantation test.* The sand shall be tested according to ASTM C117–69, "Standard Method of Test for Materials Finer Than No. 200 (75-μm) Sieve in Mineral Aggregates by Washing."

At the option of the manufacturer, the clay content and sand grain fineness may be determined by using the American Foundrymen's Society procedure, described in the *Foundry Sand Handbook,* Seventh Edition, Section 5. By this latter method, the total percentage finer than No. 200 sieve, as defined in ASTM Designation C119–71, is equal to the AFS percentage of clay plus the percentage passing through the No. 200 sieve.

No more than 2 per cent shall be lost in the decantation test.

The decantation tests of sand from an established source of supply shall be made once each 6 months. For sand from a new source, these tests shall be made not less than once a month for a period of 6 months.

4–3.2.5. *Test records.* The requirements of Sec. 4–3.2.2, 4–3.2.3, and 4–3.2.4 shall be met, and the records shall be filed for reference for 1 year.

Sec. 4–4—Water

The water used for tempering the mortar shall meet the requirements of the United States Public Health Service Drinking Water Standards 1962.

Sec. 4–5—Mortar

Mortar for the lining shall be composed of cement, sand, and water. The mortar shall be well mixed and of proper consistency to produce a dense, homogeneous lining that will adhere firmly to the pipe or fitting surface. Admixtures may be used, provided the linings meet all the requirements of this standard. The cement mortar shall contain not less than one part of cement to two parts of sand, by volume.

Sec. 4–6—Preparation of Pipe and Fittings for Lining

The surface to be lined shall be free from foreign material, which would adversely affect the lining adhesion or cause inclusions, blisters, or voids in the lining. The surface shall be free from projections of iron which may protrude through the lining.

Sec. 4–7—Method of Lining

4–7.1. *Lining of pipe and fittings.* Pipe shall be lined by the centrifugal process. Fittings shall be lined by a process that will produce linings meeting the requirements of this standard.

4–7.2. *Mortar.* The waterway surfaces of pipe and fittings shall be completely covered with the specified mortar. The mortar shall be entirely free from holidays or visible bubbles of air and shall be thoroughly compacted throughout. The consistency of the mortar and the time and speed of spinning of the pipe shall be so adjusted as to minimize the segregation of the sand from the cement and to deliver the finished lining substantially free of laitance.

4–7.3. *Repair of defective or damaged areas of linings.* Defective or damaged areas of linings may be patched by cutting out the defective or damaged lining to the metal so that the edges of the lining not removed are perpendicular or slightly undercut. A stiff mortar shall be prepared in accordance with Sec. 4–5. The cut-out area and the adjoining lining shall be thoroughly wetted, and the mortar applied and troweled smooth with the adjoining lining. After any surface water has evaporated, but while the patch is still moist, it shall be cured as specified in Sec. 4–12.

Sec. 4–8—Socket

The socket shall be free of mortar.

Sec. 4–9—Protection of Work

The lined pipe and fittings shall be protected from extreme heat due to direct rays of the sun, from impact of rainfall, and from freezing temperatures until the linings have cured sufficiently to withstand these conditions.

Sec. 4–10—Thickness of Lining

4–10.1. *Standard thickness.* The thickness of linings for pipe and fittings, as determined in Sec. 4–11, shall be not less than $\frac{1}{16}$ in. for 3–12 in. pipe, $\frac{3}{32}$ in. for 14–24 in. pipe, and $\frac{1}{8}$ in. for 30–54 in. pipe.

4–10.2. *Double thickness.* Linings with thicknesses twice those specified in Sec. 4–10.1 shall be furnished if specified by the purchaser.

4–10.3. *Taper of linings.* Lining thickness may taper to less than the specified minimum thickness at the ends of the pipe or fitting. The length of the taper shall be as short as practicable and shall not exceed 2 in.

4–10.4. *Permitted tolerances.* A thickness tolerance of $+\frac{1}{8}$ in. shall be permitted on pipe and $+\frac{1}{4}$ in. on fittings.

Sec. 4–11—Determination of Thickness

Lining thickness shall be determined at intervals frequent enough to assure compliance. Thickness of lining may be determined by means of spear measurement, with a hardened-steel point not larger than $\frac{1}{16}$ in. in diameter. The inspector shall pierce the lining immediately after it is placed in the pipe or fitting and before the mortar has set. The lining shall be pierced at four equidistant points on two cross sections of the barrel at each end of the pipe or fitting. The first set shall be not more than 4 in. from the respective ends of the pipe or fitting. The second set shall be made as far into the interior of the pipe or fitting as can be readily reached without injuring the lining.

Sec. 4–12—Curing

The lining shall be cured in such a manner as to produce a properly hydrated mortar lining that is hard and durable and will otherwise meet the requirements of Sec. 4–13. The cure may be effected by the application of a seal coat to the still-moist lining.

Sec. 4–13—Lining Quality

The lining shall be free from voids, ridges, or corrugations that reduce the thickness of lining to less than the specified thickness.

Unbonded areas of cement lining in a pipe or fitting are acceptable if the dimension of any single area does not exceed the nominal diameter in the circumferential direction and in longitudinal direction does not exceed the nominal diameter or 12 in., whichever is greater.

Longitudinal cracks less than 9 in. in length or less than the nominal diameter, whichever is greater, are acceptable. Circumferential cracks of any length are acceptable. Surface crazing is acceptable.

Repair of any unacceptable condition is permitted in the field, in accordance with Sec. 4–7.3.

Sec. 4–14—Seal Coat

4–14.1. *General.* Unless otherwise specified, the cement lining shall be given a seal coat of bituminous material. Other seal coat materials may be used, but they shall be agreed upon at the time of purchase and shall be specified on the purchase order.

4–14.2. *Seal coat characteristics.* The seal coat shall be continuous and shall adhere to the mortar lining at all points. The seal coat, after drying for at least 48 hr, shall have no deleterious effect upon the quality, color, taste, or odor of potable water.

4–14.3. *Limit of toxic substances.*

4–14.3.1. *Requirements.* The seal coat material shall not yield chloroform–soluble extractives, corrected for zinc extractives as zinc oleate, in excess of 18 mg per sq in. of surface exposed or of 50 ppm by weight of the water capacity of the test container.

4–14.3.2. *Frequency of test.* The seal coat material shall be tested at sufficiently frequent intervals to determine that it meets the requirements prescribed in Sec. 4–14.3.1.

4–14.3.3. *Method of testing.* The procedure used in the determination of the amount of toxic substances shall be in accordance with the FDA "Method of Testing for Toxicity of Coating Material Intended for Use in Transporting or Holding Food or Potable Water," as described in *Food Additives Amendment and Code of Federal Regulations,* pp. 13.0 through 13.8, April 16, 1963, Food and Drug Administration, US Department of Health, Education and Welfare. The seal coat material shall be extracted with distilled or demineralized water at 120F for 24 hr.

4–14.4. *Leaching resistance.*

4–14.4.1. *Requirements.* The seal-coated pipe shall impart to the water during any 24-hr test period no more than 25 ppm of hardness or 25 ppm of total alkalinity, and shall impart no caustic alkalinity.

4–14.4.2. *Frequency of test and records.* Leaching tests shall be made at sufficiently frequent intervals to assure compliance. The results of one test each month shall be filed for reference for 1 year.

4–14.4.3. *Method of testing.* The seal-coated pipe shall be tested as follows:

The test specimen shall be at least 6 in. in length, either cut or isolated by suitable closure pieces. When a cut section is used, it shall be bedded on end in a shallow pan of molten paraffin. After the paraffin has cooled, the cut section shall be filled nearly to the top with distilled or demineralized water at laboratory temperature. The top shall be covered with a glass plate and sealed with petroleum jelly. If an isolated section is used, it shall be filled through a tap in the closure device with distilled or demineralized water at laboratory temperature.

In either case, the water in the specimen shall be changed and tested after 24-hr contact on each of 3 successive days. The methods and procedures used in the determination of hardness and alkalinity shall be those prescribed in *Standard Methods for the Examination of Water and Wastewater,* APHA, AWWA, and WPCF, thirteenth edition, 1971.

END OF STANDARD

External Corrosion

The resistance of gray iron pipe to exterior corrosion has long since been established. Comparative corrosion tests between ductile and gray iron have now shown that the corrosion resistance of ductile iron pipe is greater than that of gray iron pipe. In a survey of water utility officials throughout the United States, it was found that about 5 percent of the soils in water distribution systems are corrosive to these pipe materials. CIPRA has developed a system for evaluation of soil corrosivity based on many years of experience. The system is described in the Appendix to ANSI/AWWA Standard C105–72 (R1977). Most common causes of corrosion involve naturally corrosive soils or soil contaminants such as peat, muck, cinders, mine wastes, deicing salts and stray direct current.

The Cast Iron Pipe Research Association has completed soil surveys along hundreds of miles of proposed pipeline installation using the standard soil test procedures. These procedures involve environmental characteristics listed in Table A1 of the Appendix to ANSI/AWWA C105. With the additional consideration of potential corrosion due to stray direct current, these procedures have proven to be highly dependable since their establishment in 1968 and proper application of polyethylene encasement where the environment is determined to be corrosive has likewise proven to be dependable.

Polyethylene encasement for protection of gray and ductile iron pipe in corrosive soils and locations of potential stray current influence is a universal practice in the USA as well as most European countries and industrial nations throughout the world. Its success is attributed to the uniform environment it provides for the pipe, its dielectric strength, and the barrier it provides between the pipe and the corrosive environment. No attempt is made to make the encasement absolutely watertight. Water, upon entering the annular space between the pipe and the encasement, may possess the corrosive characteristics of the surrounding soil; however, there is no ongoing ingress and egress of water and the results of the initial corrosion reaction create a stable situation which further enhances the polyethylene encasement protection system. It is estimated that about 4,000 miles of gray and ductile iron pipe are protected with polyethylene encasement in the USA alone. Failures are so few as to be negligible over a period exceeding 20 years.

406

AMERICAN NATIONAL STANDARD

for

POLYETHYLENE ENCASEMENT FOR GRAY AND DUCTILE CAST-IRON PIPING FOR WATER AND OTHER LIQUIDS

4
0
7

ADMINISTRATIVE SECRETARIAT

AMERICAN WATER WORKS ASSOCIATION

CO-SECRETARIATS

AMERICAN GAS ASSOCIATION
NEW ENGLAND WATER WORKS ASSOCIATION

NOTICE

This Standard has been especially printed by the American Water Works Association for incorporation into this volume. It is current as of December 1, 1975. It should be noted, however, that all AWWA Standards are updated at least once in every five years. Therefore, before applying this Standard it is suggested that you confirm its currency with the American Water Works Association.

First edition approved by American National Standards Institute, Inc., Dec. 27, 1972

PUBLISHED BY

AMERICAN WATER WORKS ASSOCIATION
6666 West Quincy Avenue, Denver, Colorado 80235

4
0
8

Table of Contents

4
0
9

iii

Committee Personnel

Subcommittee No. 4, which reviewed and recommended reaffirmation of this standard without revision, had the following personnel at that time:

W. HARRY SMITH, *Chairman*
KENNETH W. HENDERSON, *Vice-Chairman*

User Members	*Producer Members*
STANLEY C. BAKER	FRANK J. CAMEROTA
RAYMOND C. HOLMAN	W. D. GOODE
DAVID A. LINCOLN	JAMES H. SALE
RICHARD E. MORRIS JR.	EDWARD C. SEARS

Standards Committee A21—Cast Iron Pipe and Fittings, which reviewed and reaffirmed this standard without revision had the following personnel at that time.

LLOYD W. WELLER, *Chairman*
EDWARD C. SEARS, *Vice-Chairman*
PAUL A. SCHULTE, *Secretary*

Organization Represented	*Name of Representative*
American Gas Association	LEONARD ORLANDO JR.
American Society of Civil Engineers	KENNETH W. HENDERSON
American Society of Mechanical Engineers	JAMES S. VANICK
American Society for Testing and Materials	BEN C. HELTON *
	JOSEPH J. PALMER *
American Water Works Association	J. PORTER HENNINGS
	RAYMOND J. KOCOL
	ROBERT L. LEE
	ARNOLD M. TINKEY
	LLOYD W. WELLER
Cast Iron Pipe Research Association	THOMAS D. HOLMES
	HAROLD M. KENNEDY JR.
	EDWARD C. SEARS
	W. HARRY SMITH
Individual Producer	ALFRED F. CASE
Manufacturers' Standardization Society of the Valve and Fittings Industry	ABRAHAM FENSTER
New England Water Works Association	WALTER AMORY
Naval Facilities Engineering Command	STANLEY C. BAKER
Underwriters' Laboratories, Inc.	LEE J. DOSEDLO
Canadian Standards Association	W. F. SEMENCHUK *

* Liaison representative without vote.

Foreword

This foreword is provided for information and is not a part of ANSI/AWWA C105.

In 1926, ASA (now ANSI) Committee A21—Cast-Iron Pipe and Fittings was organized under the sponsorship of AGA, ASTM, AWWA, and NEWWA. The current sponsors are AGA, AWWA, and NEWWA, and the present scope of Committee A21 activity is

Standardization of specifications for cast-iron and ductile-iron pressure pipe for gas, water, and other liquids, and fittings for use with such pipe; these specifications to include design, dimensions, materials, coatings, linings, joints, accessories, and methods of inspection and test.

In 1958, Committee A21 was reorganized. Subcommittees were established to study each group of standards in accordance with the review and revision policy of ASA (now ANSI). The present scope of Subcommittee No. 4—Coatings and Linings is

To review the matter of interior and exterior corrosion of gray and ductile-iron pipe and fittings; and to draft standards for the interior and exterior protection of gray and ductile-iron pipe and fittings.

In accordance with this scope, Subcommittee No. 4 was charged with the responsibility for

1. Development of standards on polyethylene encasement materials and their installation as corrosion protection, when required, for gray and ductile cast-iron pipe and fittings.
2. Development of procedures for the investigation of soil to determine when polyethylene protection is indicated.

In response to these assignments, Subcommittee No. 4 has

1. Developed Standard ANSI A21.5–1972 (AWWA C105–72) for Polyethylene Encasement for Gray and Ductile Cast-Iron Piping for Water and Other Liquids.
2. Developed Appendix A outlining soil-investigation procedures.

In 1976 Subcommittee No. 4 reviewed the 1972 edition and submitted a recommendation to Committee A21 that the standard be reaffirmed without change from the 1972 edition, except for the updating of this foreword.

History

Loose polyethylene encasement was first used experimentally in the US for protection of cast-iron pipe in corrosive environments in 1951. The first field installation of polyethylene wrap on cast-iron pipe in an operating water system was in 1958, and consisted of about 600 ft of 12-in. pipe installed in a waste dump fill area. Since that time, hundreds of installations have been made in severely corrosive soils throughout the US in pipe sizes ranging from 4–54 in. in diameter. Polyethylene encasement has been used as a soil-corrosion preventative in Canada, England, France, Germany, and several other countries since development of the procedure in the United States.

Research

Research by the Cast Iron Pipe Research Assn. (CIPRA) on several severely corrosive test sites has indicated that polyethylene encasement provides a high degree of protection and results in minimal and generally insignificant exterior surface corrosion of gray and ductile cast-iron pipe thus protected.

Investigations of many field installations in which loose polyethylene encasement has been used as protection for gray and ductile cast-iron pipe against soil corrosion have confirmed CIPRA's findings with the experimental specimens. These field installations have further indicated that the dielectric capability of polyethylene provides shielding for gray and ductile cast-iron pipe against stray direct current at most levels encountered in the field.

Useful Life of Polyethylene

Tests on polyethylene used in the protection of gray and ductile cast-iron pipe have shown that after 19 years of exposure to severely corrosive soils, strength loss and elongation reduction are insignificant. Studies by the Bureau of Reclamation of the US Dept.

of the Interior * on polyethylene film used underground showed that tensile strength was nearly constant in a 7-yr test period and that elongation was only slightly affected. The Bureau's accelerated soil burial testing (acceleration estimated to be five to ten times that of field conditions) showed polyethylene to be highly resistant to bacteriological deterioration.

Exposure to Sunlight

Prolonged exposure to sunlight will eventually deteriorate polyethylene film. Therefore, such exposure prior to backfilling the wrapped pipe should be kept to a minimum. If several weeks of exposure prior to backfilling is anticipated, Class C material should be used (see Sec. 5–3.1.1).

Options

This standard includes certain options, which, if desired, must be specified. These are

1. Class of polyethylene material (Sec. 5–3).
2. Installation method A, B, or C (Sec. 5–4) if there is a preference.

---

* US Dept. of the Interior, Bureau of Reclamation, "Laboratory and Field Investigations of Plastic Films," Rept. No. ChE-82, Sep. 1968.

American National Standard for

Polyethylene Encasement for Gray and Ductile Cast-Iron Piping for Water and Other Liquids

Sec. 5–1—Scope

This standard covers materials and installation procedures for polyethylene encasement to be applied to underground installations of gray and ductile cast-iron pipe. This standard also may be used for polyethylene encasement of fittings, valves, and other appurtenances to gray and ductile cast-iron pipe systems.

Sec. 5–2—Definition

5–2.1. *Polyethylene encasement.* The encasement of piping with polyethylene film in tube or sheet form.

Sec. 5–3—Materials

5–3.1. *Polyethylene.* Polyethylene film shall be manufactured of virgin polyethylene material conforming to the following requirements of ASTM Standard Specification D-1248-68—Polyethylene Plastics Molding and Extrusion Materials:

5–3.1.1. *Raw material used to manufacture polyethylene film.*

Type I
Class .... A (natural color) or C (black)
Grade E-1
Flow rate (formerly melt index) ...... 0.4 maximum
Dielectric strength .. Volume resistivity, minimum ohm-cm^3=10^{15}

5–3.1.2. *Polyethylene film.*

Tensile strength .... 1,200 psi minimum
Elongation .... 300 per cent minimum
Dielectric strength .. 800 V/mil thickness minimum

5–3.2. *Thickness.* Polyethylene film shall have a minimum nominal thickness of 0.008 in. (8 mils). The minus tolerance on thickness shall not exceed 10 per cent of the nominal thickness.

5–3.3. *Tube size or sheet width.* Tube or sheet size for each pipe diameter shall be as listed in Table 5.1.

TABLE 5.1

Tube and Sheet Sizes

Nominal Pipe Diameter *in.*	Minimum Polyethylene Width *in.*	
	Flat Tube	Sheet
3	14	28
4	16	32
6	20	40
8	24	48
10	27	54
12	30	60
14	34	68
16	37	74
18	41	82
20	45	90
24	54	108
30	67	134
36	81	162
42	95	190
48	108	216
54	121	242

Sec. 5–4—Installation

5–4.1. *General.* The polyethylene encasement shall prevent contact between the pipe and the surrounding backfill and bedding material but is not intended to be a completely air- and watertight enclosure. Overlaps shall be secured by the use of adhesive tape, plastic string, or any other material capable of holding the polyethylene encasement in place until backfilling operations are completed.

5–4.2. *Pipe.* This standard includes three different methods for the installation of polyethylene encasement on pipe. Methods A and B are for use with polyethylene tubes and method C is for use with polyethylene sheets.

5–4.2.1. *Method A.* Cut polyethylene tube to a length approximately 2 ft longer than that of the pipe section. Slip the tube around the pipe, centering it to provide a 1-ft overlap on each adjacent pipe section, and bunching it accordion-fashion lengthwise until it clears the pipe ends.

Lower the pipe into the trench and make up the pipe joint with the preceding section of pipe. A shallow bell hole must be made at joints to facilitate installation of the polyethylene tube.

After assembling the pipe joint, make the overlap of the polyethylene tube. Pull the bunched polyethylene from the preceding length of pipe, slip it over the end of the new length of pipe, and secure in place. Then slip the end of the polyethylene from the new pipe section over the end of the first wrap until it overlaps the joint at the end of the preceding length of pipe. Secure the overlap in place. Take up the slack width to make a snug, but not tight, fit along the barrel of the pipe, securing the fold at quarter points.

Repair any rips, punctures, or other damage to the polyethylene with adhesive tape or with a short length of polyethylene tube cut open, wrapped around the pipe, and secured in place. Proceed with installation of the next section of pipe in the same manner.

5–4.2.2. *Method B.* Cut polyethylene tube to a length approximately 1 ft shorter than that of the pipe section. Slip the tube around the pipe, centering it to provide 6 in. of bare pipe at each end. Make polyethylene snug, but not tight; secure ends as described in Sec. 5–4.2.1.

Before making up a joint, slip a 3-ft length of polyethylene tube over the end of the preceding pipe section, bunching it accordion-fashion lengthwise. After completing the joint, pull the 3-ft length of polyethylene over the joint, overlapping the polyethylene previously installed on each adjacent section of pipe by at least 1 ft; make snug and secure each end as described in Sec. 5–4.2.1.

Repair any rips, punctures, or other damage to the polyethylene as described in Sec. 5–4.2.1. Proceed with installation of the next section of pipe in the same manner.

5–4.2.3. *Method C.* Cut polyethylene sheet to a length approximately 2 ft longer than that of the pipe section. Center the cut length to provide a 1-ft overlap on each adjacent pipe section, bunching it until it clears the pipe ends. Wrap the polyethylene around the pipe so that it circumferentially overlaps the top quadrant of the pipe. Secure the cut edge of polyethylene sheet at intervals of approximately 3 ft.

Lower the wrapped pipe into the trench and make up the pipe joint with the preceding section of pipe. A shallow bell hole must be made at joints to facilitate installation of the polyethylene. After completing the joint, make the overlap as described in Sec. 5–4.2.1.

Repair any rips, punctures, or other damage to the polyethylene as described in Sec. 5–4.2.1. Proceed with installation of the next section of pipe in the same manner.

5–4.3. *Pipe-shaped appurtenances.* Cover bends, reducers, offsets, and other pipe-shaped appurtenances with polyethylene in the same manner as the pipe.

5–4.4. *Odd-shaped appurtenances.* When valves, tees, crosses, and other odd-shaped pieces cannot be wrapped practically in a tube, wrap with a flat sheet or split length of polyethylene tube by passing the sheet under the appurtenance and bringing it up around the body. Make seams by bringing the edges together, folding over twice, and taping down. Handle width and overlaps at joints as described in Sec. 5–4.2.1. Tape polyethylene securely in place at valve stem and other penetrations.

5–4.5. *Openings in encasement.* Provide openings for branches, service taps, blow-offs, air valves, and similar appurtenances by making an X-shaped cut in the polyethylene and temporarily folding back the film. After the appurtenance is installed, tape the slack securely to the appurtenance and repair the cut, as well as any other damaged areas in the polyethylene, with tape.

5–4.6. *Junctions between wrapped and unwrapped pipe.* Where polyethylene-wrapped pipe joins an adjacent pipe that is not wrapped, extend the polyethylene wrap to cover the adjacent pipe for a distance of at least 2 ft. Secure the end with circumferential turns of tape.

5–4.7. *Backfill for polyethylene-wrapped pipe.* Use the same backfill material as that specified for pipe without polyethylene wrapping, exercising care to prevent damage to the polyethylene wrapping when placing backfill. Backfill material shall be free from cinders, refuse, boulders, rocks, stones, or other material that could damage polyethylene. In general, backfilling practice should be in accordance with AWWA Standard C600-64, or the latest revision thereof.

4
1
5

Appendix A

Notes on Procedures for Soil Survey Tests and Observations and Their Interpretation to Determine Whether Polyethylene Encasement Should Be Used

This appendix is for information and is not a part of ANSI/AWWA C105.

In the appraisal of soil and other conditions that affect the corrosion rate of gray and ductile cast-iron pipe, a minimum number of factors must be considered. They are outlined here. A method of evaluating and interpreting each factor and a method of weighting each factor to determine whether polyethylene encasement should be used are subsequently described.

Soil Survey Tests and Observations

1. Earth resistivity
   (a) Four-pin Vibroground
   (b) Single-probe Vibroground
   (c) Saturated sample Vibroground
2. pH
3. Oxidation-reduction (Redox) potential
4. Sulfides
   (a) Azide (qualitative)
5. Moisture content (relative)
   (a) Prevalence
6. Soil description
   (a) Particle size
   (b) Uniformity
   (c) Type
   (d) Color
7. Potential stray direct current
   (a) Nearby cathodic protection utilizing rectifiers
   (b) Railroads (electric)
   (c) Industrial equipment, including welding
   (d) Mine equipment
8. Experience with existing installations in the area

1. *Earth resistivity.* There are three methods for determining earth resistivity: four-pin, single probe, and soil box. In the field, a four-pin determination should be made with pins spaced at approximate pipe depth. This method yields an average of resistivity from the surface to a depth equal to pin spacing. However, results are sometimes difficult to interpret where dry top soil is underlain with wetter soils and where soil types vary with depth. The Wenner configuration is used in connection with a Vibroground which is available with varying ranges of resistance. For all-around use, a unit with a capacity of up to 10^4 ohms is suggested because of its versatility in permitting both field and laboratory testing in most soils.

Because of the afore-mentioned difficulty in interpretation, the same unit may be used with a single probe that yields resistivity at the point of the probe. A boring is made into the subsoil so that the probe may be pushed into the soil at the desired depth.

Inasmuch as the soil may not be typically wet, a sample should be removed for resistivity determination, which may be accomplished with any one of several laboratory units that permit the introduction of water to saturation, thus simulating saturated field conditions. Each of these units is used in conjunction with the Vibroground.

Interpretation of resistivity results is extremely important. To base an opinion on a four-pin reading with dry top soil averaged with wetter subsoil would probably result in an inaccurate premise. Only by reading the resistivity in soil at pipe depth can an accurate interpretation be made. Also, every effort should be made to determine the local situation concerning ground-water table, presence of shallow ground water, and approximate percentage of time the soil is likely to be water saturated.

With gray and ductile cast-iron pipe, resistance to corrosion through products of corrosion is enhanced if there are dry periods during each year. Such periods seem to permit hardening or toughening of the corrosion scale or products, which then become impervious and serve as better insulators.

In making field determinations of resistivity, temperature is important. The result obtained increases as temperature decreases. As the water in the soil approaches freezing, resistivity increases greatly, and, therefore, is not reliable. Field determinations under frozen soil conditions should be avoided. Reliable results under such conditions can be obtained only by collection of suitable subsoil samples for analysis under laboratory conditions at suitable temperature.

Interpretation of resistivity. Because of the wide variance in results obtained under the methods described, it is difficult specifically to interpret any single reading without knowing which method was used. It is proposed that interpretation be based on the lowest reading obtained with consideration being given to other conditions, such as normal moisture content of the soil in question. Because of the lack of exact correlation between experiences and resistivity, it is necessary to assign ranges of resistivity rather than specific numbers. In Table A1 (p. 7) points are assigned to various ranges of resistivity. These points, when considered along with points assigned to other soil characteristics, are meaningful.

2. *pH*. In the pH range of 0.0 to 4.0, the soil serves well as an electrolyte, and total acidity is important. In the pH range of 6.5 to 7.5, soil conditions are optimum for sulfate reduction. In the pH range of 8.5 to 14.0, soils are generally quite high in dissolved salts, yielding a low soil resistivity.

A Beckman Electromate portable pH meter is suitable for both field and laboratory pH determinations. In this test, glass and reference electrodes are pushed into the soil sample and a direct reading is made, following suitable temperature setting on the instrument. Normal procedures are followed for standardization.

3. *Oxidation-reduction (Redox) potential.* The oxidation-reduction (Redox) potential of a soil is significant because the most common sulfate-reducing bacteria can live only under anaerobic conditions. A Redox potential greater than +100 mV shows the soil to be sufficiently aerated so that it will not support sulfate reducers. Potentials of 0 to +100 mV may or may not indicate anaerobic conditions; however, a negative Redox potential definitely indicates anaerobic conditions under which sulfate reducers thrive. This test also is accomplished using a Beckman meter, with platinum and reference electrodes inserted into the soil sample, which permits a reading of potential between the two electrodes. It should be noted that soil samples removed from a boring or excavation can

undergo a change in Redox potential on exposure to air. Such samples should be tested immediately on removal from the excavation. Experience has shown that heavy clays, muck, and organic soils are often anaerobic, and these soils should be regarded as potentially corrosive.

4. *Sulfides.* The sulfide determination is recommended because of its field expediency. A positive sulfide reaction reveals a potential problem due to sulfate-reducing bacteria. The sodium azide–iodine qualitative test is used. In this determination, a solution of 3 per cent sodium azide in a 0.1 N iodine solution is introduced into a test tube containing a sample of the soil in question. Sulfides catalyze the reaction between sodium azide and iodine, with the resulting evolution of nitrogen. If strong bubbling or foaming results, sulfides are present, and the presence of sulfate-reducing bacteria is indicated. If very slight bubbling is noted, sulfides are probably present in small concentration and the result is noted as a trace.

5. *Moisture content.* Since prevailing moisture content is extremely important to all soil corrosion, every effort must be made to determine this condition. It is not proposed, however, to determine specific moisture content of a soil sample, because of the probability that content varies throughout the year, but to question local authorities who are able to observe the conditions many times during the year. (Although mentioned under item 1, this variability factor is being reiterated to emphasize the importance of notation.)

6. *Soil description.* In each investigation, soil types should be completely described. The description should include color and physical characteristics, such as particle size, plasticity, friability, and uniformity. Observation and testing will reveal whether the soil is high in organic content; this should be noted. Experience has shown that in a given area, corrosivity may often be reflected in certain types and colors of soil. This information is valuable for future investigations or for determining the most likely soils to suspect. Soil uniformity is important because of the possible development of local corrosion cells due to the difference in potential between unlike soil types, both of which are in contact with the pipe. The same is true for uniformity of aeration. If one segment of soil contains more oxygen than a neighboring segment, a corrosion cell can develop from the difference in potential. This cell is known as a differential aeration cell.

There are several basic types of soils that should be noted: sand, loam, silt, clay, muck. Unusual soils, such as peat or soils high in foreign material, also should be noted and described.

7. *Potential stray direct current.* Any soil survey should include consideration of possible stray direct current with which the proposed gray or ductile cast-iron pipe installation might interfere. The widespread use of rectifiers and ground beds for cathodic protection of underground structures has resulted in a considerable threat from this source. Proximity of such cathodic protection systems should be noted. Among other potential sources of stray direct current are electric railways, industrial equipment, including welding, and mine transportation equipment.

8. *Experience with existing installations.* The best information on corrosivity of soil with respect to gray and ductile cast-iron pipe is the result of experience with these materials in the

area in question. Every effort should be made to acquire such data by questioning local officials and, if possible, by actual observation of existing installations.

Soil-Test Evaluation

Using the soil-test procedures described herein, the following tests are considered in evaluating corrosivity of the soil: resistivity, pH, Redox potential, sulfides, and moisture. For each of these tests, results are categorized according to their contribution to corrosivity. Points are assigned based on experience with gray and ductile cast-iron pipe. When results of these five observations are available, the assigned points are totaled. If the sum is equal to ten or more, the soil is corrosive to gray or ductile cast-iron pipe and protection against exterior corrosion should be provided. This system is limited to soil corrosion and does not include consideration of stray direct current. Table A1 lists points assigned to the various test results.

General. These notes deal only with gray and ductile cast-iron pipe, the soil environment in which they will serve, and methods of determining need for polyethylene encasement.

TABLE A1

Soil-Test Evaluation*

Soil Characteristics	Points
Resistivity—*ohm-cm* (based on single probe at pipe depth or water-saturated Miller soil box):	
<700	10
700–1,000	8
1,000–1,200	5
1,200–1,500	2
1,500–2,000	1
>2,000	0
pH:	
0–2	5
2–4	3
4–6.5	0
6.5–7.5	0†
7.5–8.5	0
>8.5	3
Redox potential:	
> +100 mV	0
+50 to +100 mV	3.5
0 to +50 mV	4
Negative	5
Sulfides:	
Positive	3.5
Trace	2
Negative	0
Moisture:	
Poor drainage, continuously wet	2
Fair drainage, generally moist	1
Good drainage, generally dry	0

* Ten points = corrosive to gray or ductile cast-iron pipe; protection is indicated.
† If sulfides are present and low or negative Redox potential results are obtained, three points shall be given for this range.

Expanding Soils & Their Effect on Underground Pipe

Expanding Soils & Their Effect on Underground Pipe

First, a definition.

An expansive soil might be defined as any soil that swells or shrinks significantly upon wetting or (especially) upon drying, thus showing a volume change. Expansive soils are, in fact, usually clay soils containing large fractions of colloid-sized particles ranging from less than 1 micron to 2 microns in size. Such clays found in the continental United States are silicate clays: kaolinite, montmorillonite, hydrous micas or illites, chlorite, vermiculite, and attapulgite.

Field identification of expansive soils.

Positive field determination of an expansive soil and its potentially adverse effect on underground pipe is nearly impossible without proper test equipment. Tentative identification can be made by observing soil characteristics.

A potentially expansive soil may: (1) become very hard on drying and crack or fissure extensively; (2) become very sticky when wetted; (3) be plastic over a wide range of water content; (4) have a soapy or slick feeling when rubbed between the fingers; (5) be fine-grained, with little sand or coarse material; or (6) simply be known as a clay.

Experience with other structures in or on the soil can be helpful. Thus expansive soil may be present where there are apparently unexplainable beam breaks in the piping system, plus heaved or cracked pavement, curbs, etc., which cannot be explained by frost action or poor preparation of the support soil.

The sure way—laboratory testing.

The soil consolidometer is normally used to determine the swell pressure value of an expansive soil. However, a simpler and faster device with reasonable accuracy is also available—the Soil Potential Volume Change Meter. Readings from this Meter are converted to a swell index value in pounds per square foot (psf). This value can then be converted to Potential Volume Change rating and approximate plasticity index.

Cast Iron Pipe Research Association (CIPRA) has used this Soil Meter for nearly four years, testing hundreds of samples from many areas of the U.S., including Alaska and Hawaii.

Interpreting the test.

A Potential Volume Change Number is assigned to each swell index value obtained from tests with the meter. Here are the ranges:

Swell Index (psf)	PVC Number	Category of Expansive Soil
0-1700	0-2	Noncritical
1700-3200	2-4	Marginal
3200-4700	4-6	Critical
4700 and above	6 and above	Very Critical

Swell index values of 3000 psf and above (at Potential Volume Change Numbers above 2) are presently considered to indicate soils that may require special pipe selection.

Distribution of expansive soils.

The swell index tests performed by CIPRA and those reported in the **Guide to the Use of the FHA Soil Potential Volume Change Meter, FHA 595.** Indicate a rather wide distribution of expansive soils across the United States. Values from a series of samples follow:

Location	Swell Index (psf)	Location	Swell Index (psf)	Location	Swell Index (psf)
Arkansas	3,069	Michigan	3,968	Ohio	2,890
California	6,070	Missouri	8,604	Pennsylvania	3,726
California	4,958	Missouri	7,666	South Dakota	8,604
California	4,790	Mississippi	17,260	South Dakota	5,043
Hawaii (Oahu)	10,061	Minnesota	2,800	Texas	10,061
Illinois	6,245	North Carolia	3,248	Texas	8,989
Illinois	4,048	Oklahoma	4,460	Texas	7,042
Kansas	2,851				

Damage by expansive soils.

These soils may severely damage structures on or in them, usually more often by swelling than by shrinking. There is little documentation on such damage to pipelines. Several utilities in areas with highly expansive soils indicate that damage does occur: beam breaks in stronger rigid pipes and both beam breaks and ring crush in weaker rigid pipe. Pipe fabricated from strong ductile material appears to be little affected, and assumes a slightly new position to relieve stress.

Because of short-term experience with new pipe materials, such as plastic, it is not now possible to judge their performance in this environment.

What to watch for.

Moisture change in expansive soils is responsible for their volume change. So, even though a soil may have a high potential for swell or shrinkage, its actual volume change will depend on the amount of moisture fluctuation. Long, wet periods followed by long, dry periods favor high volume changes. And dry climates (with periodic high ground water tables) or relatively dry soil conditions with a source of moisture (such as a water leak) can also create problems.

The installation trench itself can, for a long period, attract water. Covering the trench with pavement, etc. also invites water invasion. Soil-volume changes are likely.

Shallow cover over pipe in expansive soils with relatively long drying periods can damage the pipe as differential soil pressures develop. Deeper burial of the pipe may place it in a more stable moisture condition and permit more earth weight to offset swell pressures from below.

Bedding and backfill around the pipe with selected granular material may also help, although better pipe support to sustain backfill load may be the only benefit.

About corrosion problems:

With more expansive soil, corrosion potential may increase, requiring special corrosion protection. Shrinkage of the soil may damage pipe coatings—particularly tar and asphalt coatings. It appears that loose-fitting polyethylene tube or wrap is well suited for protection in shrinking soil conditions, in that some movement around the pipe by both soil and wrap wll not result in damage of this protective system.

Our ongoing involvement.

CIPRA is now conducting laboratory tests to help answer the following questions on the effect of expansive soils: What is the range of stress applied to the pipe at different swell pressures? How is the stress applied and in what direction? What are the best methods for pipe installation?

Frost Penetration - Loads on Underground Pipe

A seasonal story.

It is common, particularly in northern areas, for municipal water systems to experience a number of pipe failures during the winter months. And one cause is frost penetration. It is a known fact that soil moisture, when frozen, expands the soil. And that causes increased vertical forces on buried pipe during the winter. Until now, there were no published data on this problem. The whole subject needed study.

A cooperative effort.

Recently an investigation was started in the Portland, Maine Water District. Ideally located for such research. The District and the Cast Iron Pipe Research Association (CIPRA) entered into a joint study on the effects of frost penetration. Test data for two winters in the Portland District have been collected and analyzed. Here is a summary of the results.

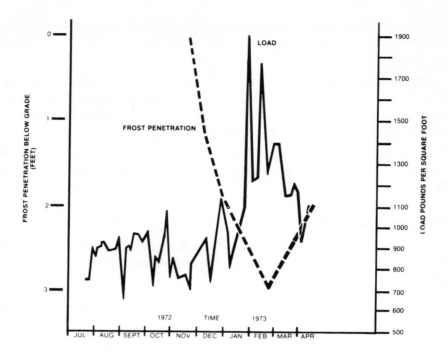

The test site.

An 8-inch cast iron pipe buried at a depth of 3.75 feet was selected as the test subject. The data collection system consisted of:

Four electronic load cells placed beneath the pipe.

A separate cell independent of the pipe, measuring direct soil load.

Five temperature sensors placed at 1-foot intervals to a depth of 5 feet.

Strip-charts for recording all load and temperature sensors.

The data.

To save space and also because the results of the second winter's study were more precise, we have reproduced here a plot of the data obtain during the second winter.

As the curve of earth load (the higher solid line) dramatically shows, load began to increase as frost penetration began in December. And as frost penetration increased to almost 3 feet in depth, the load had practically doubled the December 5 value. Also note the effect of increased moisture content on earth load. Points on the curve where significant rainfall is indicated represent marked decreases in earth load. This change in load with precipitation may be due to a change in soil friction angle or in soil cohesion.

What happens when a live load is added.

Another significant finding in the Portland CIPRA study was the effect of live loads on the test pipe. In a summer month, a loaded gravel truck of known weight caused an increase of about 2100 pounds on one of the load cells. And when an equivalently weighted truck passed over the pipe when winter's frost had reached 3 feet of penetration, the live load was only about 5% less. The difference was most likely due to wider distribution of the live load by the frost layer.

A clearer picture of frost penetration effects.

Live loads added to earth loads when frost is present are a logical explanation for the greater incidence of water main breaks during winter. Of course, the presence of much colder water in the main also causes some stresses on the pipe in wintertime.

In "normal" pipeline installations, it is also unlikely that uniform bedding can be achieved beneath the pipe. Increased external loading from frost penetration can cause beam breaks where bedding is not uniform. A piece of stone or rock beneath the pipe could act as a fulcrum. Soft portions of the bedding may permit excessive deflection caused by frost or live loads (or both).

The solution lies in the pipe.

Cold climates, where frost penetration is a recurring menace, require pipe of sufficient ductility, beam strength, and joint flexibility to withstand winter's stresses. Ductile iron pipe fits all these requirements with strength to spare. Serious consideration should therefore be given to ductile iron water systems for frost-prone areas.

Our ongoing involvement.

The Portland Water District and CIPRA have agreed to continue studying frost penetration. Even more severe winters, where frost reaches down to 4 feet, will hopefully yield further data.

SECTION IX
ENGINEERING DATA

EQUATION OF PIPE

It is frequently desired to know what numbers of pipe of a given size are equal in carrying capacity to one pipe of a larger size. At the same velocity of flow the volume delivered by two pipe of different sizes is proportional to the squares of their diameters; thus one 4-inch pipe will deliver the same volume as four 2-inch pipe. With the same head, however, the velocity is less in the smaller pipe, and the volume delivered varies about as square root of the fifth power. This table is calculated on this basis. The figures opposite the smaller sized pipe required to equal two sizes is the number of the smaller sized pipe required to equal one of the larger; thus, one 4-inch equals 5.7 two-inch.

Diam. In.	1/2	3/4	1	2	3	4	5	6	7	8	10	12	14	16	18	20	24	30	36	42	48	54
2	32.0	11.7	5.7	1.0	—																	
3	88.2	32.0	15.6	2.8	1.0	—																
4	181.	65.7	32.0	5.7	2.1	1.0	—															
5	316.	115.	55.9	9.9	3.6	1.7	1.0	—														
6	499.	181.	88.2	15.6	5.7	2.8	1.6	1.0	—													
7	733.	266.	130.	22.9	8.3	4.1	2.3	1.5	1.0	—												
8		372.	181.	32.0	11.7	5.7	3.2	2.1	1.4	1.0	—											
10		649.	316.	55.9	20.3	9.9	5.7	3.6	2.4	1.7	1.0	—										
11			401.	70.9	25.7	12.5	7.2	4.6	3.1	2.2	1.3	—										
12			499.	88.2	32.0	15.6	8.9	5.7	3.8	2.8	1.6	1.0										
13			609.	108.	39.1	19.0	10.9	7.1	4.7	3.4	1.9	1.2	—									
14			733.	130.	47.1	22.9	13.1	8.3	5.7	4.1	2.3	1.5	1.0									
15			787.	154.	55.9	27.2	15.6	9.9	6.7	4.8	2.8	1.7	1.2									
16				181.	65.7	32.0	18.3	11.7	7.9	5.7	3.2	2.1	1.4	1.0								
17				211.	76.4	37.2	21.3	13.5	9.2	6.6	3.8	2.4	1.6	1.2								
18				243.	88.2	43.0	24.6	15.6	10.6	7.6	4.3	2.8	1.9	1.3	1.0							
19				278.	101.	49.1	28.1	17.8	12.1	8.7	4.8	3.2	2.1	1.5	1.1							
20				316.	115.	55.9	32.0	20.3	13.8	9.9	5.7	3.6	2.4	1.7	1.3	1.0						
22				401.	146.	70.9	40.6	25.7	17.5	12.5	7.2	4.6	3.1	2.2	1.7	1.3						
24				499.	181.	88.2	50.5	32.0	21.8	15.6	8.9	5.7	3.8	2.8	2.1	1.6	1.0					
30										27.2	15.6	10.0	6.7	4.8	3.6	2.8	1.7	1.0				
36										—	24.6	15.6	10.6	7.6	5.7	4.3	2.8	1.6	1.0			
42										—	36.2	22.9	15.6	11.2	8.3	6.4	4.1	2.3	1.5	1.0		
48										—	50.5	32.0	21.8	15.6	11.7	8.9	5.7	3.2	2.1	1.4	1.0	
54										—	67.8	43.0	29.2	20.7	13.2	12.8	7.6	4.3	2.8	1.9	1.3	1.0

Pressures in Pounds per Square Inch, Corresponding to Heads of Water in Feet

Head Ft.	0	1	2	3	4	5	6	7	8	9
0	....	0.433	0.866	1.299	1.732	2.165	2.598	3.031	3.464	3.987
10	4.330	4.763	5.196	5.629	6.062	6.495	6.928	7.361	7.794	8.277
20	8.660	9.093	9.526	9.959	10.392	10.825	11.258	11.691	12.124	12.557
30	12.990	13.423	13.856	14.289	14.722	15.155	15.588	16.021	16.454	16.887
40	17.320	17.753	18.186	18.619	19.052	19.485	19.918	20.351	20.784	21.217
50	21.650	22.083	22.516	22.949	23.382	23.815	24.248	24.681	25.114	25.547
60	25.980	26.413	26.846	27.279	27.712	28.145	28.578	29.011	29.444	29.877
70	30.310	30.743	31.176	31.609	32.042	32.475	32.908	33.341	33.774	34.207
80	34.640	35.073	35.506	35.939	36.372	36.805	37.238	37.671	38.104	38.537
90	38.970	39.403	39.836	40.269	40.702	41.135	41.568	42.001	42.436	42.867

Heads of Water in Feet, Corresponding to Pressures in Pounds per Square Inch

Pressure Lbs. per Sq. In.	0	1	2	3	4
0	....	2.309	4.619	6.928	9.238
10	23.095	25.404	27.714	30.023	32.333
20	46.189	48.499	50.808	53.118	55.427
30	69.284	71.594	73.903	76.213	78.522
40	92.379	94.688	96.998	99.307	101.62
50	115.47	117.78	120.09	122.40	124.71
60	138.57	140.88	143.19	145.50	147.81
70	161.66	163.97	166.28	168.59	170.90
80	184.76	187.07	189.38	191.69	194.00
90	207.85	210.16	212.47	214.78	217.09

	5	6	7	8	9
0	11.547	13.857	16.166	18.476	20.785
10	34.642	36.952	39.261	41.570	43.880
20	57.737	60.046	62.356	64.665	66.975
30	80.831	83.141	85.450	87.760	90.069
40	103.93	106.24	108.55	110.85	113.16
50	127.02	129.33	131.64	133.95	136.26
60	150.12	152.42	154.73	157.04	159.35
70	173.21	175.52	177.83	180.14	182.45
80	196.31	198.61	200.92	203.23	205.54
90	219.40	221.71	224.02	226.33	228.64

At 62° F., 1 foot head = 0.433 lb. per square inch; 0.433 × 144 = 62.355 lb. per cubic foot. 1 lb. per square inch = 2.30947 feet head. 1 atmosphere = 14.7 lb. per square inch = 33.94 feet head.

CONTENTS OF PIPE

Capacities in Cubic Feet and in United States Gallons
(231 Cubic Inches) per Foot of Length

Diameter, Inches	Diameter, Feet	For 1 Foot Length Cubic Feet, Also Area in Sq. Ft.	U.S. Gal. (231 Cu. In.)	Diameter, Inches	Diameter, Feet	For 1 Foot Length Cubic Feet, Also Area in Sq. Ft.	U.S. Gal. (231 Cu. In.)	Diameter, Inches	Diameter, Feet	For 1 Foot Length Cubic Feet, Also Area in Sq. Ft.	U.S. Gal. (231 Cu. In.)
¼	.0208	.0003	.0026	6.75	.5625	.2485	1.859	19.0	1.583	1.969	14.73
5/16	.0260	.0005	.0040	7.00	.5833	.2673	1.999	19.5	1.625	2.074	15.52
3/8	.0313	.0008	.0057	7.25	.6042	.2868	2.144	20.0	1.666	2.182	16.32
7/16	.0365	.0010	.0078	7.50	.6250	.3068	2.295	20.5	1.708	2.292	17.15
½	.0417	.0014	.0102	7.75	.6458	.3275	2.450	21.0	1.750	2.405	17.99
9/16	.0469	.0017	.0129	8.00	.6667	.3490	2.611	21.5	1.792	2.521	18.86
5/8	.0521	.0021	.0159	8.25	.6875	.3713	2.777	22.0	1.833	2.640	19.75
11/16	.0573	.0026	.0193	8.50	.7083	.3940	2.948	22.5	1.875	2.761	20.65
¾	.0625	.0031	.0230	8.75	.7292	.4175	3.125	23.0	1.917	2.885	21.58
13/16	.0677	.0036	.0270	9.00	.7500	.4418	3.305	23.5	1.958	3.012	22.53
7/8	.0729	.0042	.0312	9.25	.7708	.4668	3.492	24.0	2.000	3.142	23.50
15/16	.0781	.0048	.0359	9.50	.7917	.4923	3.682	25.0	2.083	3.409	25.50
1.00	.0833	.0055	.0408	9.75	.8125	.5185	3.879	26.0	2.166	3.687	27.58
1.25	.1042	.0085	.0638	10.00	.8333	.5455	4.081	27.0	2.250	3.976	29.74
1.50	.1250	.0123	.0918	10.25	.8542	.5730	4.286	28.0	2.333	4.276	31.99
1.75	.1458	.0168	.1250	10.50	.8750	.6013	4.498	29.0	2.416	4.587	34.31
2.00	.1667	.0218	.1632	10.75	.8958	.6303	4.714	30.0	2.500	4.909	36.72
2.25	.1875	.0276	.2066	11.00	.9167	.6600	4.937	31.0	2.583	5.241	39.21
2.50	.2083	.0341	.2550	11.25	.9375	.6903	5.163	32.0	2.666	5.585	41.78
2.75	.2292	.0413	.3085	11.50	.9583	.7213	5.395	33.0	2.750	5.940	44.43
3.00	.2500	.0491	.3673	11.75	.9792	.7530	5.633	34.0	2.833	6.305	47.17
3.25	.2708	.0576	.4310	12.00	1.000	.7854	5.876	35.0	2.916	6.681	49.98
3.50	.2917	.0668	.4998	12.50	1.042	.8523	6.375	36.0	3.000	7.069	52.88
3.75	.3125	.0767	.5738	13.00	1.083	.9218	6.895	37.0	3.083	7.468	55.86
4.00	.3333	.0873	.6528	13.50	1.125	.9940	7.435	38.0	3.166	7.876	58.92
4.25	.3542	.0985	.7370	14.00	1.167	1.069	7.997	39.0	3.250	8.296	62.06
4.50	.3750	.1105	.8263	14.50	1.208	1.147	8.578	40.0	3.333	8.728	65.29
4.75	.3958	.1231	.9205	15.00	1.250	1.227	9.180	41.0	3.416	9.168	68.58
5.00	.4167	.1364	1.020	15.50	1.292	1.310	9.801	42.0	3.500	9.620	71.96
5.25	.4375	.1503	1.124	16.00	1.333	1.396	10.44	43.0	3.583	10.084	75.43
5.50	.4583	.1650	1.234	16.50	1.375	1.485	11.11	44.0	3.666	10.560	79.00
5.75	.4792	.1803	1.349	17.00	1.417	1.576	11.79	45.0	3.750	11.044	82.62
6.00	.5000	.1963	1.469	17.50	1.458	1.670	12.50	46.0	3.833	11.540	86.32
6.25	.5208	.2130	1.594	18.00	1.500	1.767	13.22	47.0	3.916	12.048	90.12
6.50	.5417	.2305	1.724	18.50	1.542	1.867	13.97	48.0	4.000	12.566	94.02

1 Cubic foot of water weighs 62.35 pounds; 1 gallon (U. S.) weighs 8.335 pounds.

LINEAR EXPANSION OF DUCTILE IRON PIPE

The coefficient of linear expansion of ductile iron may be taken as 0.0000062 per degree Fahrenheit. The expansion or contraction in inches that will take place in a line of given length with various temperature changes is shown in the following table:

Temp. Difference °F	LENGTH OF LINE IN FEET			
	100	500	1000	5280
5	0.037	0.19	0.37	1.96
10	0.074	0.37	0.74	3.93
20	0.149	0.74	0.15	7.86
30	0.223	1.12	2.23	11.78
40	0.298	1.49	2.98	15.71
50	0.372	1.86	3.72	19.64
60	0.446	2.23	4.46	23.57
70	0.520	2.60	5.20	27.50
80	0.595	2.98	5.95	31.43
90	0.670	3.35	6.70	35.35
100	0.744	3.72	7.44	39.28
120	0.893	4.46	8.93	47.14
150	1.116	5.58	11.16	58.92

Linear Expansion of Cast Iron Pipe

The coefficient of linear expansion of cast iron may be taken as 0.0000058 per degree Fahrenheit. The expansion or contraction in inches that will take place in a line of given length with various temperature changes is shown in the following table:

Temp. Difference °F	Length of Line in Feet			
	100	500	1000	5280
5	0.035	0.17	0.35	1.83
10	0.070	0.35	0.70	3.67
20	0.139	0.70	1.39	7.34
30	0.209	1.04	2.09	11.01
40	0.278	1.39	2.78	14.70
50	0.348	1.74	3.48	18.35
60	0.418	2.09	4.18	22.04
70	0.487	2.44	4.87	25.72
80	0.557	2.79	5.57	29.39
90	0.626	3.13	6.26	33.05
100	0.696	3.48	6.96	36.71
120	0.835	4.17	8.35	44.10
150	1.043	5.22	10.43	55.10

TABLE 1 Rounding Tolerances

Inches to Millimetres

Original Tolerance, in.		Fineness of Rounding, mm
at least	less than	
0.000 01	0.000 1	0.000 01
0.000 1	0.001	0.000 1
0.001	0.01	0.001
0.01	0.1	0.01
0.1	1	0.1

TABLE 2 Inch-Millimetre Equivalents

NOTE—All values in this table are exact, based on the relation 1 in. = 25.4 mm. By manipulation of the decimal point any decimal value or multiple of an inch may be converted to its exact equivalent in millimetres.

in.	0	1	2	3	4	5	6	7	8	9
					mm					
0		25.4	50.8	76.2	101.6	127.0	152.4	177.8	203.2	228.6
10	254.0	279.4	304.8	330.2	355.6	381.0	406.4	431.8	457.2	482.6
20	508.0	533.4	558.8	584.2	609.6	635.0	660.4	685.8	711.2	736.6
30	762.0	787.4	812.8	838.2	863.6	889.0	914.4	939.8	965.2	990.6
40	1016.0	1041.4	1066.8	1092.2	1117.6	1143.0	1168.4	1193.8	1219.2	1244.6
50	1270.0	1295.4	1320.8	1346.2	1371.6	1397.0	1422.4	1447.8	1473.2	1498.6
60	1524.0	1549.4	1574.8	1600.2	1625.6	1651.0	1676.4	1701.8	1727.2	1752.6
70	1778.0	1803.4	1828.8	1854.2	1879.6	1905.0	1930.4	1955.8	1981.2	2006.6
80	2032.0	2057.4	2082.8	2108.2	2133.6	2159.0	2184.4	2209.8	2235.2	2260.6
90	2286.0	2311.4	2336.8	2362.2	2387.6	2413.0	2438.4	2463.8	2489.2	2514.6
100	2540.0	...	...	...	...	...	...	...	...	...

TABLE 3 Conversion of Temperature Tolerance Requirements

Tolerance, deg F	±1	±2	±5	±10	±15	±20	±25
Tolerance, K or deg C	±0.5	±1	±3	±5.5	±8.5	±11	±14

TABLE 4 Pressure and Stress Equivalents–Pounds-Force per Square Inch and Thousand Pounds-Force per Square Inch to Kilopascals (Kilonewtons per Square Metre) and Megapascals (Meganewtons per Square Metre)

NOTE 1—This table may be used to obtain SI equivalents of values expressed in psi or ksi. SI values are usually expressed in kPa (kN/m²) when original value is in psi and in MPa (MN/m²) when original value is in ksi.

NOTE 2—This table may be extended to values below 1 or above 100 psi (ksi) by manipulation of the decimal point and addition as illustrated in 4.7.1.

Conversion Relationships:
1 in. = 0.0254 m (exactly)
1 lbf = 4.448 221 615 260 5 N (exactly)

psi / ksi	0	1	2	3	4	5	6	7	8	9
					kPa (kN/m²) MPa (MN/m²)					
0	0.0000	6.8948	13.7895	20.6843	27.5790	34.4738	41.3685	48.2633	55.1581	62.0528
10	68.9476	75.8423	82.7371	89.6318	96.5266	103.4214	110.3161	117.2109	124.1056	131.0004
20	137.8951	144.7899	151.6847	158.5794	165.4742	172.3689	179.2637	186.1584	193.0532	199.9480
30	206.8427	213.7375	220.6322	227.5270	234.4217	241.3165	248.2113	255.1060	262.0008	268.8955
40	275.7903	282.6850	289.5798	296.4746	303.3693	310.2641	317.1588	324.0536	330.9483	337.8431
50	344.7379	351.6326	358.5274	365.4221	372.3169	379.2116	386.1064	393.0012	399.8959	406.7907
60	413.6854	420.5802	427.4749	434.3697	441.2645	448.1592	455.0540	461.9487	468.8435	475.7382
70	482.6330	489.5278	496.4225	503.3173	510.2120	517.1068	524.0015	530.8963	537.7911	544.6858
80	551.5806	558.4753	565.3701	572.2648	579.1596	586.0544	592.9491	599.8439	606.7386	613.6334
90	620.5281	627.4229	634.3177	641.2124	648.1072	655.0019	661.8967	668.7914	675.6862	682.5810
100	689.4757									

TABLE 5 Pressure and Stress Equivalents—Metric Engineering to SI Units

NOTE 1—This table may be used for obtaining SI equivalents of quantities expressed in kgf/cm² by multiplying the given values by 10^{-2}, that is, by moving the decimal point two places to the left.

NOTE 2—This table may be extended to values below 1 or above 100 kgf/cm² by manipulation of the decimal point and addition as illustrated in 4.7.1.

Conversion Relationships: 1 mm = 0.001 m (exactly)
1 kgf = 9.80665 N (exactly)

kgf/mm²	MPa (MN/m²)									
	0	1	2	3	4	5	6	7	8	9
0		9.8066	19.6133	29.4200	39.2266	49.0332	58.8399	68.6466	78.4532	88.2598
10	98.0665	107.8731	117.6798	127.4864	137.2931	147.0998	156.9064	166.7130	176.5197	186.3264
20	196.1330	205.9396	215.7463	225.5530	235.3596	245.1662	254.9729	264.7796	274.5862	284.3928
30	294.1995	304.0062	313.8128	323.6194	333.4261	343.2328	353.0394	362.8460	372.6527	382.4594
40	392.2660	402.0726	411.8793	421.6860	431.4926	441.2992	451.1059	460.9126	470.7192	480.5258
50	490.3325	500.1392	509.9458	519.7524	529.5591	539.3658	549.1724	558.9790	568.7857	578.5924
60	588.3990	598.2056	608.0123	617.8190	627.6256	637.4322	647.2389	657.0456	666.8522	676.6588
70	686.4655	696.2722	706.0788	715.8854	725.6921	735.4988	745.3054	755.1120	764.9187	774.7254
80	784.5320	794.3386	804.1453	813.9520	823.7586	833.5652	843.3719	853.1786	862.9852	872.7918
90	882.5985	892.4052	902.2118	912.0184	921.8251	931.6318	941.4384	951.2450	961.0517	970.8584
100	980.6650									

SELECTED CONVERSION FACTORS

To convert from	to	multiply by
atmosphere (760 Hg)	pascal (Pa)	$1.013\ 25 \times 10^5$
board foot	metre³ (m³)	$2.359\ 737 \times 10^{-3}$
Btu (International Table)	joule (J)	$1.055\ 056 \times 10^3$
Btu (International Table)/hour	watt (W)	$2.930\ 711 \times 10^{-1}$
Btu (International Table) in./s·ft²·°F (k, thermal conductivity)	watt/metre-kelvin (W/m-K)	$5.192\ 204 \times 10^2$
calorie (International Table)	joule (J)	$4.186\ 800*$
centipoise	pascal-second (Pa-s)	$1.000\ 000* \times 10^{-3}$
centistokes	metre²/second (m²/s)	$1.000\ 000* \times 10^{-6}$
circular mil	metre² (m²)	$5.067\ 075 \times 10^{-10}$
degree Fahrenheit	degree Celsius	$t_{oc} = (t_{oF}-32)/1.8$
foot	metre (m)	$3.048\ 000* \times 10^{-1}$
foot²	metre² (m²)	$9.290\ 304* \times 10^{-2}$
foot³	metre³ (m³)	$2.831\ 685 \times 10^{-2}$
foot-pound-force	joule (J)	$1.355\ 818$
foot-pound-force/minute	watt (W)	$2.259\ 697 \times 10^{-2}$
foot/second²	metre/second² (m/s²)	$3.048\ 000* \times 10^{-1}$
gallon (U.S. liquid)	metre³ (m³)	$3.785\ 412 \times 10^{-3}$
horsepower (electric)	watt (W)	$7.460\ 000* \times 10^2$
inch	metre (m)	$2.540\ 000* \times 10^{-2}$
inch²	metre² (m²)	$6.451\ 600* \times 10^{-4}$
inch³	metre³ (m³)	$1.638\ 706 \times 10^{-5}$
inch of mercury (60° F)	pascal (Pa)	$3.376\ 85 \times 10^3$
inch of water (60° F)	pascal (Pa)	$2.488\ 4 \times 10^2$
kilogram-force/centimetre²	pascal (Pa)	$9.806\ 650* \times 10^4$
kip (1000 lbf)	newton (N)	$4.448\ 222 \times 10^3$
kip/inch² (ksi)	pascal (Pa)	$6.894\ 757 \times 10^6$
ounce (U.S. fluid)	metre³ (m³)	$2.957\ 353 \times 10^{-5}$
ounce-force (avoirdupois)	newton (N)	$2.780\ 139 \times 10^{-1}$
ounce-mass (avoirdupois)	kilogram (kg)	$2.834\ 952 \times 10^{-2}$
ounce-mass/ft²	kilogram/metre² (kg/m²)	$0.305\ 152$
ounce-mass/yard²	kilogram/metre² (kg/m²)	$3.390\ 575 \times 10^{-2}$
ounce (avoirdupois)/gallon (U.S. liquid)	kilogram/metre³ (kg/m³)	$7.489\ 152$
pint (U.S. liquid)	metre³ (m³)	$4.731\ 765 \times 10^{-4}$
pound-force (lbf avoirdupois)	newton (N)	$4.448\ 222$
pound-mass (lbm avoirdupois)	kilogram (kg)	$4.535\ 924 \times 10^{-1}$
pound-mass/second	kilogram/second (kg/s)	$4.535\ 924 \times 10^{-1}$
pound-force/inch² (psi)	pascal (Pa)	$6.894\ 757 \times 10^3$
pound-mass/inch³	kilogram/meter³ (kg/m³)	$2.767\ 990 \times 10^4$
pound-mass/foot³	kilogram/meter³ (kg/m³)	$1.601\ 846 \times 10$
quart (U.S. liquid)	metre³ (m³)	$9.463\ 529 \times 10^{-4}$
ton (short, 2000 lbm)	kilogram (kg)	$9.071\ 847 \times 10^2$
torr (mm Hg)	pascal (Pa)	$1.333\ 22 \times 10^2$
watt-hour	joule (J)	$3.600\ 000* \times 10^3$
yard	metre (m)	$9.144\ 000* \times 10^{-1}$
yard²	metre² (m²)	$8.361\ 274 \times 10^{-1}$
yard³	metre³ (m³)	$7.645\ 549 \times 10^{-1}$

---
* Exact

Units of Measurement—Conversion Factors*

*All boldface figures are exact; the others generally are given to seven significant figures.

In using conversion factors, it is possible to perform division as well as the multiplication process shown here. Division may be particularly advantageous where more than the significant figures published here are required. Division may be performed in lieu of multiplication by using the reciprocal of any indicated multiplier as divisor. For example, to convert from centimeters to inches by division, refer to the table headed "To Convert from *Inches*" and use the factor listed at "centimeters" (*2.54*) as divisor.

Units of Length

To Convert from **Centimeters**	
To	Multiply by
Inches	0.393 700 8
Feet	0.032 808 40
Yards	0.010 936 13
Meters	**0.01**

To Convert from **Inches**	
To	Multiply by
Feet	0.083 333 33
Yards	0.027 777 78
Centimeters	**2.54**
Meters	**0.025 4**

To Convert from **Meters**	
To	Multiply by
Inches	39.370 08
Feet	3.280 840
Yards	1.093 613
Miles	0.000 621 37
Millimeters	**1 000**
Centimeters	**100**
Kilometers	**0.001**

To Convert from **Feet**	
To	Multiply by
Inches	**12**
Yards	0.333 333 3
Miles	0.000 189 39
Centimeters	**30.48**
Meters	**0.304 8**
Kilometers	**0.000 304 8**

To Convert from **Yards**	
To	Multiply by
Inches	**36**
Feet	**3**
Miles	0.000 568 18
Centimeters	**91.44**
Meters	**0.914 4**

To Convert from **Miles**	
To	Multiply by
Inches	**63 360**
Feet	**5 280**
Yards	**1 760**
Centimeters	**160 934.4**
Meters	**1 609.344**
Kilometers	**1.609 344**

Units of Measurement—Conversion Factors*

*All boldface figures are exact; the others generally are given to seven significant figures.

Units of Mass

To Convert from Grams

To	Multiply by
Grains	15.432 36
Avoirdupois Drams	0.564 383 4
Avoirdupois Ounches	0.035 273 96
Troy Ounces	0.032 150 75
Troy Pounds	0.002 679 23
Avoirdupois Pounds	0.002 204 62
Milligrams	**1 000**
Kilograms	**0.001**

To Convert from Metric Tons

To	Multiply by
Avoirdupois Pounds	2 204.623
Short Hundredweights	22.046 23
Short Tons	1.102 311 3
Long Tons	0.984 206 5
Kilograms	**1 000**

To Convert from Grains

To	Multiply by
Avoirdupois Drams	0.036 571 43
Avoirdupois Ounces	0.002 285 71
Troy Ounces	0.002 083 33
Troy Pounds	0.000 173 61
Avoirdupois Pounds	0.000 142 86
Milligrams	**64.798 91**
Grams	**0.064 798 91**
Kilograms	**0.000 064 798 91**

To Convert from Avoirdupois Pounds

To	Multiply by
Grains	**7 000**
Avoirdupois Drams	**256**
Avoirdupois Ounces	**16**
Troy Ounces ..	14.583 33
Troy Pounds ..	1.215 278
Grams	453.592 37
Kilograms	**0.453 592 37**
Short Hundredweights	**0.01**
Short Tons	**0.000 5**
Long Tons	0.000 446 428 6
Metric Tons ..	**0.000 453 592 37**

To Convert from Kilograms

To	Multiply by
Grains	15 432.36
Avoirdupois Drams	564.383 4
Avoirdupois Ounces	35.273 96
Troy Ounces	32.150 75
Troy Pounds	2.679 229
Avoirdupois Pounds	2.204 623
Grams	**1 000**
Short Hundredweights	0.022 046 23
Short Tons	0.001 102 31
Long Tons	0.000 984 2
Metric Tons	**0.001**

Units of Measurement—Conversion Factors*

*All boldface figures are exact; the others generally are given to seven significant figures.

Units of Mass

To Convert from **Avoirdupois Ounces**	
To	Multiply by
Grains	437.5
Avoirdupois Drams	16
Troy Ounces	0.911 458 3
Troy Pounds	0.075 954 86
Avoirdupois Pounds	0.062 5
Grams	28.349 523 125
Kilograms	0.028 349 523 125

To Convert from **Troy Pounds**	
To	Multiply by
Grains	5 760
Avoirdupois Drams	210.651 4
Avoirdupois Ounces	13.165 71
Troy Ounces	12
Avoirdupois Pounds	0.822 857 1
Grams	373.241 721 6

To Convert from **Short Hundredweights**	
To	Multiply by
Avoirdupois Pounds	100
Short Tons	0.05
Long Tons	0.044 642 86
Kilograms	45.359 237
Metric Tons	0.045 359 237

To Convert from **Short Tons**	
To	Multiply by
Avoirdupois Pounds	2 000
Short Hundredweights	20
Long Tons	0.892 857 1
Kilograms	907.184 74
Metric Tons	0.907 184 74

To Convert from **Troy Ounces**	
To	Multiply by
Grains	480
Avoirdupois Drams	17.554 29
Avoirdupois Ounces	1.097 143
Troy Pounds	0.083 333 3
Avoirdupois Pounds	0.068 571 43
Grams	31.103 476 8

To Convert from **Long Tons**	
To	Multiply by
Avoirdupois Ounces	35 840
Avoirdupois Pounds	2 240
Short Hundred-Weights	22.4
Short Tons	1.12
Kilograms	1 016.046 908 8
Metric Tons ..	1.016 046 908 8

438

Units of Measurement—Conversion Factors*

*All boldface figures are exact; the others generally are given to seven significant figures.

Units of Capacity, or Volume, Liquid Measure

To Convert from **Milliliters**	
To	**Multiply by**
Minims	16.230 73
Liquid Ounces	0.033 814 02
Gills	0.008 453 5
Liquid Pints	0.002 113 4
Liquid Quarts	0.001 056 7
Gallons	0.000 264 17
Cubic Inches	0.061 023 74
Liters	**0.001**

To Convert from **Minims**	
To	**Multiply by**
Liquid Ounces	0.002 083 33
Gills	0.000 520 83
Cubic Inches	0.003 759 77
Milliliters	0.061 611 52

To Convert from **Gills**	
To	**Multiply by**
Minims	**1 920**
Liquid Ounces	**4**
Liquid Pints	**0.25**
Liquid Quarts	**0.125**
Gallons	**0.031 25**
Cubic Inches	7.218 75
Cubic Feet	0.004 177 517
Milliliters	118.294 118 25
Liters	0.118 294 118 25

To Convert from **Cubic Meters**	
To	**Multiply by**
Gallons	264.172 05
Cubic Inches	61 023.74
Cubic Feet	35.314 67
Liters	**1 000**
Cubic Yards	1.307 950 6

To Convert from **Liters**	
To	**Multiply by**
Liquid Ounces	33.814 02
Gills	8.453 506
Liquid Pints	2.113 376
Liquid Quarts	1.056 688
Gallons	0.264 172 05
Cubic Inches	61.023 74
Cubic Feet	0.035 314 67
Milliliters	**1 000**
Cubic Meters	**0.001**
Cubic Yards	0.001 307 95

To Convert from **Liquid Ounces**	
To	**Multiply by**
Minims	**480**
Gills	**0.25**
Liquid Pints	**0.062 5**
Liquid Quarts	**0.031 25**
Gallons	**0.007 812 5**
Cubic Inches	1.804 687 5
Cubic Feet	0.001 044 38
Milliliters	29.573 53
Liters	0.029 573 53

Units of Measurement—Conversion Factors*

*All boldface figures are exact; the others generally are given to seven significant figures.

Units of Capacity, or Volume, Liquid Measure

To Convert from Liquid Pints

To	Multiply by
Minims	7 680
Liquid Ounces	16
Gills	4
Liquid Quarts	0.5
Gallons	0.125
Cubic Inches	28.875
Cubic Feet	0.016 710 07
Milliliters	473.176 473
Liters	0.473 176 473

To Convert from Cubic Inches

To	Multiply by
Minims	265.974 0
Liquid Ounces	0.554 112 6
Gills	0.138 528 1
Liquid Pints	0.034 632 03
Liquid Quarts	0.017 316 02
Gallons	0.004 329 0
Cubic Feet	0.000 578 7
Milliliters	16.387 064
Liters	0.016 387 064
Cubic Meters	0.000 016 387 064
Cubic Yards	0.000 021 43

To Convert from Liquid Quarts

To	Multiply by
Minims	15 360
Liquid Ounces	32
Gills	8
Liquid Pints	2
Gallons	0.25
Cubic Inches	57.75
Cubic Feet	0.033 420 14
Milliliters	946.352 946
Liters	0.946 352 946

To Convert from Cubic Feet

To	Multiply by
Liquid Ounces	957.506 5
Gills	239.376 6
Liquid Pints	59.844 16
Liquid Quarts	29.922 08
Gallons	7.480 519
Cubic Inches	1 728
Liters	28.316 846 592
Cubic Meters	0.028 316 846 592
Cubic Yards	0.037 037 04

To Convert from Gallons

To	Multiply by
Minims	61 440
Liquid Ounces	128
Gills	32
Liquid Pints	8
Liquid Quarts	4
Cubic Inches	231
Cubic Feet	0.133 680 6
Milliliters	3 785.411 784
Liters	3.785 411 784
Cubic Meters	0.003 785 411 784
Cubic Yards	0.004 951 13

To Convert from Cubic Yards

To	Multiply by
Gallons	201.974 0
Cubic Inches	46 656
Cubic Feet	27
Liters	764.554 857 984
Cubic Meters	0.764 554 857 984

Units of Measurement—Conversion Factors*

*All boldface figures are exact; the others generally are given to seven significant figures.

Units of Capacity, or Volume, Dry Measure

To Convert from Liters

To	Multiply by
Dry Pints	1.816 166
Dry Quarts	0.908 082 98
Pecks	0.113 510 4
Bushels	0.028 377 59
Dekaliters	**0.1**

To Convert from Cubic Meters

To	Multiply by
Pecks	113.510 4
Bushels	28.377 59

To Convert from Dekaliters

To	Multiply by
Dry Pints	18.161 66
Dry Quarts	9.080 829 8
Pecks	1.135 104
Bushels	0.283 775 9
Cubic Inches	610.237 4
Cubic Feet	0.353 146 7
Liters	**10**

To Convert from Dry Pints

To	Multiply by
Dry Quarts	**0.5**
Pecks	**0.062 5**
Bushels	**0.015 625**
Cubic Inches	**33.600 312 5**
Cubic Feet	0.019 444 63
Liters	0.550 610 47
Dekaliters	0.055 061 05

To Convert from Dry Quarts

To	Multiply by
Dry Pints	**2**
Pecks	**0.125**
Bushels	**0.031 25**
Cubic Inches	67.200 625
Cubic Feet	0.038 889 25
Liters	1.101 221
Dekaliters	0.110 122 1

To Convert from Bushels

To	Multiply by
Dry Pints	**64**
Dry Quarts	**32**
Pecks	**4**
Cubic Inches	2 150.42
Cubic Feet	1.244 456
Liters	35.239 07
Dekaliters	3.523 907
Cubic Meters	0.035 239 07
Cubic Yards	0.046 090 96

To Convert from Pecks

To	Multiply by
Dry Pints	**16**
Dry Quarts	**8**
Bushels	**0.25**
Cubic Inches	537.605
Cubic Feet	0.311 114
Liters	8.809 767 5
Dekaliters	0.880 976 75
Cubic Meters	0.008 809 77
Cubic Yards	0.011 522 74

Units of Measurement—Conversion Factors*

*All boldface figures are exact; the others generally are given to seven significant figures.

Units of Capacity, or Volume, Dry Measure

To Convert from
Cubic Inches

To	Multiply by
Dry Pints	0.029 761 6
Dry Quarts	0.014 880 8
Pecks	0.001 860 10
Bushels	0.000 465 025

To Convert from
Cubic Feet

To	Multiply by
Dry Pints	51.428 09
Dry Quarts	25.714 05
Pecks	3.214 256
Bushels	0.803 563 95

To Convert from
Cubic Yards

To	Multiply by
Pecks	86.784 91
Bushels	21.696 227

Units of Area

To Convert from
Square Centimeters

To	Multiply by
Square Inches	0.155 000 3
Square Feet	0.001 076 39
Square Yards	0.000 119 599
Square Meters	**0.000 1**

To Convert from
Hectares

To	Multiply by
Square Feet	107 639.1
Square Yards	11 959.90..
Acres	2.471 054
Square Miles	0.003 861 02
Squire Meters	**10 000**

To Convert from
Square Feet

To	Multiply by
Square Inches	**144**
Square Yards	0.111 111 1
Acres	0.000 022 957
Square Centimeters	929.030 4
Square Meters	**0.092 903 04**

To Convert from
Acres

To	Multiply by
Square Feet	**43 560**
Square Yards	**4 840**
Square Miles	0.001 562 5
Square Meters	4 046.856 422 4
Hectares	0.404 685 642 24

Units of Measurement—Conversion Factors*

*All boldface figures are exact; the others generally are given to seven significant figures.

Units of Area

To Convert from Square Meters

To	Multiply by
Square Inches	1 550.003
Square Feet	10.763 91
Square Yards	1.195 990
Acres	0.000 247 105
Square Centimeters ..	**10 000**
Hectares	**0.000 1**

To Convert from Square Inches

To	Multiply by
Square Feet	0.006 944 44
Square Yards	0.000 771 605
Square Centimeters	6.451 6
Square Meters	**0.000 645 16**

To Convert from Square Miles

To	Multiply by
Square Feet	**27 878 400**
Square Yards	**3 097 600**
Acres	**640**
Square Meters ..	2 589 **988.110 336**
Hectares	258.998 **811 033 6**

To Convert from Square Yards

To	Multiply by
Square Inches	**1 296**
Square Feet	**9**
Acres	0.000 206 611 6
Square Miles	0.000 000 322 830 6
Square Centimeters	8 361.273 6
Square Meters	0.836 127 36
Hectares ..	0.000 083 612 736

INDEX

—A—

—D—

—E—

—F—

—T—

4
4
8

—W—

—X, Y, Z—